Life in the Universe

Jeffrey Bennett
University of Colorado at Boulder

Seth Shostak
SETI Institute

Bruce Jakosky
University of Colorado at Boulder

Addison
Wesley

San Francisco Boston New York
Capetown Hong Kong London Madrid Mexico City
Montreal Munich Paris Singapore Sydney Tokyo Toronto

Acquisitions Editor: *Adam Black*
Project Editor: *Nancy Benton*
Market Developer: *Chalon Bridges*
Marketing Manager: *Christy Lawrence*
Publishing Associate: *Liana Allday*
Developmental Editor: *Margot Otway*
Production Coordination: *Joan Marsh*
Production: *Mary Douglas, Rogue Valley Publications*
Photo Research: *Lili Weiner, Myrna Engler*
Artists: *John and Judy Waller*
Copyeditor: *Mary Roybal*
Text Designers: *Blakeley Kim, Andrew Ogus*
Partial Cover Photo: Upper left: "NGC 2080, nicknamed
'The Ghost Head Nebula'" NASA, ESA
& Mohammad Heydari-Malayeri (Obser-
vatoire de Paris, France). Lower portion:
"Valles Marineris—Point Perspective"
Jody Swann, Tammy Becker, and Alfred
McEwen of U.S. Geological Survey;
Flagstaff, AZ
Cover Designers: Quade Paul/fiVth.com and Emiko Rose
Koike / fiVth.com
Composition: *TBH Typecast, Inc.*
Cover Printer: *Phoenix Color Corporation*
Printer and Binder: *Von Hoffmann*

Library of Congress Cataloging-in-Publication Data--
Life in the Universe / Jeffrey O. Bennett . . . [et al.].
     p.   cm.
   Includes index.
   ISBN 0-8053-8577-0
   1. Astronomy.   I. Bennett, Jeffrey O.

4 5 6 7 8 9 10—VH—05

All this world is heavy with the promise of greater things, and a day will come, one day in the unending succession of days, when beings, beings who are now latent in our thoughts and hidden in our loins, shall stand upon this earth as one stands upon a footstool, and shall laugh and reach their hands amidst the stars.

H. G. WELLS, 1902

Dedication

THE QUEST TO UNDERSTAND LIFE ON EARTH and the prospects for life elsewhere in the universe touches upon the most profound questions of human existence. It sheds light on our origins, teaches us to appreciate how and why our existence on Earth became possible, and inspires us to wonder about the incredible possibilities that may await us in space. We dedicate this book to all who wish to join in this quest, with the sincere hope that knowledge will help our species act wisely and responsibly.

BRIEF CONTENTS

DETAILED CONTENTS

PREFACE

To the Reader

Few questions have so inspired humans through the ages as the mystery of whether we are alone in the universe. Many ancient Greek philosophers were confident that intelligent beings could be found far beyond Earth. When the first telescopes were trained on the Moon in the seventeenth century, some eminent astronomers interpreted lunar features as proof of an inhabited world. Barely a century ago, belief in a civilization on Mars became so widespread that the term *martian* became synonymous with *alien*. But despite this historical interest in the possibility of extraterrestrial life, until quite recently few scientists devoted any effort to understanding the issues surrounding it, let alone to making a serious search for life.

In the past couple of decades, however, a remarkable convergence of biology, geology, astronomy, and other sciences has suddenly placed the issue of extraterrestrial life at the forefront of research. Advances in our understanding of the origin of life on Earth are helping us predict the conditions under which life might arise in other places. Discoveries of microbes thriving under extreme conditions (at least by human standards) on Earth have raised hopes that life might survive even in some of the harsh environments found elsewhere in our own solar system. Proof that planets exist around other stars—first obtained in the 1990s—has given added impetus to the study of the conditions that might allow for life in other star systems. Technological advances are making it possible for us to engage in unprecedented, large-scale scrutiny of the sky for signals from other civilizations, spurring heightened interest in the search for extraterrestrial intelligence (SETI). Perhaps most important, scientists have found the inter-disciplinary study of issues related to the search for life beyond Earth to have intrinsic value, independent of whether the search is ultimately successful.

Given the intense research efforts being undertaken by the scientific community and the long-standing public fascination with the search for life, it should be no surprise that the study of life in the universe—also known as *astrobiology*—has become one of the most publicly visible sciences. But, as is often the case, scientific discovery is moving much more quickly than innovations in education. As a result, most people have had little opportunity to learn about the remarkable scientific adventure now under way to search for answers to fundamental questions about life on Earth and beyond. This book aims to help improve this situation by offering an introduction to the broad science of life in the universe in a way that is fairly comprehensive but still accessible to readers who have little or no scientific training.

Although this book is written in a format that makes it especially suitable as a textbook for introductory courses on life in the universe, it is designed to be of interest to *anyone* with a desire to learn about the current state of research in astrobiology. No special scientific training or background is assumed, and all necessary scientific concepts are reviewed as they arise. If you have a basic high school education and a willingness to learn, you are capable of understanding every topic covered in this book. We wish you well in your efforts.

Jeff Bennett
Seth Shostak
Bruce Jakosky

To Current or Prospective Instructors

The rest of this preface is aimed primarily at current or prospective instructors of courses on life in the universe. Students and general readers might still find it useful, because it explains some of the motivation behind the pedagogical features and organization of this book.

Why Teach a Course on Life in the Universe?

By itself, the rapid rise of research interest in astrobiology might not be enough to justify the creation of new courses for nonscience majors. But the subject has at least three crucial features that together make a very strong case for adding it to the standard science offerings:

1. For students who take only one or a few required science courses, the interdisciplinary nature of the study of life in the universe offers a broader understanding of the range of scientific research than can a course in any single discipline.

2. Public fascination with UFOs and alien visitation offers a unique opportunity to use life in the universe courses as a vehicle for teaching about the nature of science and how to distinguish true science from pseudoscience.

3. The science of life in the universe considers many of the most profound questions we can ask, including: What is life? How did life begin on Earth? Are we alone? Could we colonize other planets or other star systems? Students are virtually always interested in these questions, making it easy to motivate even those students who study science only because it is required.

These features probably also explain the growing number of life in the universe courses being offered at colleges around the world as well as at the high school level. It's worth noting that, besides being fascinating to students, a course on life in the universe can be a great experience for instructors. The interdisciplinary nature of the subject means that no matter what your specific scientific background you are sure to learn something new when you teach an astrobiology course at any level.

Using This Book for Your Course

While courses on life in the universe are of great interest to students, teaching them has been challenging, in part because of the lack of suitable texts. This book was designed expressly to overcome this problem both by giving current instructors of life in the universe courses a resource to match their course content and by making it easier for other instructors to start teaching such courses.

This book was not written in isolation but rather was shaped by careful study of the needs of students in our own courses and by detailed consultation with other instructors teaching life in the universe courses. Before embarking on the actual writing, we undertook a survey of nearly all existing life in the universe courses in the United States, soliciting input on what would be most useful in a textbook. The content of this book represents what we believe is a broad consensus about what is needed to teach a successful course on life in the universe. This consensus is reflected both in the content selection and in the pedagogical features designed to make the content understandable to nonscience majors.

Course Types This book is designed primarily for use in courses for nonscience majors, such as core course requirements in natural science or elective follow-on courses for students who lack the preparation needed for more technical offerings in astrobiology. It can also be used at the senior high school level, especially for integrated science courses that seek to break down the traditional boundaries separating individual science disciplines.

Overall Structure We've developed this book with a four-part structure that matches the content of most courses on life in the universe. The table of contents gives more detail; a brief outline of the structure follows:

I. Introducing Life in the Universe (Chapters 1 and 2). Chapter 1 offers an overview of why the science of life in the universe has moved to the forefront of research. Chapter 2 discusses the nature of science based on the assumption that this is many students' first real exposure to how scientific thinking differs from other modes of thinking.

II. Life on Earth (Chapters 3–5). This is the first of three parts devoted to in-depth study of astrobiology issues. Here we discuss the current state of knowledge about life on Earth. Chapter 3 explores the nature of life on Earth. Chapter 4 discusses the geological history of Earth and how this history has made life possible. Chapter 5 discusses current ideas about the origin and subsequent evolution of life on Earth.

III. Life in the Solar System (Chapters 6–9). We next use what we've learned about life on Earth in Part II to explore the possibilities for life elsewhere in our solar system. Chapter 6 discusses the environmental requirements for life and the methods of searching for life in the solar system and then offers a brief tour of various worlds in our solar system, exploring their potential habitability. Chapters 7 and 8 focus on the places that seem most likely to offer possibilities for life: Mars (Chapter 7) and the jovian moons Europa, Ganymede, Callisto, and Titan (Chapter 8). Chapter 9 discusses how habitability evolves over time in the solar system, with emphasis on the past and present habitability of Venus, Earth, and Mars and the future habitability of Earth; this chapter also introduces the concept of a habitable zone around a star, setting the stage for the discussion of life beyond our solar system in Part IV.

IV. Life Among the Stars (Chapters 10–14). This final set of chapters deals with the question of life beyond our solar system. Chapter 10 focuses on recent discoveries of extrasolar planets and what they might tell us about the likelihood of finding habitable planets elsewhere; it also discusses techniques used to search for extrasolar planets and ways we might soon be able to search for signatures of life. Chapter 11 covers the search for extraterrestrial intelligence (SETI). Chapter 12 discusses the challenge of and prospects for interstellar travel and uses these ideas to help explain why most claims of alien visitation seem implausible. Chapter 13 addresses the issues behind the so-called Fermi paradox ("Where is everybody?"), the potential solutions to the paradox, and the implications of the considered solutions. Finally, Chapter 14 offers a brief philosophical wrap-up, discussing the profound implications of discovering extraterrestrial life, whether microbial or intelligent.

Pace of Course Coverage Although the chapters are not all of equal length, it should be possible to cover them at an average rate of approximately one chapter per week in a typical 3-hour college course. Thus, the 14 chapters in this book should provide about the right amount of material for a typical one-semester college course. If you are teaching a one-quarter course, you might need to be selective in your coverage, perhaps dropping some topics entirely. If you are teaching a year-long course, as might be the case at the high school level, you can spread out the material to cover it at an average rate of about one chapter every two weeks.

Pedagogical Features

This book includes a number of pedagogical features that should be of use to instructors. Here is a brief list of the features, with notes on how they are intended to be used.

End-of-Chapter Questions, Problems, and Projects
The end-of-chapter material is designed for self-study, class discussion, and assignment. It is divided into four sets of questions at the end of each chapter:

- *Review Questions.* The review questions replace a standard chapter summary. Like a chapter summary, they review the important concepts from the chapter. However, they are more effective than a standard summary as a study tool because they require students to think rather than mindlessly wield a highlight pen. While it is possible to assign review questions as homework, keep in mind that these questions generally can be answered with little more than a quick rereading of the relevant sections of the text.

- *Discussion Questions.* These questions are meant to be particularly thought-provoking and generally do not have objective answers. As such, they are ideal for class discussions.

- *Problems.* Each chapter includes several problems of varying levels of difficulty designed to be assigned as written homework. Note that every problem set is structured roughly as follows:

  - The problem set begins with questions designed to test student understanding of basic chapter concepts. These questions, which carry a label such as Sense or Nonsense? or Surprising Discoveries?, make a simple statement that students are asked to evaluate. They are designed so that they can be answered quickly in just a few sentences by students who understand the concepts, while being almost impossible for students to answer reasonably if they have not studied the chapter material adequately.

  - The initial set of problems is followed by somewhat more in-depth questions, some requiring longer essay answers.

  - Most problem sets end with a few problems that require mathematical manipulation; these are marked by an asterisk to help you identify them. All the necessary equations are given, as are hints on the necessary calculations. Keep in mind that these questions are designed to help reinforce key chapter concepts through mathematics, not to test students' mathematical ability.

- *Web Projects.* Each chapter ends with one or more Web projects designed for independent research. The projects typically ask students to learn more about a topic of relevance to the chapter, such as a current or planned space mission.

Think About It This feature, which consists of short, thought-provoking questions within the main text narrative, gives readers the opportunity to reflect on important new concepts. The Think About It questions can also serve as a starting point for classroom discussions; in particular, instructors may use them to make lectures interactive by posing them to the class and asking students to spend a few minutes discussing them in groups of two or three.

Boxed Material The book includes a number of "boxed" segments designed to be read separately from the main narrative flow. These boxes are always placed near the narrative material to which they are most relevant. Note that boxes are used for several distinct purposes:

- *Review Boxes* (example: Basic Properties of Atoms, p. 16). These boxes review concepts that students have almost certainly encountered in prior high school science courses but that are critical to understanding the chapter material. Thus, they serve as a quick review for those students who may have forgotten (or never learned) the concepts.

- *Definition Summary Boxes* (example: Key Biological Definitions, p. 64). These boxes summarize the definitions of key terms introduced in the text. The terms are grouped categorically, rather than alphabetically, and thus should be a useful study aid. Students looking for a particular term can consult the alphabetical glossary that appears at the back of the book.

- *Movie Madness* (example: *Star Wars*, p. 248). These boxes, essentially brief critical reviews of popular movies, are designed to help students see where Hollywood sometimes leads viewers astray in terms of what is scientifically plausible.

- *General Interest Boxes* (example: Greek Thought on Extraterrestrial Life, p. 31). These boxes contain supplementary discussion of topics related to the chapter material but not prerequisite to the continuing discussion.

The Big Picture This feature, which appears at the end of the narrative for each chapter, is designed to help students put the details they've learned in the chapter into the broader context of life in the universe.

Cross-References When we discuss a concept that is covered in greater detail elsewhere in the book, we include a cross-reference in brackets. For example, "[Section 5.2]" means that the concept being discussed is covered in greater detail in Section 5.2.

Glossary A detailed glossary at the back of the book makes it easy for students to look up unfamiliar terms. Every boldface term in the book is included in the glossary.

Appendixes The appendices include a number of useful references, including a list of key constants (Appendix A), a brief review of key mathematical skills such as working with scientific notation and operating with units (Appendix B), the periodic table of the elements (Appendix C), a brief description of the chemical reactions that can provide energy to life (Appendix D), planetary data for our solar system (Appendix E), and a list of useful Web sites (Appendix F).

References/Bibliography A list of references and suggested readings is organized by chapter and topic. Those marked with an asterisk should be accessible to students and general readers; the rest may be of special interest to instructors seeking technical references.

Supplements and Resources

In addition to the book itself, a number of supplements are available to help you as an instructor. The following is a brief summary; contact your local Addison Wesley representative for more information.

- *Life in the Universe Companion Web Site* (www.astronomyplace.com). This free accompanying Web site provides students with an out-of-class study aid. It includes quizzes, further reading, and extensive Web links. The site also provides resources to help instructors set up and design a course, including links to courses currently taught at other schools and advice from experts in the field.

- *Life in the Universe Activities Manual,* by Ed Prather, Erika Offerdahl, and Tim Slater (ISBN 0-8053-8735-8). This manual provides 15 creative projects that explore a wide range of concepts in astrobiology. It can be used as a laboratory component for a life in the universe course or as a source for group activities in the classroom.

Acknowledgments

A textbook may carry the names of its authors, but it is the result of the hard work of a long list of committed individuals. We could not possibly name everyone who has had a part in this book, but would like to call attention to a few people who have played particularly important roles. First, we thank the friends and family members who put up with us during the long hours that we worked on this book, with special thanks to Lisa, Grant, Brooke, RND, Mamie, Fers, and Merry. Without their support, this book would not have been possible.

Next, we thank three people who have gone to extraordinary effort to make this book possible: our editor, Adam Black; our production manager, Mary Douglas; and Margot Otway, who carefully reviewed every word in the text for scientific accuracy and pedagogical efficacy. Many other people have also played crucial roles in the book's development and production, including Joan Marsh, Nancy Benton, Chalon Bridges, Claire Masson, Blakeley Kim, and Hassan Herz.

We've also been fortunate to be able to draw on the expertise of several other Addison Wesley authors, in some cases drawing ideas and artwork directly from their outstanding texts. For their gracious help, we thank

- Neil Campbell and Jane Reece, authors of *Biology* (6th ed.) and *Essential Biology*, both from Benjamin Cummings, an imprint of Addison Wesley Longman, Inc.

- Megan Donahue, Nicholas Schneider, and Mark Voit, coauthors of *The Cosmic Perspective* (2d ed.) and *The Essential Cosmic Perspective* (2d ed.), also available from Addison Wesley.

Finally, we thank the many people who have carefully reviewed portions of the book in order to help us make it both as scientifically up-to-date and as pedagogically useful as possible:

Wayne Anderson, Sacramento City College
Timothy Barker, Wheaton College
Sukanta Bose, Washington State University
Greg Bothun, University of Oregon
Paul Braterman, University of North Texas
Juan Cabanela, Haverford College
Leo Connolly, San Bernardino State
Steven J. Dick, U.S. Naval Observatory
James Dilley, Ohio University
Jack Farmer, Arizona State University
Richard Frankel, California Polytechnic State University
Tracy Furutani, California Polytechnic State University
Bob Garrison, University of Toronto
Harold Geller, George Mason University
Bob Greeney, Holyoke Community College
Bruce Hapke, University of Pittsburgh
James Kasting, Pennsylvania State University
Jim Knapp, Holyoke Community College
Bruce Margon, Space Telescope Science Institute
Lori Marino, Emory University
Ken Nealson, University of Southern California
Norm Pace, University of Colorado, Boulder
Stacy Palen, University of Washington
Eugenie Scott, National Center for Science Education
David Thomas, Lyon College
Gianfranco Vidali, Syracuse State University
John Wernegreen, Eastern Kentucky University
William Wharton, Wheaton College
Ben Zuckerman, University of California, Los Angeles

HOW TO SUCCEED IN
A COURSE ON LIFE IN THE UNIVERSE

Many readers of this book are enrolled in a college course on life in the universe. If you are one of these readers, we offer you the following hints to help you succeed in your course.

Using This Book

Before we address general strategies for studying, we offer a few guidelines that will help you use *this* book most effectively.

- Read assigned material twice.

  - Make your first pass *before* the material is covered in class. Use this pass to get a "feel" for all the material and to identify concepts you might want to ask about in class.

  - Read the material for the second time shortly after it is covered in class. This will help solidify your understanding and allow you to make notes that will help you study for exams later.

- Take advantage of the features that will help you study.

  - Always read the *Think About It* features, and use them to help you absorb the material and to identify areas where you may have questions.

  - Use the *cross-references* indicated in brackets in the text and the *glossary* at the back of the book to find more information about terms or concepts that you don't recall.

  - Go to The Astronomy Place at **www. astronomyplace.com** to find additional study aids.

- It's your book, so don't be afraid to make notes in it that will help you study later.

- Don't highlight—underline! Using a pen or pencil to underline material requires greater care than highlighting and therefore helps keep you alert as you study. And be selective in your underlining—for purposes of studying later, it won't help if you've underlined everything.

- There's plenty of "white space" in the margins and elsewhere, so use it to make notes as you read. Your own notes will be very valuable when you are doing homework or studying for exams.

- After you complete the reading, and again when you study for exams, make sure you can answer the *review questions* at the end of each chapter. If you are having difficulty with a review question, reread the relevant portions of the chapter until the answer becomes clear.

Budgeting Your Time

One of the easiest ways to ensure success in any college course is to make sure you budget enough time for studying. A general rule of thumb for college classes is that you should expect to study about 2 to 3 hours per week *outside* of class for each unit of credit. For example, based on this rule of thumb, a student taking 15 credit hours should expect to spend 30 to 45 hours each week studying outside of class. Combined with time in class, this works out to a total of 45 to 60 hours spent on academic work—not much more than the time a typical job requires, and you get to choose your own hours. Of course, if you are working while you attend school, you will need to budget your time carefully.

As a rough guideline, your studying time in the course might be divided as shown in the table at the top of p. xv. If you find that you are spending fewer

If Your Course Is:	Time for Reading the Assigned Text (per week)	Time for Homework Assignments (per week)	Time for Review and Test Preparation (average per week)	Total Study Time (per week)
3 credits	2 to 4 hours	2 to 3 hours	2 hours	6 to 9 hours
4 credits	3 to 5 hours	2 to 4 hours	3 hours	8 to 12 hours
5 credits	3 to 5 hours	3 to 6 hours	4 hours	10 to 15 hours

hours than these guidelines suggest, you can probably improve your grade by studying more. If you are spending more hours than these guidelines suggest, you may be studying inefficiently; in that case, you should talk to your instructor about how to study more effectively.

General Strategies for Studying

- Don't miss class. Listening to lectures and participating in discussions is much more effective than reading someone else's notes. Active participation will help you retain what you are learning.

- Budget your time effectively. An hour or two each day is more effective, and far less painful, than studying all night before homework is due or before exams.

- If a concept gives you trouble, do additional reading or studying beyond what has been assigned. And if you still have trouble, ask for help. You surely can find friends, colleagues, or teachers who will be glad to help you learn.

- Working together with friends can be valuable in helping you understand difficult concepts. However, be sure that you learn *with* your friends and do not become dependent on them.

- Be sure that any work you turn in is of *collegiate quality*: neat and easy to read, well organized, and demonstrating mastery of the subject matter. Although it takes extra effort to make your work look this good, the effort will help you solidify your learning and is also good practice for the expectations that your future professors and employers will have.

Preparing for Exams

- Study the review questions, and rework problems and other assignments; try additional questions to be sure you understand the concepts. Study your performance on assignments, quizzes, or exams from earlier in the term.

- Check the resources available on the text Web site at **www.astronomyplace.com.**

- Study your notes from lectures and discussions. Pay attention to what your instructor expects you to know for an exam.

- Reread the relevant sections in the book, paying special attention to notes you have made on the pages.

- Study individually *before* joining a study group with friends. Study groups are effective only if every individual comes prepared to contribute.

- Don't stay up too late before an exam. Don't eat a big meal within an hour of the exam (thinking is more difficult when blood is being diverted to the digestive system).

- Try to relax before and during the exam. If you have studied effectively, you are capable of doing well. Staying relaxed will help you think clearly.

ABOUT THE AUTHORS

Jeffrey Bennett received a B.A. in biophysics from the University of California, San Diego (1981) and a Ph.D. in astrophysics from the University of Colorado, Boulder (1987). He currently spends most of his time as a teacher, speaker, and writer. He has taught extensively at all levels, including having founded and run a science summer school for elementary and middle school children. At the college level, he has taught more than fifty classes in subjects ranging from astronomy, physics, and mathematics, to education. He served two years as a visiting senior scientist at NASA headquarters, where he helped create numerous programs for science education. He also proposed the idea for and helped develop the *Voyage* Scale Model Solar System, which opened in 2001 on the National Mall in Washington, D.C. (He is pictured here with the model Sun.) In addition to this textbook on life in the universe, he has written college-level textbooks in astronomy, mathematics, and statistics, and a book for the general public, *On the Cosmic Horizon* (Addison Wesley, 2001). He also recently completed his first children's book, *Max Goes to the Moon* (Big Kid Science, 2003). When not working, he enjoys participating in masters swimming and in the daily adventures of life with his wife Lisa, his children Grant and Brooke, and his dog, Max.

Seth Shostak earned his B.A. in physics from Princeton University (1965) and a Ph.D. in astronomy from the California Institute of Technology (1972). He is currently a senior astronomer at the SETI Institute, in Mountain View, California, where he helps press the search for intelligent cosmic company. For much of his career, Seth conducted radio astronomy research on galaxies and investigated the fact that these massive objects contain large amounts of unseen mass. He has worked at the National Radio Astronomy Observatory in Charlottesville, Virginia as well as the Kapteyn Astronomical Institute, in Groningen, The Netherlands (where he learned to speak bad Dutch). Seth also founded and ran a company that produced computer animation for television. He has written several hundred popular articles on various topics in astronomy, technology, film, and television and teaches courses at the California Academy of Sciences and elsewhere. A frequent fixture on the lecture circuit, Seth gives approximately seventy talks annually at both educational and corporate institutions. He is a Distinguished Speaker for the American Institute of Aeronautics and Astronautics and a frequent commentator on astronomical matters for radio and television. His book *Sharing the Universe* (Berkeley Hills Books, 1998) details the rationale and expectations of the scientific search for intelligent life. When he's not trying to track down the aliens, Seth can often be found in the darkroom, where he continues to hope that something interesting will develop.

Bruce Jakosky received his B.S. degree in geophysics and space physics from U.C.L.A. (1977), and his Ph.D. in planetary science and geophysics from the California Institute of Technology (1982). He has been at the University of Colorado in Boulder since 1982 and is now a Professor in the Department of Geological Sciences and the Laboratory for Atmospheric and Space Physics. He teaches undergraduate and graduate courses in geology, planetary science, and astrobiology, and his research emphasizes planetary geology, the evolution of the surface and atmosphere of Mars, and the potential for life on Mars and elsewhere in the solar system. In addition, he is exploring the connections between society and science, especially astrobiology. He has been involved in analysis of data from the *Viking, Solar Mesosphere Explorer, Clementine, Mars Observer, Mars Global Surveyor,* and *Mars Odyssey* spacecraft missions. He heads up the University of Colorado's astrobiology program, which is a part of the NASA Astrobiology Institute. He has served on numerous advisory committees within NASA and the National Research Council and has been editor or a member of the editorial board of a number of planetary science and astrobiology journals. He is a co-editor of the book *Mars* (Univ. Arizona Press, 1992), and author of the book *The Search for Life on Other Planets* (Cambridge Univ. Press, 1998). He doesn't really like gardening but does it to keep his yard from becoming completely overgrown with weeds, and he appears to love flying judging by the amount of time he spends traveling to meetings.

CHAPTER 1

A Universe of Life?

Gazing upward at night, we can't help but be amazed at the stars glittering in the sky. Stars are suns, much like our own Sun, each shining brightly for millions or billions of years. Many have planets, some of which may be much like the planets in our own solar system. Among these countless worlds, it is hard to imagine that Earth is the only home for life. Nevertheless, we do not yet have proof that life exists beyond Earth, and in the absence of evidence we cannot be sure that it does.

Learning whether the universe is full of life holds great significance for the way we view ourselves and our planet. If life is rare or nonexistent elsewhere, we will view our planet Earth with added wonder. If life is common, we'll know that the Earth is not quite as special as it may seem. If civilizations are common, we'll be forced to accept that we ourselves are just one of many intelligent species throughout the universe. The profound implications of finding—or not finding—extraterrestrial life are what makes the question of life beyond Earth such an exciting topic of study. In this first chapter, we'll learn why scientists suspect that the universe *might* be full of life. We'll also explore just exactly what we mean by the study of life on Earth and beyond.

FIGURE 1.1 Aliens have become a part of modern culture, as illustrated in this 1956 movie advertisement.

1.1 Life Elsewhere

It may seem as if aliens are everywhere (Figure 1.1). Television starships like the *Enterprise* or *Voyager* are on constant prowl throughout the galaxy, seeking out new life and hoping it speaks English (or something close enough to English for the "universal translator"). In *Star Wars*, aliens from many planets gather at bars to share drinks and stories, and presumably to marvel at the fact that they have greater similarity in their level of technology than do different nations on Earth. Closer to home, tabloids like the *Weekly World News* routinely carry headlines about the latest alien atrocities committed against humans or about which candidate aliens support for president. Even serious newspapers and magazines carry occasional articles about UFO sightings or about claims that the U.S. government is hiding frozen alien corpses at "Area 51." Given this intense media interest, perhaps it's unsurprising that close to half of all Americans claim to believe that we have already been visited by beings from other worlds.

Scientists are interested in aliens too, although most scientists remain deeply skeptical about reports of aliens on Earth [Sections 2.3, 12.6]. In the past few decades, the question of life beyond Earth has changed from one primarily philosophical in nature to one of serious scientific research. Most of this research is directed toward the possibility of finding microbial life, but some also seeks to learn about the potential existence of intelligent aliens. Fuel for this growing scientific interest has come from startling discoveries in many fields of research.

Prior to the twentieth century, many scientists assumed that other planets in our own solar system would harbor life. For example, some guessed that the clouds of Venus might hide a tropical paradise, and others thought they saw evidence of seasonal changes in vegetation on Mars. A few even claimed to see canals on Mars, which some (notably Percival Lowell [Section 7.1]) interpreted as evidence of a martian civilization. However, better data gradually

---

Movie Madness: Cinema Aliens

Aliens have invaded Hollywood.

Helped by new computer graphics technology that has simplified the creation of weird-looking extraterrestrials, the silver screen is now crawling with cosmic critters eager to chow down on us, abduct us, or just tick us off as they total our planet. A few, like loveable little E.T., have kinder intentions: that big-eyed, wrinkled little extraterrestrial came two million light-years (the distance from the Andromeda galaxy) merely to pick some plants and hang with the kids.

Why are aliens suddenly infesting the local multiplex? Partly it's because after the collapse of the Soviet Union, Hollywood had to hunt around for a new source of bad guys. But our own space program has also convinced many among the popcorn-eating public that visiting other worlds will be a walk in the park for any advanced species. The movie moguls quietly ignore the fact (which you'll encounter later on in this book) that traveling between the stars is enormously more difficult than checking out the planets of your own solar system. The aliens won't do it just to abduct you for unauthorized breeding experiments.

But the really big problem with Hollywood aliens, other than the fact that they seldom wear any clothes, is that they are inevitably portrayed as being close to our own level of development. This is so that we can take them on in aerial dogfights, or challenge them to a manly light-saber duel. But the reality is somewhat different. As we'll discuss in Chapter 13, if we ever make contact with real aliens, their culture will probably be thousands, millions, or billions of years beyond ours.

Of course, an invasion by hostile aliens with a million-year head start on *Homo sapiens* wouldn't make for an interesting movie. It would be Godzilla versus the chipmunks. But you don't mistake the movies for reality, do you?

FIGURE 1.2 Today, the surface of Mars is too cold and the atmosphere too thin for liquid water to flow. But this photograph from NASA's Pathfinder lander (which landed on Mars in 1997) shows an ancient floodplain, part of the evidence that Mars once had flowing water. Scientists wonder whether life may have evolved on Mars during its warmer, wetter past.

FIGURE 1.3 This photograph shows Jupiter and two of its moons: Io is the moon in front of Jupiter's Great Red Spot, and Europa is to the right. Scientists suspect that Europa has a deep ocean beneath its surface of ice, making it a prime target in the search for life in our solar system.

proved these guesses incorrect. By the late-1970s, when the Viking missions to Mars found no clear evidence of life [Section 7.3], even the prospect of finding microbes seemed dim. However, discoveries made in the past three decades have rejuvenated hopes that life might exist elsewhere in our solar system.

By and large, we have not yet been able to search for actual living organisms on other worlds. Nevertheless, we are learning more and more about the potential habitability of other worlds. Note that, when we ask whether a world is **habitable,** we are asking whether it offers environmental conditions under which life of some kind could arise or survive, not whether it actually harbors life. Today, growing evidence for the past habitability of Mars (Figure 1.2) and for an ocean under the icy surface of Jupiter's moon Europa (Figure 1.3) has made these two worlds prime targets in the search for extraterrestrial life. Equally important, new discoveries in biology have shown that some forms of life on Earth can survive under an astonishing range of conditions. These discoveries expand the range of environmental conditions that qualify as habitable and make it more likely that life might exist on other planets or moons.

If life is discovered elsewhere in our solar system, it's unlikely to be anything with which you could carry on a conversation—unless you enjoy talking to microbes. Now that we have detailed pictures of most of the planets in our solar system, there seems little chance that we'll find intelligent life nearby. If we want new friends, we'll have to pin our hopes on planets around other stars. Here, too, recent discoveries give at least some reason for optimism. As re-

cently as 1995, no one was sure whether planets even existed around other stars like the Sun. Since that time, discoveries of such **extrasolar planets** have become commonplace, and the known planets around other stars now far outnumber the planets of our own solar system. (These discoveries have been made by detecting gravitational tugs of the planets on their stars; we have not taken actual pictures of any of these planets [Section 10.3].) Although our technology is not yet capable of discovering planets as small as Earth, the existence of so many large planets suggests that small planets might also be common. Moreover, it's conceivable that some of the newly discovered planets could have moons with Earth-like surface conditions. These possibilities are spurring astronomers to develop new technologies for

FIGURE 1.4 This 140-foot radio telescope in West Virginia was used in 1996 to search for signals from extraterrestrial civilizations.

searching for Earth-size planets, while biologists, chemists, and geologists are considering how we might use telescopic observations to determine whether life actually exists on such planets.

The growing optimism for finding life has also spawned renewed enthusiasm for attempts to detect other civilizations that might be out there. These efforts, known as the **search for extraterrestrial intelligence** (or **SETI**), involve scanning the skies for signals broadcast by aliens (Figure 1.4). Although we cannot know whether the search will meet with success, we can be sure that the unambiguous receipt of an alien message would be one of the most significant discoveries in human history.

All in all, the possibility of finding aliens of any kind, from microbes to advanced civilizations, is bringing together researchers from almost every field of science and engineering. In the rest of this book, we will examine the current state of this research in some detail. Let's begin by getting a better idea of why the question of life elsewhere now seems so worthy of study.

1.2 Worlds Beyond Imagination

Until about four hundred years ago, most people assumed that the Earth was the center of the universe and that the Sun, Moon, planets, and stars belonged to an entirely separate realm known as "the heavens." This *geocentric* (Earth-centered) view of the universe gave the Earth a unique place in the cosmos and implied a clear distinction between Earth and anyplace else. Although the geocentric belief did not prevent people from speculating about inhabitants of the heavens (often imagined to be godlike), it certainly limited the possibilities for Earth-like life. Our modern view couldn't be more different.

Today, we know that our planet is just one of nine planets in our **solar system,** which consists of the Sun, the planets and their moons, and countless smaller objects including asteroids, comets, and specks of interplanetary dust. Our solar system, in turn, is just one of more than 100 billion star systems that make up the **Milky Way Galaxy.** And our galaxy is one of some 100 billion galaxies in our universe (Figure 1.5).

Numbers like 100 billion are truly astronomical; it takes some effort to conceive of their size. Let's start by considering a galaxy of 100 billion stars. Imagine that, tonight, you are having difficulty falling asleep, perhaps because you are contemplating the vastness of the Milky Way Galaxy. Instead of counting sheep, you decide to count stars. If you count about one star each second, how long would it take to count 100 billion stars? Clearly, the answer is about 100 billion seconds. But how long is that? You can get the answer by dividing 100 billion seconds by 60 seconds per minute, 60 minutes per hour, 24 hours per day, and 365 days per year. If you do this calculation, you'll find that 100 billion seconds is more than 3,000 years. In other words, you would need thousands of years just to *count* the stars in the Milky Way Galaxy, let alone study them or search their planets for signs of life. And this assumes you never take a break—no sleeping, no eating, and absolutely no dying!

The number of stars in the universe is even more incredible. Just as it would take thousands of years to count the 100 billion (or more) stars in the Milky Way, it would take thousands of years to count the 100 billion galaxies in our universe. How can we conceive of the total number of stars in the universe? Visit a beach. Run your hands through the fine-grained sand. Try to imagine counting every tiny grain of sand as it slips through your fingers (Figure 1.6). Then imagine continuing to scoop up and count the grains until you have counted every grain of sand on the beach. Next think about visiting every beach on Earth and counting every grain of dry sand you

FIGURE 1.5 Our place in the universe. Earth is one of nine planets in our solar system. Our solar system is one of more than 100 billion star systems in the Milky Way Galaxy. The Milky Way is one of some 100 billion galaxies in our universe.

the Milky Way Galaxy

the solar system (not to scale)

the local group

the local supercluster

the universe

Earth

FIGURE 1.6 The number of stars in our universe is roughly the same as the number of grains of dry sand on all the beaches on Earth.

can find. Of course, you could never actually complete this task. But if you could, you'd eventually know the total number of grains of sand on all the beaches on Earth. Incredibly, this number is roughly the same as the number of stars in our universe.

The total number of *worlds*—by which we mean any reasonably large bodies in space, such as planets, moons, or even large asteroids—may be even greater.

If planetary systems are as common as recent discoveries suggest, many or even most stars may have at least a few planets or moons, some of which could potentially harbor life. Clearly, our universe contains worlds beyond imagination.

THINK ABOUT IT . . . *Contemplate the fact that there are as many stars as grains of sand on all the beaches on Earth and that each star is a potential sun for a system of planets. With so many possible homes for life, do you think it is conceivable that life exists only on Earth? Why or why not?*

The Vast Distances Between Worlds

The sheer number of possible planets may make it seem as if it should be easy to search for life on at least some of them, but the vast distances between planets and stars make the search extremely challenging. One of the best ways to understand the challenge is to imagine our solar system shrunk down to a scale on which you could walk through it, such as the scale of the *Voyage* scale model solar system in Washington, D.C. (Figure 1.7). The *Voyage* model shows the Sun and the planets, and the distances between them, at *one ten-billionth* the actual sizes and distances.

Basic Astronomical Definitions

This box summarizes a few key astronomical definitions introduced in this chapter and used throughout the book.

star: Our Sun and other ordinary stars are large, glowing balls of gas that generate heat and light through *nuclear fusion*—the smashing together of light nuclei to make heavier nuclei—in their cores. (The term *star* is also applied to objects that are in the process of becoming true stars, such as protostars, and to the remains of stars that have died, such as neutron stars.)

planet: A planet is a moderately large object—either rocky or gaseous—that orbits a star. There is no official minimum or maximum size for planets, which can lead to disagreement about what counts as a planet. On the minimum side, for example, some astronomers argue that Pluto is too small to count as a planet. On the maximum side, astronomers disagree on the dividing line between large planets (up to several times the size of Jupiter) and "failed stars" (called *brown dwarfs*) that are starlike but too small to sustain significant nuclear fusion in their cores.

extrasolar planet: A planet orbiting a star other than our Sun.

habitable planet (or *habitable world*): A planet (or world) with environmental conditions under which life could potentially arise or survive. (Note that the term *world* is used generically to refer to any large body in space, such as a planet, moon, or large asteroid.)

moon (or *satellite*): An object that orbits a planet. The term *satellite* is also used more generally to refer to any object orbiting another object.

asteroid: A relatively small and rocky object that orbits a star; asteroids are sometimes called *minor planets* because they are similar to planets but smaller.

comet: A relatively small and icy object that orbits a star.

star system: One or more stars and any planets and other material that orbit them. The term *solar system* technically refers to our own star system (because *solar* means "of the Sun") but is sometimes used to describe other star systems.

galaxy: A great island of stars in space, containing from a few hundred million to a trillion or more stars, all held together by gravity and orbiting a common center.

universe (or *cosmos*): The sum total of all matter and energy; that is, everything within and between all galaxies.

On the *Voyage* scale, the Sun is about the size of a grapefruit, Jupiter is about the size of a marble, and Earth is about the size of a pinhead (Figure 1.8a). Compared to their small sizes, the distances between the planets are substantial (Figure 1.8b). The inner planets (Mercury, Venus, Earth, and Mars) all lie within just a few steps of the Sun on this scale. The outer planets (Jupiter, Saturn, Uranus, Neptune, and Pluto) extend out to a distance of about 600 meters (about ⅓ mile). When we view it on this scale, we are struck by the emptiness of our solar system. Over an area the size of hundreds of football fields, the only objects large enough to be seen by the naked eye are the grapefruit-size Sun, the nine planets, and a few moons.

Seeing the solar system to scale can help us understand why the search for life in the solar system is only just beginning. The Moon, the only other world on which humans have ever stepped, lies only about 4 centimeters away from Earth in the model. Thus, on this scale, the palm of your hand can cover the entire region of the universe in which humans have so far traveled. Our robotic spacecraft have visited all the planets except Pluto, but with current technology these journeys can take many years. For example, while you can walk from the Sun to Pluto in just a

FIGURE 1.7 The Sun and the locations of the inner planets in the *Voyage* scale model solar system on the National Mall in Washington, D.C.; the planets themselves are too small to see in this photograph.

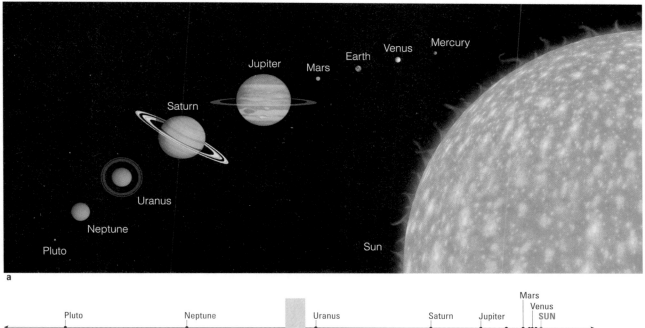

FIGURE 1.8 The *Voyage* model shows the Sun and planets, and the distances between them, at *one ten-billionth* the actual sizes and distances. (**a**) This painting shows the sizes (but not distances) of the Sun and planets in the *Voyage* model. (**b**) This map shows the locations of the Sun and planets in the *Voyage* model. The total distance from the Sun to Pluto in this model is about 600 meters. Remember that, while *Voyage* lays out the planets in a line from the Sun, in reality each planet orbits the Sun independently. Thus, the "emptiness" of the solar system is even greater than it appears when walking through the *Voyage* model.

few minutes on the *Voyage* scale, the real journey would take more than a decade for our fastest current spacecraft.

Distances to the Stars

Where are other stars on the scale of our model solar system? Amazingly, on the same scale that allows you to walk from the Sun to Pluto in minutes, even the nearest stars would be more than 4,000 kilometers (2,500 miles) away. If you started at the grapefruit-size Sun in the *Voyage* model in Washington, D.C., you'd have to walk to California to find the next grapefruit-size star. Now you can see why it is so difficult to discover planets around other stars. Trying to see the Earth from the nearest star (besides the Sun) would be like looking from San Francisco to try to see a pinhead in Washington, D.C.

The vast distances to the stars also show why interstellar travel remains far beyond our current technology. We have so far sent four spacecraft on paths that will eventually carry them to the realm of other stars: *Pioneer 10, Pioneer 11, Voyager 1,* and *Voyager 2.* These four spacecraft flew past and studied the outer planets of our solar system during the 1970s and 1980s and are now continuing their outward journeys. Traveling at speeds close to 50,000 kilometers per hour—about 100 times faster than a speeding bullet—these spacecraft are about the fastest vehicles we have ever built. But even at such speeds, these spacecraft will take some 100,000 years to reach the distances of the nearest stars. Crossing the Milky Way Galaxy would take such spacecraft more than a billion years. Clearly, our current technology is nowhere near what we see in science fiction shows like *Star Trek,* in which humans just a couple of hundred years in the future can travel rapidly among the stars. In fact, there's good reason to believe travel like that shown in *Star Trek* is impossible, although we cannot be certain. We'll discuss the prospects for interstellar travel in Chapter 12.

THINK ABOUT IT . . . *Try to combine the distance scale with the number of stars to get a sense of the immensity of the Milky Way Galaxy. Stars are typically separated like grapefruits thousands of miles apart, but at the same time the galaxy contains so many stars that it would take thousands of years just to count them. Can you think of other ways to describe the vastness of our galaxy? Do you think a civilization could ever fully explore the Milky Way Galaxy? Defend your opinion.*

Communicating Between Star Systems

Because interstellar travel will remain impractical for the foreseeable future, our only hope for communicating with civilizations on distant planets lies with signals carried by radio waves or other forms of light. Light travels extremely fast; in fact, according to Einstein's theory of relativity, nothing can travel faster than light. The speed of light is about 300,000 kilometers per second (186,000 miles per second), which means that light could circle the Earth nearly eight times in just 1 second. But despite its awesome speed, even light requires years to cross the vast chasms between the stars. As a result, it can be useful to think of interstellar distances in terms of light travel times.

If you multiply the speed of light (300,000 kilometers per second) by the number of seconds in a year (about 31.5 million seconds), you'll find that light can travel about 10 trillion kilometers (6 trillion miles) in 1 year. We therefore call this distance a **light-year;** that is, 1 light-year is about 10 trillion kilometers (more precisely, 9.46 trillion kilometers). Note that a light-year is a unit of distance, not of time.

The nearest star besides the Sun is about 4 light-years away, meaning that light from this star takes about 4 years to reach us. Thus, if we sent a radio message to a planet around this star, it would take about 4 years for the message to get there. (Note that radio waves are a form of light; see box on p. 103.) It would take another 4 years for us to receive a reply from anyone who happened to be living there. Clearly, it would require a lot of patience to carry on a true conversation, because we'd have to wait 8 years for a reply to each message we sent.

How Far Is a Light-Year?

It's easy to calculate the distance represented by a light-year if you recall that

$$distance = speed \times time$$

For example, if you travel at a speed of 50 kilometers per hour for 2 hours, you will travel 100 kilometers. A light-year is the distance covered by light, traveling at a speed of 300,000 kilometers per second, in a time of 1 year. In the process of multiplying the speed and the time, you must convert the year to seconds in order to arrive at a final answer in units of kilometers.

$$1 \text{ light-year} = (\text{speed of light}) \times (1 \text{ yr})$$
$$= \left(300,000 \, \frac{km}{s}\right) \times \left(1 \, yr \times \frac{365 \, days}{1 \, yr}\right.$$
$$\left. \times \frac{24 \, hr}{1 \, day} \times \frac{60 \, min}{1 \, hr} \times \frac{60 \, s}{1 \, min}\right)$$
$$= 9,460,000,000,000 \text{ km}$$

That is, "1 light-year" is just an easy way of saying "9.46 trillion kilometers" or "almost 10 trillion kilometers."

FIGURE 1.9 The Orion Nebula (photo below) is a giant cloud of gas and dust in which new stars are forming. It is located about 1,500 light-years away, which makes it relatively nearby within the Milky Way Galaxy. The diagram shows the location of the Orion Nebula within the constellation Orion; it is visible to the naked eye, and you can see some detail with a good pair of binoculars.

The situation quickly becomes more complicated for more distant stars. Consider the Orion Nebula, an interstellar cloud of gas and dust that lies about 1,500 light-years away and appears to the naked eye in the sword of the constellation Orion (Figure 1.9). Light from the Orion Nebula, including any signals sent by aliens living there, takes about 1,500 years to reach us. Thus, if we were to receive a message from the Orion Nebula, it would have to have been sent some 1,500 years ago, around the end of the Roman Empire on Earth. If we sent a message in return, we couldn't expect to hear a reply for at least 3,000 years.

The Milky Way Galaxy stretches about 100,000 light-years from end to end, and we lie about midway out on one side, about 28,000 light-years from the center (Figure 1.10). Thus, it takes tens of thousands of years for light (or messages) from most of the Milky Way's star systems to reach Earth. Even if there are billions of habitable planets in our galaxy, we have no hope of carrying on two-way conversations with most of them. In essence, then, SETI efforts listen primarily for signals that may have been sent in the distant past.

The difficulty of interstellar communication may be disheartening in some ways, but being aware of it can help us determine search strategies. For example, because most talk would be one-way communication, broadcasting civilizations would presumably go to great lengths to make sure they could be heard and understood by others. We will discuss SETI search strategies in Chapter 11.

1.3 The Stuff of Life

In our discussions so far, we've assumed that planets should be common around stars throughout the universe. However, all the extrasolar planets discovered to date are quite nearby on the scale of the Milky Way Galaxy. Why, then, do we believe that planets should be common everywhere? Part of the answer lies in the fact that, in science, we always assume that our location in the universe is typical of many other places, and not special in any way. In that case, the discovery of numerous extrasolar planets nearby must imply that there are many others waiting to be discovered at greater distances. In addition to this observational evidence, our understanding of how stars and planets are made gives us good reason to think that planets must be common. To see why, we must look briefly at the history of matter in the universe.

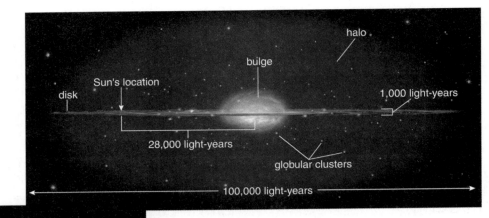

a Artist's conception of the Milky Way viewed from its outskirts.

halo

bulge

Sun's location

disk

28,000 light-years

1,000 light-years

globular clusters

100,000 light-years

b Edge-on schematic view of the Milky Way.

FIGURE 1.10 The Milky Way Galaxy. The disk of the galaxy, marked by the majestic spiral arms, contains most of the visible stars, mixed with substantial quantities of interstellar gas and dust. The disk rotates, carrying our solar system in a complete orbit around the galactic center every 230 million years. The region above and below the disk is called the galaxy's *halo*. The halo contains a much sparser population of stars than the disk and virtually no gas or dust. Many of the halo stars are grouped together in *globular clusters*, which typically contain a few hundred thousand stars.

globular clusters

halo

bulge

disk

spiral arms

halo

Strong evidence now points to the idea that our universe was born somewhere between about 12 and 15 billion years ago, in an event we call the **Big Bang** (see the box "The Extent of the Universe," p. 12). The universe began in a state of extremely high density and temperature and has been expanding and cooling ever since. However, even as the universe as a whole expands, on smaller scales the force of gravity has drawn matter together to make galaxies. That is, galaxies represent localized places where gravity has won out against the overall expansion.

Within galaxies, gravity drives the collapse of clouds of gas and dust to form stars (and their planets). Stars are certainly not living organisms, but they nonetheless go through "life cycles." After its birth in a giant cloud of gas and dust, a star shines for millions or billions of years by carrying out nuclear reactions in its core. A star dies when it exhausts its useable fuel. When a star dies, it blows much of its gas back out into space. The returned matter mixes with other interstellar matter, eventually forming new clouds of gas from which new generations of stars can be born. Thus, the Milky Way Galaxy is in many ways like a giant recycling plant, recycling matter

from dead stars into new generations of living stars. Our own Sun is a product of many generations of such recycling—the Milky Way Galaxy predates our Sun by at least 5 billion years.

Planets like Earth must be made from material that has been cycled through generations of stars. Based on evidence we'll discuss shortly, we have good reason to believe that the early universe contained only the simplest chemical elements: hydrogen and helium (and a trace amount of lithium).[1] But we and the Earth are made primarily of "other" elements such as carbon, nitrogen, oxygen, and iron. Where did all these other elements come from? Remarkably, modern science tells us that all these elements were manufactured inside stars (or during stellar explosions that occur at the end of the lives of massive stars). This means that most of the atoms from which we and the Earth are made were manufactured inside stars that lived and died long ago. In the words of noted astronomer Carl Sagan (1934–1996), we are "star stuff."

Figure 1.11 summarizes the basic history we have just described. In the rest of this section, we'll investigate stellar lives and planet formation in a bit more detail. As we'll see, the same processes that led to the existence of our solar system probably have created planetary systems around many or most other stars.

---

[1] Interestingly, studies of the distribution of mass in the universe (based on observing gravitational effects) suggest that the vast majority of the matter in the universe is *not* made from atoms. We do not yet know the nature of this mysterious matter, which is usually referred to as *dark matter*.

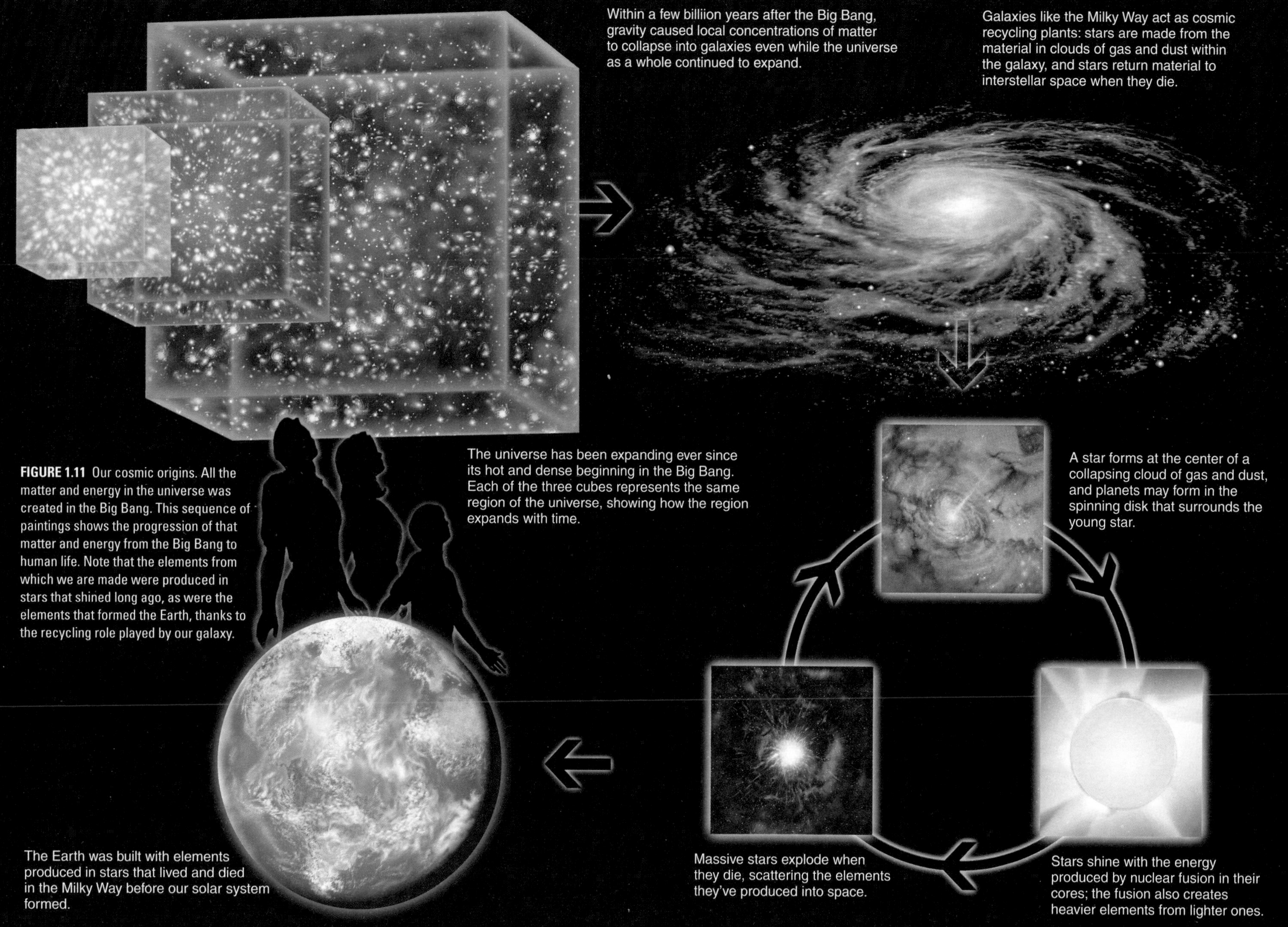

Galaxies like the Milky Way act as cosmic recycling plants: stars are made from the material in clouds of gas and dust within the galaxy, and stars return material to interstellar space when they die.

A star forms at the center of a collapsing cloud of gas and dust, and planets may form in the spinning disk that surrounds the young star.

Stars shine with the energy produced by nuclear fusion in their cores; the fusion also creates heavier elements from lighter ones.

Massive stars explode when they die, scattering the elements they've produced into space.

Within a few billion years after the Big Bang, gravity caused local concentrations of matter to collapse into galaxies even while the universe as a whole continued to expand.

The universe has been expanding ever since its hot and dense beginning in the Big Bang. Each of the three cubes represents the same region of the universe, showing how the region expands with time.

FIGURE 1.11 Our cosmic origins. All the matter and energy in the universe was created in the Big Bang. This sequence of paintings shows the progression of that matter and energy from the Big Bang to human life. Note that the elements from which we are made were produced in stars that shined long ago, as were the elements that formed the Earth, thanks to the recycling role played by our galaxy.

The Earth was built with elements produced in stars that lived and died in the Milky Way before our solar system formed.

The Extent of the Universe

The universe is the sum total of all matter and energy, and no one knows the precise size of the entire universe. How, then, can we say that there are some 100 billion galaxies in "our universe"? The answer comes down to a fundamental limitation on the portion of the universe that we can see, even in principle. The limitation has to do with the age of the universe. To understand how we know this age (at least approximately), we must first investigate how we learned that the universe is expanding.

At the dawn of the last century, many astronomers assumed that the universe as a whole was permanent and largely unchanging. However, thanks to work started in the 1920s by Edwin Hubble (for whom the Hubble Space Telescope is named), we now know that the universe is expanding. That is, average distances between galaxies in the universe are increasing with time, and space itself is growing to account for these larger distances. (The galaxies themselves are *not* expanding, nor are star systems or planets expanding; it is only the space between galaxies that is growing.)

Hubble discovered the universal expansion by observing many galaxies and the speeds at which they appear to move relative to Earth. He found that all galaxies (except for a few nearby members of our *Local Group* of galaxies) are moving away from the Milky Way, and the farther away they are, the faster they are going. Figure 1.12 uses a simple analogy to show why this observation tells us that the universe is expanding. Imagine that you make a raisin cake in which the distance between adjacent raisins is 1 centimeter. You place the cake in an oven, where it expands as it bakes. After 1 hour, you remove the cake, which has expanded so that the distance between adjacent raisins has increased to 3 centimeters. From the outside, the expansion of the cake is fairly obvious. But what would you see if you lived in, say, the "Local Raisin" inside the cake? The table in Figure 1.12 shows the answer. Notice, for example, that Raisin 1 starts out at a distance of 1 centimeter before baking and ends up at a distance of 3 centimeters after baking, which means it moves a distance of 2 centimeters away from the Local Raisin during the hour of baking. Hence, its speed as seen from the Local Raisin is 2 centimeters per hour. Raisin 2 moves from a distance of 2 centimeters before baking to a distance of 6 centimeters after baking, which means it moves a distance of 4 centimeters away from the Local Raisin during the hour—making its speed 4 centimeters per hour, or twice as fast as the speed of Raisin 1. Generalizing, the fact that the cake is expanding means that all raisins are moving away from the Local Raisin, with more distant raisins moving away faster. Thus, Hubble's discovery that more distant galaxies move away from our galaxy faster than nearer ones must imply that the universe in which we live is expanding, much like the raisin cake.

The fact that distances between galaxies are increasing implies that these distances must have been smaller in the past. If we go back far enough, there must have been a time when galaxies (or the matter from which they were made) were all on top of one another, marking the beginning of the universal expansion. This idea that expansion

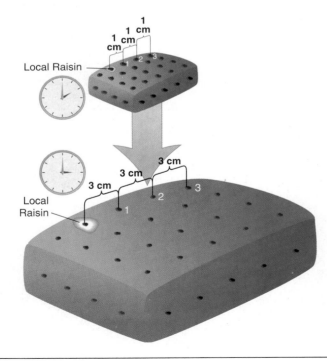

Distances and Speeds As Seen from the Local Raisin

Raisin Number	Distance Before Baking	Distance After Baking (1 hour later)	Speed
1	1 cm	3 cm	2 cm/hr
2	2 cm	6 cm	4 cm/hr
3	3 cm	9 cm	6 cm/hr
⋮	⋮	⋮	⋮

FIGURE 1.12 From the outside, the raisin cake appears to expand uniformly. But from the inside, anyone living in one of the raisins would find that all other raisins are moving away as the cake expands, with more distant raisins moving away faster. This analogy shows why the fact that more distant galaxies move away from us faster than nearer ones implies that our universe is expanding.

implies a precise moment of beginning for the universe was first enunciated by the astronomer and Catholic priest Georges Lemaître. Today, we call this beginning the Big Bang. Although Lemaître based his original idea solely on the logic of working the current expansion backward in time, today the Big Bang is supported by two other lines of strong, independent evidence. First, we have detected radiation in space left over from the Big Bang (called the *cosmic microwave background*), and its characteristics precisely match those predicted by calculations based on the theory of the Big Bang. Second, the Big Bang theory allows us to do calculations in which we essentially run the expansion backward to determine the temperature and density of the universe in the past. Based on the temperature and density, we can determine when nuclear fusion occurred in the early universe (remarkably, during only about the first 3 minutes after the Big Bang) and thereby predict the chemical composition of the universe. These predictions tell us that the chemical composition of the universe should have ended up about three-fourths hydrogen and one-fourth helium—which is precisely what we observe for the actual chemical composition of the universe. This excellent agreement between prediction and observation gives strong support to the Big Bang theory.

Hubble's original measurements of the universal expansion were fairly crude and have been greatly improved upon since his time. Based on the best available data for the current separations of galaxies and the speeds at which galaxies are moving apart, astronomers now calculate that the Big Bang must have occurred somewhere between about 12 and 15 billion years ago.

The age of the universe limits how far we can see because light takes time to travel through space (Figure 1.13). To understand why, note that looking to great distances means looking far back into the past. For example, if we look at a galaxy that is 1 billion light-years away, its light has taken 1 billion years to reach us—which means we are seeing it as it looked 1 billion years ago. (This assumes we have properly accounted for expansion during the billion years; more technically, we are talking about a *lookback time* of 1 billion years.) If we look at a galaxy that is 6 billion light-years away, its light has taken 6 billion years to reach us—which means we are seeing it as it looked 6 billion years ago, when the universe was about half its current age. Now suppose the universe is 12 billion years old. In that case, we cannot possibly see anything that is more than 12 billion light-years away, because its light has not yet had enough time to reach us. Thus, if the universe is 12 billion years old, our *observable universe*—the portion of the entire universe that we can potentially observe—consists only of objects that lie within 12 billion light-years of us. (A similar argument applies if the universe is some other age.) Note that this fact does not put any limit on the size of the entire universe, which is almost certainly far larger than our observable universe. But we have no hope of seeing or studying anything beyond the bounds of our observable universe.

Based on counts of the number of galaxies we can see when we photograph small pieces of the sky with powerful telescopes, astronomers estimate that the observable universe contains about 100 billion galaxies. For simplicity, in this book we will refer to the observable universe simply as "our universe."

We see this galaxy as it was 6 billion years ago, when the universe was only about half its current age.

We see this galaxy as it was 10 billion years ago—so if the universe is 12 billion years old today, we are seeing this galaxy as it looked when the universe was only 2 billion years old.

Light from this distance shows us how the universe looked very shortly after the Big Bang.

If the universe is 12 billion years old and we try to look to a distance of, say, 13 billion light-years, we are looking to a time before the universe existed—which means we cannot see anything at this distance, even in principle.

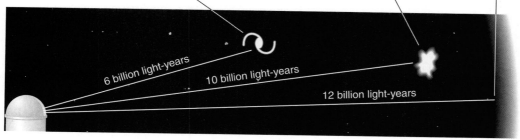

6 billion light-years

10 billion light-years

12 billion light-years

FIGURE 1.13 Because light travels at a finite speed, looking farther away in space means looking further back in time. This fact limits the extent of our observable universe. This figure assumes the universe is 12 billion years old; its actual age may be anywhere between about 12 and 15 billion years.

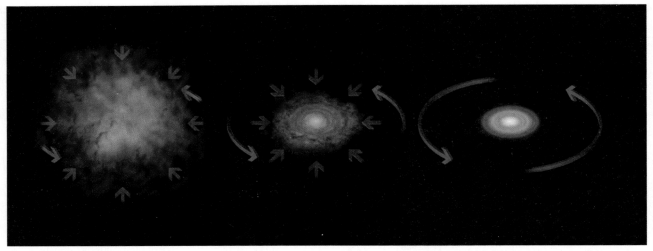

a The original cloud is large and diffuse, and its rotation is imperceptibly slow.

b The cloud heats up and spins faster as it contracts.

c The result is a spinning, flattened disk, with mass concentrated near the center. Temperature is also highest near the center.

FIGURE 1.14 This sequence of paintings shows how the gravitational collapse of a large cloud of gas causes it to become a spinning disk of matter. The hot, dense central bulge becomes a star, while planets can form in the surrounding disk.

The Birth of Stars and Planets

Stars are born when gravity pulls together a large, interstellar gas cloud. These clouds are made almost entirely of hydrogen and helium—roughly three-quarters hydrogen and one-quarter helium, by mass. Only about 2% of the cloud material consists of heavier elements (this fraction was smaller in the past, before stellar recycling made these heavier elements). Because new stars are made from this interstellar material, they are born with the same chemical composition of roughly 98% hydrogen and helium and 2% other elements.

A typical star-forming cloud, such as the Orion Nebula (see Figure 1.9), gives birth to several thousand stars. However, the stars are not all born at once; the individual stars may be born over a time span of many millions of years, depending on how rapidly gravity can make small pieces of the cloud collapse into stars.

Stars form not as isolated spheres but rather as the central objects within broad, spinning disks of gas. Planets can be born within these disks as part of the star formation process. The disks form because of three critical processes that occur when gravity collapses a cloud of gas (Figure 1.14):

- As the cloud shrinks in size, it gradually spins faster and faster. This spin-up occurs for the same reason that ice skaters spin faster when they pull in their arms (a phenomenon known in physics as "conservation of angular momentum").

The original cloud is so large and diffuse that its rotation may be imperceptible, but as it shrinks in size the rotation becomes noticeable, just as is the case with an ice skater.

- As the rate of spin increases, the cloud flattens into a disk. The flattening occurs because any gas particles that are not moving in the plane of rotation tend to collide with each other and with particles in the disk, which gradually forces all the particles into the same plane.

- Near the cloud center, the temperature rises as the cloud density and pressure increase. (This occurs because as the cloud shrinks its gravitational potential energy is converted into thermal energy.) When the temperature grows hot enough, nuclear reactions can begin and the central object ignites as a star.

Because all three processes must occur in any collapsing cloud, scientists believe that virtually all stars are surrounded by spinning disks as they are born. Observations confirm that many young stars are surrounded by such disks (Figure 1.15). There are many possible ways in which the disks can later be destroyed. For example, in many star systems the central object rotates so fast that it splits into two stars (making a binary star system), and the competing gravitational tugs from the two stars may disrupt any disk. Nevertheless, the existence of our own solar system (and other known planetary systems)

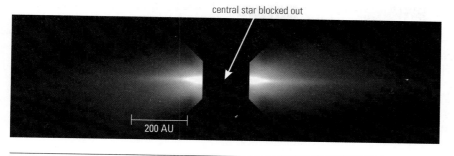

central star blocked out

200 AU

FIGURE 1.15 This Hubble Space Telescope photo shows an edge-on view of a disk of material around a star called Beta Pictoris. Similar disks have been detected around many other young stars, providing observational evidence in support of the idea that spinning disks surround most or all stars when they are born.

proves that planets do form within these disks in at least some cases. Before we discuss how planets form, let's first investigate the lives of stars to see where planet-building material comes from.

Stellar Lives and the Origin of Elements

Although elements besides hydrogen and helium make up only about 2% of the chemical constituents of our galaxy, these heavier elements are crucial to the existence of life. Rocky planets like Earth are made almost entirely of these elements. (Hydrogen and helium make up less than 0.1% of the Earth's mass.) Life itself is also made primarily of these other elements (especially carbon, oxygen, and nitrogen); living cells contain some hydrogen, as part of molecules of water and other compounds, but no helium. As we noted earlier, these elements of life were created in stars. Let's investigate the evidence for—and the implications of—this amazing fact.

Key observational evidence for the fact that elements are manufactured in stars comes from studying the chemical composition of stars of different ages. Careful observations show that only the *disk* of the Milky Way Galaxy contains substantial amounts of interstellar gas; the *halo* of the galaxy does not (see Figure 1.10). This suggests that the galaxy as a whole formed much as stars form but on a far larger scale: A giant cloud of gas collapsed under gravity, becoming the spinning disk of the Milky Way Galaxy. Because all the gas resides in the disk, the recycling of stellar matter that gives birth to new generations of stars can occur only in the disk. As a result, no new stars have been born in the halo since the time when the gas first collapsed into the disk, which means that stars in the halo must be very old. (Other, independent evidence also supports the old age of halo stars but is beyond the scope of this book.) The chemical composition of these old stars turns out to be much closer to pure hydrogen and helium; instead of containing 2% heavier elements, these stars typically contain only 0.1% heavier elements. Thus, we

have clear evidence that the abundance of heavier elements has increased with time, suggesting that the early universe must have contained only the elements hydrogen and helium and that all other elements were manufactured later.

THINK ABOUT IT . . . *The Milky Way is typical of many but not all galaxies. Some galaxies lack a gas-filled disk, so all their stars are old like the stars in the Milky Way's halo. (Galaxies with disks like the Milky Way are called* spiral galaxies; *galaxies without disks are called* elliptical galaxies.) *Would you expect the overall composition of these galaxies to be closer to 0.1% heavier elements or closer to 2% heavier elements? Why?*

A second piece of evidence for the changing composition of the universe comes from theoretical studies of the Big Bang (see box, p. 12–13). These studies suggest that the early universe should have contained only hydrogen and helium (and a trace of lithium), in which case all other elements must have been manufactured at later times.

The third and perhaps strongest piece of evidence for the stellar manufacturing of elements comes from the study of how stars produce energy. Deep in a star's core, the temperature and pressure are so high that pairs of atomic nuclei frequently bang together hard enough so that they stick together, or *fuse*, forming a single larger nucleus. This process, called **nuclear fusion,** releases the energy that makes the star shine. During most of a star's life, nuclear fusion combines hydrogen nuclei to make helium nuclei. It takes four hydrogen nuclei to make one helium nucleus (the process involves several steps); energy is released because a helium nucleus has slightly less mass than the four hydrogen nuclei. This means that a small amount of the mass of the hydrogen has disappeared and become energy in accord with Einstein's famous formula, $E = mc^2$

(text continues on p. 18)

Basic Properties of Atoms

Atoms and molecules are the basic building blocks of stars, planets, and life. You are probably familiar with atoms and molecules from high school science courses, but they are so important to understanding the search for life that we review their key characteristics here.

Atoms are incredibly small: Millions could fit end to end across the period at the end of this sentence, and the number in a single drop of water (10^{22}–10^{23} atoms) exceeds the number of stars in our universe. Yet atoms are composed of even smaller particles: **protons, neutrons, and electrons.** (Protons and neutrons are, in turn, made of even smaller particles called *quarks.*) Protons and neutrons are found in the tiny **nucleus** at the center of the atom. The rest of the atom's volume contains only electrons (Figure 1.16). Note that, although the nucleus is very small compared to the atom as a whole, it contains most of the atom's mass. An atom's size is dictated primarily by its electrons, which form a "smeared out" cloud around the nucleus.

The properties of an atom depend mainly on the **electrical charge** of its protons and electrons. Each proton has one unit of positive charge (a charge of +1), and each electron has one unit of negative charge (a charge of −1). Neutrons are electrically neutral; they have no charge. Opposite charges attract each other, and negatively charged electrons remain bound in atoms because of their attraction to the positive charge of the protons in the nucleus. Most of the atoms in and around you contain the same number of electrons as protons and thus are electrically neutral overall. However, atoms often lose or gain electrons, in which case they obtain a net electrical charge. We call such atoms **ions.** A *positive ion* is an atom that has lost one or more electrons so that it has more positive than negative charge overall; a *negative ion* is an atom that has gained one or more electrons, giving it a net negative charge.

There are more than 100 different kinds of atoms, called **chemical elements** (or **elements** for short). Hydrogen, carbon, iron, and gold are examples. The number of protons in an atom's nucleus—the atom's **atomic number**—determines what element that atom represents. For example, all atoms with a single proton in the nucleus have atomic number 1 and represent the element hydrogen. Similarly, all helium atoms have two protons in the nucleus and thus atomic number 2, and atoms with six protons in the nucleus (atomic number 6) are carbon atoms. The complete set of known elements is listed in the **periodic table of the elements** (see Appendix C).

The combined number of protons and neutrons in an atom is its **atomic mass number.** For example, the mass number of ordinary hydrogen is 1 because its nucleus is just a single proton, while helium usually has a mass number of 4 because its nucleus contains two neutrons in addition to its two protons. (Note: Protons and neutrons have nearly equal mass, and this mass is roughly 2,000 times greater than an electron's mass.) While every atom of a given element contains exactly the same number of protons, the number of neutrons may vary. For example, all carbon atoms have six protons, but they may have six, seven, or eight neutrons. Versions of an element with different numbers of neutrons are called **isotopes** of the element. Isotopes are extremely useful to many important studies related to life in the universe, including those determining the ages of rocks and fossils, so they will come up many times in this book.

To name the isotopes of an element, we use their mass numbers. For example, the most common isotope of carbon, with six neutrons, has a mass number of 6 protons + 6 neutrons = 12. We call it carbon-12. The other isotopes of carbon are carbon-13 (because its six protons and seven neutrons give it mass number 13) and carbon-14 (six protons and eight neutrons give it mass number 14). We can also write isotopes by writing the mass number as a superscript to the left of the element symbol: ^{12}C, ^{13}C, ^{14}C. We read ^{12}C as "carbon-12." Figure 1.17 summarizes some of this basic atomic terminology.

The number of different substances is far greater than the number of chemical elements because atoms can combine with each other to form more complex materials. Many substances consist of **molecules,** in which two or more atoms are chemically bound together in a way that gives them properties distinct from those of the individual atoms they contain. For example, the oxygen in the air we breathe consists not of O atoms (which would be highly toxic) but of O_2 molecules. Other familiar molecules include water

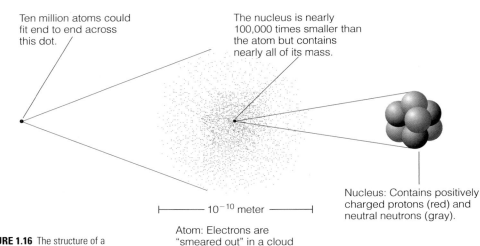

Ten million atoms could fit end to end across this dot.

The nucleus is nearly 100,000 times smaller than the atom but contains nearly all of its mass.

Nucleus: Contains positively charged protons (red) and neutral neutrons (gray).

10^{-10} meter

Atom: Electrons are "smeared out" in a cloud around the nucleus.

FIGURE 1.16 The structure of a typical atom.

(H_2O), carbon dioxide (CO_2), and ammonia (NH_3). Life on Earth is based on the chemistry of molecules containing carbon, which are called **organic molecules.** In diagrams, molecules are often represented with ball-and-stick models that show how the atoms are arranged (Figure 1.18a).

Other complex substances consist of atoms (or molecules) arranged in **crystals;** that is, atoms arranged in precise geometrical patterns. For example, table salt consists of crystals made from a precise, alternating arrangement of sodium (Na) and chlorine (Cl) ions; hence, we say that table salt is made from crystals of sodium chloride (Figure 1.18b). In general, any substance made from more than one element is called a **compound.** Thus, for example, water (H_2O) is a molecular compound, but oxygen (O_2) is not; table salt is a crystalline compound, but diamond (which consists of crystals made from only carbon atoms) is not.

FIGURE 1.17 Terminology of atoms.

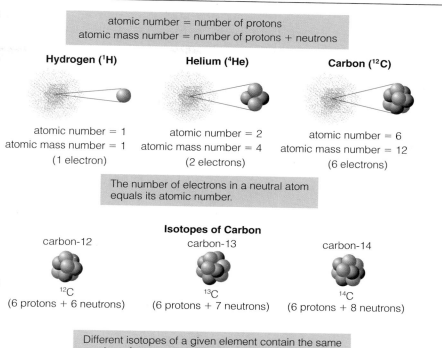

atomic number = number of protons
atomic mass number = number of protons + neutrons

Hydrogen (^{1}H)

atomic number = 1
atomic mass number = 1
(1 electron)

Helium (^{4}He)

atomic number = 2
atomic mass number = 4
(2 electrons)

Carbon (^{12}C)

atomic number = 6
atomic mass number = 12
(6 electrons)

The number of electrons in a neutral atom equals its atomic number.

Isotopes of Carbon

carbon-12

^{12}C
(6 protons + 6 neutrons)

carbon-13

^{13}C
(6 protons + 7 neutrons)

carbon-14

^{14}C
(6 protons + 8 neutrons)

Different isotopes of a given element contain the same number of protons but different numbers of neutrons.

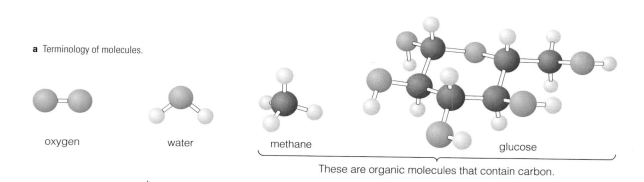

a Terminology of molecules.

oxygen

water

methane

glucose

These are organic molecules that contain carbon.

These are compounds (molecules made from atoms of two or more different elements).

These are all molecules made from two or more atoms.

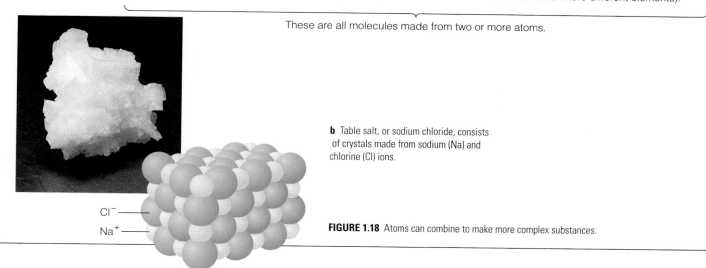

Cl⁻

Na⁺

b Table salt, or sodium chloride, consists of crystals made from sodium (Na) and chlorine (Cl) ions.

FIGURE 1.18 Atoms can combine to make more complex substances.

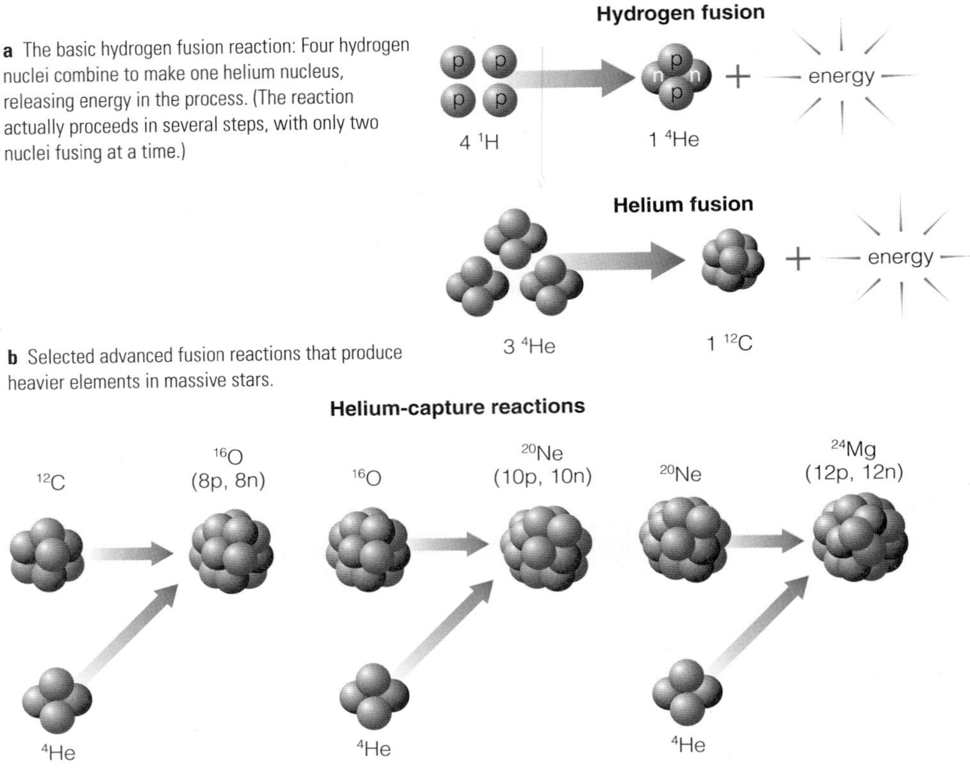

a The basic hydrogen fusion reaction: Four hydrogen nuclei combine to make one helium nucleus, releasing energy in the process. (The reaction actually proceeds in several steps, with only two nuclei fusing at a time.)

Hydrogen fusion

4 ^{1}H → 1 ^{4}He + — energy —

FIGURE 1.19 These diagrams represent nuclear fusion reactions; protons are shown in red and neutrons in gray.

Helium fusion

3 ^{4}He → 1 ^{12}C + — energy —

b Selected advanced fusion reactions that produce heavier elements in massive stars.

Helium-capture reactions

^{12}C → ^{16}O (8p, 8n) ^{4}He

^{16}O → ^{20}Ne (10p, 10n) ^{4}He

^{20}Ne → ^{24}Mg (12p, 12n) ^{4}He

(Figure 1.19a). Indeed, that is how our Sun shines today—with energy generated deep in its core by the fusion of hydrogen into helium.

Our Sun has been fusing hydrogen into helium and shining fairly steadily for more than 4 billion years, and it will continue to do so for billions of years to come. The Sun will eventually use up all the available hydrogen in its core, and at that point the Sun's "life" will be near its end. It will continue to generate energy for a few hundred million years more, partly by fusing helium into carbon,[2] but then its central power plant will shut down, leaving our solar system cold and dark.

More massive stars end their lives quite differently. Massive stars burn through their supply of hydrogen much more rapidly than low-mass stars like the Sun, in some cases exhausting their core hydrogen in just a few million years (as opposed to a few billion years). Near the ends of their lives, the cores of these stars become so hot and dense that they can fuse, successively, many heavier elements. They fuse helium into carbon, carbon into oxygen, oxygen into neon, and so on (Figure 1.19b). Then, at the ends of their lives, these massive stars die in titanic explosions—called *supernovae*—that scatter the manufactured elements into space (Figure 1.20). The supernova itself also leads to the creation of some new elements, notably those heavier than iron. In

this way, massive stars manufacture and release to space the elements from which we and the Earth are made.

Astronomers test this theory by observing the relative abundances of various chemical elements. The theory of nuclear fusion in massive stars makes specific predictions about relative abundances; for example, it says that the elements carbon and oxygen should be more abundant than nitrogen and that neon should be more abundant than fluorine. Figure 1.21 shows the observed relative abundances of elements. The observations agree well with the predicted pattern, suggesting that we really do understand the origin of the elements. Further confirmation comes from direct observations of massive stars that have exploded, as the observed abundance of elements within the debris agrees with the theoretical models.

In summary, three lines of evidence together make an overwhelming case for the fact that all elements besides hydrogen and helium were manufactured in stars: the observational evidence for increasing abundance of these elements over time, the theoretical calculations suggesting that the early universe began with only the chemical elements hydrogen and helium, and the theoretical and observational evidence that stars manufacture elements through nuclear fusion. Moreover, the nature of the element production results in substantial quantities of the most important elements for life. For example, Figure 1.21 shows that three of the four most abundant elements besides hydrogen and helium are carbon, nitrogen, and oxygen—which also happen to be three primary ingredients of living organisms on

---

[2] In fact, much of the carbon in the universe is thought to have come from relatively low-mass stars like our Sun.

FIGURE 1.20 This photograph shows the Crab Nebula, the remains of a massive star whose explosion was witnessed on Earth in A.D. 1054. The glowing gas has been moving outward from the point of the explosion for about a thousand years. In a few tens of thousands of years more, it will have fully dispersed, mixing the elements forged in the exploded star with other gas in the Milky Way Galaxy.

Earth. We are indeed made of "star stuff," and the fact that stars are found throughout the universe means that the stuff of life must also be common throughout the universe.

Planet Formation

The very first generation of stars in the universe must have been made entirely of hydrogen and helium, because other chemical elements did not yet exist. These first-generation stars must have lived and died before the birth of the old stars in the galactic halo, producing the relatively small amount of heavier elements that we find in halo stars. As time passed, many generations of massive, short-lived stars added heavier elements to the galaxy. By the time our solar system formed, about 4.6 billion years ago, interstellar gas in the Milky Way Galaxy already contained close to its present value of 2% heavier elements. Although 2% may not sound like much, it is more than enough to trigger the process of planet formation and to make small, rocky planets like Earth. Indeed, it's probably possible for planetary systems to form even around stars with much smaller abundances of heavy elements, such as around the stars in the halo, though we don't yet know for sure.

According to present theory, the process of planet formation begins as tiny solid particles condense from the gas in the spinning disk around a forming star. (In cooler regions of the disk, some of the solid particles may actually be preexisting grains of dust that are commonly found in cool interstellar clouds.) These particles condense for much the same reason that raindrops or snowflakes condense in clouds: When the temperature is low enough, some atoms or molecules in the gas will bond together. When the solid particles collide gently, they stick together, thus growing larger. Eventually, they can become large enough so that their own gravity begins to attract more matter, enabling them to become larger still. If nothing interrupts this process, gravity will eventually make full-fledged planets from what started as tiny solid "seeds."

Because pure hydrogen and helium do not solidify, most of the material in the spinning disk always remains gaseous. Only a small fraction of the material in the disk can condense to make the solid seeds. These seeds come in two basic types:

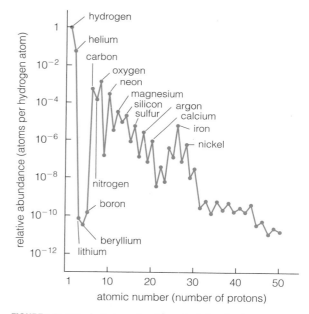

FIGURE 1.21 This graph shows the observed relative abundances of elements in the galaxy. For example, the relative abundance of 10^{-4} for nitrogen means that nitrogen is only about $10^{-4} = 0.0001$ as abundant as hydrogen. The observed abundances agree well with theoretical predictions based on the assumption that elements besides hydrogen and helium are manufactured in stars.

1. Some of the seeds are made purely from heavier elements that can take solid form as metals (such as iron and nickel) or rocks (such as silicon-based minerals). These seeds can form and remain solid even at relatively high temperatures, and therefore can be present throughout most of the spinning disk around the star.

2. Other seeds are solid grains of ice that include hydrogen as well as a heavier element. The most common ices are water ice (H_2O), methane ice (CH_4), and ammonia ice (NH_3). These ices can condense from the gas only at relatively low temperatures and therefore can be present only in the cold outer regions of the spinning disk.

This theory makes it easy to explain why the inner planets of our solar system (Mercury, Venus, Earth, and Mars) are so different in character from the giant outer planets (Jupiter, Saturn, Uranus, and Neptune). In the inner regions of a spinning disk, near the central star, it is too hot for ices to condense. But metal and rock can solidify at fairly high temperatures, so in these regions bits of solid metal and rock condense from the warm gas (Figure 1.22). Because the heavy elements that make metals and rocks are so rare, there's not enough solid material in these inner regions to make planets much larger than Earth. Moreover, these small planets lack the gravitational strength needed to hold on to the abundant gas around them. These inner planets therefore end up being made almost entirely of metal and rock. We call them **terrestrial planets;** *terrestrial* means "Earth-like."

Bits of metal and rock also condense farther from the central star, but here the temperatures are cold enough for ices to condense as well. Because the ices are made from more abundant elements (see Figure 1.21) than are rocks and metals, ices actually make up most of the solid material in these regions of the spinning disk. Thus, the growing chunks of solid material in an outer solar system are made mostly of ice, mixed with smaller amounts of metal and rock. These iceballs can grow fairly large—perhaps 10 times the mass of the Earth or more. At that point, their gravity begins to attract the surrounding hydrogen and helium gas, which makes them grow even bigger. By the time the process is complete, these outer planets have become giants made mostly of hydrogen and helium. We call them **jovian planets;** *jovian* means "Jupiter-like." The gas drawn into a jovian planet tends to form its own spinning disk, rather like a miniature version of the spinning disk around the star. The same general processes then occur within these planetary disks, leading to the formation of moons. This is one reason why the jovian planets tend to have many moons.

THINK ABOUT IT . . . *Use the ideas described above to explain why the moons of jovian planets tend to contain much more ice than the terrestrial planets.*

Not every solid seed present during the formation of our solar system became a planet; in fact, the vast majority did not. Only those that grew fastest and avoided being shattered in collisions became the planets that we see today. The rest either eventually crashed into a planet, becoming part of it, or are still orbiting the Sun today as "leftovers" from the planetary formation process. These leftovers are the asteroids and comets; asteroids are the leftovers made of metal and rock, while comets are the leftovers made primarily of ice. These leftovers still sometimes crash into planets, and, as we'll discuss in Chapter 5, they have played a major role in the evolution of life on Earth.

The process of planet formation in our solar system is thought to have taken a few tens of millions of years, which may seem like a long time but is only a tiny fraction of the 4.6-billion-year age of our solar system. As the planets formed, the Sun also formed. Studies of other stars suggest that the young Sun would have had a strong **solar wind,** consisting of particles blown off its surface and out into space. This wind would have swept any remaining gas into interstellar space. At this point, the formation of our solar system was essentially complete; no further substantial planetary growth could occur. Figure 1.23 summarizes the process of solar system formation.

FIGURE 1.22 In the warm inner regions of a forming star system, only bits of rock and metal can condense from the gas. In the cooler outer regions, ices (which consist of hydrogen compounds) can also condense.

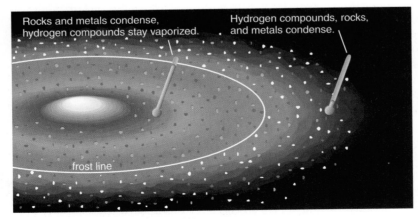

Rocks and metals condense, hydrogen compounds stay vaporized.

Hydrogen compounds, rocks, and metals condense.

frost line

FIGURE 1.23 (Opposite page) A summary of the process by which our solar system formed.

Large, diffuse interstellar gas cloud (solar nebula) contracts under gravity.

As it contracts, the cloud heats, flattens, and spins faster, becoming a spinning disk of dust and gas.

Sun will be born in center.

Planets will form in disk.

Hydrogen and helium remain gaseous, but other materials can condense into solid "seeds" for building planets.

Warm temperatures allow only metal/rock "seeds" to condense in inner solar system.

Cold temperatures allow "seeds" to contain abundant ice in outer solar system.

Solid "seeds" collide and stick together. Larger ones attract others with their gravity, growing bigger still.

Terrestrial planets are built from metal and rock.

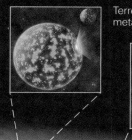

The seeds of jovian planets grow large enough to attract hydrogen and helium gas, making them into giant, mostly gaseous planets; moons form in disks of dust and gas that surround the planets.

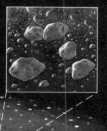

Solar wind blows remaining gas into interstellar space.

Terrestrial planets remain in inner solar system.

Jovian planets remain in outer solar system.

"Leftovers" from the formation process will become asteroids (metal/rock) and comets (mostly ice).

Not to scale

Should We Expect Other Star Systems to Be Like Ours?

As we've discussed, virtually all stars should be born within spinning disks of gas, suggesting that the raw material for planets is nearly always present. Moreover, the processes of condensation and planetary growth that we've described also follow from basic laws of nature, suggesting that other solar systems should form much as ours did. As a result, prior to the discoveries of extrasolar planets, astronomers generally assumed that other planetary systems would be laid out much like ours, with terrestrial planets close in and jovian planets farther out. But recent discoveries have shown that the reality may be much more complex.

As we'll discuss further in Chapter 10, many of the newly discovered planets appear to be jovian in nature though they orbit quite close to their stars. Planetary scientists are still struggling to explain this seemingly anomalous fact. The leading explanation suggests that the surprising planets probably did form in the cold outer regions of the star system but migrated inward due to friction with the gas in the spinning disk. (Once a jovian planet forms, it can remain intact even if it migrates to regions of much warmer temperature.) In such systems, the close-in jovian planets may have disrupted the formation of terrestrial planets or perhaps even "kicked out" any terrestrial planets that had already formed. In addition, some theoretical work suggests that it may not always be easy to grow planets in the first place. In some cases, the growing solid seeds may collide and shatter rather than continue to grow into full-fledged planets.

We do not yet know whether planets—or our solar system's layout—will prove to be common or rare. Nevertheless, given the enormous number of stars in our galaxy, it seems almost inevitable that at least some other systems like ours exist. The stuff of life is everywhere and must be gathered on many planets or moons that would make suitable homes for life. We are now ready to turn to the question of whether life itself should be expected in such places.

1.4 Could Biology Be Universal?

Wherever we look in the Milky Way Galaxy or elsewhere in the universe, we find clear evidence that the same laws of nature are operating. Many galaxies have disks and halos like the Milky Way, and we've discovered many stars that are surrounded by spinning disks of gas or bona fide planets. These facts tell us that the same physical processes that created the disks of the Milky Way and of our own solar system must be operating throughout the universe. In situations where we can observe orbital motions, we find that they agree with what we expect from the law of gravity. Thus, we can be confident that the basic laws of physics that we've discovered here on Earth also hold throughout the universe.

We can be similarly confident that the laws of chemistry are universal. Observations of distant stars show that they are made of the same chemical elements we find here in our own solar system and that interstellar gas clouds contain many of the same molecules we find on Earth. This provides conclusive evidence that atoms come in the same types and combine in the same ways throughout the universe.

The universality of physics and chemistry is what makes us confident that we will find planets, including many that are terrestrial in nature, throughout the galaxy and the universe. Could biology also be universal? That is, could the biological processes we find on Earth be common throughout the universe? If the answer is yes, then the search for life elsewhere should be exciting and fruitful. If the answer is no, then life may be a rarity. We do not yet know whether biology is universal; however, we have good reason to be optimistic. Laboratory experiments suggest that the chemical constituents that made up the early Earth would have combined readily into complex organic (carbon-based) molecules, including many of the building blocks of life [Section 5.2]. Indeed, scientists have found organic molecules in meteorites (chunks of rock that fall to Earth from space) and, through spectroscopy [Section 6.2], in clouds of gas between the stars. The fact that such molecules form even under the extreme conditions of space suggests that they form quite readily.

Of course, the mere presence of organic molecules does not necessarily mean that life will arise, but the history of life on Earth gives us good reason to think that the step from chemistry to biology is not especially difficult. As we'll discuss in Chapter 5, geological evidence tells us that life on Earth arose almost as early as it possibly could have after the Earth's formation. This early origin of life suggests that the process was not that difficult, because a difficult process would probably have required much more time.

THINK ABOUT IT . . . *While life arose early in the Earth's history, humans did not come on the scene until quite recently—some 4 billion years after the earliest living organisms. Do you think this fact tells us anything about the likelihood of finding intelligent life, as opposed to finding any life, on planets around other stars? Explain.*

If life can emerge easily under the right conditions, as the evidence from Earth suggests, the only remaining question is the prevalence of those "right" conditions. Here, too, recent discoveries give us reason for optimism. In particular, biologists have found that life can survive and prosper under a much wider range of conditions than was believed only a few decades ago [Section 3.5]. For example, we now know that life exists in extremely hot water near deep-sea volcanic vents, in the frigid conditions of Antarctica, and inside rocks buried a kilometer or more beneath the Earth's surface. Indeed, if we could export these strange organisms from Earth to other planets in our solar system, it seems likely that at least some of them would survive. This suggests that the range of "right" conditions for life may be quite broad, in which case it might be possible to find life even on planets that are quite different in character from Earth.

In summary, we have no reason to think that life ought to be rare and several reasons to expect that it may be quite common (Figure 1.24). Moreover, if life is indeed common, studying it will give us new insights into life on Earth, even if we don't find other intelligent civilizations. These enticing prospects have captured the interest of scientists from many disciplines and from around the world, giving birth to a new science devoted to the study of life in the universe.

1.5 The Science of Astrobiology

The science of life in the universe has been given a number of names, including "exobiology" and "bioastronomy," but in this book we follow the lead of NASA and call it **astrobiology.** This term is meant to invoke the combination of astronomy (the study of the universe) and biology (the study of life), so *astrobiology* literally means "the study of life in the universe."

Because astrobiology is a young science, scientists are still working to define it more precisely and to decide where to focus their research efforts. One major player in this effort is the NASA Astrobiology Institute, a collaboration involving scientists from NASA and more than a dozen other research institutions across the United States. Similar efforts are under way in other countries, including the United Kingdom, Sweden, France, Spain, and Australia. Generally speaking, astrobiology is about studying the conditions conducive to the formation of life, looking for such conditions on other planets in our solar system and around other stars, and looking for the actual occurrence of life elsewhere.

Note that this field involves much more than simply searching for extraterrestrial life or civilizations. Astrobiology research seeks to reveal the fundamental connections between living organisms and the places where they reside. In this sense, finding

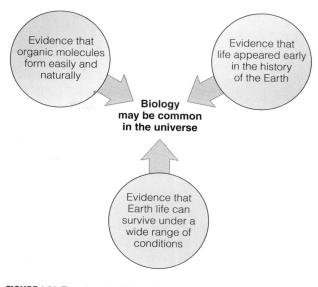

FIGURE 1.24 Three important lines of evidence suggest that biology may be quite common in the universe.

no life (on Mars, for example) is just as significant a result as finding life, because either result tells us about the conditions that can result in the development of life, how life evolves in conjunction with planets, and whether life is likely to be rare or widespread throughout the universe. In a document intended as a "road map" for scientists, NASA has identified three fundamental questions to be addressed by astrobiology research:

1. How does life begin and develop?

2. Does life exist elsewhere in the universe?

3. What is life's future on Earth and beyond?

THINK ABOUT IT . . . *Briefly describe how these three questions relate to one another. For example, why is question 1 relevant to question 2, and how might the answer to question 2 influence the answer to question 3? Explain clearly.*

These three questions require not only searching for life elsewhere, but also learning much more about the origin and evolution of life on Earth. Thus, a great deal of the research that comes under the auspices of astrobiology involves biologists, geologists, and chemists searching to understand the development of life on our own planet. In the rest of this book, we will focus largely on the same three questions. After discussing the nature of science in greater detail in Chapter 2, we'll turn our attention to the nature and origin of life on Earth in Chapters 3 through 5. Then we'll discuss prospects for life elsewhere in our solar system in Chapters 6 through 9, followed by discussion of the prospects for finding life—including intelligent life—beyond our solar system in Chapters 10 through 13. Along the way, we'll learn what science can presently say about the future of life on Earth and consider possible futures for our own species. Finally, in Chapter 14, we'll discuss the philosophical implications of the search for—and potential discovery of—life beyond Earth.

THE BIG PICTURE

In this first chapter, we have explored why the science of astrobiology seems so promising and exciting. As you continue in your study, keep in mind the following "big picture" ideas:

■ Despite the abundance of aliens in popular media, we do not yet have any convincing evidence for life beyond Earth. Nevertheless, we have good reason to think that life may be widespread, which makes the question of life in the universe an exciting topic of scientific research.

■ It's conceivable that life may exist on any of several worlds in our own solar system, but it's extremely unlikely that any of this life is intelligent. The search for intelligent life requires looking to planets in other star systems. Though the search may be difficult, the possibilities are incredible given that there are as many stars in our universe as grains of sand on all the beaches on Earth.

■ We are "star stuff" in that we are made of elements that were manufactured in stars. The same elements are available to make planets and life throughout the universe, and we already have both theoretical and observational evidence that planets are common. Thus, while we cannot be sure that life exists elsewhere, we know that the necessary raw materials are available.

■ The prospect that life may be common has given rise to the science of astrobiology, which involves the study of the origin and evolution of life on Earth, the search for life elsewhere, and the study of possible futures for life on Earth.

Review Questions

1. Briefly describe why scientists believe that life could exist on other worlds in our *solar system* while also believing that we are unlikely to find intelligent life in our solar system. What do we mean by a *habitable* planet?

2. What are *extrasolar planets*? When were the first ones discovered?

3. What is *SETI*?

4. Briefly explain what we mean by a *star system* (like our solar system), a *galaxy*, and the *universe*. How many stars are in the *Milky Way Galaxy*? How many are in our universe?

5. Use the idea of a scale model to describe the separations between *planets* in our own solar system and the distances to the nearest other *stars*. How does this model show the difficulty of detecting planets around other stars?

6. What is a *light-year*? Explain why the distances to the stars make interstellar travel impractical with current technology, and why even radio conversations with other civilizations may be difficult to achieve.

7. Briefly explain what it means to say that we are "star stuff."

8. Describe the three processes that lead to spinning disks around forming stars.

9. Briefly describe the structure of *atoms* and *molecules*. What do we mean by *atomic number* and *atomic mass number*? What are *isotopes*? What are *molecules*, *crystals*, and *compounds*?

10. Describe the three lines of evidence supporting the idea that the universe began with only the *chemical elements* hydrogen and helium (and trace amounts of lithium) and that all other elements have been produced in stars.

11. What is *nuclear fusion*? Explain how it creates heavier elements and how those elements can be recycled into later generations of stars.

12. Briefly explain how temperature differences in the spinning disk that formed our solar system led to different types of solid seeds in different regions. Describe the two basic types of solid seeds and where they formed.

13. Describe how the different types of solid seeds led to the existence of two basic planet types: *terrestrial* and *jovian*. Why do jovian planets tend to have many moons? Also explain the nature of *asteroids* and *comets*.

14. Briefly discuss whether we should expect other solar systems to be like ours.

15. Briefly describe the evidence that suggests life could be common.

16. What do we mean by *astrobiology*? What other terms are sometimes used to describe it? What fundamental questions does it study?

Discussion Questions

1. *Aliens Among Us.* Take an informal poll of your friends or classmates. How many believe we have already been visited by aliens? On what basis do they hold their beliefs? How strong are their convictions on this issue? In light of your findings and what you've learned in this chapter, discuss whether public interest in aliens visiting Earth has any bearing on the scientific study of astrobiology.

2. *Conducting the Search.* Given the large number of possible places to look for life, how would you prioritize the search? For example, how would you prioritize the search for life on other worlds in our own solar system, and how would you come up with a search strategy for other star systems? Explain your priorities and strategies clearly.

3. *Funding for Astrobiology.* Imagine that you are a member of Congress, so it is your job to decide how much government funding goes to research in astrobiology. What factors would influence your decision? Do you think you would increase or decrease such funding from the current level? Explain.

Problems

(Quantitative problems are marked with an asterisk.)

Sense or Nonsense? In **problems 1–8,** evaluate the given statement and decide whether it makes sense. Explain your reasoning clearly.

Example: I walked east from our base camp at the North Pole.

Solution: This statement does not make sense because *east* has no meaning at the North Pole; all directions are south from the North Pole.

1. For all we know, there could be great cities on the surface of Mars, because our current technology would not yet be able to detect them.

2. If life exists on about one out of a million planets, then the Milky Way Galaxy must be teeming with life.

3. If we could build a spaceship that would reach Pluto in a year, then the same spaceship could take us to nearby stars in only about a decade.

4. Based on our understanding of solar system formation, there's no reason why Jupiter couldn't have formed at the Earth's distance from the Sun.

5. It will take me light-years to complete this homework assignment.

6. Even if we discover a civilization around other stars, we will never be able to talk with them with the same ease with which we carry on conversations with people on Earth.

7. If the universe did not contain stars more massive than our Sun, we couldn't be here.

8. I may not be a big star myself, but I owe my existence to big stars.

9. *Common Levels of Technology.* In *Star Wars,* aliens from many worlds share approximately the same level of technological development. Explain why this would be extremely unlikely in reality.

10. *Planetary Composition.* Study Figure 1.21. Based on the abundances shown, is it likely or unlikely that we would find a terrestrial planet made mostly of gold (atomic number 79)? Explain.

11. *Fission and Fusion.* Stars make energy through nuclear fusion, but nuclear power plants on Earth make energy through *nuclear fission*—the splitting of the nuclei of heavy elements, such as uranium. Do a bit of research to learn more about nuclear fission and why power plants use it, rather than nuclear fusion, at present. Also learn why fusion would be a more abundant and cleaner source of power if we learned how to use it in power plants. Write a one- to two-page paper summarizing your findings.

*12. *Scale of the Solar System.* The real diameters of the Sun and the Earth are approximately 1.4 million km and 6,400 km, respectively. The Earth–Sun distance is approximately 150 million km. Calculate the sizes of the Earth and Sun, and the distance between them, on a scale of 1 to 10 billion. Show your work clearly.

*13. *Counting Stars.* Suppose there are 400 billion stars in the Milky Way Galaxy. How long would it take to count them if you could count continuously at a rate of 1 per second? Show your work clearly.

*14. *Interstellar Travel.* Our fastest current spacecraft travel away from Earth at a speed of roughly 50,000 km/hr. At this speed, how long would it take to travel the 4.3-light-year distance to Alpha Centauri (the nearest star system to our own)? Show your work clearly. (*Hint:* Recall that a light-year is approximately 9.5×10^{12} kilometers.)

Web Projects

1. *Astrobiology News.* Go to NASA's Astrobiology home page and read some of the recent news from astrobiology research. Choose one recent news article, and write a one- to two-page summary of the research and how it fits into astrobiology research in general.

2. *The NASA Astrobiology Institute.* Go to the home page for the NASA Astrobiology Institute (NAI) and learn more about how it is organized and the type of research it supports. Also learn whether your school or any nearby institutions participate in the NAI. In about one page or less, describe the NAI and its work and discuss the particular contributions of any institutions located near you.

3. *International Astrobiology.* Search the Web for information on astrobiology efforts outside the United States. Learn about the effort in one particular country or group of countries. What areas of research are emphasized? How do these areas compare to the areas that NASA addresses with the three questions described in Section 1.5? How do the researchers involved in the effort collaborate with other international astrobiology efforts? Write a one- to two-page report on your findings.

4. *The Search for Extraterrestrial Intelligence.* Go to the home page for the SETI Institute. Learn more about how SETI is funded and carried out. In about one page or less, describe the SETI Institute and its work.

CHAPTER 2

The Science of Life in the Universe

Ideas of life beyond Earth go far back in human history. Although many of these ideas were mythical in nature, imagining gods or other supernatural beings living among the constellations, some ancient ideas were much more modern in character. For example, many ancient Greek scholars imagined other worlds populated with intelligent beings like us. But until quite recently, all these ideas remained purely speculative, because there was no way to study the question of life beyond Earth scientifically.

Given that we don't yet know of any life beyond Earth, you might wonder how we can make a science of studying it. The answer is that we use science to help us understand the conditions under which we might expect to find life, the likely characteristics of life elsewhere, and methods we can use to search for it. Without this science, the search for life would be a shot in the dark. With it, we can focus our efforts in a way that will help us know where to look and how to interpret our results whether we do or do not find life.

In this chapter, we will look at the story of how we moved from ancient speculation to the methods of modern science we use to search for life in the universe. We will also examine the nature of science and how science differs from other sorts of knowledge and from the impostor we call "pseudoscience."

2.1 The Ancient Roots of Science

The roots of science go deep into human history, and one of the best ways to understand modern science is to explore its ancient roots. Let's begin by trying to view the world as our ancestors did. Imagine living in ancient times, looking up at the sky without the benefit of any of our modern knowledge. What would you see?

Every day, the Sun rises in the east and sets in the west, though its precise path through the sky varies with the seasons. Every night, the stars likewise move across the sky, but with different constellations visible at different times of year. The Moon goes through its monthly phases, from new to full and back again. Meanwhile, in contrast to the constant motion you observe for these objects in the sky, the Earth beneath you feels steady and looks flat. It would be quite natural to assume—as did people of many early cultures—that the Earth is a flat, motionless disk surmounted by a domelike sky across which the heavenly bodies move.

The story of how we progressed from this simple, intuitive view of the Earth and heavens to our modern understanding of the Earth as a tiny planet in a vast cosmos is in many ways the story of the development of science. Our ancestors were surely curious about many aspects of the world around them, but astronomy held special interest. The Sun clearly plays a central role in our lives, governing daylight and darkness and the progression of the seasons. The Moon's connection to the tides would have been obvious to people living near the sea. The evident power of these celestial bodies probably explains why they attained prominent roles in many early religions and may be one reason why it seemed so important to know the sky. Careful observations of the sky also served a practical need by enabling ancient peoples to keep track of the time and the seasons—crucial skills for agricultural societies.

As civilizations rose, astronomical observations became more careful and elaborate. In some cases, the results were recorded in writing. The ancient Chinese kept detailed records of astronomical observations beginning some 5,000 years ago. By about 2,500 years ago, written records allowed the Babylonians (in the region of modern-day Iraq) to predict eclipses with great success. Halfway around the world (and a few centuries later), the Mayans of Central America independently developed the same ability.

These ancient, recorded observations of astronomy represented databases of facts—the raw material of science. But, as far as we know, in most cases these facts were never put to use for anything beyond meeting immediate religious and practical needs. One clear exception was ancient Greece, where scholars began to use these facts in an attempt to understand the architecture of the cosmos. It took a couple thousand years, but the Greek ideas ultimately blossomed into the methods of modern science.

The Greek Setting

The civilization of ancient Greece, which extended well beyond the boundaries of modern-day Greece in Europe, rose as a power in the Middle East around 500 B.C. Its geographical location placed it at a crossroads for travelers, merchants, and armies of Europe, Asia, and Africa. Its location in time allowed it to absorb the mathematical and astronomical knowledge developed by prior civilizations of the Middle East, including those of ancient Egypt and Mesopotamia.

Philosophers in ancient Greece began asking basic questions about the nature of the universe. We generally credit the philosopher Thales (c. 624–546 B.C.; pronounced "thay-lees") as the founder of Greek science. Among his many accomplishments, Thales was the first person known to have addressed the question, "What is the universe made of?" without resorting to supernatural explanations. (Our knowledge of Thales comes from second- and third-hand accounts by later Greeks, so it is difficult to know whether all of his supposed accomplishments were real or were embellished by legend.) His own guess—that the universe fundamentally consisted of water and that the Earth was a flat disk on an infinite ocean—was not widely accepted even in his own time, but his mere asking of the question helped set the stage for all later science. For the first time, someone had suggested that the world was inherently understandable and not just the result of arbitrary or incomprehensible events.

The scholarly tradition begun by Thales was carried on by many others, perhaps most famously Plato (427–347 B.C.) and his student Aristotle (384–322 B.C.). Each Greek philosopher introduced new ideas, sometimes in contradiction to the ideas of others. None of these ideas rose quite to the level of modern science, primarily because the Greeks tended to rely more on pure thought and intuition than on observations or experimental tests. Nevertheless, with hindsight we can see at least three major innovations in Greek thought that helped pave the way for science.

First, the Greek philosophers developed a tradition of thinking and reasoning independent of any preconceived beliefs. For example, whereas earlier Greeks might simply have accepted that the Sun moves across the sky because it is pulled by the god Apollo in his chariot—an idea whose roots were already lost in antiquity—the philosophers sought a natural explanation that caused them to speculate

FIGURE 2.1 This photograph, taken at Arches National Park with a 6-hour exposure, shows how the stars appear to circle around the pole star each day. Today, other stars appear to circle nearly around the star we call Polaris, or the North Star. The basic circling looked the same in ancient times, but with a different star near the center. (This change occurs because of a phenomenon called *precession*, which causes the Earth's axis to change where it points in space with an approximately 26,000-year cycle.)

anew about the construction of the heavens. They were free to think creatively because they were not merely trying to prove preconceived ideas, and they recognized that new ideas should be open to challenge. As a result, they often worked communally, debating and testing each other's ideas. This tradition of challenging virtually every new idea remains one of the distinguishing features of scientific work today.

Second, the Greeks developed mathematics in the form of geometry. They valued this discipline for its own sake, and they understood its power, using geometry to solve both engineering and scientific problems. Without their mathematical sophistication, they would not have gone far in their attempts to make sense of the cosmos. Like the Greek tradition of challenging ideas, the use of mathematics to help explore the implications of new ideas remains an important part of modern science.

Third, while much of their philosophical activity consisted of subtle debates grounded only in thought and thus was not scientific in the modern sense, the Greeks also saw the power of reasoning from observations. They understood that an explanation about the world could not be right if it disagreed with observed fact. This willingness to discard explanations that simply don't work is also a crucial part of modern science.

The Geocentric Model

Perhaps the greatest Greek contribution to science came from the way they synthesized all three innovations into the idea of creating **models** of nature. Just as a model airplane seeks to represent a real airplane, a scientific model seeks to represent some aspect of nature. However, scientific models usually are conceptual models rather than miniature representations. The purpose of a scientific model is to explain and predict real phenomena. For example, a model of the solar system seeks to explain and predict the observed motions of the Sun, Moon, and planets, and a model of an earthquake seeks to explain and predict the occurrence of earthquakes. Of course, just as the model airplane may or may not be faithful in its representation of the real airplane, a scientific model may or may not capture the true, underlying essence of nature.

In astronomy, the Greeks constructed conceptual models of the universe in an attempt to explain what they observed in the sky, which quickly led them past simplistic ideas of a flat Earth under a dome-shaped sky to a far more sophisticated view of the cosmos. We will focus on how human thought about the heavens developed from the time of the Greeks because this story is so intimately tied to the development of science as a whole.

One of the first crucial steps was taken by a student of Thales, Anaximander (610–c. 547 B.C.). In an attempt to explain the way the sky appears to turn around the pole star each day (Figure 2.1), Anaximander suggested that the heavens must form a complete sphere around the Earth. Moreover, based on how the sky changes with travel north and south, he concluded that the Earth must not be flat. Because the sky does not change with east-west travel, he guessed that the Earth might be a cylinder curved only in the north-south direction. By about 500 B.C., for reasons that are not completely clear, Pythagoras was teaching that the Earth was a sphere. (A little over a century later, Aristotle cited the Earth's curved shadow on the Moon during lunar eclipses as observational support for a spherical Earth.) Thus, early Greek philosophers adopted a basic **geocentric** (Earth-centered) **model** of the universe consisting of a spherical Earth surrounded by the sphere of the heavens.

THINK ABOUT IT . . . *A widespread myth holds that Columbus proved the Earth to be round rather than flat when he sailed to America in 1492. It is probably true that most people in Columbus's day believed the Earth to be flat, but scholars were well aware of the then 2,000-year-old Greek idea of a round Earth. Why do you suppose the idea of a flat Earth persisted so long after the Greeks discovered otherwise?*

If you watch the sky closely, you'll notice that, while the patterns of the constellations seem not to change, the Sun, the Moon, and the five planets visible to the naked eye (Mercury, Venus, Mars, Jupiter, and Saturn) gradually move among the constellations from one day to the next. Indeed, the word *planet* comes from the Greek for "wanderer," and it originally referred to the Sun and Moon as well as to the five visible planets. Our seven-day week is directly traceable to the fact that seven "planets" are visible in the heavens (Table 2.1).

To the Greek philosophers, the motions of the "planets" among the stars soon suggested that these other objects could not be part of the same sphere as the stars. Moreover, each of these objects moves independently and at a different rate. For example, the Sun appears to make a complete circuit around the sphere of the stars (passing through the constel-

FIGURE 2.2 This model represents the Greek idea of the heavenly spheres (c. 200 B.C.). The Earth is a sphere that rests in the center. The Sun, the Moon, and each of the planets moves on its own sphere, and the outermost sphere holds the stars.

lations of the zodiac) once each year, while the Moon makes a similar circuit in only about a month (think "moonth"). The Greek geocentric model had to become more complex to account for the differing motions of different objects, and it came to include a separate nested sphere for the stars, the Sun, the Moon, and each of the five visible planets (Figure 2.2).

By allowing the different spheres to rotate at different rates, this model of nested spheres could account reasonably well for the observed motions of the Sun, Moon, and stars. However, it was far less successful for the motions of the planets. The problem is that while the Sun and Moon follow simple

(*text continues on p. 32*)

Table 2.1 The Seven Days of the Week and the Astronomical Objects They Honor
The correspondence between objects and days is easy to see in French and Spanish. In English, the correspondence becomes clear when we look at the names of the objects used by the Teutonic tribes who lived in the region of modern-day Germany.

Object	Teutonic Name	English	French	Spanish
Sun	Sun	Sunday	dimanche	domingo
Moon	Moon	Monday	lundi	lunes
Mars	Tiw	Tuesday	mardi	martes
Mercury	Woden	Wednesday	mercredi	miércoles
Jupiter	Thor	Thursday	jeudi	jueves
Venus	Fria	Friday	vendredi	viernes
Saturn	Saturn	Saturday	samedi	sábado

The ancient Greek philosophers began a spirited debate about the possibility of other worlds and of life on those worlds. The roots of the debate go back to Thales' question of what the universe was made of. Whereas Thales claimed the fundamental element to be water, his student Anaximander imagined a more mystical element that he called *apeiron*, meaning "infinite." He suggested that all material things arose from and returned to the apeiron. This allowed him to imagine that worlds might be born and die repeatedly through eternal time. (By "world," the Greeks meant both the Earth and the heavenly spheres assumed to surround the Earth; note that this is different from our modern idea of a "world" as a single body such as a planet or moon.) Thus, while he made no known claim of life existing elsewhere in the present, he essentially suggested that other Earths and other beings might exist at other times.

Other Greeks took the debate in a slightly different direction, and eventually a general consensus emerged in favor of the world's having been built from four elements: fire, water, earth, and air. However, two distinct schools of thought emerged concerning the nature and extent of these elements: the *atomists* held that the universe was made from an infinite number of indivisible *atoms* of each of the four elements; the *Aristotelians* (after Aristotle) held that the four elements (not necessarily made from atoms) were confined to the realm of the Earth, while the heavens were made of a distinct fifth element, often called the *aether* or the *quintessence* (literally, "the fifth essence").

The atomist doctrine naturally led to the idea of life beyond Earth, because it held that the universe was infinite and contained other, similar worlds. Even before the doctrine was fully formulated, some of the Greek philosophers understood this connection. For example, Xenophanes (c. 570–480 B.C.) suggested that the Earth might not be unique and that the Moon might be inhabited. Shortly thereafter, Anaxagoras (c. 500–428 B.C.) hit upon the idea that the heavens and Earth were made of the same elements, suggesting that the Sun was a flaming rock. He further suggested that the universe arose from an infinite number of "seeds" and that the same processes that created the Earth could have occurred elsewhere, leading to the idea of other worlds.

The actual idea of indivisible atoms is attributed to Leucippus (c. 490 B.C.–?), about whom we know very little. He supposedly was the teacher of Democritus (c. 470–380 B.C.), who developed atomism more fully. Democritus held a number of strikingly modern beliefs. For example, he claimed that the Moon had mountains and valleys, and he was the first person known to suggest that the Milky Way was in fact a vast conglomeration of individual stars. Democritus argued that the Earth and other worlds were created by the random motions of infinite atoms. Because this view lacked any intelligent Creator, atomism became intimately tied with atheism, which may be one reason why the later debate about life on other worlds became so intertwined with debates about religion.

None of Democritus's original writings survive, so we know of the connection between atomism and views of extraterrestrial life only through later writers. For example, after describing the doctrine of atomism, Epicurus (341–270 B.C.) wrote in his *Letter to Herodotus*:

> There are infinite worlds both like and unlike this world of ours . . . we must believe that in all worlds there are living creatures and plants and other things we see in this world.

These ideas were further extended by the Roman philosopher Leucretius (c. 95–55 B.C.), who wrote in his book *On the Nature of Things*:

> It is in the highest degree unlikely that this earth and sky is the only one to have been created. . . . This follows from the fact that our world has been made by the spontaneous and casual collision and the multifarious, accidental, random and purposeless congregation and coalescence of atoms whose suddenly formed combinations could serve [to produce] . . . earth and sky and the races of living creatures. . . . So we must realize that there are other worlds in other parts of the universe, with races of different men and different animals.

On the other side of the debate, Aristotle argued forcefully against the atomist beliefs. Aristotle held that all elements had their own natural motion and place. For example, he believed that the element earth moved naturally toward the center of the universe, from which he concluded that the Earth must reside at the center. The element fire, he claimed, naturally rose away from the center, which is why flames jut upward into the sky. These incorrect ideas about physics, which were not disproved until the time of Galileo and Newton almost 2,000 years later, caused Aristotle to reject the atomist idea of many worlds. If there was more than one world, there would be more than one natural place for the elements to go, which would be a logical contradiction. Aristotle concluded: "The world must be unique. . . . There cannot be several worlds."

Interestingly, Aristotle's philosophies were not particularly influential until many centuries after his death. His books were preserved and valued—in particular, by Islamic scholars of the late first millennium—but they were unknown in Europe until they were translated into Latin in the twelfth and thirteenth centuries. St. Thomas Aquinas (1225–1274) integrated Aristotle's philosophy into Christian theology. At this point, the contradiction between the Aristotelian notion of a single world and the atomist notion of many worlds became a subject of great concern to Christian theologians. Even today, the theological issues are not fully settled, and echoes of the ancient Greek debate between the atomists and the Aristotelians still reverberate in our time.

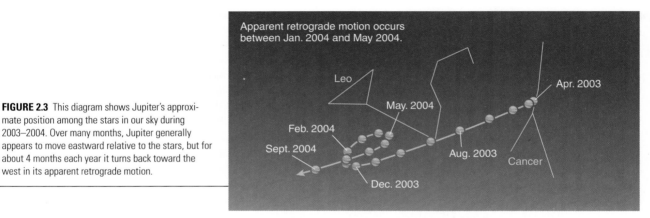

Apparent retrograde motion occurs between Jan. 2004 and May 2004.

Leo

May. 2004

Feb. 2004

Sept. 2004

Dec. 2003

Apr. 2003

Aug. 2003

Cancer

FIGURE 2.3 This diagram shows Jupiter's approximate position among the stars in our sky during 2003–2004. Over many months, Jupiter generally appears to move eastward relative to the stars, but for about 4 months each year it turns back toward the west in its apparent retrograde motion.

Dots represent Jupiter's approximate position at 1-month intervals. (Jupiter not to scale.)

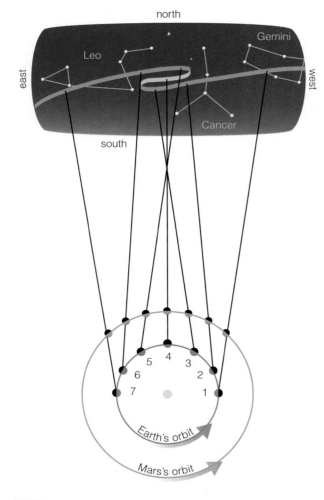

north

Gemini

Leo

east

west

Cancer

south

5 4 3

6 2

7 1

Earth's orbit

Mars's orbit

FIGURE 2.4 The explanation for apparent retrograde motion. The numbered dots represent positions of Earth and Mars at various times in their respective orbits of the Sun. Follow the lines of sight from Earth to Mars in numerical order, noting where Mars appears among the stars. Seen from Earth, Mars usually appears to move toward the east, but it shifts toward the west in apparent retrograde motion during the period when it is nearly opposite the Sun in our sky.

paths that carry them eastward relative to the stars at fairly steady rates (about 1° per day for the Sun and about 12° per day for the Moon), planetary motion is much more complex. If you chart the position of a planet (such as Mars or Jupiter) relative to the stars over the course of a year, you'll find not only that its speed varies considerably during the year, but that for part of the year the planet seems to go backward compared to its motion the rest of the year (Figure 2.3). The period during which the planet moves backward is called its period of **apparent retrograde motion.**

Today, we know that the explanation for apparent retrograde motion is really quite simple: It occurs as the Earth passes another planet in its orbit. For example, by tracing lines of sight from Earth to Mars as shown in Figure 2.4, you can see how Mars appears to move backward during the period when it is nearly opposite the Sun in our sky. Note that Mars never really goes backward in its orbit; it only *appears* to go backward as seen from our perspective on the moving Earth.

Of course, this simple explanation of apparent retrograde motion works only if you accept the fact that Earth and the other planets orbit the Sun. Although the Greeks considered this possibility (as we'll discuss shortly), they ultimately rejected it. Part of the reason for this rejection came from untested philosophical ideas. For example, Aristotle outlined a number of arguments for the Earth's having a central position, but in retrospect his arguments were grounded in his misconceptions about physics. Another and more valid (at least, by modern scientific standards) reason why the Greeks held on to the geocentric system was because the alternative seemed inconsistent with other observations.

In particular, the Greeks recognized that their model would have a clear problem if they placed the Sun rather than the Earth at the center of their spheres. In that case, Earth would be closer to different portions of the sphere of the stars at different times of year. At times of year when we were closer to a particular part of the sphere, the stars on that part of the sphere would appear more widely separated than when we were farther from that part of the sphere (just as the spacing between the two headlights on a car looks greater when you are closer to the car). This would create annual shifts in the separations of stars—and the Greeks observed no such shifts. They knew that there are only two possible ways to account for the lack of an observed shift: Either the Earth is at the center of the universe, or the stars are so far away that the shift is undetectable to the naked eye. They did not consider it possible that the stars could be *that* far away and therefore concluded that the Earth must be the center of the universe. Significantly, this basic argument still holds even when we allow for the stars to be at different distances rather than all on the same sphere. In that case, as Earth orbits the Sun we would look at particular stars from slightly different positions at different times of year, and this change would affect the apparent positions of nearby stars more than distant stars. Today, we do indeed detect such shifts (called stellar parallax)—though they are too small to see with the naked eye—providing concrete proof that the Earth really does go around the Sun (Figure 2.5).

The Greeks remained wedded to their geocentric model, making it far more difficult for them to account for the apparent retrograde motion of the planets. That they managed to do so at all is a tribute to human ingenuity. The Greeks came up with a variety of ways to model the mysterious retrograde motion of the planets, but the idea that eventually stuck was this: Each planet was imagined to move around the Earth on a small circle that turned upon a larger circle (Figure 2.6). This circle-upon-circle motion makes a planet trace a loop as seen from Earth, with the backward portion of the loop mimicking apparent retrograde motion.

The Greeks gradually refined their model in an effort to better match their observations of planetary positions among the stars. The final Greek model was that of the astronomer Ptolemy (c. A.D. 100–170; pronounced *tol′-e-mee*); we often call it the **Ptolemaic model** to distinguish it from earlier geocentric models of the ancient Greeks. Ptolemy found that a simple application of the circle-upon-circle idea was not enough to produce satisfactory predictions of planetary positions. He therefore devised a number of other mathematical tricks (such as adding even more circles and not having the Earth at the precise center of all the circles). But one trick he did not try was any deviation from circles. Following Greek tradition, Ptolemy assumed that all heavenly motion must proceed in perfect circles. In the end,

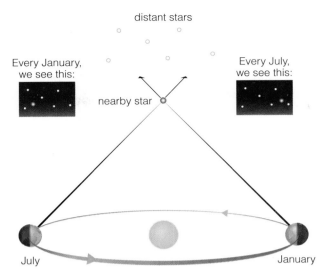

FIGURE 2.5 If the Earth orbits the Sun, then over the course of each year we should see nearby stars shift slightly back and forth relative to more distant stars (stellar parallax). The Greeks could not detect any shift, which meant either that the stars were too far away for it to be visible to the naked eye or that the Earth does not orbit the Sun. They chose the latter possibility. Today, we *can* detect the shift with telescopic observations, which proves that the Earth does orbit the Sun. (The figure is greatly exaggerated; the actual shift is far too small to be noticeable to the naked eye.)

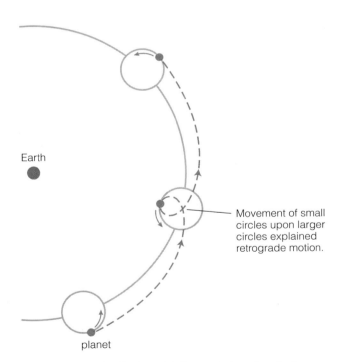
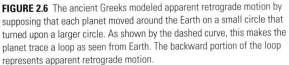

FIGURE 2.6 The ancient Greeks modeled apparent retrograde motion by supposing that each planet moved around the Earth on a small circle that turned upon a larger circle. As shown by the dashed curve, this makes the planet trace a loop as seen from Earth. The backward portion of the loop represents apparent retrograde motion.

Ptolemy succeeded in creating a model that could forecast future planetary positions to within a few degrees of arc—roughly equivalent to holding your hand at arm's length against the sky—which was considered sufficiently accurate at the time. Indeed, his model generally worked so well that it remained in use for the next 1,500 years.

Before we leave the ancient Greeks, it's worth noting that we know of at least one case in which a Greek philosopher rejected the geocentric idea outright. In about 260 B.C., Aristarchus (c. 310–230 B.C.) argued that the Earth goes around the Sun, rather than vice versa. Little of Aristarchus's work survives to the present day, so we do not know why he suggested a Sun-centered solar system. We do know that he made measurements that convinced him that the Sun is much larger than the Earth, so perhaps he simply concluded that it was more natural for the smaller Earth to orbit the larger Sun. In addition, he probably recognized that a Sun-centered system offers a simple explanation for apparent retrograde motion (see Figure 2.4). His idea did not gain wide acceptance in Greece, but it never fully died. We know that Copernicus was aware of Aristarchus's views when he resurrected the idea of a Sun-centered solar system some 1,800 years later.

Toward a Scientific Renaissance

One reason why Greek thought gained such influence was that the Greeks proved to be as adept at politics and war as they were at philosophy. In about 330 B.C., Alexander the Great began a series of conquests that expanded the Greek Empire throughout the Middle East, absorbing all the former empires of Egypt and Mesopotamia. Alexander was more than just a military leader; he also had a keen interest in science and education. Indeed, as a teenager, Alexander's personal tutor had been none other than Aristotle. Alexander encouraged the pursuit of knowledge and respect for foreign cultures. On the Nile delta in Egypt, he founded the city of Alexandria, which became a center of world culture.

Although Alexander did not live to see it, the heart of Alexandria was a great library and research center that was completed in about 300 B.C. The Library of Alexandria was the world's preeminent center of research for the next 700 years. At its peak, the library held more than a half million books, handwritten on papyrus scrolls. Most of the scrolls probably were original manuscripts or the single copies of original manuscripts. When the library was destroyed during a time of anti-intellectual fervor in the fifth century A.D., most of its storehouse of knowledge was lost forever.

Much more would have been lost if not for the rise of a new center of intellectual achievement in Baghdad (in present-day Iraq). While European civilization fell into the Dark Ages, scholars of the new religion of Islam sought knowledge of mathematics and astronomy in hopes of better understanding the wisdom of Allah. The Islamic scholars translated and thereby saved many of the ancient Greek works. Building on what they learned from the Greek manuscripts, they went on to develop the mathematics of algebra as well as many new instruments and techniques for astronomical observation.

The Islamic world of the Middle Ages was in frequent contact with Hindu scholars from India, who in turn brought ideas and discoveries from China. Hence, the intellectual center in Baghdad achieved a synthesis of the surviving work of the ancient Greeks, the Indians, and the Chinese and the contributions of its own scholars. This accumulated knowledge spread throughout the Byzantine Empire (the eastern part of the former Roman Empire). When the Byzantine capital of Constantinople (modern-day Istanbul) fell in 1453, many Eastern scholars headed west to Europe, carrying with them the knowledge that helped ignite the European Renaissance and led directly to the development of modern science.

2.2 The Copernican Revolution

The principles of modern science developed during the European Renaissance. Historians attribute much of their development to a new spirit of inquiry that blossomed during this period. Technological developments also fueled the change. The most significant new technology was the printing press with movable type, invented by Johannes Gutenberg around 1450 (Figure 2.7). Prior to its invention, books had to be laboriously copied by hand or printed from hand-carved entire pages of type, making books expensive and rare. Indeed, one of the main reasons why most people remained illiterate at the time was that without readily available reading material there was little point in investing the effort required to learn to read. The printing press changed all that. By 1500, some 9 million printed copies of some 30,000 works were in circulation. With books cheap and widely available, many more people learned to read. This had the effect of democratizing knowledge and naturally led to a much larger pool of scholars. The stage was set for a dramatic rethinking of our place in the universe.

In 1543, Nicholas Copernicus published *De Revolutionibus Orbium Caelestium* ("Concerning the Revolutions of the Heavenly Spheres"), launching what we now call the **Copernican revolution.** In

FIGURE 2.7 This colored engraving shows Johannes Gutenberg holding a proof sheet from his printing press, which is visible to the left.

his book, Copernicus made the radical suggestion that the Earth is just one of the planets going around the Sun, thereby removing the Earth from any central place in the cosmos. Over the next century and a half, philosophers and scientists (who were often one and the same) debated and tested the Copernican idea. The debate was in many ways a clash between competing models of the universe, each designed to explain the observed motions of the planets and stars in our sky. Many of the ideas that now form the foundation of modern science first arose as this debate played out. Indeed, the Copernican revolution had such a profound impact on philosophy that we cannot understand modern science without first understanding the key features of this revolution.

Nicholas Copernicus—The Revolution Begins

Nicholas Copernicus was born in Torún, Poland, on February 19, 1473. His family was wealthy, and he received a first-class education, studying mathematics, medicine, and law. He began studying astronomy in his late teens. By that time, tables of planetary motion based on the Ptolemaic model were noticeably inaccurate. Copernicus concluded that planetary motion could be more simply explained in a Sun-centered solar system, and he began developing a Sun-centered model for predicting planetary positions. Copernicus was hesitant to publish his work, fearing that his suggestion that the Earth moved would be considered absurd. Nevertheless, he discussed his system with other scholars, generating great interest in his work. At the urging of some of these scholars, including some high-ranking officials of the Church, he finally agreed to publish his work. Copernicus saw the first printed copy of his book on the same day that he died—May 24, 1543.

In his book, Copernicus laid out a detailed model of planetary motion in which the Earth was just one of the planets going around the Sun. Many of his early supporters recognized the aesthetic advantages of his model, particularly in its more natural explanation for apparent retrograde motion. However, his model did not make substantially better predictions than Ptolemy's, largely because he still believed that heavenly motion must be in perfect circles. Because the true orbits of the planets are *not* circles, Copernicus found it necessary to add circles upon circles to his system, just as in the Ptolemaic system. As a result, the Copernican model was no more accurate and no less complex than Ptolemy's model, and it won relatively few converts in the 50 years after it was published. After all, why throw out thousands of years of tradition for a new model that worked equally poorly?

Tycho and Kepler

Part of the difficulty faced by astronomers who sought to improve either the Ptolemaic or the Copernican model was a lack of quality data. The telescope had not yet been invented, and existing naked-eye observations were not very accurate. In the late 1500s, Danish nobleman Tycho Brahe (1546–1601), usually known simply as Tycho, set about correcting this problem.

Tycho was an eccentric genius who, at age 20, lost part of his nose in a sword fight with another student over who was the better mathematician. Taking advantage of his royal connections, he built large naked-eye observatories (which worked much like giant protractors) with which he could measure planetary positions with unprecedented accuracy—to within 1 minute of arc, about the thickness of a fingernail held at arm's length. He collected such measurements over a period of more than three decades. Although Tycho never succeeded in coming up with a fully satisfactory explanation for his observations, he found someone who did. In 1600, he hired a young German astronomer named Johannes Kepler (1571–1630). Kepler and Tycho had a strained relationship, but Tycho recognized the talent of his young apprentice. In 1601, as he lay on his deathbed, Tycho begged Kepler to find a system that would make sense of his observations so "that it may not appear I have lived in vain."

It took Kepler most of a decade to find a model that could successfully reproduce Tycho's observations. Part of the difficulty was that Kepler still believed strongly in the Greek idea that planets must move in perfect circles. But, unable to match Tycho's data to planets moving in circles, Kepler tried other shapes and eventually discovered that planetary orbits around the Sun actually are ellipses rather than perfect circles (Figure 2.8).

Kepler based his final model, which gave a perfect match to Tycho's data, on what we now call

Kepler's laws of planetary motion (the first two laws published in 1610, and the third in 1618). Kepler's first law of planetary motion states that the orbits of all planets are ellipses with the Sun at one focus (Figure 2.9a). His second law states that a planet's speed varies in such a way that it sweeps out equal areas in equal times, which means it moves faster when near the Sun in its orbit and slower when farther from the Sun (Figure 2.9b). His third law relates a planet's orbital period to its average distance from the Sun and shows that more distant planets orbit the Sun more slowly.

Galileo

Despite the success of Kepler's model, many scientists objected to the idea that the Earth could move around the Sun, because it contradicted accepted ideas of physics that dated back to Aristotle. Such objections were overcome almost single-handedly by the great Italian scientist Galileo Galilei (1564–1642), nearly always known by only his first name. Galileo answered the objections with a combination of experiments and, beginning in 1609, observations through telescopes that he built. (The telescope was invented in 1608 by Hans Lippershey, but Galileo took what was little more than a toy and made improvements that turned it into a scientific instrument.)

Galileo overcame the deeply rooted objections to the idea of a Sun-centered solar system by perform-

---

FIGURE 2.8 An ellipse is a special type of oval. This figure shows how an ellipse differs from a circle and how different ellipses vary in their eccentricity.

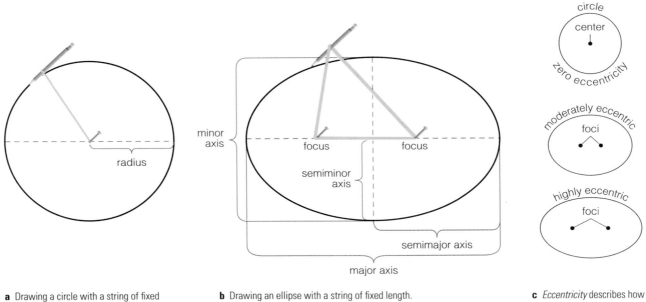

a Drawing a circle with a string of fixed length.

b Drawing an ellipse with a string of fixed length.

c *Eccentricity* describes how much an ellipse deviates from a perfect circle.

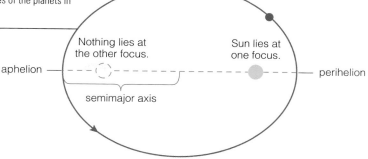

FIGURE 2.9 These diagrams illustrate Kepler's first two laws of planetary motion. (The eccentricity shown here is exaggerated compared to the actual eccentricities of the planets in our solar system.)

a Kepler's first law: The orbit of each planet about the Sun is an ellipse with the Sun at one focus.

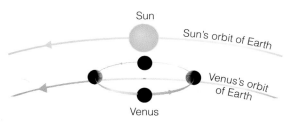

The areas swept out in 30-day periods are all equal.

b Kepler's second law: As a planet moves around its orbit, it sweeps out equal areas in equal times. This means that it moves faster when near the Sun and slower when farther from the Sun.

ing experiments that revealed the errors of Aristotelian physics. For example, Aristotle had argued that the Earth could not be moving because, if it were, objects such as birds and clouds would be left behind as the Earth moved along its way. Galileo's experiments showed that objects do not change their motion unless a force acts upon them (an idea now codified in Newton's first law of motion), thereby proving that birds and clouds would naturally stay with a moving Earth.

Galileo's telescopic observations may have been even more important in making a case for the Copernican revolution. Three sets of observations were especially crucial. First, he found clear evidence of mountains and craters on the Moon and of sunspots on the Sun. These findings contradicted the ancient belief that the heavenly bodies must be "perfect" and made the idea of elliptical orbits (as opposed to "perfect" circles) more acceptable. Second, he observed that Venus goes through phases rather like the phases of our Moon, but in a way that can be explained only by its orbiting the Sun rather than the Earth (Figure 2.10). Third, he discovered four moons orbiting Jupiter, clearly demonstrating that not everything orbits the Earth. Galileo's experiments and observations, combined with the success of Kepler's model, provided an exceedingly strong case for the Copernican idea of a Sun-centered solar system.

Although our historical hindsight shows clearly that Galileo won the day, the story was more complex in his own time, when Catholic Church doctrine still held the Earth to be the center of the universe. On June 22, 1633, Galileo was brought before a Church inquisition in Rome and ordered to recant his claim that the Earth orbits the Sun. Already nearly 70 years old and fearing for his remaining life, Galileo did as ordered and his life was spared. However, legend has it that as he rose from his knees he whispered under his breath *Eppur si muove*—Italian for "And yet it moves." (Given the likely consequences if Church officials had heard him say this, most historians doubt the veracity of this legend.) Galileo was not formally vindicated by the Church until 1992 (in a statement by Pope John Paul II), but the Church gave up the argument long before that. Galileo's book, *Dialogue Concerning the Two Chief World Systems,* was removed from the Church's index of banned books in 1824. Today, Catholic scientists are at the forefront of much astronomical research, and

FIGURE 2.10 Galileo observed phases of Venus through his telescope, providing conclusive evidence that Venus orbits the Sun and not the Earth.

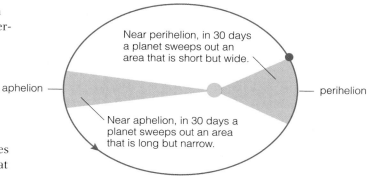

a In the Ptolemaic system, Venus follows a circle upon a circle that keeps it close to the Sun in our sky. Therefore, its phases would range only from new to crescent.

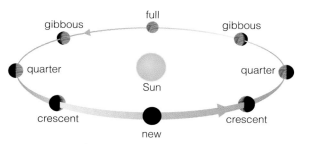

b Galileo observed Venus go through a complete set of phases and therefore proved that it orbits the Sun.

The case of Galileo, in which the Catholic Church held tightly to a geocentric belief and was unwilling to consider the evidence of science, is often portrayed as having exposed a deep conflict between science and religion. However, even a cursory look at the history of the debate over geocentrism in the Church shows that the reality was much more complex, with deep divisions within the Church hierarchy.

Perhaps the clearest evidence for a more open-minded Church comes from the case of Copernicus, in that numerous Church officials voiced strong support for his revolutionary work. A less well known but equally interesting case occurred a century earlier with the work of Nicholas of Cusa (1401–1464). In *De Docta Ignorantia* ("On Learned Ignorance"), published in 1440, Nicholas argued that the Earth goes around the Sun and even weighed in on the subject of extraterrestrial life. He wrote:

> *Rather than think that so many stars and parts of the heavens are uninhabited and that this earth of ours alone is peopled . . . we will suppose that in every region there are inhabitants, differing in nature by rank and allowing their origin to God, who is the centre and circumference of all stellar regions.*

Nicholas was ordained a priest in the same year his book was published, and he was later elevated to cardinal. Clearly, his views caused no problems for Church officials of the time. (Although Nicholas of Cusa published his book more than a century ahead of Copernicus, Copernicus gets credit for starting the revolution for two reasons: First, Nicholas's book was not widely disseminated, and

there's no evidence that Copernicus was even aware of it. Second, Nicholas merely expressed opinions, while Copernicus offered a detailed model that could be tested against observations.)

Many other scientists received similar support from within the Church. Indeed, for most of his life, Galileo counted cardinals (and even the pope who later excommunicated him) among his friends. Some historians suspect that Galileo got into trouble less for his views than for being somewhat insensitive to the political realities created by internal divisions within the Church. In 1632, just a year before his famous trial, he published a book laying out the evidence for a Sun-centered system in the form of a dialog between two fictional characters on the two sides of the debate. For the character taking the geocentric position, he chose the name Simplicio, which pretty much says it all in terms of the simplistic views the character held. The noted science and science fiction author Isaac Asimov described the ramifications of Galileo's book as follows:

> *The pope was persuaded that Simplicio . . . was a deliberate and insulting caricature of himself. The book was all the more damaging to those who felt themselves insulted, because it was written in vigorous Italian for the general public (and not merely for the Latin-learned scholars) and was quickly translated into other languages—even Chinese!*

Another case that almost certainly reached its tragic conclusion because of a clash of personalities concerned the Italian philosopher Giordano Bruno (1548–1600).

official Church teachings are compatible not only with the Earth's planetary status but also with the theories of the Big Bang and the subsequent evolution of the cosmos and of life.

THINK ABOUT IT . . . *Although the Catholic Church today teaches that science and the Bible are compatible, not all religious denominations hold the same belief. Do you think that science and the Bible are compatible? Defend your opinion.*

Newton—The Revolution Concludes

The closing chapter to the Copernican revolution was delivered by Sir Isaac Newton (1642–1727). Perhaps the greatest unanswered question after the work of Kepler and Galileo was *why* the planets would move in the seemingly strange ways described by Kepler's laws. Newton answered this question by finding mathematical formulations for how motion works in general (with his three laws of motion) and

how gravity works in particular (with his law of universal gravitation). In 1687, Newton published a famous book known as *Principia* (short for *Philosophiae Naturalis Principia Mathematica*, or "Mathematical Principles of Natural Philosophy"). In this book he showed mathematically how Kepler's laws are natural consequences of the law of gravity. In essence, Newton had created a new model for the inner workings of the universe in which motion is governed by clear laws and the force of gravity. The model explained so much about the nature of motion in the everyday world, as well as about the movements of the planets, that the geocentric idea simply could no longer be taken seriously.

The Copernican Perspective on Life in the Universe

The Copernican revolution made it clear that the Moon and the planets really were other *worlds*, not mere lights in the sky. This dramatic change in the human perspective on Earth's place in the universe caused scientists and philosophers to take seriously

Once a Dominican monk, Bruno later became quite extreme in his views concerning the truth of the Copernican system and the prospect of extraterrestrial life. In his book *On the Infinite Universe and Worlds*, published in 1584, Bruno wrote:

> For it is impossible that a rational being . . . can imagine that these innumerable worlds, manifest as like to our own or yet more magnificent, should be destitute of similar or even superior inhabitants.

Note that, rather than posing a possibility, Bruno was adamant not only that other worlds must be inhabited but also that no "rational being" could disagree with him. This and other, similar claims placed Bruno in direct, personal conflict with conservative Church officials, which led to his being branded a heretic and burned at the stake on February 17, 1600.

Perhaps the main lesson to be drawn from these stories is that while science has advanced dramatically in the past several centuries, people remain much the same. The Church was never a monolithic entity, and just as different people today debate the meaning of words in the Bible or other religious texts, so did different Church scholars at the time of the Copernican revolution. The political pendulum swung back and forth—or perhaps even chaotically—between the geocentric and Copernican views. Even when the evidence became overwhelming, a few die-hards never gave in, and only the passing of generations finally ended the antagonism that had accompanied the great debate.

the possibility of extraterrestrial life. Some even tried to make observations of the Moon and planets that might reveal any life present on them. Galileo suggested that the lunar features he saw through his telescope might be land and water much like that on Earth. Kepler agreed and went further, suggesting that the Moon had an atmosphere and was inhabited by intelligent beings. Kepler even wrote a science fiction story, *Somnium* ("The Dream"), in which he imagined a trip to the Moon and described the lunar inhabitants. Unfortunately, as shown by these incorrect conclusions about the existence of water and air on the Moon (which has neither), the telescopes of the day were not sufficiently powerful to make observations useful to the search for life.

Because the observations were open to differing interpretations, scientists also debated the issue of extraterrestrial life on theoretical grounds. In particular, Newton's laws of physics suggested that any physical processes that had happened here on Earth also could have happened elsewhere. However, just because they could have happened doesn't prove that they did, and no one had any way of determining

whether other planets in our solar system were similar to the Earth or whether other stars were similar to the Sun. Moreover, even if other worlds existed, Newton's physical laws did not seem to give any insight into whether life could exist on those worlds. Thus, real study of life in the universe had to await more modern inventions. Before discussing some of these inventions in Section 2.4, we will first discuss the nature of science explicitly by using what we've learned from the Copernican revolution.

2.3 The Nature of Modern Science

The story just told, of how our ancestors gradually figured out the basic architecture of the cosmos, shows many of the features of what today is considered "good science." For example, we saw how models were formulated and tested against observations, and modified or replaced when they failed these tests. The story also shows some classic mistakes, such as the failure of everyone before Kepler to question the belief that orbits must be circles. The ultimate success of the Copernican revolution led scientists, philosophers, and theologians to reassess the various modes of thinking that played a role in the 2,000-year debate. The principles of modern science emerged from this reassessment.

Perhaps surprisingly, it turns out to be quite difficult to define the term *science* precisely. The word itself comes from the Latin *scientia*, meaning "knowledge"; it is also the root of the words *conscience* (related to the idea of self-knowledge or self-awareness) and *omniscience* (the quality of being all-knowing). But not all knowledge is science; for example, you may know what music you like best, but your taste is not the result of scientific study. In the rest of this section, we'll explore in some detail the nature of modern science.

Approaches to Science

One reason why it is difficult to define science is that not all science works in the same way. For example, philosophers of science often distinguish between two main scientific approaches: discovery science and hypothesis-driven science.

Discovery science involves going out and looking at nature in a general way in hopes of learning something new and unexpected. Once we gather the observations, we try to interpret and explain them, and we often find that they suggest new questions to be studied in greater depth. Sending a spacecraft to a world that we have not studied in depth is an example of discovery science. Consider the case of Europa, the moon of Jupiter suspected of having a subsurface ocean. Prior to sending the Voyager spacecraft there, we had only vague hints of this possibility. Thus, the act of sending the spacecraft to make general observations helped us discover an exciting possibility.

Hypothesis-driven science involves proposing an idea and then performing experiments or making observations that put it to the test. As an example of the process, suppose your flashlight suddenly stops working. This observation may lead you to question why it has stopped working, and you may *hypothesize* that the reason is that the batteries have died. In other words, you've created a tentative explanation, or **hypothesis,** for the flashlight's failure. A hypothesis is sometimes called an *educated guess*—here, it is "educated" because you must know that flashlights need batteries in order to construct your hypothesis of battery failure. Your hypothesis allows you to make a simple prediction: If you replace the batteries with new ones, the flashlight should work. You can test this prediction by replacing the battery. If the flashlight now works, you've confirmed your hypothesis. If it doesn't, you must revise or discard your hypothesis, hopefully in favor of some other one that you can then test (such as that the bulb is dead). Figure 2.11 illustrates the basic flow of this process, which is often referred to as "the scientific method." The scientific method is a useful idealization of some of the process of science. However, as the development of human ideas about the universe shows, science rarely progresses in quite this way.

Most scientific progress occurs through some combination of discovery science and hypothesis-driven science. Moreover, neither approach is unique to science. You use the discovery approach when you listen to new music to discover whether you like it. As in the flashlight example, you probably use the hypothesis-driven approach to solve many everyday problems. So while the two approaches help us understand *how* we do science, they still do not tell us what constitutes science. For that we must look a little deeper to find a few distinguishing characteristics of science.

Before we leave this topic, however, note that there are other ways to categorize different types of science. For the study of life in the universe, it's often useful to distinguish between what we call *historical science* and *experimental science*. Experimental science seeks to uncover general principles of physics, chemistry, or other sciences, usually by carrying out specific experiments in a laboratory. Experiments are repeatable and can also be modified to test slightly different ideas. In contrast, historical science involves looking at present-day evidence to try to discover something about past events. Because we cannot repeat or vary the past, historical science

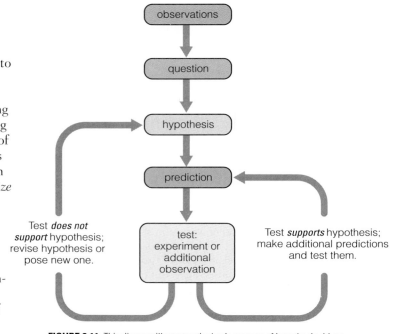

FIGURE 2.11 This diagram illustrates the basic process of hypothesis-driven science.

involves piecing together plausible descriptions of past events by careful examination of the evidence they have left behind. For example, the study of the history of life on Earth is a historical science, because we learn about now-extinct organisms not by observing them directly but by piecing together their story from an examination of fossils and other evidence that we can find today. The history of the Earth is a historical science for the same reason—we cannot do an experiment to see how the Earth would evolve under particular circumstances, but we can learn what occurred by studying the record that past events have left behind in rocks and fossils.

THINK ABOUT IT . . . *Notice that the standard view of "the scientific method," as shown in Figure 2.11, describes hypothesis-driven, experimental science. Does this description of the scientific method also work for historical science? Explain.*

Hallmarks of Science

If the debate over the Earth's place in the universe had been held only on philosophical grounds, we might still believe in an Earth-centered universe today. But the debate actually involved competition among different models of the universe, notably the models of Ptolemy, Copernicus, and Kepler. Each model was based on observations of the motions of the Sun, Moon, planets, and stars, and each could

Is there an "absolute truth" in science, with scientists discovering increasingly better approximations to this truth, or is scientific interpretation influenced by one's cultural biases?

In the early and middle 1900s, Karl Popper argued that there is an absolute truth. He said, however, that one cannot actually prove any theory true in science. A theory can only be demonstrated again and again as being consistent with observations. In this view, the single distinguishing hallmark of science is "falsifiability": It takes only one observation that is inconsistent with a theory to show that the theory is wrong. Incorrect theories are abandoned, leaving only those that are most consistent with the observations. As theories get better and better with time, they provide increasingly accurate representations that are increasingly closer approximations to the absolute truth.

In the late 1950s, Thomas Kuhn challenged this view of science. He argued that, occasionally, observations accumulate that are so inconsistent with an accepted theory that the theory has to be thrown out completely and replaced with a new one. Such shifts in *paradigm,* or a basic view of the world, have occurred many times—for example, with the development of the idea that the universe is not centered on the Earth or that living species evolve over time rather than remain static. Other paradigm shifts accompanied the theories of relativity, quantum physics, and continental drift (plate tectonics). In each of these cases, the interpretation we give to observations of the world around us depends on the particular paradigm we accept; in one sense, this is another way of saying that there really is a cultural influence on scientific interpretation.

So, is there an absolute truth in science? As we dig deeper both into how the world works and into the nature of science, we continually find surprises that make it difficult to know for sure. There is no answer today to the question of absolute truth, and in fact there may never be one.

make predictions about the future motions of those objects in our sky.

It's tempting to think that Kepler's model gained acceptance simply because it proved far superior to the other models in its ability to match the data. However, as we've discussed, that is not the whole story. For example, given that the original model of Copernicus was not noticeably better at matching the data than the Ptolemaic model, a purely data-driven decision might have caused scientists to reject the Sun-centered idea immediately. Instead, many scientists found elements of the Copernican model appealing—such as the simplicity of its explanation for apparent retrograde motion—and therefore kept it alive until Kepler finally found a way to make it match the data.

At that point, a data-driven decision would have led everyone to accept Kepler's model immediately, but widespread acceptance came only after the additional work of Galileo and Newton. Indeed, if data alone were the criterion for judgment, we could imagine a modern-day Ptolemy adding millions or billions of circles and complexities to the geocentric model in an effort to force it to match the observed data. In principle, a sufficiently complex geocentric model could reproduce the observations with arbitrary precision—but it would not convince us that the Earth is the center of the universe. We would still choose the Copernican view over the geocentric view because, in light of everything else we now know about the universe, it simply makes much more sense than could any convoluted, complex, and arbitrary model that managed to force-fit the data.

In many ways, the challenge of defining the nature of science comes down to listing the elements that enable us to choose wisely between competing models. Historians and philosophers of science have examined (and continue to examine) this issue in great depth, and entire books are devoted to questions about how to define science. And, of course, different people may set out the criteria that define science somewhat differently. Nevertheless, it seems fairly clear that everything we consider to be science today shares at least the following three major characteristics, which we will refer to as the "hallmarks" of science:

1. *Modern science is driven by observations and a belief that the world is inherently understandable.* Science always begins by assuming that the world is inherently understandable and that we can learn how it works by observing it and examining the processes that affect it. All of science, therefore, is based on observations of the world around us. This is true whether the observations are in the form of laboratory experiments or of field observations of the Earth's geology, the planets, or the universe around us.

2. *Science progresses through the creation and testing of models designed to explain our observations, and these models should be as simple as possible and be consistent with a rational understanding of other aspects of science.* This hallmark plays out in a variety of ways. For example, when two (or more) models are capable of explaining the same observations, we generally prefer the model that is simpler and based on fewer assumptions. This idea, called *Occam's razor* after the medieval English scholar William of Occam (1285–1349),

explains why we would keep our current model of the universe even if a modern-day Ptolemy came up with a more complex model that also matched the data. Moreover, models should not have to assume any special or unique circumstances—such as supernatural intervention—to explain what we see. In geology, for example, we assume that past events can be understood in terms of the same geological processes that we observe operating today (an idea called *uniformitarianism*). Perhaps most important, in science we let the observations direct our thinking as we construct a model rather than building a model to support our preconceived notions. This idea was crucial in the Copernican revolution; Kepler succeeded with his model only when he let the observations point him to ellipses and finally let go of his preconceived notion that orbits must be circles.

3. *A scientific model should be testable so that we can always envision further ways to check its validity or, if necessary, to conclude that it is incorrect.* In many ways, this notion of testability holds the key to distinguishing science from nonscience. For example, people may choose to believe in God because of their observations of the world around them, and they may construct a model of the universe based on their beliefs. But the existence (or nonexistence) of God is not testable and therefore does not constitute science. (That is why thousands of scientists have no difficulty at all reconciling their religious beliefs with their scientific work and why you should be very suspicious when anyone claims to have found scientific evidence either for or against the existence of God.) In contrast, a scientific model, such as Kepler's model, must ultimately be testable. If the planets suddenly began to move in a way that is inconsistent with Kepler's elliptical orbits, we would be forced to revisit the model. Moreover, because a model must always be testable, we can never consider a scientific model to be correct beyond all doubt; at best, we can conclude that it is extremely likely that the model is correct.

THINK ABOUT IT . . . *Choose a scientific theory that you know something about, such as the theory of evolution or the theory of global warming. For the case you choose, discuss whether the theory exhibits the three hallmarks of science.*

Note that while all science should exhibit these three hallmarks, they are not the only features that show up in scientific work. Indeed, scientists often work in ways that may seem rather "unscientific," such as engaging in pure speculation before embarking on a new project. For example, scientists speculated about life in the universe long before the data

allowed anyone even to come up with a testable hypothesis. Such speculation is natural for curious humans and often leads to real scientific advances. Nevertheless, it is important to distinguish between mere speculation and true science, and looking for the three hallmarks can help.

Theories in Science

Many people associate facts with science, but accumulating facts is not what science is primarily about. A telephone book is an impressive catalog of factual information, for example, but it has little to do with science. It is true that facts, in the form of verifiable observations and repeatable experiments, are the prerequisites of science. What really advances science, however, is discovering underlying principles that tie together a number of observations that previously seemed unrelated. People like Newton, Charles Darwin, and Albert Einstein stand out in the history of science not because they discovered a great many facts but because their theories had such broad explanatory power.

What is a scientific theory, and how is it different from a hypothesis? Theories differ from hypotheses in two important ways. First, and most important, a hypothesis is only a *tentative* explanation of a specific process or event, whereas a **theory** is supported by a wide range of extensive and varied evidence. When first proposed, a hypothesis might have little or no observational support. We then subject the hypothesis to experimental or observational tests. A hypothesis may become a theory *only* if it has been repeatedly subjected to real tests and found to be both consistent with the observations and more consistent with them than other hypotheses. Second, theories tend to be broader in scope than the ideas suggested as hypotheses, providing explanations for a wide range of observations and tying together broad issues in science.

As an example, consider Kepler's discovery of his first law of planetary motion, which states that planets move in elliptical orbits around the Sun with the Sun at one focus (see Figure 2.9a). When Kepler first proposed this idea, it qualified only as a hypothesis; he couldn't make circular orbits match Tycho's observational data and thought that ellipses might help. His hypothesis gained support only when he calculated what the positions of the planets in the sky would be if each were on an elliptical orbit (combined with the orbital velocity properties he described with his second law) and got an excellent match with the data. This match and the consistency of elliptical orbits with a huge amount of data obtained subsequently give us enormous confidence that the idea is correct and allow us to elevate it to the level of theory. However, because the fact that planets move in elliptical orbits is only one aspect of the more general issue of planetary motion, we usually don't think of it as a theory by itself. Instead, we think of it as a part of our theory of planetary motion. (Note that we call the fact that planets move in elliptical orbits a *law*, though this term is not always used in precisely this way.) Newton's discoveries broadened our understanding even further, showing planetary motion to be a natural outgrowth of his theory of gravity. Still later, Einstein showed that Newton's theory of gravity is an outgrowth of an even more general theory—Einstein's general theory of relativity.

Note that this scientific use of the term *theory* to denote a comprehensive explanation supported by abundant evidence contrasts with our everyday usage, which equates theories more with speculations or hypotheses. In everyday life, someone might get a new idea and say, for example, "I have a new theory about why people enjoy the beach." But without the support of a broad range of evidence that has been tested and confirmed by others, this idea is really only a hypothesis. Newton's theory of gravity qualifies as a scientific theory because of its broad application and because it has been validated by a great many observations and experiments. Similarly, the theory of evolution has been continually validated by observations, and it ties together so many seemingly disparate facts that, as we'll discuss shortly, it is now considered the unifying theme of all biology. In accord with our third hallmark of science, a scientific theory can never be proved true beyond all doubt, but anything that qualifies as a real theory must be supported by an enormous body of evidence.

THINK ABOUT IT . . . *When someone claims that something is "only a theory," what do you think they mean? Does this meaning agree with the definition of a theory in science? Explain.*

Nonscience and Pseudoscience

People often seek knowledge in ways that do not represent science. For example, while science seeks to explain how life arose and evolved on Earth, it says nothing about the meaning and value of life—topics about which you probably hold strong convictions. Similarly, suppose you are learning to play drums or write history papers, or deciding what you think about a complex political issue. In each case, you will make observations, exercise logic, and form and test hypotheses. You will also employ intuition (as do working scientists), and you may call on—and perhaps in the long run modify—your values and beliefs. However, these examples are clearly not science. Remember that science is an attempt to provide rational, testable explanations for observed natural phenomena. Thus, science focuses on particular types of questions and uses a particular set of methods. Most types of nonscientific knowledge seeking, such as the examples

above, address different types of questions than does science. These types of knowledge are generally easy to distinguish from scientific knowledge.

In other cases, distinguishing science from nonscience can be tricky. It can also be very important, especially when we are dealing with proposed explanations of natural phenomena. For example, suppose someone tells you that life didn't actually originate on Earth but arrived here from outer space on meteorites. That is a real possibility that scientists consider [Section 5.2]. But is it a scientific hypothesis, or mere speculation? A few decades ago, it was speculation because we had no way to test it. Today it is at least partially testable, because we can now examine meteorites in great detail.

For a case that is more likely to be in the news, consider the widespread belief that aliens regularly visit the Earth in UFOs ("unidentified flying objects"). Is that a scientific theory—or even a fact, as many assert? Why or why not? For alien visitation to count as an observed fact, we would have to be able to reliably observe evidence that could not be better explained by anything else. If aliens land in Paris, hold press conferences, shake hands with crowds, and provide artifacts, most people will decide that their visit is a fact. In reality, however, scientists have studied many UFO sightings, and all of them can be credibly explained as something other than an alien visit—from an unfamiliar natural phenomenon to a hoax or a delusion or our own technology (airplanes, balloons, rockets, etc.). A person who has seen a UFO may truly believe that it was due to aliens, but the rest of us cannot accept the assertion as fact without unambiguous evidence.

THINK ABOUT IT . . . *Have you ever seen anything in the sky that you could not explain? Was it "flying" in the sense of qualifying as an "unidentified flying object"? If so, can you think of explanations besides aliens for what you saw? In general, what do you think of the common tendency to associate "unidentified" objects with aliens?*

For a more specific example, consider the famous claim that an alien vessel crashed near Roswell, New Mexico, in 1947. As the story is told by believers, this crash left not just the wrecked spacecraft but actual alien bodies—exactly the sort of evidence that would prove beyond a doubt that aliens visit us. Yet none of this supposed evidence has ever become available for public study. Believers explain that the military collected the material and that the government has maintained a cover-up ever since. Consider what that would entail. For more than 50 years, through administrations with widely varying political agendas, hundreds of scientists and other military personnel would have had access to such a secret. Yet not a shred of actual evidence has emerged to public view, despite the fact that anyone who produced such evidence would become instantly rich and famous on the talk-show circuit. This claim of a cover-up is, at the minimum, very improbable. Moreover, a far more likely explanation is available: Government records speak to a crash in Roswell in 1947—but of a then-secret military balloon experiment, not an alien spacecraft (Figure 2.12). Believers may continue to *believe* that an alien spaceship crash took place and that the real evidence is being hidden by a conspiracy, but their belief is untestable and therefore not science. Indeed, this example illustrates a basic difference between science and belief. A scientist is obliged to allow the evidence to discredit even a favorite theory. A believer can start by asserting that the belief in question is true and "explain away" any evidence to the contrary (in this case, by claiming a conspiracy).

Note that the nonscientific nature of beliefs in alien visitation does not mean that such visits don't occur. Scientists will always acknowledge that "absence of evidence is not evidence of absence," meaning that the alien visitation idea could be true even in the absence of scientific evidence to back it up. But from the standpoint of science, the beliefs offer no hard evidence, they make less sense than other explanations for the claimed alien visits, and, even in rare cases (such as Roswell) in which claims can be at least partially put to a test, the true believers won't

accept the results. These shortcomings explain why most scientists remain extremely skeptical of reports of aliens having visited Earth. (We will discuss these issues further in Chapter 12.)

The UFO example also illustrates an important distinction between different types of nonscience. If everyone simply agreed that belief in alien visits was nonscientific, then there'd be no reason to distinguish these beliefs from other nonscientific beliefs. But at least some UFO adherents claim that their beliefs *are* scientific—even to the point of demanding funding for their "research" from scientific agencies. The fact that none of the hallmarks of science are evident in the alien visitation beliefs marks such claims as **pseudoscience,** which literally means "false science."

Generally speaking, pseudoscience refers to claims that *can* be tested scientifically but for which the adherents do not pay attention to the results of the tests. For example, at the beginning of each year, you can find tabloid newspapers offering predictions made by people who claim to be able to "see" the future. Because they make specific predictions, we can test their claims by checking whether their predictions come true. Numerous studies have shown that their predictions come true no more often than would be expected by pure chance, but these seers seem unconcerned by such studies of their abilities. The fact that they make testable claims but then

Movie Madness: Alien Autopsy—Film of a Dead Extraterrestrial?

In 1995, an American TV network broadcast a "documentary" purporting to be the filmed record of an alien dissection. "Alien Autopsy," as the show was called, consisted of black-and-white film footage supposedly made by U.S. government agents in 1947. This would have been shortly after the famed Roswell incident, in which the military collected crash debris discovered on a New Mexico ranch. Beginning in the 1970s, some people began insisting that what the military had hauled away were the remains of an alien spacecraft, including the bodies of extraterrestrial crash victims. The "Alien Autopsy" film was offered as proof of this claim, as it showed white-coated technicians cutting up a short humanoid body, presumably one of the unfortunate aliens.

But was this documentary for real? Or was it a hoax—a staged, recently made film, in which actors slice up a synthetic dummy? The incentive to make and broadcast a faked film would be obvious: enormous public interest in the Roswell incident guaranteed a large, lucrative audience.

The show generated much debate. Those who suspected a government cover-up of evidence for

visiting aliens felt that the program gave weight to their beliefs. But skeptics quickly assembled a laundry list of things that were suspicious about the film. The alien itself looks like a human female, which on the basis of the diversity of life on our own planet is more than a little surprising. Why would extraterrestrials resemble us? In addition, the camera work is amateurish, which doesn't seem appropriate for such an important assignment. No tripod is used, and the close-ups are badly out of focus.

Perhaps the greatest reason to suspect that "Alien Autopsy" was a fabrication was that the producers refused to let the film be analyzed by Kodak or other companies that could verify that the original film stock actually dates from 1947. While snippets of film leader (not showing any aspect of the autopsy) were offered for inspection, and were indeed from that era, they clearly proved nothing about the film itself.

In the late 1990s, all remaining suspicions about "Alien Autopsy" evaporated when the broadcast network admitted that it was a hoax.

FIGURE 2.12 Debris recovered in Roswell, New Mexico, in 1947, claimed by some to be the remnants of an extraterrestrial craft. In this photo, made by newspaper photographer James Johnson, General Roger Ramey is showing the debris to reporters in Fort Worth, Texas.

ignore the results of the tests marks their claimed ability to see the future as pseudoscience.

In the rest of this book, we will focus almost exclusively on the scientific study of life in the universe. However, in public discussions and in the media, it's almost impossible to avoid hearing nonscientific or pseudoscientific claims of life beyond Earth (or of why life beyond Earth is not possible). We shouldn't be surprised by such claims—after all, even scholarly discussion of the question of life beyond Earth was nonscientific for most of its long history. But it is very important to be able to distinguish between claims that stand up to scientific scrutiny and those that don't.

2.4 The Tools and Methods of the Search

The study of life in the universe has become a real science because we now have the ability to challenge hypotheses concerning such things as how life evolved on Earth and whether conditions for life exist on other worlds. In this section, we'll discuss in broad terms a few of the key tools and methods that apply to the three fundamental questions in astrobiology listed in Section 1.5; we'll save details of some of the more complex tools for later chapters.

Studying the Origin and Evolution of Life

As we'll discuss in Chapter 3, virtually everything we know today about the evolution of life on Earth rests on the foundation of Charles Darwin's *theory* of evolution, in which he proposed a specific model to describe how organisms evolve (through *natural selection* [Section 3.1]) and cited a tremendous amount of evidence in support of his model. Thus, Darwin's theory provided biologists with the intellectual framework needed to try to understand the origin and evolution of life. However, gathering precise data to back up claims about early life or about the conditions under which life arose required new technological tools to address questions such as these:

■ When did fossil organisms live?

■ What was the Earth like at the time that ancient organisms lived?

■ How are modern organisms related to one another and to their ancient ancestors?

Addressing the first question requires technologies for age-dating rocks and fossils. A number of techniques can be used to estimate such ages, but for specific ages we rely almost exclusively on **radiometric dating.** This technique, which we will discuss in detail in Chapter 4, makes it possible to measure ages of rocks and fossils with great precision. Radiometric dating, developed in the mid-twentieth century, has allowed us to put together a detailed timeline for life on Earth.

The second question requires that we have a way to determine the prevailing conditions on Earth—such as the oxygen content of the atmosphere and the average temperature—at various times in the past. This also requires studying ancient rocks and fossils, but we look for clues beyond their ages. For example, chemical analysis of a rock may tell us whether it has been in the presence of liquid water, and the nature of a fossil organism (such as whether it required oxygen) may tell us about past atmospheric composition. By combining results from many parts of the world and many different times, scientists can put together a picture of how the Earth has changed through time.

Until very recently, the third question was studied primarily by comparing visible characteristics of different organisms; for example, it's fairly obvious

FIGURE 2.13 The summit of Mauna Kea, on the Big Island of Hawaii, is home to many of the world's most powerful telescopes. The high altitude and clear skies of Mauna Kea make it an almost ideal observing site.

that humans are more closely related to chimpanzees than to fish. Today, the powerful new tool of **genetic analysis** can be applied to this question. All living organisms contain genetic material in the form of DNA [Section 3.4]. Genetic analysis allows us to compare the genetic material of different species in order to determine what makes one species different from another. This knowledge, in turn, allows us to map relationships among species much more clearly. We'll discuss the method in more detail in Chapter 5; suffice it to say that genetic analysis has revolutionized our understanding of species relationships.

In summary, scientific tools have enabled us to learn details of how the Earth and life have evolved through time, making it possible for us to understand what characteristics of the Earth are important for supporting life and whether conditions on other worlds are suitable for life.

Searching for Life Beyond Earth

Besides understanding whether life could have arisen on other worlds, we'd also like to search for actual living organisms. We can carry out this search either by going to another world to look for evidence of past or present life or by studying other worlds remotely with telescopes.

Within our own solar system, we now have the technology to look for life in both ways. We can observe planets and moons in our solar system either by studying them telescopically from Earth or by sending spacecraft to observe them more closely. Today, these observations are of sufficiently high quality that we can be quite confident that no other world in our solar system is home to a variety of surface life like that found on Earth. However, it remains quite possible that life is hidden in ways that we may someday uncover by visiting other worlds. For example, we might find microscopic life tucked away underground on Mars or in an ocean beneath the icy crust of Europa [Section 1.1].

Beyond our solar system, the vast distances to the stars [Section 1.2] mean that going to other worlds remains well beyond our present technology. Therefore, our present hopes for finding life among the stars rest solely with telescopic technologies. Here, too, we are in the midst of a revolution. For example, the 5-meter Hale telescope on Mount Palomar (outside San Diego) was the world's most powerful for nearly 50 years after it was built in 1947. Then, beginning in the 1990s, a renaissance in mirror-crafting techniques initiated a deluge of telescope building, and the Hale telescope is no longer even among the world's 10 largest (Figure 2.13). Space-based telescopes, such as the Hubble Space Telescope and the Chandra X-Ray Observatory, have further improved our observing abilities by overcoming the blurring caused by the Earth's atmosphere and allowing us to study forms of light (such as ultraviolet rays and X-rays) that don't penetrate the atmosphere.

FIGURE 2.14 This image of the Earth at night is dotted with the lights of civilization (mostly lights from cities and from agricultural, oil, and gas fires). If aliens on a distant world had a telescope large enough to capture an image of night lights on Earth, they would know that we have an advanced civilization.

Telescopes can help us search for inhabited planets in two basic ways. First, we can make *images* (photographs) of distant worlds. High-resolution images could allow us to see evidence of life such as seasonal changes in surface colors caused by vegetation or the lights of a civilization (Figure 2.14). Unfortunately, no existing telescopes are capable of making images of extrasolar planets, and it will probably be many decades before we can build a telescope large enough to obtain even moderately high resolution images of planets around other stars. Second, through the remarkable techniques of *spectroscopy,* we can in principle determine key characteristics of a distant world such as its atmospheric composition, surface temperature, and rotation rate. Spectra might even reveal evidence of life if we detect combinations of atmospheric gases that can exist only if life is present. We are much closer to having the technology needed for spectroscopy than for imaging of extrasolar planets, and scientists are already developing plans for telescopes that might enable us to search for spectral signatures of life on planets around nearby stars. We will discuss both imaging and spectroscopy in more detail in Chapter 6.

Finally, we can search for advanced civilizations through SETI efforts, in which we use telescopes to seek signals sent deliberately. As we will discuss in detail in Chapter 11, we already have the technology needed to begin modest SETI efforts, and more advanced efforts will be possible in coming decades.

Investigating the Future of Life on Earth and Beyond

The third major question in astrobiology concerns the future of life on Earth and beyond. As with the other major questions, research on this question relies largely on technologies developed only recently. Several technologies can be brought to bear on the question of life's future on Earth. Scientists use computers to create models of how the Sun and the Earth will change through time. For example, models of the Sun show that it will gradually become hotter and brighter with time, so the Earth will eventually become uninhabitable—but probably not for at least a billion years [Section 9.4]. Of more immediate concern, models of the Earth's climate can be used to understand how human activity is changing the Earth. For example, climatologists use computer models to try to understand how global warming will affect the Earth, while biologists use models to try to understand the impact of deforestation on ecosystems.

THINK ABOUT IT . . . *While the Earth's long-term future is clearly of interest scientifically, should we be concerned about the fact that the Earth may become uninhabitable in a billion years? Why or why not? (Hint: Consider the fact that the earliest hominids [human ancestors] walked the earth a few million years ago, while the earliest civilization [such as Egypt] emerged a few thousand years ago.)*

A related question concerns the possibility that we might be able to spread our own civilization beyond Earth. We can study this question by exploring things such as how our bodies respond to space travel, how we might "terraform" other planets (such as Mars) by changing their environments to allow us to survive on them [Section 7.5], and how we might someday develop spacecraft capable of interstellar travel.

For the longest time scales, astronomers are trying to understand how the ultimate fate of life is tied up with the eventual fate of stars, galaxies, and the universe. Is life a passing phase in the history of the universe, or might it somehow survive as the universe grows older? On this question, we remain in the realm of speculation rather than science. However, modern science is giving us an increasingly accurate picture of what the large-scale future of the universe will be like, and we can hope that we'll someday be able to answer this deep philosophical question.

2.5 The Golden Age of Astrobiology

It seems almost a part of human nature to wonder about the nature of life and whether life exists anywhere besides Earth. Nearly every culture has developed at least some mythology concerning these questions, and they were the subject of intense debate among ancient peoples such as the Greeks and Romans. But no matter how deeply our ancestors considered these questions, until recently there was no way to address them scientifically. As we've discussed, part of the reason was that science itself developed only slowly through history. Then, even after the methods of science were available in principle, we lacked the technology to address these questions in practice.

Today, many new technologies are being brought to bear on questions of life in the universe. Scientists from a variety of disciplines now work together in astrobiological research. In this sense, we are living at the beginning of what might be a "golden age of astrobiology" in which the ancient dream of searching for life beyond Earth is finally coming to fruition. It may take decades, centuries, or even millennia to play out, but we are for the first time on a path that should ultimately yield answers to ancient questions about life in the universe.

THE BIG PICTURE

In this chapter, we've explored the nature of science and its role in the study of life in the universe. As you continue your study, keep in mind the following "big picture" ideas:

- The questions that drive research about life in the universe have been debated for thousands of years. However, until recently the debate was purely speculative rather than scientific in a modern sense, both because it took a long time for the modern scientific method to develop and because we lacked the necessary technology for real research into questions of the origin of life and the existence of life elsewhere.

- Many of the hallmarks of modern science developed as a result of the Copernican revolution, which changed our view of the universe from one with the Earth at its center to one in which Earth is just one planet orbiting the Sun. This fundamental change in human perspective had a dramatic impact not only on science in general but also on the particular question of life in the universe, because it showed that planets really are other *worlds* and not mere lights in the sky.

- Although the methods of science developed gradually, they are what make it possible for us to learn so much about the universe today. Science always begins by assuming that the world is inherently understandable and that we can learn how it works by observing it and by examining the processes that affect it. All of science, therefore, is based on observations of the world around us.

- Even after the methods of modern science developed, it took time for the study of life in the universe to become a science because of the lack of a sound intellectual framework and the lack of sufficient technologies for research. We are now entering what might be called a "golden age of astrobiology" in which new ideas and new technologies are helping scientists from many disciplines work together to try to answer fundamental questions about life in the universe.

Review Questions

1. Briefly discuss why the ancient Greeks are so intimately connected with modern science and why Greece provided a setting in which early scientific ideas could develop.

2. What do we mean by a *model* of nature?

3. Summarize the development of the Greek *geocentric model*, from Thales through Ptolemy.

4. What is *apparent retrograde motion*, and why was it so difficult to explain with the geocentric model? What is its real explanation?

5. What was the *Copernican revolution*, and how did it change the human view of the universe? Briefly describe the major players and events in the Copernican revolution.

6. How did the Copernican revolution impact scholarly thought regarding the question of life beyond Earth?

7. Briefly contrast *discovery science* and *hypothesis-driven science*. What is historical science?

8. Describe the three hallmarks of science and why they are so important.

9. What is a *theory* in science, and how does it differ from a *hypothesis*? How is the scientific definition of a theory different from the way the term is used in everyday life?

10. Briefly explain why the belief that UFO sightings mean alien visitation does not meet the standards of science. Does this make such beliefs wrong? Explain.

11. What do we mean by *pseudoscience*? How does it differ from more general nonscience?

12. Briefly describe key tools used to study the three major questions in astrobiology research.

13. In what sense are we entering a "golden age" in astrobiology? Explain.

Discussion Questions

1. *Religion and Life Beyond Earth.* Choose one religion (your own or another) and investigate its beliefs with regard to the possibility of life on other worlds. If scholars of this religion have made any definitive statements about this possibility, what did they conclude? If there are no definitive statements, discuss whether the religious beliefs are in any way incompatible with the idea of extraterrestrial life. Discuss your findings with those of students who have investigated other religions.

2. *Science and Religion.* Science and religion are often claimed to be in conflict. Do you believe this conflict is real and hence unreconcilable, or is it a result of misunderstanding the differing natures of science and religion? Defend your opinion.

3. *A Science of UFOs?* Is it possible to test scientifically the idea that UFOs are bringing aliens to visit Earth? If so, how? If not, why not? If you think that the idea can be tested scientifically, how much effort (e.g., time, money) is it worth contributing to such research? If you do not think it can be tested scientifically, should it be part of the debate over whether we are alone in the universe? Defend your opinions.

4. *Answers in Our Lifetimes?* Discuss the current state of technology for astrobiology research. How far do you think we are from being able to answer the fundamental questions? Do you think we will have answers to any of these questions within your lifetime, or will the answers be available only to future generations? Explain.

5. *Absolute Truth.* An important issue in the philosophy of science is whether science deals with absolute truth. We can think about this issue by imagining the science of other civilizations. For example, would aliens necessarily discover the same laws of physics that we have discovered, or would the laws they observe depend on the type of culture they have? How does the answer to this question relate to the idea of absolute truth in science? Overall, do you believe that science is concerned with absolute truth? Defend your opinion.

Problems

Testable Claims? Each of **problems 1–8** lists some claim that may or may not be true. In each case, state whether the claim is testable, both in principle and in practice. If it is, suggest at least one way it could be tested. If it is not, explain why not.

1. Mars was home to an ancient, advanced civilization, but the civilization vanished without a trace.

2. Several kilometers below its surface, Europa has an ocean of liquid water.

3. There is no liquid water on the surface of Venus today.

4. Our Sun is a star.

5. Bacteria from Earth can survive on Mars.

6. Children born when Jupiter is in the constellation Taurus are more likely to be musicians than other children.

7. Aliens can manipulate time so that they can abduct people and perform experiments on them without the people ever realizing they were taken.

8. A huge fleet of alien spacecraft will land on Earth and introduce an era of peace and prosperity on January 1, 2020.

9. *Testing UFOs.* Consider at least one claim that you've heard about alien visitation (such as a claim about the Roswell crash, about an alien abduction, or about aliens among us). Based on what you've heard, can the claim be tested scientifically? If so, how? If not, why not? Do you think the claim should be considered more seriously or more skeptically? Defend your opinion.

10. *Science or Nonscience?* Find a recent news report from "mainstream" media (i.e., not from a supermarket tabloid) that makes some type of claim about extraterrestrial life. Analyze the report and decide whether the claim is scientific or nonscientific. Write two or three paragraphs explaining your conclusion.

11. *Tabloid Astrobiology.* Find a recent article from a supermarket tabloid that makes a claim about alien visitation to Earth. Analyze the article and write two or three paragraphs explaining whether the claim is scientific.

12. *Biographical Research: Post-Copernican Viewpoints on Life in the Universe.* Many seventeenth- and eighteenth-century writers expressed interesting opinions on extraterrestrial life. Each individual listed below wrote a book that discussed this topic; book titles (and original publication dates) follow each name. Choose one or more individuals and research their arguments about extraterrestrial life. Write a one- to two-page summary of the person's arguments, and discuss which (if any) parts of these arguments are still valid in the current debate over life on other worlds.

    Bishop John Wilkins, *Discovery of a World in the Moone* (1638).

    René Descartes, *Philosophical Principles* (1644).

Bernard Le Bovier De Fontenelle, *Conversations on the Plurality of Worlds* (1686).

Richard Bentley, *A Confutation of Atheism from the Origin and Frame of the World* (1693).

Christiaan Huygens, *Cosmotheros, or, Conjectures Concerning the Celestial Earths and Their Adornments* (1698).

William Derham, *Astro-Theology: Or a Demonstration of the Being and Attributes of God from a Survey of the Heavens* (1715).

Thomas Wright, *An Original Theory or New Hypothesis of the Universe* (1750).

Thomas Paine, *The Age of Reason* (1793).

Web Projects

1. *The Galileo Affair.* In recent years, the Vatican has devoted a lot of resources to learning more about the trial of Galileo and understanding past actions of the Church in the Galileo case. Learn more about such studies, and write a short report about the current Vatican view of the case.

2. *Science or Pseudoscience.* Choose some pseudoscientific claim that has been in the news recently, and learn more about it and how scientists have "debunked" it. Write a short summary of your findings.

3. *UFOlogy.* You can find an amazing amount of material about UFOs on the Web. Search for some such sites. Choose one that looks particularly interesting or entertaining and, in light of what you have learned about science, evaluate the site critically. Write a short review of the site.

There is grandeur in this view of life, with its several powers, having been originally breathed by the Creator into a few forms or into one; and that, whilst this planet has gone cycling on according to the fixed law of gravity, from so simple a beginning endless forms most beautiful and most wonderful have been, and are being, evolved.

Charles Darwin, *The Origin of Species*, 1859

CHAPTER 3
The Nature of Life

What is life? This seemingly simple question lies at the heart of research into life in the universe. After all, if we are interested in the possibility of life elsewhere, we must know what it is that we are looking for. Unfortunately, defining life has proved to be surprisingly difficult, even when we consider only life on Earth. It is all the more difficult when we consider life elsewhere, because we cannot be sure that life on other worlds would resemble life on Earth physically, chemically, or functionally. Given the difficulty of defining life, the only sensible way to proceed is by studying the one example of life that we know, hoping it will yield fundamental insights into how life operates and into the environmental conditions required to support life. For this reason, a significant part of research about life in the universe focuses on understanding the nature of life on Earth.

In this chapter, we'll explore our current understanding of life on Earth. Along the way, we'll discover that life elsewhere almost certainly shares at least a few characteristics with life on Earth; understanding the characteristics may provide clues to how we might search for life on other worlds. We'll also explore some of the remarkable conditions under which life thrives on Earth, including some conditions that closely resemble those found on other worlds in our solar system. Thus, we'll leave this chapter with greater reason to think that life might exist elsewhere in our own solar system, perhaps even in environments that might seem unduly harsh by human standards.

3.1 What Is Life?

A cat and a car have much in common. Both require energy to function—the cat gets energy from food, and the car gets energy from gasoline. Both can move at varying speeds and can turn corners. Both expel waste products. But a cat clearly is alive, while a car clearly is not. What's the difference?

In the case of a cat and a car, we can find many important differences without looking too far. For example, cats reproduce themselves, while cars must be built in factories. But as we look deeper into the nature of life, it becomes increasingly difficult to decide what characteristics separate living organisms from rocks and other nonliving materials. Indeed, the question can be so difficult to answer that we may be tempted to fall back on the famous words of Supreme Court Justice Potter Stewart, who, in avoiding the difficulty of defining pornography, wrote: "I shall not today attempt further to define [it]. . . . But I know it when I see it."[1] If living organisms on other worlds turn out to be much like those on Earth, it may prove true that we'll know them when we see them. But if the organisms are fairly different from those on Earth, we'll need clearer guidelines in order to decide whether or not they are truly "living." In this section, we'll investigate some of the distinguishing features of life.

Properties of Life

One way to seek distinguishing features of life is to study living organisms, looking for common characteristics. Given the difficulty of defining life, you probably won't be surprised to learn that there are exceptions to almost any "rule" we think of. Nevertheless, biologists have identified at least six key properties that appear to be shared by most or all living organisms on Earth (Figure 3.1). Let's briefly investigate each property.

- *Order* (Figure 3.1a). The materials in living organisms always exhibit some type of order. For example, the molecules in living cells are not scattered randomly about but instead are arranged in specific patterns to make cell structures. These structures, in turn, make possible all the other properties of life shown in Figure 3.1. Note that order alone does not make something living; a book also has order, because words are not scattered randomly on the pages, but it is not alive. However, it seems reasonable to expect that all living things will show order. In logical terms, we

say that order is a *necessary condition* for life, because something cannot be alive without order. However, order is not a *sufficient condition* for life, because order alone does not make something alive.

THINK ABOUT IT . . . *The idea of necessary and sufficient conditions is very important in science. To make sure you understand it, decide whether each of the following conditions is necessary or sufficient (or neither or both) for the given effect: (1) condition: breathing; effect: human survival while sleeping; (2) condition: living in New York City; effect: living in the United States; (3) condition: meeting all requirements for a college degree; effect: receiving a college degree.*

- *Reproduction* (Figure 3.1b). Living organisms reproduce their own kind. Simple life-forms, such as bacteria, reproduce by dividing to make nearly exact copies of themselves. More complex organisms may reproduce in more sophisticated ways —including sexual reproduction, in which offspring have genetic material from two individuals. In general, reproduction seems necessary to a definition of life, though with some caveats. For example, a mule (the hybrid offspring of a horse and a donkey) is sterile and cannot reproduce, but it is clearly alive. In addition, reproduction exposes some borderline cases in defining life, such as *viruses*. A virus can reproduce only by infecting a living organism and commandeering the organism's cellular machinery to make copies of itself. Thus, it meets the condition of reproduction when it infects another organism but not when it is on its own.

- *Growth and development* (Figure 3.1c). Living organisms grow and develop in patterns directed at least in part by **heredity**—traits passed to an organism from its parent(s). The property of growth and development appears necessary to life in that all organisms grow or develop during at least some periods in their life cycles, but it is not sufficient to constitute life. For example, fire grows and develops as it spreads through a forest, but a fire is not alive.

- *Energy utilization* (Figure 3.1d). Living organisms use energy to fuel all other properties of life, such as to create and maintain patterns of order within their cells, to reproduce, and to grow. Life without energy utilization is simply not possible (though some organisms can survive temporarily in dormant states, in which they don't use energy). Of course, energy utilization is not sufficient to constitute life; any electrical or gas-powered appliance uses energy to function.

---

[1] From Potter Stewart's concurring opinion in *Jacobellis v. Ohio*, 378 U.S. 184, 198 (1964).

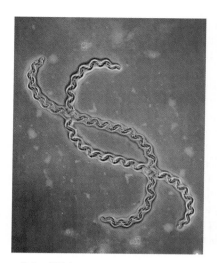

a Order: All living organisms exhibit order in their internal structure, as is apparent in this microscopic view of spiral patterns in two single-celled organisms.

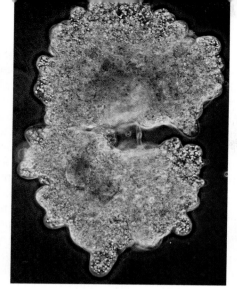

b Reproduction: Organisms reproduce their own kind. Here, a single-celled organism (an amoeba) has already copied its genetic material (DNA) and is now dividing into two cells, each of which will be genetically identical to each other and to the original cell.

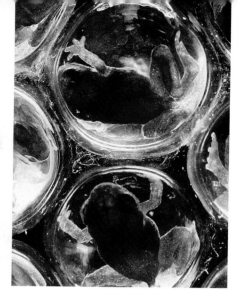

c Growth and development: Living organisms grow and develop in patterns determined at least in part by heredity. Here, we see growing embryos of a Costa Rican frog.

d Energy utilization: Living organisms use energy to fuel their many activities. Here, we see tube worms living near a deep sea volcanic vent. These organisms obtain much of their energy from chemical reactions made possible in part by the heat released from the volcanic vent.

e Response to the environment: Life actively responds to changes in its surroundings. Here, we see a blacktail jackrabbit's ears flush with blood; the blood flow adjusts automatically to help the animal maintain a constant internal temperature by adjusting the heat loss from the ears.

f Evolutionary adaptation: Life evolves in a way that leads to organisms that are adapted to their environments. Here, the white feathers of the white-tail ptarmigan in winter plumage make it nearly invisible against the animal's snowy surroundings.

FIGURE 3.1 Six key properties of life.

■ *Response to the environment* (Figure 3.1e). All living organisms interact with their surroundings and actively respond in at least some ways to environmental changes. For example, some simple organisms may alter their chemistry in the presence of a food source, and warm-blooded mammals may sweat, pant, or adjust blood flow to maintain a constant internal temperature. Like all the other properties on our list, response to

the environment is a necessary but not sufficient condition for life. Many human-made devices also respond to changes in the environment; for example, the mercury in a thermometer rises as the surrounding temperature rises.

■ *Evolutionary adaptation* (Figure 3.1f). Life has changed dramatically over time as the primitive organisms that lived billions of years ago have gradually evolved into the great variety of organisms found on Earth today. Life evolves as a result of the interactions between organisms and their environments, leading over time to evolutionary adaptations that make species better suited to their environments. When the adaptations are significant enough, organisms carrying the adaptations may be so different from their ancestors that they constitute an entirely new species.

Evolution as Biology's Unifying Theme

All six properties of life that we have discussed are important, but biologists today regard evolution as the most fundamental and unifying of all these properties. Evolutionary adaptation is the only property that can explain the great diversity of life on Earth. Moreover, understanding how evolution works allows us to understand how all the other properties of life came to be. Because the theory of evolution is so central to modern biology, let's briefly investigate the origin of the theory and the evidence that supports it.

An Ancient Idea The idea that life might evolve through time goes back more than 2,500 years. The Greek scientist Anaximander (c. 610–547 B.C.) promoted the idea that life originally arose in water and gradually evolved from simpler to more complex forms. A century later, Empedocles (c. 492–432 B.C.) suggested that creatures poorly adapted to their environments would perish, foreshadowing the modern idea of evolutionary adaptation. Many of the early Greek atomists [Section 2.1] probably held similar beliefs, though the evidence is sparse. Aristotle, however, held that species are fixed and independent of one another and do not evolve. This Aristotelian view eventually became entrenched within the theology of Christianity, and evolution was not taken seriously

CREATIONISM AND EVOLUTION

If you live in the United States, you've almost certainly heard about public battles between scientists who argue in favor of evolution and people who believe in creationism—in particular, in the doctrine that God created the Earth just a little over 6,000 years ago in precisely the manner described in the biblical story of Genesis. Indeed, while the scientific community is near-unanimous in support of the theory of evolution, polls consistently show that a large fraction of the population in the United States believes in creationism or closely related ideas.

Given the widespread debate, at some point you are almost certain to be asked, Do you believe in creationism or evolution? However, from a scientific standpoint, the question is poorly phrased. Science is about evidence, not beliefs. A tremendous weight of evidence, some of it discussed in this and subsequent chapters, supports the idea that the Earth is some 4½ billion years old (and resides in an even older universe), that life on Earth has existed for nearly 4 billion years and has undergone dramatic change during that time, and that this change has occurred through the mechanism of evolution by natural selection. Nevertheless, as discussed in Chapter 2, all the evidence in the world can never prove a theory to be true beyond any doubt. To take a somewhat extreme case, you could accept the overwhelming scientific evidence for evolution and still believe in creationism by, for example, assuming that God created the Earth and universe a mere 6,000 years ago as described in Genesis but did so with all the evidence for evolution in place. Because in this case the evidence would look the same as if evolution actually occurred, science cannot say anything about the validity of such a belief. (Note, however, that this evidence would have to include not only having all the rocks and fossils in place on Earth with the ages we find from radiometric

dating, but also other things such as having starlight already en route from stars and galaxies more than 6,000 light-years away, having stars that appear to be of various much older ages, and so on.) Thus, from the standpoint of science, the important issue is not what individuals choose to believe but rather how we interpret scientific evidence—and, more to the point of the U.S. debate, what we teach in science classes in schools. Here, the issues are much more clear-cut.

One clear difference between the scientific study of evolution and a belief in creationism is how evidence is gathered and interpreted. One hallmark of science is that hypotheses must be testable, and these tests are conducted through experiments or observations of the world around us [Section 2.3]. The tests must be repeatable by anybody (at least in principle) so that we can always check hypotheses for ourselves and need never take anybody's word for what is correct. Evolution rose to the status of a scientific theory only because it has been so thoroughly tested by so many people and has been continually found to be fully consistent with the observations. Moreover, like any scientific theory, the theory of evolution is subject to constant reexamination and modification. For example, Darwin thought that evolution would be a gradual, steady-paced process occurring in a large population of a given species, but more recent evidence suggests that evolution often occurs in small, isolated populations and often in episodic bursts. This type of ongoing debate about the precise nature of evolution sometimes makes nonscientists think there is scientific controversy over the basic theory. But while no theory is ever set in stone, the basic principle that life evolves through natural selection is so strongly supported by direct evidence that it is unlikely to be overturned and indeed has remained intact since it was again for some 2,000 years. In the mid-1700s, scientists began to suspect that many fossils represented extinct ancestors of living species. Then, in the early 1800s, French naturalist Jean Baptiste Lamarck suggested that the best explanation for the relationship between fossils and living organisms is that life-forms evolve by gradually adapting to perform successfully in their environments.

Lamarck's idea of evolution by adaptation represented the first clear statement of what we now consider the "observed facts" of evolution. That is, observations of fossils and of relationships between living species make it quite clear that life has changed over time. However, Lamarck was unable to come up with a successful theory to explain *how* evolution occurs. His hypothesis concerning the mechanism of evolution, called "inheritance of acquired characteristics," suggested that organisms develop new characteristics during their lives and then pass these characteristics on to their offspring.

For example, Lamarck would have imagined that weightlifting would enable a person to create an adaptation of great strength that could be genetically passed to his or her children. While this hypothesis may have seemed quite reasonable at the time, it has not stood up to scientific scrutiny and therefore has been discarded as a model of how evolution occurs. It has been replaced by a different model, proposed by the British naturalist Charles Darwin.

Darwin's Inescapable Conclusion Charles Darwin described his theory of evolution in his book *The Origin of Species,* first published in 1859. In this book, Darwin laid out the case for evolution in two fundamental ways. First, he described his observations of living organisms (made during his voyages on the HMS *Beagle*) and showed how they supported the idea that evolutionary change really does occur. Second, he put forth a new model of *how* evolution occurs, backing up his model with a wealth of

first put forth by Darwin. In contrast to this constant reexamination of evolution by scientists, creationists generally do not look for evidence that might cause them to change their beliefs. There is nothing right or wrong in this—it just means that creationism is based on faith, not science.

Another clear difference between the science of evolution and beliefs in creation is that while the former follows the accepted methods of science as discussed in Chapter 2, we can't even define "creationism" clearly enough to test it against the hallmarks of science. Many religions incorporate stories of creation, and these stories are often quite different from one another. For example, many Native American religious beliefs speak of creation in terms that bear little resemblance to the story in Genesis. Even among Christians who claim a literal belief in the Bible, there are differences of interpretation about creation. Some biblical literalists argue that the creation must have occurred in just 6 days, as the first chapter of Genesis seems to say, while others suggest that the term *day* in Genesis does not necessarily mean 24 hours and therefore that the story in Genesis is compatible with a much older Earth. (This latter argument arises, at least in part, because the second chapter of Genesis describes the events of creation in a different order from that in the first chapter.)

Of course, some creationists have tried to find scientific support for their beliefs. For example, some people who believe in creation have recently begun advocating a doctrine known as "intelligent design." This doctrine attempts to use scientific arguments to prove that the complexity of life is too great to have arisen naturally and therefore that there must have been divine intervention. Intelligent design can look much like science, especially because most of its proponents generally accept both the idea of an old Earth and the idea that some living things

have changed through time as indicated by the fossil record. However, as we will discuss in Chapter 5, science can now offer plausible scenarios for how life arose and evolved naturally, making the underpinnings of intelligent design look quite suspect. While proponents of intelligent design will surely continue to look for evidence in support of their belief, their idea has certainly not risen to the level of a scientific theory that could be taught alongside evolution.

Note that science does *not* argue against divine intervention—but neither does science offer any specific reason to think that such intervention occurred and science is consistent with the possibility that it didn't. Thus, within limits, a belief in divine intervention and the theory of evolution are quite compatible. The current doctrine of the Catholic Church, for example, holds that there is no contradiction between Christianity and the theory of evolution by allowing for the idea that God has "directed" the course of evolution without requiring this idea to be backed by any scientific evidence.

The bottom line is that science and faith are very different things. One is not necessarily any more valid, useful, or important than the other; they are different subjects of study. (Of course, science clearly plays a greater role in technological developments—for example, there would be no cars or computers without science—but whether this is "useful" is a value judgment.) As a society, we must choose what subjects we will teach in our schools. If we choose to teach religion and faith, then creationism should be a part of this teaching. But when we choose to teach science, the abundance of evidence supporting the theory of evolution makes the study of evolution a clear and crucial part of the curriculum.

Charles Robert Darwin was born into a wealthy and educated family in England on February 12, 1809—also the birth date of Abraham Lincoln. His father was a physician, and he had two famous grandfathers. His paternal grandfather, Erasmus Darwin (1731–1802), was a renowned physician and scientist who was a strong proponent of the idea that life evolved gradually. His maternal grandfather, Josiah Wedgwood, started the famous Wedgwood Pottery and China company that still bears his name. Darwin's mother died when he was just 8 years old, but his father and his extended family provided him with a generally happy childhood.

At his father's urging, Darwin enrolled in medical school at age 16. However, he was so horrified by the sight of operations, then done without anesthesia, that he left after just 2 years. He next enrolled in Christ College at Cambridge University, intending to become a minister. While there, he began to indulge a childhood love for the study of nature. Shortly after graduating in 1831, Darwin was offered the opportunity to serve as the naturalist aboard a ship of exploration—HMS *Beagle*. Darwin was 22 years old when the *Beagle* set sail on December 27, 1831. The voyage lasted nearly 5 years, and proved to be the defining event of Darwin's life.

The *Beagle* spent much of its voyage exploring the coasts of South America and nearby islands. While the crew conducted surveys, Darwin went ashore to observe the geology and life and collected numerous specimens to take back to England. He also read extensively during the voyage, and one book that had recently been published proved particularly influential: Charles Lyell's *Principles of Geology*, which presented the case for an ancient Earth sculpted by gradual geological processes. Darwin was given the book by a friend who expected Darwin to disagree with its conclusions; instead, Darwin found that his own observations of geology only gave further credence to Lyell's theory.

Meanwhile, Darwin became intrigued by the many adaptations he observed for species in varied environments. He was particularly impressed by the animal life he observed during a 5-week stay on the Galápagos Islands, which lie approximately 1000 kilometers (600

Charles Darwin and his son William posed for this photograph in 1842.

miles) due west of the South American coast of Equador. He focused special attention on the Galápagos finches (now called "Darwin's finches"), concluding that the different bird species must have evolved from a common, mainland ancestor (see Figure 3.3). However, at the time he returned to England, he still did not understand how the evolutionary changes occurred.

In 1838, Darwin read Thomas Malthus's *An Essay on the Principle of Population,* in which Malthus famously argued that populations are capable of growing faster than food supplies can support. The essay helped Darwin crystallize the idea of natural selection by making clear that individuals within a population must compete for survival (the idea embodied in fact 1, below). He then began intensive study of how humans bred domestic plants and animals, which helped him understand the variation in populations (fact 2, below). By 1842, Darwin was con-

evidence and unassailable logic. As evolutionary biologist Stephen Jay Gould (1941–2002) put it, Darwin built the case for his model on "two undeniable facts and an inescapable conclusion." The logic proceeds like this:

- *Fact 1: overproduction and struggle for survival.* Any population of a species has the potential to produce far more offspring than the environment can possibly support with resources such as food and shelter. (By a "population" we mean a group of individuals of the same species living in a particular area.) This overproduction leads to a struggle for survival among the individuals of a population.

- *Fact 2: individual variation.* Individuals in a population of any species vary in many heritable traits (that is, traits passed from parents to offspring). No two individuals are exactly alike, and some individuals possess traits that make them better able to compete for food and other vital resources.

vinced that natural selection held the key to evolution, and he began to draft the text that would eventually be published as *The Origin of Species*.

Darwin is said to have been a pleasant man who wished never to offend others. He was an ardent opponent of slavery, and while he was no feminist he believed that women should be treated with dignity and respect. He was deeply concerned with the impact his theory would have on those who believed in the biblical story of creation. That is probably why he did not publish his theory immediately—he wanted to take time building his case, in hopes that his theory would be so strong that it would be accepted by all without anyone taking offense. Indeed, Darwin might have delayed publication indefinitely if not for a manuscript he received from another scientist, Alfred Russel Wallace, on June 18, 1858.

Wallace had been observing geology and life in Indonesia and had independently come to the same conclusion as Darwin: that life evolves through natural selection. Upon reading Wallace's draft paper, Darwin worried that "all my originality will be smashed." Fortunately, both Darwin and Wallace were willing to share credit, and they jointly published the first major paper on the theory of evolution. It was read before a scientific society in London on July 1, 1858. A little over a year later, Darwin finally published *The Origin of Species*. All 1,250 copies in the first printing sold out on the first day. Within a decade, Darwin's theory won acceptance by the vast majority of biologists, an acceptance that has grown stronger ever since.

Darwin never had a taste for arguing about evolution, leaving that to other scientists (especially Thomas Huxley, who called himself "Darwin's bulldog"). He continued his scientific work, publishing several more books on evolution and related topics. In his personal life, Darwin married a cousin, Emma. He and Emma had 10 children, but two died in infancy and a third died at age 10. Darwin died on April 19, 1882, at the age of 73. A parliamentary petition won him burial in London's Westminster Abbey, where he lies next to Sir Isaac Newton.

a This imaginary beetle population has colonized a locale where the soil has been blackened by a recent brush fire. Initially, the population varies extensively in the coloration of individuals, from very light gray to charcoal.

b For hungry birds that prey on the beetles, it is easiest to spot the beetles that are lightest in color.

c The selective predation favors the survival and reproductive success of the darker beetles. Thus, genes for dark color are passed along to the next generation in greater frequency than genes for light color.

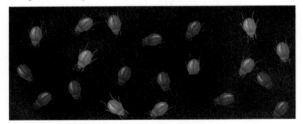

d Generation after generation, the beetle population adapts to its environment through natural selection.

FIGURE 3.2 Natural selection is the primary mechanism of evolution. Here, we see how an imaginary population of beetles of mixed color evolves into one in which all the beetles are dark in color.

→ *The inescapable conclusion: unequal reproductive success.* In the struggle for survival, those individuals whose traits best enable them to survive and reproduce will, on average, leave the largest number of offspring that in turn survive to reproduce. Therefore, in any local environment, heritable traits that enhance survival and successful reproduction will become progressively more common in succeeding generations.

It is this unequal reproductive success that Darwin called **natural selection;** that is, over time, advantageous genetic traits will naturally win out (be "selected") over less advantageous traits because they are more likely to be passed down through many generations. This process explains how species can change in response to their environment—by favoring traits that improve adaptation—and thus is the primary mechanism of evolution (Figure 3.2).

Darwin backed his logical claim that evolution proceeds through natural selection by documenting

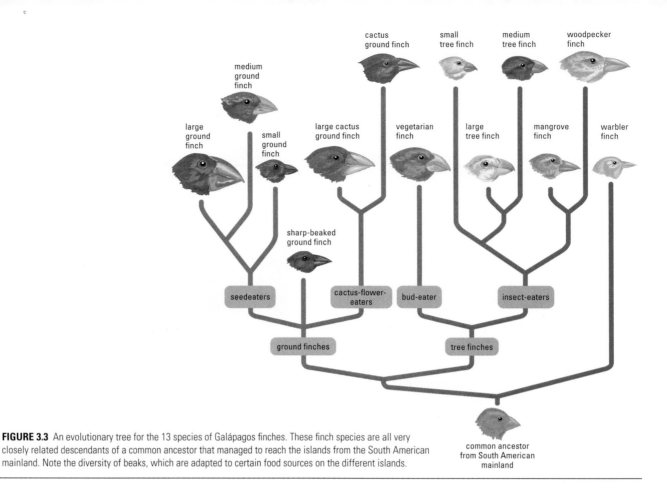

FIGURE 3.3 An evolutionary tree for the 13 species of Galápagos finches. These finch species are all very closely related descendants of a common ancestor that managed to reach the islands from the South American mainland. Note the diversity of beaks, which are adapted to certain food sources on the different islands.

cases in which related organisms are adapted to different environments or lifestyles. He found a particularly striking example among the finches of the Galápagos Islands (Figure 3.3). The 13 finch species on these young volcanic islands closely resemble one another and mainland finches. However, each species is adapted to particular microenvironments and lifestyles. Some live on the ground, and some in cacti or brush; some have beaks adapted for cracking seeds, whereas others have insect-catching beaks. One species even plays the same role that woodpeckers play on the mainland. They are found nowhere else in the world. Natural selection, as Darwin saw, easily explains this situation. We may presume that an ancestral pair of finches flew from the mainland to the Galápagos. Over millennia, local populations of island finches adapted to different environments and lifestyles. In time, these differing populations came to be distinct species.

Darwin also found strong support for his theory of evolution by looking at examples of *artificial selection*—the selective breeding of domesticated plants or animals by humans. Over the past few thousand years, our ancestors have gradually bred many plants and animals into forms that bear little resemblance to their wild ancestors. For example, Figure 3.4

shows how artificial selection has created a variety of vegetables from a single common ancestor. Dogs also are examples of the power of artificial selection: Dog breeds as different as Rottweilers and Chihuahuas were bred from a common ancestor within just a few thousand years. Darwin recognized that if artificial selection could cause such profound changes in just a few thousand years, natural selection could do far more over the millions or billions of years during which it has operated.

Today, we can observe natural selection occurring right before our eyes. In many places on Earth, species have changed in time spans as short as a few decades in response to human-induced environmental changes. On a microbial level, natural selection is what allows a population of bacteria to become resistant to specific antibiotics; those few bacteria that acquire a genetic trait of resistance are the only ones that survive in the presence of the antibiotic. Indeed, bacterial cases of natural selection pose a difficult problem for modern medicine, because bacteria can quickly develop resistance to almost any new drug we produce. As a result, pharmaceutical companies are constantly working to develop new antibiotics as bacteria become resistant to existing ones.

cabbage

brussels sprouts

cauliflower

kale

broccoli

wild mustard

kohlrabi

FIGURE 3.4 The six vegetables shown all look and taste quite different, but they were all bred by humans from the same wild ancestor (wild mustard). This is an example of artificial selection, which is much like natural selection except that humans (rather than nature) decide which individuals in each generation survive and breed offspring.

Evolution on a Molecular Level Darwin's theory of evolution by natural selection tells us that species adapt and change by passing hereditary traits from one generation to the next. However, Darwin did not know precisely how these traits were communicated across generations, nor did he know why there is always variation among individuals or how new traits can appear in a population. Today, thanks to discoveries in molecular biology made since the mid-twentieth century, we know the answers to all these questions. In particular, we now know that organisms are built from instructions contained in a molecule called **DNA** (short for *deoxyribonucleic acid*), and biologists can now trace how evolutionary adaptations are related to changes that occur through time in DNA. We'll discuss how DNA makes evolutionary adaptations possible when we discuss its structure and function in Section 3.4, but for now the key point is that our understanding of DNA means that we understand the specific mechanisms by which natural selection occurs.

Our detailed understanding of how evolution proceeds on a molecular level, coupled with all the other evidence for evolution collected by Darwin and others, puts the theory of evolution by natural selection on a rock-solid foundation. That is, it truly is a scientific theory [Section 2.3] that has withstood countless tests and challenges, each of which has only strengthened its foundation. Like any scientific theory, the theory of evolution can never be proved beyond all doubt. However, no credible *scientific* alternative to the theory of evolution has been proposed, and the evidence in the theory's favor is over-whelming. Indeed, it is difficult to imagine any aspect of biological science that can be understood without being examined in the context of the theory of evolution. That is why evolution has become the unifying theme of all modern biology.

THINK ABOUT IT . . . *The idea that life changes through time is quite ancient, and it was already well-supported by observations of fossils before Darwin was even born. Moreover, Lamarck recognized that evolution occurs as a result of adaptations about a half-century before Darwin advanced the idea of natural selection. Given these facts, explain why we credit Darwin with the theory of evolution.*

A Definition of Life?

Now that we have examined the fundamental properties of life on Earth, let's return to our original question: Can we come up with a definition of life?

Based on the central role of evolution, our simplest definition might be that life is something that can reproduce and evolve through natural selection. For most practical purposes, this definition would probably suffice. However, some cases may still challenge this definition. For example, computer scientists can now write programs (lines of computer code) that can reproduce themselves (create additional sets of identical lines of code). By adding programming instructions that allow random changes, they can even make "artificial life" that evolves on a computer. Should this "artificial life," which consists of nothing but electronic signals processed by computer chips, be considered alive?

The fact that we have such difficulty distinguishing the living from the nonliving on Earth suggests that we should be very cautious about constraining our search for life elsewhere. No matter what definition of life we choose, the possibility always remains that we'll someday encounter something that challenges our definition. Nevertheless, the properties of reproduction and evolution seem very likely to be shared by most, if not all, life in the universe. Thus, they provide a useful starting point as we consider how to explore the possibility of extraterrestrial life.

THINK ABOUT IT . . . *Do you think computer programs that can reproduce and evolve are alive? Why or why not? Would your opinion change if these programs evolved to the point where they could write their own computer code or exchange e-mail with us? What if they wrote other programs that operated machinery to build other computers? Do you think it is possible to create true life on a computer?*

3.2 Cells: The Basic Units of Life

Now that we have discussed general properties of life, we are ready to look more specifically at the nature of life on Earth. All living organisms are made of **cells**[2]—microscopic units in which the living matter inside is separated from the outside world by a barrier called a **membrane** (Figure 3.5). Thus, cells are the basic structures of life on Earth. Some organisms consist only of a single cell. Other organisms, like oak trees and people, are complex structures in which trillions of cells work cooperatively, dividing various tasks among cells of different types.

Given the great diversity of life on Earth, you may be surprised to learn that all living cells share a great many similarities. For example, all pass on their hereditary information in the same basic way with DNA, and many other chemical processes are nearly the same in all cells. These similarities, which we'll discuss in more detail later, are profoundly important to our understanding of the origin of life. As far as we know, there is no reason why all living cells must share these characteristics; that is, while life elsewhere might also be composed of cells, we should not expect those cells to have the same biochemistry

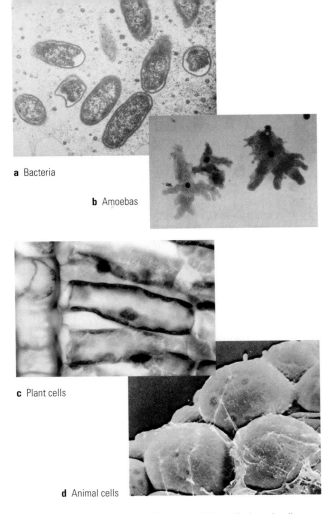

a Bacteria

b Amoebas

c Plant cells

d Animal cells

FIGURE 3.5 Microscopic views of four types of living cells. In each cell, a membrane separates the living matter inside the cell from the outside world.

as cells on Earth. Thus, the many similarities among all cells on our planet suggest a startling conclusion: All life on Earth shares a common ancestor. In other words, every living organism on Earth is related to every other one because all evolved over billions of years from the same origin of life.

Living Cells Are Carbon-Based

Life on Earth is made from more than 20 different chemical elements. However, just four of these elements—oxygen, carbon, hydrogen, and nitrogen—make up about 96% of the mass of typical living cells. Most of the remaining mass consists of just a few other elements, notably calcium, phosphorus, potassium, and sulfur (Figure 3.6).

Given that oxygen dominates Figure 3.6, you might be tempted to say that life on Earth is "oxygen-based." However, most of the oxygen in living cells is actually part of water molecules (H_2O). The molecules that account for a cell's structure and function

---

[2] Some organisms, such as some slime molds, do not perfectly fit this picture of discrete cells, because they consist of a large mass of protoplasm containing thousands of nuclei. Nevertheless, the basic idea that living tissue is contained in a package separated from the external environment still holds.

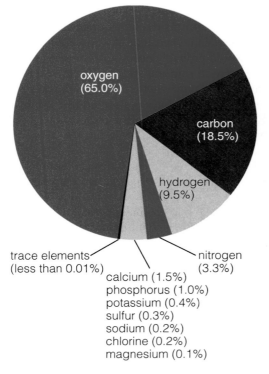

FIGURE 3.6 This pie chart shows the chemical composition of the human body by weight; this composition is fairly typical of all living matter on Earth.

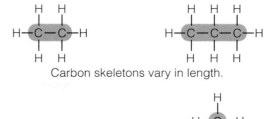
Carbon skeletons vary in length.

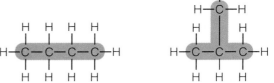
Carbon skeletons may be unbranched or branched.

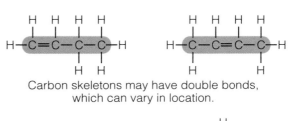
Carbon skeletons may have double bonds, which can vary in location.

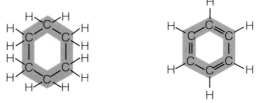
Carbon skeletons may be arranged in rings.

FIGURE 3.8 These diagrams represent several relatively simple hydrocarbons—organic molecules consisting of a carbon skeleton attached to hydrogen atoms. The carbon skeletons are highlighted in green. Each single line represents a single chemical bond; a double line represents a double bond. Note that every carbon atom has a total of four bonds (a double bond counts as two single bonds).

FIGURE 3.9 In a more complex organic molecule, at least one bond links a carbon atom to something besides hydrogen or another carbon atom. Here, one of the carbon atom's four bonds links it to an *amino group* (which consists of a nitrogen atom and two hydrogen atoms), highlighted in green.

owe their remarkable qualities to a different element: carbon. For this reason, we say that life on Earth is **carbon-based.**

Why Is Carbon So Important to Life on Earth? The primary answer to this question lies in how carbon can combine with other elements to make complex molecules. The atoms in any molecule are linked together by what we call **chemical bonds.** Different elements are capable of making chemical bonds in different ways. For example, hydrogen atoms generally can bond with only one other atom at a time, while oxygen atoms generally can bond with at most two other atoms at a time. We can see these properties in a water molecule (Figure 3.7):

FIGURE 3.7 A water molecule has a single oxygen atom bonded to two hydrogen atoms.

The oxygen atom has two chemical bonds, one to each of the two hydrogen atoms, while each hydrogen atom has only a single chemical bond to the oxygen atom.

Carbon is a particularly versatile chemical element because it can bond to from one to four atoms at a time. This allows carbon atoms to link together in an endless variety of carbon "skeletons" varying in size and branching patterns (Figure 3.8). The sim-plest carbon molecules, which we refer to generically as **organic molecules,** consist of carbon skeletons bonded only to hydrogen atoms; these simple organic molecules are often called *hydrocarbons* to reflect the fact that they contain only hydrogen and carbon. In more complex organic molecules, one or more carbon atoms are bonded to something besides hydrogen and other carbon atoms (Figure 3.9).

Could Life Elsewhere Be Based on Something Else? When we consider the possibility of extraterrestrial life, it's natural to wonder whether it might be based on an element besides carbon. In truth, we cannot say for sure whether other elements would work. However, given the importance to life on Earth of carbon's ability to form four bonds at once, we might expect that any other elemental basis for life would have to have the same bonding capability. Among the elements common on Earth's surface— and likely to be common on other planets—silicon is the only element besides carbon that can have four bonds at once. As a result, science fiction writers have often speculated about finding silicon-based life on other worlds.

Unfortunately for science fiction, silicon has at least two strikes against it as a basis for life. First and most important, the bonds formed by silicon are significantly weaker than equivalent bonds formed by carbon. As a result, complex molecules based on silicon are more fragile than those based on carbon— probably too fragile to form the structural components of living cells. In particular, complex silicon-based molecules cannot exist long in water, which is also generally thought to be necessary to life [Section 3.3]. Second, unlike carbon, silicon does not normally form double bonds; instead, it forms only single bonds. This limits the range of chemical reactions that silicon-based molecules can engage in as well as the variety of molecular structures that can form. Given the two strikes against silicon, most scientists consider it unlikely that life can be silicon-based.

A few other elements have also been suggested as possibilities for replacing carbon on other worlds, but most scientists believe carbon's natural advantages will always win out. We have found carbon-based (organic) molecules even in space (as identified in meteorites and interstellar clouds), suggesting that carbon chemistry is so easy and so common that life with any other basis is difficult to imagine.

Molecular Components of Cells

All the major components of cells are made from complex organic molecules. Today, biologists know the precise chemical structure of a great many of these molecules, and this knowledge has enabled them to gain a deep understanding of the biochemistry of life. If you take a course in biology, you will learn about much of this biochemistry; here we focus only on generalities about the molecules of life. The large molecular components of cells fall into four main classes: carbohydrates, lipids, proteins, and nucleic acids. Let's briefly investigate the properties of each class that are most important to life.

Carbohydrates You're probably familiar with carbohydrates as a source of food energy—the sugars and starches known to athletes as "carbs." In addition to providing energy to cells, carbohydrates make impor-

tant cellular structures. For example, a carbohydrate called *cellulose* forms the fibers of cotton and linen and is the main constituent of wood. Life on other worlds would probably need molecules to play these same energy-storing and structural roles. However, we do not yet know whether other molecules could potentially play the roles performed by carbohydrates on Earth, so we will have little more to say about carbohydrates in this book.

Lipids Like carbohydrates, lipids can store energy for cells. The types of lipids that store energy are more commonly known as *fats;* thus, despite the bad reputation of fat, it is actually critical to living cells. Lipids also play a variety of other roles in cells on Earth, but from the standpoint of life in the universe perhaps their most important role is as the major ingredients of cell membranes. That is, lipids form the barriers that make it possible for cells to exist. As we'll discuss in more detail in Chapter 5, lipids are thought to have played a critical role in the origin of life. Lipids can spontaneously form membranes in water and probably did so on the early Earth. Other organic molecules were trapped inside the space formed by the membranes, making what were in essence tiny chemical factories. These tiny chemical factories may have facilitated the chemical reactions that ultimately led to the origin of life.

Proteins Proteins are often considered the workhorses of cells, because they participate in such a vast array of functions. Some proteins serve as structural elements in cells. Others, called **enzymes,** are crucial to nearly all the important biochemical reactions that occur within cells—including the copying of genetic material (DNA)—because they serve as **catalysts** for these reactions. A catalyst is any substance (not necessarily a single molecule) that either facilitates a chemical reaction that would not occur otherwise or accelerates a reaction that would otherwise occur very slowly but is not itself changed by the process. Enzymes are catalysts because they greatly accelerate the reactions in which they are involved, even though they enter and leave the reactions essentially unchanged. Moreover, because an enzyme is left unchanged after it catalyzes a reaction, a single enzyme can catalyze many identical chemical reactions without needing to be rebuilt.

Proteins are extremely large molecules built from long chains of smaller molecules called **amino acids.** The nature of these chains provides important evidence supporting the idea that all life on Earth shares a common ancestor. While more than 70 different amino acids are found in nature, most life on Earth builds proteins from only 20 of them. (Two additional amino acids are known to be used in rare cases by particular microorganisms, and scientists suspect that other cases of rare amino acids may yet

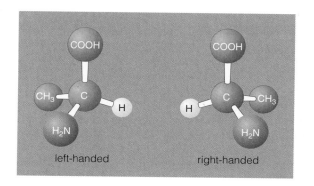

FIGURE 3.10 Any particular amino acid can come in two forms, distinguished by their handedness. These diagrams show the left- and right-handed versions of the amino acid alanine. The two versions are mirror images of each other.

be discovered.) If life on Earth had more than one common ancestor, we might expect that different organisms would use different sets of amino acids, but they don't. Moreover, naturally occurring amino acids come in two slightly different forms, distinguished by their **handedness:** The "left-handed" and "right-handed" versions are mirror images of each other (Figure 3.10). Amino acids found in nonbiological circumstances generally consist of a mix of the left- and right-handed versions, but living cells use only the left-handed versions of amino acids to build proteins. Again, the fact that all life on Earth makes use of the same versions of amino acids suggests a common ancestor. (Carbohydrates provide some similar evidence, as life on Earth uses mainly the right-handed versions of sugars.)

THINK ABOUT IT . . . *As we'll discuss in Chapter 5, some scientists hypothesize that early life actually migrated to Earth on meteorites from another world rather than arising here on its own. Suppose we discover life on another planet and, while it also has proteins, it builds them from a different set of amino acids than does life on Earth, and they are all the right-handed versions. Would this support or contradict the hypothesis that life migrated from this planet to Earth?*

Nucleic Acids Perhaps no molecule is more famous than DNA, short for *deoxyribonucleic acid*. DNA is the basic hereditary material of all life on Earth. A second important nucleic acid, RNA (short for *ribonucleic acid*), helps carry out the instructions contained in DNA. (We'll discuss both molecules in more detail in Section 3.4.) Thus, the nucleic acids DNA and RNA are responsible for allowing cells to function according to precise, heritable instructions. Changing a cell's DNA changes the very nature of an organism, and, indeed, it is changes to DNA that allow species to evolve. We do not know whether other molecules could replace nucleic acids in life elsewhere, but it is difficult to imagine life existing in any form without molecules that serve the critical functions of DNA and RNA.

Two Basic Cell Types: Prokaryotic and Eukaryotic

While all living cells on Earth are built from essentially the same molecules and share a common chemistry, some cells are clearly more complex than others. In particular, living cells on Earth appear to come in two basic types: **prokaryotic cells** and **eukaryotic cells** (Figure 3.11). Prokaryotic cells are generally much simpler and smaller than eukaryotic cells. The most obvious difference between the two cell types is that eukaryotic cells have a **cell nucleus** —that is, an internal membrane that effectively walls off the genetic material (DNA) from the rest of the cell—and prokaryotic cells do not.

prokaryotic cell

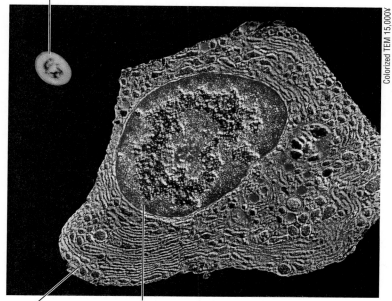

Colorized TEM 15,000×

eukaryotic cell   nucleus

FIGURE 3.11 These microscopic photographs contrast a typical prokaryotic cell and a typical eukaryotic cell. The eukaryotic cell is about 10 times larger in diameter than the prokaryotic cell (a bacterium); this size difference is typical, though prokaryotic cells come in a fairly wide range of sizes. The eukaryotic cell is also much more complex, with numerous specialized structures and a cell nucleus containing its DNA. The prokaryotic cell lacks a distinct nucleus, though its DNA is still concentrated near the center.

In general, every prokaryotic cell represents a distinct, single-celled organism (though some of these organisms form small colonies in which they live or work together). There are many different species of prokaryotic organisms, including all bacteria. For example, the well-known bacteria *E. coli* and *Salmonella*, which are common causes of food poisoning, are single-celled organisms in which the single cell is a prokaryotic cell. Organisms consisting of prokaryotic cells are often called **prokaryotes** for short.

Organisms made from eukaryotic cells—called **eukaryotes** for short—may be either single-celled or multicellular. Perhaps the best-known single-celled eukaryotes are amoebas, which you may have seen if you have ever looked at pond water under a microscope (see Figure 3.5b). Among the multicellular eukaryotes are all familiar plants and animals; even though plants and animals may be made of trillions of individual cells, every one of these cells is eukaryotic.

Because we are more familiar with plants and animals than with bacteria, most people assume that eukaryotes are more common than prokaryotes. Also, because we tend to associate bacteria with disease, many people assume that prokaryotes are generally harmful. In fact, both assumptions are wrong. The microscopic prokaryotes far outnumber and outweigh all eukaryotes combined. Moreover, most prokaryotes (including most bacteria) are harmless to humans, and many are crucial to our survival. For example, bacteria in our intestines provide us with important vitamins, and others living in our mouths prevent harmful fungi from growing there. Other prokaryotes play crucial roles in cycling carbon and other vital chemical elements between organic matter and the soil and atmosphere; for example, prokaryotes are responsible for decomposing dead plants and animals. Indeed, eukaryotic life would be doomed if prokaryotes somehow disappeared from Earth. In contrast, prokaryotes could survive just fine without eukaryotes, as they did during the early history of life on Earth [Section 5.3].

Classifying Life by Cell Type: The Three Domains

A deeper look at prokaryotes shows that they are not all as alike as they may at first seem. In fact, biologists have identified two major branches of the prokaryotes: **Bacteria** and **Archaea.** Members of Bacteria and Archaea look similar under a microscope, but they differ in quite a few basic features of

---

Key Biological Definitions

This box summarizes the definitions of a few of the most basic biological terms used in this chapter and throughout the book.

Terms related to evolution:

evolution (*biological*): The gradual change in populations of living organisms responsible for transforming life on Earth from its primitive origins to the great diversity of life today.

evolutionary adaptation: An inherited trait that enhances an organism's ability to survive and reproduce in a particular environment.

theory of evolution: The theory, first advanced by Charles Darwin, that explains *how* evolution occurs through the process of natural selection (also referred to as *Darwinian evolution*).

natural selection: The primary mechanism by which evolution proceeds. More specifically, natural selection refers to the process by which, over time, advantageous genetic traits naturally win out (are "selected") over less advantageous traits because they are more likely to be passed down through succeeding generations.

Terms related to heredity:

heredity: The characteristics of an organism passed to it by its parent(s), which it can pass on to its offspring. The term can also apply to the transmission of these characteristics from one genera-

tion to the next. Hereditary information is encoded in DNA.

gene: The basic functional unit of an organism's heredity. A single gene consists of a sequence of DNA bases (or RNA bases, in some viruses) that provide the instructions for a single cell function (such as building a protein).

genome: The complete sequence of DNA bases in an organism, encompassing all of the organism's genes.

genetic code: The specific set of rules by which the sequence of bases in DNA is "read" to provide the instructions that make up genes.

DNA (*deoxyribonucleic acid*): The basic hereditary molecule of life on Earth. A DNA molecule consists of two strands, twisted in the shape of a double helix, along each of which lies a long sequence of *DNA bases*. The four DNA bases are adenine (A), cytosine (C), guanine (G), and thymine (T), and they can be paired across the two DNA strands only so that A pairs with T and C pairs with G.

RNA (*ribonucleic acid*): A molecule closely related to DNA, but with only a single strand and a slightly different backbone and set of bases, that plays critical roles in carrying out the instructions encoded in DNA.

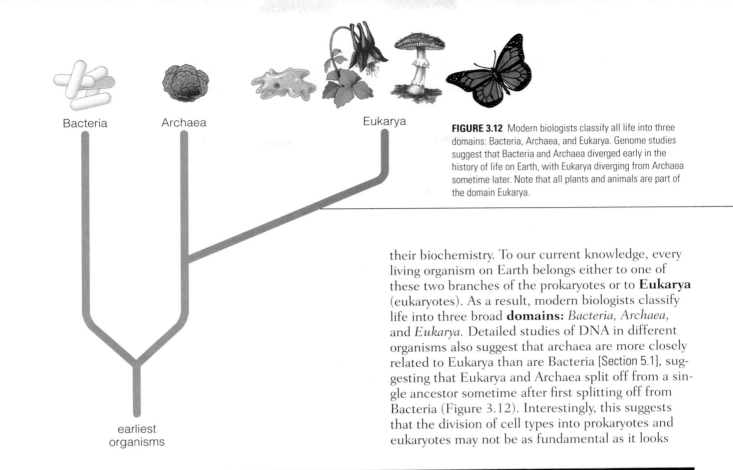

Bacteria Archaea Eukarya

earliest
organisms

FIGURE 3.12 Modern biologists classify all life into three domains: Bacteria, Archaea, and Eukarya. Genome studies suggest that Bacteria and Archaea diverged early in the history of life on Earth, with Eukarya diverging from Archaea sometime later. Note that all plants and animals are part of the domain Eukarya.

their biochemistry. To our current knowledge, every living organism on Earth belongs either to one of these two branches of the prokaryotes or to **Eukarya** (eukaryotes). As a result, modern biologists classify life into three broad **domains:** *Bacteria, Archaea,* and *Eukarya.* Detailed studies of DNA in different organisms also suggest that archaea are more closely related to Eukarya than are Bacteria [Section 5.1], suggesting that Eukarya and Archaea split off from a single ancestor sometime after first splitting off from Bacteria (Figure 3.12). Interestingly, this suggests that the division of cell types into prokaryotes and eukaryotes may not be as fundamental as it looks

Terms related to cell types:

cell: The basic structure of all life on Earth, in which the living matter inside is separated from the outside world by a barrier called a *membrane.* Cells come in two basic types: *Eukaryotic cells* contain a distinct *nucleus* that is separated from the rest of the cell by its own membrane; *prokaryotic cells* lack a nucleus.

eukaryote: A living organism made from one or more cells that contain a nucleus. All eukaryotes are members of the domain *Eukarya,* one of the three *domains* of life (the others are *Bacteria* and *Archaea*).

prokaryote: A living organism made from cells that lack a distinct, membrane-enclosed nucleus. Most prokaryotes are single-celled. Prokaryotes include all the organisms in two of the three domains of life: Bacteria and Archaea.

Terms related to cellular chemistry:

organic molecule: Generally, any molecule containing carbon, but with some exceptions among molecules that are most commonly found independent of life. For our purposes in this book, the important exceptions are carbon dioxide (CO_2) and carbonate minerals, which we do not consider to be organic.

organic chemistry: The chemistry of organic molecules (whether or not the molecules are involved in life).

biochemistry: The chemistry of life.

amino acids: The molecules that form the building blocks of proteins. Most organisms construct proteins from a particular set of 20 amino acids, although several dozen other amino acids can be found in nature. More technically, an amino acid is a molecule containing both an *amino group* (NH or NH_2) and a *carboxyl group* (COOH).

protein: A large molecule assembled from amino acids according to instructions encoded in DNA. Proteins play many roles in cells; a special category of proteins, called **enzymes,** catalyzes nearly all the important biochemical reactions that occur within cells.

catalysis: The process of causing or accelerating a chemical reaction by involving a substance or molecule that is not permanently changed by the reaction. The unchanged substance or molecule involved in catalysis is called a **catalyst.** In living cells, the most important catalysts are the proteins known as *enzymes.*

metabolism: The many chemical reactions that occur in living organisms.

under a microscope. That is, while the prokaryote/eukaryote division puts Bacteria and Archaea in the same category (prokaryotes) and Eukarya in a different category, the genetic evidence suggests that Archaea are genetically more like Eukarya than Bacteria. As we will discuss in Chapter 5, this fact has important implications to our understanding of the early evolution of life.

Note that the idea of classifying all life into the three domains by cell type is somewhat different from the older idea of classifying life into "kingdoms" (such as the plant kingdom and the animal kingdom), which you may have learned about in earlier courses. The kingdom classification system was originally based on structural and physiological differences among different species. While such differences are relatively easy to notice, until recently biologists had no way to know which ones were most fundamental. In the past two decades, as it became possible to map the DNA of organisms, biologists have focused instead on classifying organisms according to genetic similarities and differences. It is these new studies that led to the three-domain system. The kingdoms fit under the three domains, so in essence the domains have given us a new, higher level of classification. It remains possible that additional domains will someday be discovered, because the vast majority of microscopic species have not yet been studied carefully.

3.3 Metabolism: The Chemistry of Life

Why are cells so important to life on Earth? More to the point, is it possible that life elsewhere might exist without having a fundamental organizational unit like the cell? To answer these questions, we must understand the processes that take place inside living cells. These processes, which are all chemical in nature, make up what we call **metabolism.** More specifically, metabolism is a blanket term that refers to all of the many chemical reactions that occur in living organisms.

Basic Metabolic Needs

Most of the important chemical reactions that occur in cells share a common characteristic: Without the help provided by the cell itself, the reactions would occur too slowly to be useful for life. In this sense, a cell's primary purpose is to serve as a tiny chemical factory in which desired chemical reactions occur much more rapidly than they could otherwise, thereby making it possible to turn simple molecules into the great variety of complex organic molecules needed by living organisms. (As is also the case in many factories, cellular work sometimes involves

breaking down molecules as well as building them.) Like any manufacturing process, this biochemical manufacturing process requires two basic things:

1. A source of raw materials with which to build new products. In the case of living cells, the key raw materials are molecules that provide the cell with carbon and other basic elements of life.

2. A source of energy to fuel the metabolic processes that manufacture new molecules.

Given the large variety of molecules involved in metabolic processes, you might think that cells would need an equally large variety of sources of raw materials and energy to survive. However, cells have the ability to build incredible variety from very limited starting materials. Part of this ability comes from the remarkable variety of enzymes in living cells. Each enzyme is specialized to catalyze particular chemical reactions needed in cellular manufacturing. The remarkable diversity of enzymes in living organisms today is a testament to the power of evolution. The instructions for enzyme creation are encoded in DNA and hence have been evolving for billions of years.

Another reason why cells can produce so much variety from so little input is that, regardless of where they get their energy, all cells put the energy to work in nearly the same way. Every living cell uses the same molecule, called **ATP** (short for *adenosine triphosphate*), to store and release energy for nearly all its chemical manufacturing (Figure 3.13a). This usage of ATP vastly simplifies the manufacturing process, because it means that the cell needs an outside energy source only for the purpose of producing ATP, rather than for producing the full variety of organic molecules in cells. Once ATP is produced, it can be used to provide energy for any cellular reaction. Moreover, the nature of ATP makes it completely recyclable. Each time a cell draws energy from a molecule of ATP, it leaves a closely related by-product, called ADP (short for *adenosine diphosphate*), that can be easily turned back into ATP (Figure 3.13b).

Incidentally, the fact that all life on Earth uses the same molecule (ATP) for energy storage offers further evidence for a common origin of life. There's no known reason why other molecules could not fill the role of ATP. Thus, the fact that all living cells use ATP suggests that they all evolved from a common ancestor that was the first to make use of this remarkable molecule.

Carbon and Energy Sources

Living cells on Earth get their raw materials (most notably carbon, because we are dealing with carbon-

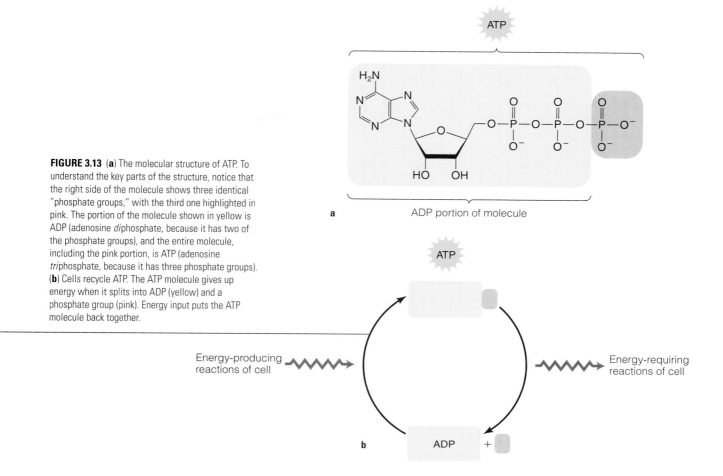

FIGURE 3.13 (**a**) The molecular structure of ATP. To understand the key parts of the structure, notice that the right side of the molecule shows three identical "phosphate groups," with the third one highlighted in pink. The portion of the molecule shown in yellow is ADP (adenosine *di*phosphate, because it has two of the phosphate groups), and the entire molecule, including the pink portion, is ATP (adenosine *tri*phosphate, because it has three phosphate groups). (**b**) Cells recycle ATP. The ATP molecule gives up energy when it splits into ADP (yellow) and a phosphate group (pink). Energy input puts the ATP molecule back together.

ADP portion of molecule

Energy-producing reactions of cell

Energy-requiring reactions of cell

Table 3.1 Metabolic Classifications of Living Organisms

Metabolic Classification	Carbon Source	Energy Source	Examples
Photoautotroph	Carbon dioxide	Sunlight	Plants, photosynthetic prokaryotes
Chemoautotroph	Carbon dioxide	Inorganic chemicals (e.g., iron, sulfur, ammonia)	Certain prokaryotes, especially in extreme environments
Photoheterotroph	Organic compounds	Sunlight	Certain prokaryotes
Chemoheterotroph	Organic compounds	Organic compounds	Animals, many prokaryotes

based life) and energy from a great variety of sources. Fortunately, we can classify these many sources quite simply by looking at them broadly. We discuss these classifications here; because they have rather long names, you may find it useful to begin with the summary in Table 3.1 and refer back to it as you read. (Note that you should understand the meanings of the metabolic classification names but it is not necessary to memorize them—whenever the names come up in this book, there will also be a reminder of what they mean, and you can always refer back to Table 3.1. The names are useful to know, however, because they appear frequently in articles about astrobiology.)

Carbon: Autotrophs and Heterotrophs Cells need a source of carbon from which to build the skeletons of their organic molecules. In the broadest sense, cells can get their carbon in either of two ways:

1. Some cells get carbon by consuming preexisting organic compounds, that is, by eating. For example, humans acquire carbon by eating plants or other animals. Any organism that gets its carbon by eating is called a **heterotroph;** the word comes from *hetero* meaning "others" and *troph* meaning "to feed." All animals are heterotrophs, as are many microscopic organisms.

2. Some cells get carbon directly from the atmosphere in the form of carbon dioxide. An organism that gets its carbon from carbon dioxide is called an **autotroph,** meaning "self-feeding." For example, trees and most other plants are autotrophs.

Energy: Light and Chemical Reactions Broadly speaking, the energy source that a living cell uses to make ATP can be one of three things (see Appendix D for more details about chemical energy for life):

1. Some cells get energy directly from sunlight, using the process we call *photosynthesis.* For example, plants acquire their energy from sunlight. Organisms that get energy from sunlight are given the prefix *photo.* Thus, heterotrophs that get energy from sunlight are called **photo-heterotrophs,** while autotrophs that get energy from sunlight are called **photoautotrophs.** The latter category is much more familiar—nearly all plants are photoautotrophs. The former category is rarer; the known photoheterotrophs are prokaryotes that get their carbon by eating organic compounds but make ATP using energy from sunlight.

2. Some cells get energy from food; that is, they take chemical energy from organic compounds they've eaten and use it to make their own ATP. All animals fall into this category, because animals cannot make energy from sunlight. Because organisms that acquire energy in this way have eaten organic compounds, they are heterotrophs. To distinguish their energy source from that of the photoheterotrophs (which get energy from sunlight), we call them **chemo-heterotrophs,** with the prefix *chemo* standing for chemical energy.

3. Some cells get energy from *inorganic* chemicals (chemicals that do not contain carbon) in the environment. For example, some prokaryotes get their energy from chemical reactions involving iron, sulfur, or ammonia. These organisms, which need neither organic food nor sunlight to survive, are classified as **chemoautotrophs.** Again the prefix *chemo* indicates that their energy source is chemical reactions, but note that these reactions are quite different in character from those used by the food-consuming chemoheterotrophs. Indeed, chemoautotrophs are often found in environments where most other organisms could not survive.

THINK ABOUT IT . . . *Classify each of the following into one of the four categories listed in Table 3.1: (1) an organism that gets its energy from chemicals near an undersea volcano and gets its carbon from carbon dioxide dissolved in the water; (2) a tomato plant; (3) you.*

Metabolism and Cells

Although the metabolic classifications may seem somewhat arbitrary at first, we'll see later that they are very useful to understanding the differing conditions under which life might survive. Thus, they will prove useful both to our understanding of the origin of life and to our consideration of what other worlds might have conditions suitable for life. For now, however, we've seen enough to answer the questions that began this section.

The four metabolic classifications we've explored are quite general; they ought to apply equally well to life elsewhere and to life on Earth. Moreover, any type of complex metabolism requires the existence of some kind of structure that allows carbon and energy to come together to manufacture (and break down) the molecules needed by life. Thus, unless we are failing to imagine an entirely different potential mode of operation, it seems likely that all living organisms must have a fundamental structure that functions much like cells on Earth. This crucial observation means that we can search for life in the universe by searching for cells rather than having to search for a much broader variety of possible structures.

The Importance of Water

One final ingredient in metabolism, which we have ignored to this point, is liquid water. On Earth, water plays three key roles in metabolism. First, metabolism requires that organic chemicals be readily available for reactions; liquid water makes this possible by essentially allowing organic chemicals to float within the cell (because the chemicals dissolve in water). Second, metabolism requires a means of transporting chemicals to and waste products away from cells, and water is the medium of this transport. Third, water plays a role in many of the metabolic reactions within cells; for example, water is involved in the reactions that store and release energy in ATP.

All living cells on Earth depend on liquid water to play these three roles, and this dependence limits the conditions under which we find life on our planet. That is, we find life only in places where it is neither too cold nor too hot for liquid water to exist. Indeed, while we've seen that life on Earth can use a variety of different carbon and energy sources, liquid water is one thing that no organism can survive without. (Some organisms can become dormant and survive temporarily in the absence of liquid water, but they cannot survive permanently in such conditions.) Does this need for liquid water also apply to life on other worlds? Certainly, some kind of liquid seems necessary, but we'll save discussion of possibilities other than water for Chapter 6.

3.4 DNA and Heredity

In the previous two sections, we studied two key features of life on Earth that are likely to be crucial to life anywhere else as well: the structural units of cells and the metabolic processes that keep cells alive. A third feature that seems generally needed for all life is some means of storing information, namely, a set of "operating instructions" for the cell and a way of passing these instructions down through the generations. This information is what we generally call an organism's *heredity.*

All living things on Earth encode their hereditary information in the molecule known as DNA. (Some viruses use RNA, but they are not usually considered "alive" and cannot reproduce unless they invade cells containing DNA.) That is, DNA holds the "operating instructions" for living organisms on Earth. DNA also allows organisms to reproduce, because it can be accurately copied. In this section, we will explore how DNA determines the nature of an organism and allows reproduction. We will also discuss how rare errors in the copying of DNA can lead to evolutionary adaptations, thus giving us the molecular-level understanding of natural selection that we first mentioned in Section 3.1.

The Double Helix

The molecular structure of DNA, a *double helix,* is probably one of the most familiar icons of our time (Figure 3.14). A helix is a three-dimensional spiral, such as you would make by extending a Slinky toy; a double helix has two intertwined strands, each in the shape of a helix. The structure looks much like a zipper twisted into a spiral. The fabric edges of the zipper represent the "backbone" of the DNA molecule, while the zipper teeth that link the two strands represent molecular components called **DNA bases.** The chemical structure of the backbone is interesting and

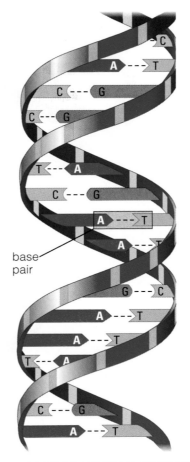

base pair

FIGURE 3.14 This diagram represents a DNA molecule, which looks much like a zipper twisted into a spiral. The important hereditary information is contained in the "teeth" linking the strands. These "teeth" are the DNA bases. Only four DNA bases are used, and they can link up between the two strands only in specific ways: T attaches only to A, and C attaches only to G. (The color coding is arbitrary and is used only to represent different types of chemical groups; in the backbone, blue and yellow represent sugar and phosphate groups, respectively.)

important in its own right, but it is the DNA bases that hold the key to heredity. Life on Earth makes use of only four DNA bases: adenine (abbreviated A), guanine (G), thymine (T), and cytosine (C).

As we'll see shortly, the key to DNA's ability to copy itself lies in the way the four DNA bases can pair up to link the two strands: T can pair up only with A, while C can pair up only with G. Figure 3.14 shows this by representing the different bases with different shapes. Note that the shape of A, which ends with an open triangle, fits only into the notch in T. Similarly, the curved end of G fits only into the curved notch in C. Although the diagrams are only schematic representations, the real chemical bases work much the same way: Their actual shapes and sizes determine how they pair up.

DNA Replication

The specific pairing of the bases in DNA explains how DNA replicates, or copies itself. The process, called **DNA replication,** is illustrated in Figure 3.15. Step 1 begins with the complete double helix. In Step 2, the two strands separate, "unzipping" the links between the paired bases. Step 3 shows how,

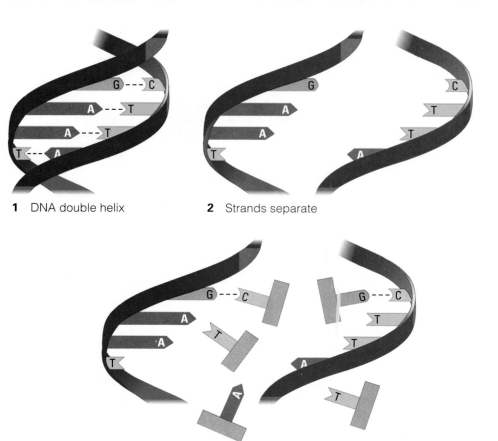

1   DNA double helix   **2**   Strands separate

3   Each strand serves as a template for a new complementary strand

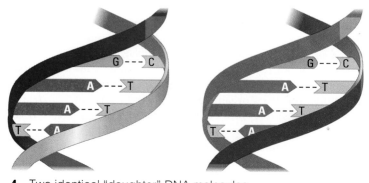

FIGURE 3.15 DNA replication. DNA copies itself by "unzipping" its two strands, each of which then serves as a template for making a new, complementary strand built in accord with the base pairing rules (A goes only with T, and C goes only with G). The end result is two identical copies of the original DNA molecule.

4   Two identical "daughter" DNA molecules

once the strands have been "unzipped," each strand can serve as a template for making a new strand. Because the "teeth" of each new strand must link to the existing strands according to the base pairing rules—T goes only with A, and C goes only with G—each new strand will be *complementary* to an existing one. (By saying that two strands are complementary, we mean that, while they are not identical, they contain essentially the same information because knowing the base sequence on one strand automatically tells us the base sequence on the other strand.) The end result, shown in Step 4, is two identical copies of the original DNA molecule. When a cell divides,

one copy goes to each daughter cell. Because cell division is the key to passing down genetic material from one generation to the next, DNA replication explains the basis of heredity.

Although the DNA copying process is easy in principle, the actual mechanics are fairly complex. More than a dozen special enzymes are involved in the various steps, performing tasks such as unwinding the double helix, making sure the correct bases pair up, checking for and correcting any errors in the copying process, and rewinding the new DNA molecules. This complexity is one reason why errors sometimes occur in DNA replication; as we'll discuss shortly, these errors are crucial to evolution. The complexity of replication also makes it extremely unlikely that DNA could have been the original hereditary molecule for life on Earth. Instead, most biologists believe that DNA evolved from a simpler self-replicating molecule (probably RNA) that carried hereditary information in the earliest living organisms [Section 5.2]. However DNA evolved, it has proved remarkably successful—it is now the hereditary material for every known organism on Earth. The basic copying process shown in Figure 3.15 explains how every organism on Earth, from the smallest bacteria to humans, passes its genetic information from one generation to the next.

Genes, Genomes, and the Genetic Code

Besides having the ability to replicate, DNA also determines the structure and function of the cells within any living organism. In essence, the "operating instructions" for a living organism are contained in the precise arrangement of chemical bases (A, T, C, and G) in the organism's DNA. Let's briefly examine a few key features of these "operating instructions."

What Is a Gene? Within a large DNA molecule, isolated sequences of DNA bases represent the instructions for a variety of cell functions. For example, a particular sequence of bases may contain the instructions for building a protein, for building a piece of RNA, or for carrying out or regulating one of these building processes. The instructions representing any individual function—such as the instructions for building a single protein—make up what we call a **gene.** A gene is the basic functional unit of an organism's heredity—a single gene consists of a sequence of DNA bases (or RNA bases, in some viruses) that provides the instructions for a single cell function.

What Is a Genome? The complete sequence of DNA bases in an organism, encompassing all of the organism's genes, is called the organism's **genome.** The human genome, for example, consists of about 3 billion DNA bases (spread among the 46 chromosomes in human cells) and contains somewhere between about 30,000 and 120,000 genes. (As this range shows, the total number of human genes is still not known, but it should be pinned down within a few years as part of the human genome project, an ongoing effort to map the complete human genome.) Note that every cell in a living organism generally contains the same set of genes. Different cell types, such as muscle cells or brain cells, differ only because they *express,* or actually use, different portions of their full set of genes. Thus, the DNA found in

any cell in any organism contains the complete instructions for building an organism of that species. This is also the basis of *cloning,* in which a single cell from a living organism is used to grow an entirely new organism with an identical set of genes.

The Genetic Code A strand of DNA contains a long, unbroken sequence of DNA bases; for example, a particular sequence might contain the bases ACTCAGCTTCAACGG. . . . In order for a sequence like this to be useful as the instructions for a cell function, there must be a set of rules for how to "read" the sequence. That is, the rules must specify how to break the long sequence into individual "words," as well as where to start reading and where to stop reading the words that represent the instructions for a single gene. The set of rules for reading DNA is called the **genetic code.** More specifically, genetic "words" consist of three DNA bases in a row. For the purpose of protein building, each word represents either a particular amino acid or a "start reading" or "stop reading" instruction.[3]

Because the genetic words consist of three DNA bases in a row and there are four DNA bases to choose from (A, C, T, G), the total number of words in the genetic code is $4^3 = 4 \times 4 \times 4 = 64$. Notice that this is significantly more than the number of amino acids used to make proteins, which is 20 for most organisms. Thus, the genetic code contains a fair amount of redundancy. For example, the genetic words ACC and ACA both represent the same amino acid. Moreover, a close examination of the genetic code also offers a hint about the likely evolution of DNA: The codes for most amino acids really depend on just the first two bases in the three-base genetic words. For example, all four of the three-base words starting with AC—ACC, ACA, ACT, and ACG—code for the same amino acid. This suggests that the genetic code once depended only on two-base words rather than three-base words. Most biologists now believe that early life-forms used only a two-base language, which later evolved into the current three-base language of the genetic code.

THINK ABOUT IT . . . *Note that a two-base language would allow only $4 \times 4 = 16$ possible words—not enough for all the amino acids used by living organisms today. What does this imply about proteins in early life-forms? Does it seem likely that even earlier organisms might have used only a one-base language? Explain.*

---

[3] You can find a simple table listing the complete genetic code—that is, the meaning of each of the 64 different three-base "words"—in any biology text or on the Web. It is not given here because it contains more details than are necessary for this book.

Another important feature of the genetic code is that it is the same in nearly all living organisms on Earth. Only a few organisms show any variations at all on this code, and the variations that do exist are minor. (Variations in the genetic code are also found in mitochondria, structures within eukaryotic cells that contain their own DNA.) Nevertheless, the fact that some variations occur tells us that not all the specifics of the genetic code were inevitable. If we think of the genetic code as a language, the fact that nearly all organisms use the same genetic code is rather as if everyone on Earth spoke the same language, even though other languages are possible. This common language of the genetic code is further evidence for a common ancestor of all life on Earth.

The Role of RNA While the sequence of bases in a gene holds the instructions for its function, the actual implementation of these instructions is quite complex. As with DNA replication, many enzymes are involved in carrying out genetic instructions. In addition, the molecule **RNA** plays a particularly important role in these functions. A molecule of RNA is quite similar in structure to a *single* strand of DNA, except that it has a slightly different backbone and uses a different set of four bases (only one of which is different from the DNA bases).

RNA participates in carrying out genetic instructions in several different ways. For example, in the process of building a protein, a molecule of RNA is first assembled along one strand of DNA, essentially transcribing the DNA instructions for use in another part of the cell. (This process is called *transcription*.) This molecule of RNA then goes to a site in the cell where amino acids are actually assembled into proteins. Other molecules of RNA collect amino acids from within the cell and bring them to this site, where yet another form of RNA helps attach the amino acids into the chains that make proteins. (This process is called *translation,* because it effectively *translates* the genetic instructions into an actual protein.) If you take a biology course, you will learn much more about the

Movie Madness: The Andromeda Strain

Most movie aliens are complex critters—"bug-eyed monsters" that come to Earth in high-speed, hi-tech spacecraft to undertake a campaign of havoc and destruction. But a different threat is offered by *The Andromeda Strain* (1971) as the U.S. military sends a satellite into space to scoop up some simple alien life—green goo attached to small meteors—hoping that microbe-sized aliens might be useful for biological warfare.

You might argue that, as a scheme for developing new weaponry, this approach is a long shot. But not in the movies. The satellite returns to Earth, and some of the alien microbes escape into the air to promptly wipe out a small, New Mexico town. (Aliens nearly always prefer landing in the American Southwest.) Unaware of the military's game plan, a team of cranky civilian scientists is brought together in a top-secret Nevada bio lab to figure out what's afoot and stop the infection from space.

Most of the movie is taken up with the step-by-step analysis performed by the scientists as they methodically try to isolate and understand the meteor-borne microbes. They find that the space critters have no nucleic acids, no amino acids, and none of the usual features of biology on Earth. In fact, these extraterrestrial entities are crystals— the entire organism is all made of the *same* molecule (like salt, except with a much more complicated molecule). This is a bit hard to swallow, but there's more: each time the alien bugs reproduce (roughly every few seconds), they also mutate. And unlike for earthly life, the mutations are not random. Each cell mutates simultaneously and identically to every other.

By the end of the film, this peculiar property makes all the weighty work of the scientists irrelevant. The alien microbes mutate into a harmless form and drift out to sea. But think about it: any species in which all individuals mutate the same way at the same time will be unable to adapt to a changing environment. Every microbe is, and remains, identical. You should feel sorry for the Andromeda strain, because it has no Darwinian evolution and, consequently, no chance of long-term survival. As a strategy for life, it strains credulity.

In real life, NASA has a program to ensure that materials brought to Earth by future spacecraft will be carefully handled to forestall any chance of contamination by alien microbes. This news should be a relief to residents of the Southwest.

many roles of RNA.[4] However, for our purposes in this book, it is not necessary to know much more about RNA except the fact that it is very important for carrying out the instructions encoded in DNA.

Mutations and Evolution

One of the most remarkable aspects of our current knowledge of DNA is that it has allowed us to confirm Darwin's theory of evolution by natural selection. In particular, whereas Darwin had to base his theory on the observation of genetic variation in populations, we now know precisely why such variation occurs. The key to this knowledge lies in understanding how DNA molecules gradually change through time. Based on what we've already said about DNA replication and protein building, we can see how and why changes in DNA occur.

Despite its complexity, the process of DNA replication proceeds with remarkable speed and accuracy. Some bacteria can copy their complete genomes in a matter of minutes, and copying the complete 3-billion-base sequence in human DNA takes a human cell only a few hours. In terms of accuracy, the copying process generally occurs with less than one error per billion bases copied. Nevertheless, errors do sometimes occur. For example, the wrong base may occasionally get attached in a base pair, such as linking C to A rather than to G. In other cases, an extra base may be accidentally inserted into a gene, a base may be deleted, or an entire sequence of bases might be duplicated or eliminated. Absorption of ultraviolet light or nuclear radiation or the action of certain chemicals (carcinogens) can also cause mistakes to occur. Any change in the base sequence of an organism's DNA is called a **mutation.**

Mutations can affect proteins in a variety of ways. Some mutations have no effect at all. For example, suppose a mutation causes the genetic word ACC to change to ACA in a gene that makes a protein. As we noted earlier, both of these words code for the same amino acid, so this mutation will not change the protein made by the gene. Other mutations in which a letter changes will change a single amino acid in a protein. In some cases, such a change will alter a protein only slightly, hardly affecting its functionality. But in other cases, the change can be much more dramatic. For example, the cause of sickle-cell dis-

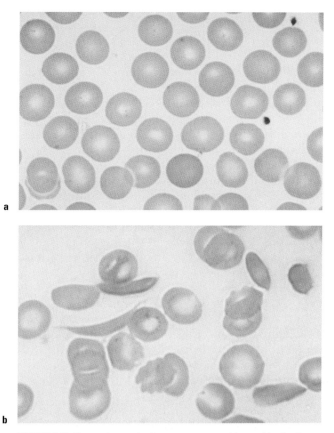

a

b

FIGURE 3.16 These microscopic views contrast normal human blood cells with the blood cells found in patients with sickle-cell disease. The sickle shape makes it easier for the blood cells to clog tiny blood vessels, which can lead to debilitating disease. Sickle-cell disease occurs in people whose gene for hemoglobin differs from the "normal" gene in just a single DNA base.

ease (Figure 3.16), which kills some 100,000 people each year worldwide, can be traced to a single mutation in the gene that makes hemoglobin in which the base A changed to the base T in just one place within the gene.[5] Mutations that add or delete a base within a gene tend to have the most dramatic effects on protein structure. The reason is that the genetic code has no "punctuation" between words; instead of saying something like "the fat cat ate the rat," for example, it says "thefatcatatetherat." Thus, if a letter (base) is added to or deleted from such a sequence, the result will be nonsense from that point on. For example, inserting an "a" so that the sequence becomes "theafatcatatetherat" would cause it to be read as "the afa tca tat eth era t."

---

[4] Readers familiar with the different types of RNA will recognize that messenger RNA (mRNA) is involved in transcription, transfer RNA (tRNA) is responsible for collecting amino acids, and both of these meet up to assemble proteins on structures called ribosomes, which are built from proteins and ribosomal RNA (rRNA).

[5] Humans have two copies of each gene, and sickle-cell disease generally occurs only in people who have the sickle-cell mutation in both copies of the gene. From an evolutionary standpoint, this mutation remains prevalent in the population because it actually confers an advantage—malaria resistance—to people with only one copy of the mutated gene.

Mutations that change proteins are sometimes lethal, because the cell may not be able to survive without the correctly structured protein. However, if the cell survives, the mutation will be copied every time its DNA is replicated. That is, the mutation makes a permanent change in the cell's hereditary information. If the cell happens to be one that gets passed to the organism's offspring—as is always the case for single-celled organisms and can be the case for animals if the mutation occurs in an egg or sperm cell—the offspring will have a gene that differs from that of the parent. It is this process of mutation that leads to variation among individuals in a species. Each of us differs slightly from all other humans because we each possess a unique set of genes with slightly different base sequences.

THINK ABOUT IT . . . *Ultraviolet radiation from the Sun can cause mutations in the DNA of skin cells. Based on what you've learned, explain why this is potentially dangerous (and, indeed, is the cause of skin cancer). How would sunscreen help prevent such mutations?*

Mutations therefore also provide the basis for evolution. Given that different individuals of a species possess slightly different genes, it is inevitable that some genes will provide advantageous adaptations to the environment. As we discussed in Section 3.1, the combination of individual variation and population pressure leads to natural selection, in which the advantageous adaptations will preferentially be passed down through the generations. Thus, what was once a random mutation in a single individual can eventually become the "normal" version of the gene for an entire species. In this way species evolve through time. Notice that, while we often associate the word *mutation* with harm, evolution actually proceeds through the occasional beneficial mutation. Although such beneficial mutations may be relatively rare compared to other mutations, natural selection allows these mutations to propagate preferentially, so tremendous positive changes can accrue over time.

Will Life Elsewhere Use DNA?

It is difficult to imagine life that does not have heredity, because it seems crucial for any form of life to have some means of storing its operating instructions and passing them on to offspring. We've seen that DNA is the carrier of heredity for all life on Earth, though as we'll discuss in Chapter 5 we have good reason to believe that very early life on Earth used RNA for this role. Should we expect DNA, or RNA, to also be the heredity molecule for life elsewhere?

We do not yet know whether other, quite different molecules might be able to carry hereditary information in the same way as DNA. However, it seems a near certainty that any form of life anywhere will have some molecule that plays the same functional role that DNA plays on Earth.

3.5 Life at the Extreme

We've discussed all the fundamental characteristics of life on Earth: the basic structure of cells, the metabolism of cells, and how cells store and pass on their heredity. We've also discussed why these characteristics seem likely to be shared, at least in a general sense, by any life we find elsewhere. In essence, our bottom-line conclusion is that life elsewhere ought to share a lot of common features with life on Earth. But don't be tempted to think this means that life elsewhere will look like us—that is, like humans or even eukaryotes. In fact, most life on Earth does not look much like "us." We've already noted that prokaryotes are far more common than eukaryotes. Perhaps even more startling, in recent decades biologists have discovered that life can survive in an astonishing variety of environments that would be lethal to humans.

Extremophiles

Deep on the ocean floor are places where volcanic activity releases hot water and rock into the surrounding ocean. Minerals and other dissolved chemicals cause the water to turn black, so the vents and their plumes of hot water are called *black smokers* (Figure 3.17). The water coming out of the black smokers is heated to temperatures as high as about 400°C (750°F), far above the normal boiling point of 100°C (212°F). However, the ocean pressure at these depths is so great that the water remains liquid despite its high temperature.

If you took any "ordinary" organism and placed it in the hot water near a black smoker, it would die quickly because the high temperature would cause many of its critical cell structures to fall apart. Yet in recent decades scientists have discovered life— mostly prokaryotes of the domain Archaea[6]—thriving near black smokers in water with temperatures above 110°C (230°F). Similar organisms thrive in hot

---

[6] Many known species of Archaea live in extreme environments, but extreme conditions are *not* a general feature of domain Archaea. Indeed, some of the most common organisms in "ordinary" environments on Earth are Archaea. For example, some 20–50% of the living cells in cool ocean water typically represent various species of Archaea.

FIGURE 3.17 This photograph shows a black smoker—a volcanic vent on the ocean floor that spews out extremely hot, mineral-rich water.

FIGURE 3.18 A hot spring in Yellowstone National Park. The different colors in the water are from different microbes that survive in water of different temperatures. To get a sense of scale, note the walkway winding along the lower right.

springs on the Earth's surface, such as in the springs around Yellowstone National Park (Figure 3.18). Organisms that survive in extremely hot water are sometimes called *thermophiles,* meaning "lovers of heat" (the suffix *phile* means "lover") or, for those living at the highest temperatures, *hyperthermophiles.* More generally, organisms that survive in extreme environments of any kind are called **extremophiles,** or "lovers of the extreme." Extremophiles are quite varied, though most of them are prokaryotes (either Bacteria or Archaea). Some can live in "normal" as well as extreme conditions, while others survive only in the extreme conditions. For example, the hyperthermophiles die when brought to "normal" temperatures because their enzymes have evolved to function only at the high temperatures in which they live. Indeed, many extremophiles are anaerobic, and they are poisoned by the oxygen on which our own lives depend.

Hot environments are not the only extreme conditions favored by some organisms. The dry valleys of Antarctica receive so little rain or snowfall that they are among the driest deserts on Earth, and temperatures in these valleys rarely rise above freezing (Figure 3.19). Nevertheless, there is life in the dry valleys living *inside* rocks. We often think of rocks as solid, but most rocks are composed of individual mineral grains packed together, leaving small spaces between the grains. Even in the dry valleys, these spaces within the rocks contain water from the rare rain or snowfall. Sunlight can penetrate up to a few

FIGURE 3.19 A dry valley in Antarctica. The valleys are deserts with extremely little rain or snowfall. The ice in the valley comes from runoff from surrounding regions.

millimeters into the rock before being completely absorbed, so the layers just below the rock's surface can have temperatures slightly above freezing despite the freezing temperatures around them. Amazingly, there are microbes that thrive in these tiny pockets of liquid water inside rocks in freezing cold valleys (Figure 3.20). Other extremophiles live in conditions far too acidic, alkaline, or salty for "ordinary" life to survive.

One particular group of extremophiles is of special interest for its possible relevance to life on other worlds. Microbes called *lithophiles* (meaning "rock lovers") live up to several kilometers below the surface of the Earth in water that fills the pore spaces

FIGURE 3.20 This photograph shows a slice of rock from a dry valley in Antarctica. The colored zones contain microbes that live inside the rock. Microscopic algae, fungi, and bacteria live in the airspaces between tiny mineral grains. During most of each year, the organisms are frozen, but sunlight warms the rock above the freezing point of water for about 500 hours each year. The piece of rock shown is 1.8 centimeters across.

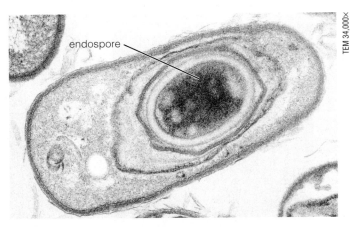

FIGURE 3.21 This microscopic photo shows an endospore created by the bacterium *Bacillus anthracis*. There are actually two cells here, one inside the other. The outer cell produced the specialized inner cell, which is the endospore. The endospore has a thick, protective coat. Its interior is dehydrated, and no metabolism occurs. Under harsh conditions, the outer cell may disintegrate, but the endospore can survive. When the environment becomes more hospitable, the endospore absorbs water and resumes growth.

within rock. One community of lithophiles, discovered in 1995, consists of bacteria living deep beneath the surface of Oregon and Washington in a rock formation known as the Columbia River Basalt. Lithophiles get their energy for metabolism from chemical reactions between the water and the surrounding rock, and they get their nutrients from chemicals within the rock itself and from carbon dioxide that has filtered down from the surface. Although these microbes clearly share the same common ancestors as the rest of terrestrial life, they tell us something very significant about the robustness of life and its ability to survive in diverse environments. In particular, the subsurface environment in which they live almost certainly exists in similar form on other planets, even on some planets that may be too cold for liquid water to exist on the surface. Mars, for example, is likely to have just this type of subsurface environment, in which liquid water exists in pore spaces within the surrounding rock. Thus, it is very likely that other planets could support life like the lithophiles on Earth, though of course we do not yet know whether life actually exists on such planets.

Perhaps one of the most amazing adaptations to extreme conditions is found in **endospores**—special "resting" cells produced by some prokaryotes (Figure 3.21). Endospores allow the organisms that create them to become dormant, neither growing nor dying in extremely inhospitable conditions. (The nondormant organisms are not necessarily extremophiles; some live under more "normal" conditions.) For example, endospores of the bacterium *Bacillus anthracis,* which causes the deadly disease anthrax, can survive a complete lack of water, extreme heat or cold, and most poisons. Some endospores can survive even in the vacuum of space, which is why planetary scientists worry that our interplanetary spacecraft could potentially contaminate other worlds with

life from Earth. Moreover, some endospores can remain dormant for centuries—and perhaps even longer—raising the possibility that life could survive journeys aboard meteorites that are blasted off one planet and land on another [Section 7.4].

Are Extremophiles Really Extreme?

From our human point of view, the environments in which extremophiles survive truly are extreme. But if an extremophile could think, it would probably claim that its environment is quite normal and that ours is the one that is extreme. So who's right, humans or the extremophiles?

In some sense, it's just a matter of opinion. Any species would naturally consider its own environment to be normal and others to be extreme. On a deeper level, we might ask which environment is more common. Surprisingly, if we look at the history of the Earth, so-called extreme environments have been much more common than an environment suitable for humans. Earth's atmosphere probably has contained oxygen at a level suitable for human life for only a few hundred million years, or about 10% of the Earth's history [Section 5.3]. That is, if you could play Russian roulette with a time machine capable of sending you to any point in the Earth's history, you'd have only about a 1 in 10 chance of being able to breathe the air. Indeed, for the first couple of billion years after life first arose on Earth, extremophiles were the only organisms that could survive. Even today, it's an open question whether extremophiles are more or less common than organisms that live in conditions favorable to humans. All in all, extreme life appears to be much more the norm than is life that lives in an environment with moderate temperatures and plentiful oxygen.

Implications for Extraterrestrial Life

The study of extremophiles has several important implications for the search for extraterrestrial life, but two are particularly important. First, the fact that extremophiles apparently evolved earlier than other forms of life [Section 5.1] suggests that we should begin the search for life elsewhere by searching for similar extreme organisms. That is, rather than looking for oxygen-breathing organisms, we should be looking for organisms that survive in a variety of more "extreme" environments. Second, the fact that extremophiles can survive such a broad range of conditions suggests that life may be possible in many more places than we would have guessed only a few decades ago. That is, any world containing an environment in which an extremophile might survive becomes a good candidate for the search for life.

THE BIG PICTURE

In this chapter, we have surveyed the nature of life on Earth and explored the implications of this survey for the search for life elsewhere. As you continue in your study, keep in mind the following "big picture" ideas:

- If we are going to search for life, it's useful to think about just what it is we are searching for. Defining life turns out to be surprisingly difficult, but at a minimum it seems that life must be capable of reproducing and evolving. Thus, evolution plays a central role in the definition of life as well as in our understanding of life on Earth.

- Life on Earth has at least three key features that are likely to be shared by any other life we find elsewhere: (1) Life has a fundamental structural unit, which we call the cell; (2) living cells undergo processes of metabolism, by which we mean chemical reactions that keep the cell alive; and (3) living cells have a heredity molecule, which is DNA for life on Earth, that allows them to store their operating instructions and to pass these instructions to their offspring.

- Life on Earth survives under a much wider range of conditions than we would have guessed a few decades ago, suggesting that life elsewhere might similarly be found in a fairly broad range of environments. This fact greatly increases the number of worlds on which we might hope to find life.

Review Questions

1. Briefly describe the six key properties that appear to be shared by most living organisms on Earth.

2. What is *natural selection?* Summarize the logic by which Darwin came to the "inescapable conclusion" that evolution occurs by natural selection. Describe some of the evidence that supports Darwin's *theory of evolution.*

3. Briefly describe the evidence that points to a single common ancestor for all life on Earth.

4. Why do we say that living cells are *carbon-based?* Briefly discuss whether life elsewhere could be based on something besides carbon.

5. Distinguish between *prokaryotic cells* and *eukaryotic cells.* Which type are we made from? Which type is more common?

6. What do we mean by the three *domains* of life?

7. What is *metabolism,* and what are the two basic metabolic needs of any organism? Explain the four metabolic classifications listed in Table 3.1.

8. Briefly describe why water is so important to life on Earth.

9. Describe the double helix structure of *DNA.* How does a DNA molecule replicate?

10. What is a *gene?* A *genome?* The *genetic code?*

11. What are *mutations,* and what effects can they have? Briefly explain why mutations represent the molecular mechanism of natural selection.

12. What are *extremophiles?* Give several examples of organisms that live in extreme environments. What are the implications of the existence of extremophiles for the search for extraterrestrial life?

Discussion Questions

1. *The Theory of Evolution.* Given the "inescapable" logic behind Darwin's theory of evolution and the extraordinary body of evidence supporting his theory, why do you think the theory remains so controversial among the general public? Do you believe the controversy concerns the theory itself, a misunderstanding of the theory, or other factors? Defend your opinion.

2. *Computer Life.* Although scientists have already developed computer programs capable of reproducing themselves and evolving, few people consider such programs to be alive. But consider future developments in computing and robotic technology. Do you think we'll ever make something out of electronics that is truly alive? Could it also be intelligent? If so, what rights should we accord to such "artificial" life?

3. *Genetic Engineering and Future Evolution.* For billions of years, evolution has proceeded through random mutations and natural selection. Today, however, we have the ability to deliberately alter DNA in what we call "genetic engineering." How do you think this ability will affect the future evolution of life on Earth? How will it affect future human evolution? Based on your answers, should we expect extraterrestrial civilizations to be naturally evolved or to be products of their own genetic engineering? Discuss and defend your opinions.

4. *Gene Transfer and GMOs.* In some cases, organisms can transfer entire genes to other organisms. This fact causes some people to worry that organisms that we have genetically engineered—commonly referred to as GMOs, for genetically *modified organisms*—may transfer their genes to other organisms in unexpected ways. For example, a crop engineered with a gene that gives it resistance to some pest may transfer its gene to weeds, giving them the same resistance. Discuss how GMOs might affect other organisms. Overall, what if any controls do you think the government should put on the use of GMOs?

Problems

Surprising Discoveries of Life on Earth? For **problems 1–8**, suppose we found an organism on Earth with the characteristics described. In light of our current understanding of life on Earth, should we be surprised that such an organism exists? Why or why not?

1. A prokaryote that builds proteins using 45 different amino acids.

2. A single-celled organism that lives deep in peat bogs, where no oxygen is available.

3. A eukaryote that reproduces without passing copies of its DNA to its offspring.

4. A prokaryote that can survive in a dormant state even in the complete absence of any liquid water.

5. A multicellular organism that can grow and reproduce even in the absence of water.

6. A bacterium that lives in rock underneath a glacier.

7. A bacterium that lives in the 1,000°C molten rock of a volcano.

8. An animal whose cells belong to the domain Archaea.

True Statements? For **problems 9–12,** decide whether the statements are true or false and explain how you know.

9. Charles Darwin was the first person to recognize that life evolves.

10. Some organisms on Earth are capable of surviving, at least temporarily, in the vacuum of space.

11. All humans have precisely the same genes, with the same DNA sequences.

12. Any early life that existed at the time when the genetic code consisted of only two-base "words" (rather than three-base "words") probably still used the same 20 amino acids used by life today.

13. *Rock Life?* How do you know that a rock is not alive? In terms of the properties of life discussed in this chapter, clearly describe why a rock does not meet the criteria for being alive.

14. *Artificial Selection.* Suppose you lived hundreds of years ago (before we knew about genetic engineering) and wanted to breed a herd of cows that provided more milk than cows in your current herd. How would you have gone about it? Explain, and describe how your breeding would have worked in terms of the idea of artificial selection. How does this breeding offer evidence in favor of the idea of natural selection?

Web Projects

1. *Darwin on Evolution.* You can find on-line the entire text of Charles Darwin's *The Origin of Species.* Read the final chapter, in which Darwin addresses potential criticisms of his theory. Evaluate how well he has presented his case. How much stronger does the theory seem today than at the time Darwin first described it in 1859? Summarize your conclusions in a one-page essay.

2. *Extreme Life.* Look for information about a recent discovery of a previously unknown type of extremophile. Describe the organism and the environment in which it lives, and discuss any implications of the finding for the search for life beyond Earth. Summarize your findings in a one-page report.

3. *The Genetic Code.* Find a Web site that gives and explains the complete genetic code. Once you understand how to read it, describe how it offers evidence for the idea that the current genetic code evolved from an earlier code that used two-base, rather than three-base, genetic "words."

CHAPTER 4

The Geological History of the Earth

In Chapter 3, we studied the nature of life on Earth, finding that it depends on the availability of liquid water, on carbon and other elements from the atmosphere and Earth, and on both chemical and solar sources of energy. These facts tell us that life is intimately connected to the physical properties of Earth. Thus, if we hope to understand the conditions under which life might exist elsewhere, we must begin by exploring how our own planet came to be a home for life.

This task involves two major steps. First, we must understand the Earth's geological history, which will allow us to understand how our planet maintains a physical environment suitable for life. Second, we must explore how life arose and evolved within this physical environment. We will focus primarily on the first step in this chapter, saving our discussion of the origin and evolution of life for the next chapter.

We will begin this chapter by exploring how scientists reconstruct the Earth's history by studying rocks and fossils. We'll next discuss current understanding of the Earth's geological history, with emphasis on how our planet formed and came to have its atmosphere and oceans. We'll also investigate geological processes and how they keep the climate hospitable for life. This discussion of various aspects of the Earth's geology will not only help us understand life on Earth, but will also help us explore the potential habitability of other worlds.

4.1 Reading the Earth's History

Human recorded history dates back only a few thousand years on a planet that has existed for about $4\frac{1}{2}$ billion years. To put this fact in perspective, imagine making a timeline to represent the history of the Earth. On a timeline the length of a football field, human civilization would appear only in a tiny sliver at the end of the line—a sliver no thicker than your fingernail. How, then, can we possibly know anything about the $4\frac{1}{2}$ billion years of Earth's history that preceded human civilization?

The answer is that this history is recorded in rocks and **fossils,** relics of organisms that lived and died long ago. Reading this history is not as easy as reading words on a page, but with proper scientific tools it can be read just as reliably. What is required is an understanding of how fossils and rocks are made, why they are found in distinct geological layers, and how we can determine their ages through radiometric dating and other techniques. Our task in this section is to explore the use of these scientific tools and what they tell us about the events of the Earth's history.

Rocks and Fossils

Rocks and fossils tell us about the Earth's past in a variety of ways. For example, a rock's structure and composition depend on the conditions under which it formed, which may offer clues to the climate at the time. Fossils may also tell us about past environmental conditions and allow us to reconstruct the evolutionary history of life. Reading the history recorded in rocks and fossils—often called the **rock record** and the **fossil record,** respectively—requires knowing a little about how rocks and fossils are formed.

Geologists classify rocks into three basic types according to how they are made (Figure 4.1). **Igneous rock** is made from molten rock that cools and solidifies. **Sedimentary rock** is made by the gradual compression of *sediments,* such as sand and silt at the bottoms of seas and swamps. **Metamorphic rock** is rock that has been structurally transformed by high pressure or heat that is not quite high enough to melt it.

Note that rock can change from one type to another. Igneous rock is often made from sedimentary or metamorphic rock that has been carried deep underground and melts under high heat and pressure. This molten rock cools and solidifies into igneous rock as it rises toward the surface or erupts from a volcano. Sedimentary rock may be made as erosion breaks up existing rock into sediments that are carried to the sea. Metamorphic rock is made from igneous or sedimentary rock that is transformed by high pressure or heat. Because rock can be recycled among the three types, the rock type is not directly related to its composition. In fact, individual rocks of any of the three types usually contain a mixture of different crystals in close contact. Each individual crystal represents a particular **mineral,** which is the word we use to describe a crystal of a particular chemical composition and structure. Thus, a rock's type (igneous, metamorphic, or sedimentary) tells us how it was made, while its mineral composition tells us what it is made of.

If you study geology in greater depth, you'll learn a variety of other terms used by geologists to subclassify rocks by their type and mineral composition. For our purposes in this book, however, you need to be aware of only one other classification. Igneous rocks may differ substantially from one another depending on exactly what melted to make them, and geologists have special names for different types of igneous rocks. For example, a type of dark, dense igneous rock commonly produced by undersea volcanoes is called **basalt;** another common type of igneous rock is **granite,** which is named for its grainy appearance and is often found in mountain ranges. (Other common igneous rocks include *andesite* and *rhyolite.*) To help you keep the classifications and definitions

FIGURE 4.1 Samples of the three basic rock types.

a Igneous rock.

b Metamorphic rock.

c Sedimentary rock.

a A dinosaur bone preserved in sandstone in Dinosaur National Monument, which straddles Utah and Colorado.

b Stone trees in the Petrified Forest in Arizona.

c These shell-like impressions are casts of dead organisms made when minerals filled the empty space left after the organisms decayed.

d This 40-million-year-old leaf still retains organic material, including DNA.

FIGURE 4.2 A gallery of fossils.

e This 30-million-year-old scorpion is embedded in hardened tree resin (often called *amber*).

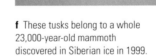

f These tusks belong to a whole 23,000-year-old mammoth discovered in Siberian ice in 1999.

g This boy is standing in a 150-million-year-old dinosaur track in Colorado.

straight, they are summarized along with other key definitions in the box on pages 85 and 86.

We generally find recognizable fossils only in sedimentary rocks, because the high heat and pressure involved in forming metamorphic or igneous rocks tends to destroy fossils. Although we tend to think of fossils as "remains" of living organisms, most fossils contain little or no organic matter. Figure 4.2 summarizes some of the ways in which fossils are made. In general, when an organism dies and gets buried in sediments, minerals dissolved in groundwater gradually replace organic material. Mineral-rich portions of organisms, such as bones, teeth, and shells, may be left behind, becoming fossils like those of the dinosaur bones displayed in many museums (Figure 4.2a). In some cases, the mineral replacement is complete and organisms literally turn to stone; the "stone trees" of Arizona's Petrified Forest formed in this way (Figure 4.2b). In many other cases, the organisms themselves decay, but in doing so they leave an empty mold that fills with minerals dissolved in water. The minerals may then make a cast in the shape of the dead organism (Figure 4.2c). More rarely, some of the organic material from a dead organism may be preserved well enough to allow at least some study. Some fossil plant leaves are still green and well enough preserved for their cells to be studied with microscopes, even though they died millions of years ago (Figure 4.2d). In other rare cases, whole organisms may be preserved in tree resin (Figure 4.2e) or frozen in ice (Figure 4.2f). One of the most interesting types of fossil is left not by a dead organism but by the activity of an organism while it was alive. For example, "coprolites" are rocks

that consist of petrified excrement, which can allow us to learn about an animal's diet. In other cases, scientists have found fossilized dinosaur footprints, made when mineral processes preserved impressions left by a dinosaur as it walked through soft soil or mud (Figure 4.2g). Such fossil tracks provide clues aout how dinosaurs walked and can help scientists hypothesize about dinosaur behavior.

THINK ABOUT IT . . . *Molecules of DNA tend to break down rapidly and easily after an organism dies, making it difficult or impossible to identify even fragments of DNA in most fossils. What types of fossils do you think are most likely to yield intact DNA? What types are least likely to yield intact DNA? Explain.*

It's important to note that very few living organisms leave fossils behind. For example, scientists have discovered only a small number of complete dinosaur skeletons despite the huge numbers of dinosaurs that once must have roamed the Earth. Fossils are rare because most dead organisms decay—becoming food for living organisms in the soil—long before any mineral replacement can occur. If you've ever read about criminologists exhuming the skeletons of people who have been dead for just a few years, you know that even bones and teeth usually decay quite rapidly after death. Nevertheless, over millions and billions of years, enough dead organisms have become fossilized to leave a substantial fossil record.

Sedimentary Strata

Sedimentary rock is particularly important to our study of the Earth's history not only because it may contain intact fossils, but also because the way in which it forms tends to preserve a record of time. The sediments that make sedimentary rock are produced primarily by erosion on land. Most of them are carried by rivers and deposited on floodplains or in the oceans. Over millions of years, sediments pile up on the seafloor, and the weight of the upper layers compresses underlying layers into rock. Fossils are made when remains of living organisms are buried along with the sediments. Aquatic organisms may be buried in sediments simply because they settle to the bottom of the sea. Some land organisms form fossils when their remains are swept into bodies of water. In other cases, remains of land organisms may be buried in place by windblown silt and later compressed by sediments when changing sea levels rise over them.

Sediments deposited at different times tend to look different as a result of changes in the rate of sedimentation, in the composition or grain size of sediments settling to the bottom, or in the type of organisms leaving fossils. As a result, sedimentary rock tends to be marked by distinct layers, or **strata** (singular, *stratum*). We can view these strata in sedimentary rocks that have been exposed by, for example, the gradual action of a river carving through the rock over millions of years or by a cut made through a mountain to make way for a road. Figure 4.3 summarizes the process by which sedimentary rock forms. Figure 4.4 shows part of the Grand Canyon, which was carved by the Colorado River to expose the beautiful colors in its strata of sedimentary rock. The rock layers of the Grand Canyon record more than 2 billion years of Earth's history.

Relative Ages and the Geological Time Scale

Because sedimentary rock builds up gradually over time, at any particular location the more deeply buried layers generally are older. This allows geologists to determine the *relative* ages of rocks and fossils buried in sediments. For example, because fossils

---

FIGURE 4.3 Formation of sedimentary rock. Each stratum, or layer, represents a particular time and place in Earth's history and is characterized by fossils of organisms that lived in that place and time.

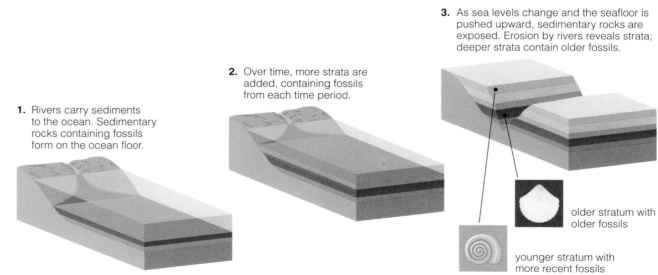

3. As sea levels change and the seafloor is pushed upward, sedimentary rocks are exposed. Erosion by rivers reveals strata; deeper strata contain older fossils.

2. Over time, more strata are added, containing fossils from each time period.

1. Rivers carry sediments to the ocean. Sedimentary rocks containing fossils form on the ocean floor.

older stratum with older fossils

younger stratum with more recent fossils

FIGURE 4.4 The walls of the Grand Canyon are exposed sedimentary rock in which the strata record more than 2 billion years of Earth's history.

of dinosaurs appear only in layers older than those in which we find fossils of primates, we conclude that dinosaurs lived before primates.

No single location contains strata from the entire history of Earth, but geologists have put together a fairly detailed fossil record by comparing sedimentary strata from many sites around the world. Scientists correlate the strata from different sites by looking for layers with similar fossils. For example, suppose an upper stratum at one location contains fossils of the same type as those found in a lower stratum at another location. In that case, the first location must contain more ancient strata than the second location (Figure 4.5).

The fossil record reveals many changes in the history of the Earth. To help organize this history, scientists divide it into a set of distinct intervals that make up what we call the **geological time scale.** Figure 4.6 shows the names of the various intervals on a timeline, along with numerous important events that we will discuss in this chapter and in Chapter 5. It is not necessary to memorize the names of the geological time intervals for the purposes of this book. However, the names are commonly used in books and articles about the Earth or astrobiology, so familiarity with them is useful.

FIGURE 4.5 In this diagram, we imagine comparing sedimentary strata at two locations. We find that the fossils found in a particular layer near the top at Location 1 are of the same type as those found in a lower layer at Location 2. We conclude that the two sets of strata represent overlapping time periods, with the strata at Location 1 going farther back in time.

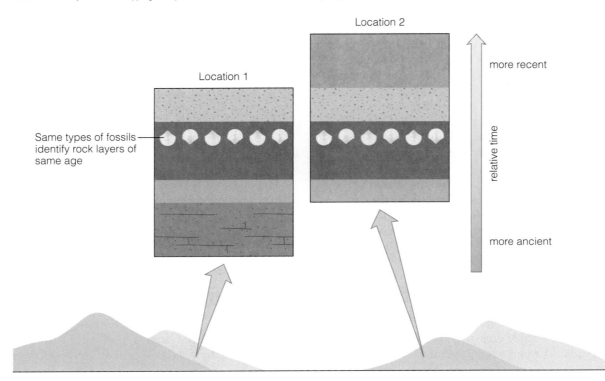

Location 2

Location 1

Same types of fossils identify rock layers of same age

more recent

relative time

more ancient

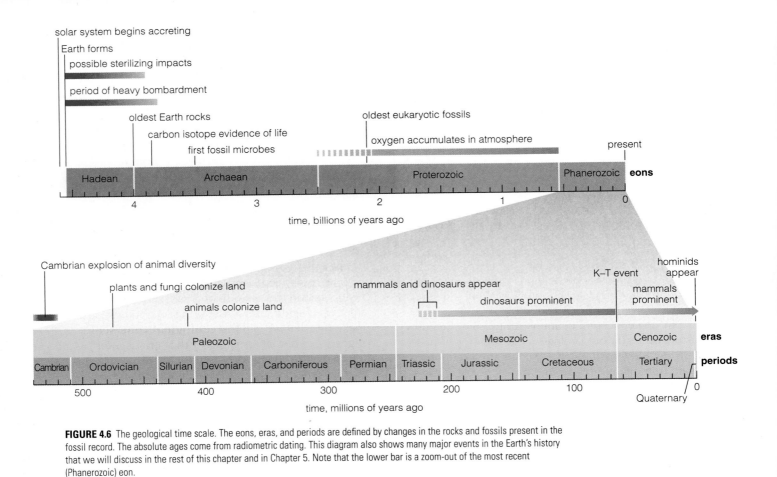

FIGURE 4.6 The geological time scale. The eons, eras, and periods are defined by changes in the rocks and fossils present in the fossil record. The absolute ages come from radiometric dating. This diagram also shows many major events in the Earth's history that we will discuss in the rest of this chapter and in Chapter 5. Note that the lower bar is a zoom-out of the most recent (Phanerozoic) eon.

The first major division of geological time is into a set of four **eons:** the Hadean, Archaean, Proterozoic, and Phanerozoic. We can understand these names by working backward in time and looking at their Greek roots. The Phanerozoic eon extends from the present back to about 540 million years ago; its name comes from the Greek for "visible life" because it is marked by the presence of fossils visible to the naked eye. The Proterozoic eon, which extends back to about 2.5 billion years ago, means the eon of "earlier life" because it shows fossils of single-celled organisms that lived before the Phanerozoic. The Archaean eon, which extends back to about 4.0 billion years ago, is named for "ancient life"; it got this name after the discovery of fossils from the first half of Earth's history. The Hadean eon gets its name from the "hellish" conditions of the environment on the early Earth before 4.0 billion years ago (Hades was the Greek mythological name for the underworld).

The most recent, or Phanerozoic, eon is subdivided into three major **eras:** the Paleozoic, Mesozoic, and Cenozoic. These names also have Greek roots and mean, respectively, "old life," "middle life," and "recent life." The three eras are further subdivided into **periods.** The periods do not follow any consis-

tent naming scheme. For example, the Cambrian period gets its name from the Roman name for Wales (in Great Britain), the Jurassic period gets its name from rocks found in the Jura mountains of Europe, and the Tertiary period simply means "third period." The Cambrian is the earliest period in the Phanerozoic eon and was once thought to be the first period in which fossil organisms could be found. For that reason, the entire time before the Phanerozoic eon— that is, the Hadean, Archaean, and Proterozoic eons —is sometimes called the *Precambrian*. (The recent geologic periods are further subdivided into *epochs* and *ages;* these are not shown in Figure 4.6.)

The divisions between the various geological time intervals mark places in the record where some fossils disappear and others begin to appear, which is why the eons, eras, and periods do not have uniform lengths. For example, the Proterozoic eon represents about three times the length of time of the Phanerozoic eon. Similarly, the Paleozoic era is longer than the Mesozoic and Cenozoic eras combined.

When geologists first began to establish the geological time scale, they could determine relative ages but not absolute ages. That is, the layering of the strata allowed them to determine which fossils were deposited earlier and which were deposited later, but they could not determine the precise ages of any of the fossils. Today, thanks primarily to the technique of *radiometric dating,* we can determine the absolute ages of rocks and fossils. We briefly discussed the importance of radiometric dating in Section 2.4. Now let's discuss how the technique works.

Absolute Ages Through Radiometric Dating

Recall that each chemical element is uniquely characterized by the number of protons in its nucleus (see the box on p. 16–17). Different *isotopes* of the same element differ only in their numbers of neutrons. A **radioactive isotope** is an isotope that is prone to spontaneous change, or *decay.* Many common elements have radioactive isotopes. For example, carbon comes in two stable isotopes: carbon-12 (98.9% of all carbon atoms), with six protons and six neutrons, and carbon-13 (1.1% of all carbon atoms), with six protons and seven neutrons. This stable carbon is generally found mixed with trace amounts of the radioactive isotope carbon-14, which has six protons and eight neutrons. Other elements, such as uranium, are always unstable—and therefore radioactive—no matter which isotope we are dealing with.

Radioactive decay can occur in a variety of ways. Sometimes a large atomic nucleus ejects a helium nucleus, which consists of two protons and two neutrons. (This process is called *alpha decay.*) In that case, the remaining, or *daughter,* nucleus has a lower atomic mass than the original, or *parent,* nucleus. For example, uranium-238 decays by ejecting a helium

---

Key Geological Definitions

This box summarizes the definitions of a few of the most basic geological terms used in this chapter and throughout the book.

Three basic rock types, distinguished by their formation process:

igneous rock: Rock made when molten rock cools and solidifies. Igneous rocks are further classified according to the types of minerals they contain; for example, *basalt* is a dark, dense igneous rock commonly produced by undersea volcanoes.

metamorphic rock: Rock made from igneous or sedimentary rock that is transformed (but not melted) by high heat or pressure.

sedimentary rock: Rock that builds up through time as sediments accumulate, often on the seafloor, and become compressed into solid rock. The sediments tend to build up in distinct layers, or *strata.*

Note: Rocks of any of these three types may be subclassified by the minerals they contain. A *mineral* has a particular chemical composition and crystal structure.

Terms related to geological time:

fossil record and *rock record:* The information about Earth's past that is recorded in fossils (fossil record) and rocks (rock record). Note that the terms are often used synonymously.

geological time scale: The time scale used to measure the history of the Earth. It is divided into four *eons* (the Hadean, Archaean, Proterozoic, and Phanerozoic). The last (Phanerozoic) eon is subdivided into three *eras* (the Paleozoic, Mesozoic, and Cenozoic), which in turn are subdivided into several *periods.* (The periods are further subdivided into *epochs* and *ages.*)

radiometric dating: The method of determining the age of a rock or fossil from study of radioactive isotopes contained within it. For a rock, it usually tells us the time since the rock solidified.

half-life: The time it takes for half of the atoms to decay in a sample of a radioactive substance.

Terms related to Earth's formation and early history:

protoplanetary disk: A spinning disk of gas in which planets may eventually form, as they did in our solar system.

solar nebula: The entire cloud of gas from which our solar system formed.

accretion (of planets): The process of building larger objects from collisions and impacts of smaller ones. As accretion proceeded in the early solar system, it first built up *planetesimals,* some of which later grew into larger *protoplanets.*

differentiation: The process by which materials separate by composition. In the Earth, differentiation led to a dense *core* made mostly of iron and nickel, a rocky *mantle* made mostly of *silicates* (minerals rich in silicon and oxygen), and a *crust* made of the lowest-density rocks.

heavy bombardment: The period of time during which the planets were heavily bombarded by leftover planetesimals, starting from the time the planets first formed and likely ending some 3.8–4.0 billion years ago.

(continued on next page)

nucleus, leaving thorium-234 as its daughter (Figure 4.7a). However, this isotope is not the final daughter of uranium-238, because thorium-234 is also radioactive. Through a chain of individual decays, eight of which involve the emission of a helium nucleus, uranium-238 ultimately decays into lead-206, which is stable and decays no further.

In other cases, radioactive decay occurs when a nucleus spontaneously emits or absorbs an electron, causing one of its neutrons to turn into a proton, or vice versa. The processes are called *beta decay* and *electron capture,* respectively. (An absorbed electron is one of the atom's own electrons that gets captured by the nucleus.) In these cases, the parent and daughter nuclei will both have the same atomic mass number (the same total number of protons and neutrons), but they will represent different elements because they have different numbers of protons. For example, carbon-14 decays by emitting an electron as one of its neutrons becomes a proton (Figure 4.7b). Thus, the daughter of carbon-14 decay has

seven protons and seven neutrons and therefore is nitrogen-14.

Regardless of the decay process, radioactive decay always occurs at a specific and measurable rate that is different for every radioactive isotope. Thus, the basic idea behind radiometric dating is to determine the age of a rock or fossil from the ratio of parent and daughter atoms within it, which depends only on the decay rate and the length of time over which the decay has been occurring. To fully understand how this technique works, we must explore the nature of radioactive decay in a bit more detail.

The Probabilistic Nature of Radioactive Decay

Radioactive decay is governed by the laws of quantum physics, which means it is a probabilistic process, much like coin tossing. If you toss a single coin, you cannot determine for sure whether it will land heads or tails. All you can say is that it has a 50% chance of landing heads and a 50% chance of landing tails. However, while the probabilistic nature

---

Key Geological Definitions (continued)

outgassing: The process of releasing gases trapped in a planetary interior into the atmosphere.

Terms related to Earth's active geology:

lithosphere: The layer of cooler, more rigid rock that sits upon the warmer, softer mantle rock below. It encompasses both the crust and the uppermost portion of the mantle, extending to a depth of about 100 kilometers. On Earth, the lithosphere is broken into a set of large *plates*.

seafloor crust: The relatively dense, thin, young crust found on Earth's sea floors, composed largely of the igneous rock called *basalt*.

continental crust: The thicker, lower-density crust that makes up the Earth's continents. It is made when remelting of seafloor crust allows lower-density rock to separate, and typically consists of granite. Continental crust ranges in age from very young to as old as about 4.0 billion years.

plate tectonics: The geological process in which lithospheric plates move around the surface of the Earth. It acts like a conveyor belt, with new seafloor crust erupting and spreading outward from mid-ocean ridges and then being recycled back into the mantle by subduction at ocean trenches. It also explains *continental drift,* because plates carry the continents with them as they move.

mantle convection: The flow pattern in which hot mantle material expands and rises while cooler material contracts and falls. Individual "loops" of rising and falling material make *convection cells*.

mantle plume (or *hot spot*): A place where a plume of hot material rises from deep within the mantle, not at the boundary between plates.

Terms related to climate and climate regulation:

greenhouse effect: The effect that makes the surface and lower atmosphere warmer than they would be in the absence of an atmosphere. It is caused by the presence of *greenhouse gases*—such as carbon dioxide (CO_2), water vapor (H_2O), and methane (CH_4)—that can absorb infrared light emitted by the planetary surface (after the surface is heated by sunlight).

carbon dioxide cycle (CO_2 cycle): The cycle that keeps the amount of carbon dioxide in Earth's atmosphere small and nearly steady and hence keeps the Earth habitable. Over time, this cycle has locked up most of the Earth's carbon dioxide in *carbonate* rocks (rocks rich in carbon and oxygen) such as limestone.

ice ages: Periods of time during which the Earth becomes unusually cold, so that water from the oceans freezes out as ice and covers a substantial portion of the continents.

snowball Earth: Refers to recently identified periods of extreme ice ages that may have occurred several times between 750 and 580 million years ago.

global warming: Usually refers to the current warming of the Earth caused by human input of greenhouse gases into the atmosphere.

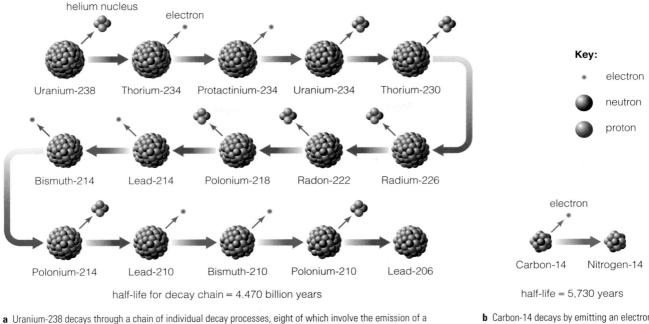

Key:
- · electron
- ● neutron
- ● proton

half-life for decay chain = 4.470 billion years

half-life = 5,730 years

a Uranium-238 decays through a chain of individual decay processes, eight of which involve the emission of a helium nucleus, ultimately leaving lead-206 as its stable daughter isotope. The half-life for the decay chain as a whole is 4.470 billion years.

b Carbon-14 decays by emitting an electron, leaving nitrogen-14 as its stable daughter, with a half-life of 5,730 years.

FIGURE 4.7 Examples of radioactive decay.

of coin tossing means you cannot predict the outcome for a single coin toss, it allows you to predict the outcome accurately for many coin tosses. For example, if you toss a fair coin 1 million times, you can be quite certain that heads will come up close to 500,000 times (but you would not expect the number to be exactly 500,000).

THINK ABOUT IT . . . *To convince yourself that probability can be used to make predictions involving large numbers of atoms or coins, toss a coin 100 times (or toss a set of 100 coins all at once) and record your results. Do you get a result close to 50, as we would expect? Compare your results with those of other students, and discuss what they tell you about the nature of probability.*

In the case of radioactive decay, the relevant probability describes the chance of a single nucleus decaying within a specific amount of time (such as within a year). However, because atoms are so tiny, we nearly always deal with such huge numbers of them that we can ignore the individual probabilities and focus instead on the rate at which large numbers of the radioactive nuclei decay.

For example, suppose we study a sample of 1 microgram of a radioactive substance, which despite its small weight (about one-millionth the weight of a paperclip) contains trillions of individual atoms. If we find that 1% of the atoms in the sample undergo radioactive decay within 1 year—meaning that at

end of the year we have 0.99 microgram of the parent substance remaining—we can conclude that any individual atom has a 1% probability of decaying in a 1-year period.

Note that, like coin tosses, results from the past do not affect results for the future. Just as a coin landing heads on one toss does not change the 50% probability of its landing heads on the next toss, the past history of the radioactive sample does not affect the future probability of decay. In our example, we would find that 1% of the atoms in the sample decay in any 1-year period, regardless of how we obtained the sample, how many atoms it contains, or how long we have been studying it. We can therefore measure the decay rate for any radioactive substance by studying a sample of the substance in a laboratory.

The Concept of a Half-life Although the probability of decay in a 1-year period is a perfectly good way to describe a substance's decay rate, we usually describe the decay rate by the substance's **half-life**—the time it would take for half the atoms in a sample of the substance to decay. That is, the amount of radioactive substance in any size sample drops by half with each half-life.

Consider the radioactive decay of potassium-40, which undergoes spontaneous change into argon-40 when its nucleus absorbs an electron to change one of its protons into a neutron. Laboratory measurements show that this decay process has a half-life of

1.25 billion years. Suppose a rock originally contained 1 microgram of potassium-40 and no argon-40. The half-life of 1.25 billion years means that half of the original potassium-40 will have decayed into argon-40 after 1.25 billion years. Thus, after 1.25 billion years, the rock will contain $\frac{1}{2}$ microgram of potassium-40 and $\frac{1}{2}$ microgram of argon-40. Half again will have decayed by the end of the next 1.25 billion years, leaving $\frac{1}{4}$ microgram of potassium-40 and $\frac{3}{4}$ microgram of argon-40 after 2×1.25 billion = 2.5 billion years. After three half-lives, or 3×1.25 billion = 3.75 billion years, only $\frac{1}{8}$ microgram of potassium-40 remains, while $\frac{7}{8}$ microgram has become argon-40. Figure 4.8 shows the gradual decline in the amount of potassium-40 and the corresponding rise in the amount of argon-40 in the rock over a period of 5 billion years.

THINK ABOUT IT . . . *Briefly describe how it is possible for us to know that potassium-40 has a half-life of 1.25 billion years, even though we have been capable of studying potassium-40 in the laboratory for only a few decades.*

The Essence of Radiometric Dating The potassium-40 example shows the essence of radiometric dating. Suppose you find a rock that contains equal amounts of potassium-40 and argon-40. As long as you can be confident that the rock contained no argon-40 when it formed (which you can, as we'll discuss shortly), you can conclude that the rock is one half-life, or 1.25 billion years, old. If instead the rock contains seven times as much argon-40 as potassium-40—that is, only 1/8 of the original potassium-40 remains—you can conclude that it is three half-lives, or 3.75 billion years, old. More generally, you can get the age of any such rock from the relative amounts of potassium-40 and argon-40 and the graphs in Figure 4.8.

Radiometric dating with other substances works the same way. By comparing the amounts of parent and daughter isotopes in the rock or fossil, we can determine the fraction of parent isotopes that has decayed since the rock or fossil formed.[1] Then, based on laboratory measurements of the decay rate (half-life), we can determine the age of the rock or fossil. Table 4.1 lists several of the parent–daughter isotope pairs commonly used in radiometric dating. Notice that the half-lives vary dramatically, so differ-

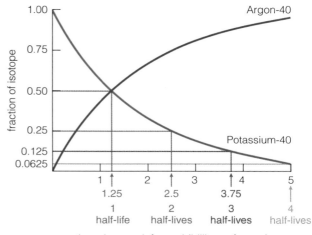

FIGURE 4.8 Potassium-40 is radioactive, decaying into argon-40 with a half-life of 1.25 billion years. The red line shows the decreasing amount of potassium-40, and the blue line shows the increasing amount of argon-40. Note that the remaining amount of potassium-40 drops in half with each successive half-life. Note also that some atoms decay every year, which enables us to measure the half-life through laboratory measurements.

ent pairs are useful for dating materials of different ages. For example, carbon-14 is useful only for dating objects less than about 50,000 years old, while uranium-238 is useful for dating the oldest rocks on Earth.

Only two major issues remain for us to consider in understanding radiometric dating. First, what do we mean by the "age" of a rock or fossil? Because the concept of a half-life applies to a collection of many atoms, radiometric dating allows us to measure the time since a particular collection of atoms came together. In the case of an igneous rock it means *the time since the rock last solidified,* because any melting would allow atoms to separate by composition (for example, gaseous argon-40 can escape, and other parent/daughter isotopes can separate by density). A sedimentary rock cannot be dated as a whole, but we can constrain its age by dating igneous rocks buried above or below it. (We can sometimes date mineral grains found within a sedimentary rock, but this tells us when the grains originally formed as opposed to when the sediments were deposited or compressed into a solid rock.) Most metamorphic rocks are altered sedimentary rocks and therefore are not readily datable.

The second major issue concerns how we can know the original composition of a rock or fossil. If we're going to get the age by comparing the amounts of a radioactive isotope and its daughter product, we must have some way of knowing how much of the daughter was originally present in the rock or fossil.

---

[1]The specific methodology varies in some cases. Carbon-14, for example, is continually produced in the Earth's atmosphere by interactions between nitrogen-14 and high-energy particles coming from the Sun. An estimate of the original carbon-14 content of a fossil is based on the atmospheric production rate rather than on a parent–daughter isotope ratio.

Table 4.1 Selected Isotopes Used for Radiometric Dating of Rocks and Fossils Some of the isotopes decay in several stages of parent–daughter pairs, and only the final daughter product is shown.

Parent Isotope	Daughter Isotope	Half-Life
Uranium-238	Lead-206	4.470 billion years
Uranium-235	Lead-207	704 million years
Thorium-232	Lead-208	14 billion years
Potassium-40	Argon-40	1.250 billion years
Rubidium-87	Strontium-87	48.8 billion years
Aluminum-26	Magnesium-26	700,000 years
Carbon-14	Nitrogen-14	5,730 years

In the case of potassium-40, this is quite easy. The daughter product, argon-40, is a gas that rarely combines with other elements. Thus, in general, argon-40 cannot be part of a rock when it solidifies, so any argon-40 trapped inside a rock must have come from radioactive decay of potassium-40. We therefore can be confident that no argon-40 is ever present when a rock first forms. Radiometric dating with other substances is not always as easy, but detailed study of several isotope ratios or several different mineral grains within a single rock often allows geologists to determine a rock's original composition.

Confirming Radiometric Ages Scientists today have great confidence in ages measured through radiometric dating. The underlying processes that govern radioactive decay are well understood, and the measurements needed to determine half-lives are straightforward when done with care. Perhaps most important, in many cases scientists can use several different radioactive isotopes to measure the age of rocks and fossils from a particular stratum of sedimentary rock, and the ages found using different isotopes all agree.

The consistency of the ages found for various strata around the world also gives us great confidence that we now have an accurate picture of the geological time scale. A further check comes from the fact that, at least in some cases, we can estimate ages in ways that are independent of radiometric dating. For example, if radiometric ages are correct, then they should confirm the ordering of the relative ages in sedimentary strata—and they do. A more precise check can be made in cases in which we can identify amino acids in fossils. Recall that all amino acids in living organisms are of the "left-handed" variety [Section 3.2]. After an organism dies, the remaining amino acids gradually convert to a mixture of left- and right-handed types. Measuring a fossil's ratio of left- to right-handed amino acids therefore gives an estimate of the fossil's age. The ages determined from the amino acids agree with the radiometric ages. For the

The Mathematics of Radioactive Decay

The amount of a radioactive substance decays by half with each half-life, so we can express the decay process with a simple formula relating the current amount of a radioactive substance in a rock to the original amount:

$$\frac{\text{current amount}}{\text{original amount}} = \left(\frac{1}{2}\right)^{t/t_{\text{half}}}$$

where t is the time since the rock formed and t_{half} is the half-life of the radioactive material. We can solve this equation for the age t by taking the logarithm of both sides and rearranging the terms. The resulting general equation for the age is:

$$t = t_{half} \times \frac{\log_{10}\left(\dfrac{\text{current amount}}{\text{original amount}}\right)}{\log_{10}\left(\dfrac{1}{2}\right)}$$

Example You heat and chemically analyze a small sample of a meteorite. Potassium-40 and argon-40 are present in a ratio of approximately 0.85 unit of potassium-40 atoms to 9.15 units of gaseous argon-40 atoms. (The units are unimportant, because only the relative amounts of the parent and daughter materials matter.) How old is the meteorite?

Solution Because no argon gas could have been present in the meteorite when it formed, the 9.15 units of argon must originally have been potassium. The sample must therefore have started with 0.85 + 9.15 = 10 units of potassium-40, of which 0.85 unit remains. Thus, for calculating the age, the *current amount* of potassium-40 is 0.85 unit, and the *original amount* is 10 units. With these values and the above equation, the age of the meteorite is:

$$t = 1.25\,\text{billion yr} \times \frac{\log_{10}\left(\dfrac{0.85}{10}\right)}{\log_{10}\left(\dfrac{1}{2}\right)}$$

$$= 1.25\,\text{billion yr} \times \left(\frac{(-1.07)}{(-0.301)}\right)$$

$$= 4.45\,\text{billion yr}$$

This meteorite solidified about 4.45 billion years ago. (*Note:* This example captures the essence of radiometric dating, but real cases generally require more detailed analysis.)

age of the Earth and the solar system as a whole [Section 4.2], we can roughly confirm the 4.6-billion-year age found from radiometric dating by comparing it to an age based on detailed study of the Sun. Theoretical models of the Sun, along with observations of other stars, show that stars slowly expand and brighten as they age. The current state of the Sun agrees with what we expect for a star of its mass at an age of about $4\frac{1}{2}$ billion years.

Taken together, our theoretical understanding of radioactive decay and the various ways of confirming ages found by radiometric dating leave virtually no room for doubt about the reliability of the technique. Today, scientists routinely use radiometric dating to determine ages of rocks and fossils, and the overall uncertainty in radiometric age dates is usually less than about 1–2%.

It is important to note that while radiometric dating is straightforward in principle, it can be complex in practice. For example, if a rock has undergone partial melting or a major shock during its history, or if some atoms have been removed from the rock by water, we may find somewhat different ages when we date the rock with different parent–daughter isotope pairs. Nevertheless, decades of experience in working with these complications have given geologists a precise understanding of how they affect radiometric dating. As a result, it is nearly always possible to know whether an age is correct or uncertain.

How Far Back Does the Record Go?

The rock and fossil records allow us to study much, but not all, of the Earth's history. The oldest rocks that have been found (excluding meteorites) date to about 4.0 billion years ago. Apparently, rocks that formed prior to this time have been remelted or re-shaped so much that we cannot obtain an age through radiometric dating. However, in a few cases scientists have found tiny crystal grains that are much older (Figure 4.9). These grains, found embedded within much younger sedimentary rocks, date to as far back as 4.3–4.4 billion years ago. The grains discovered to date are crystals of zirconium silicate, or zircons, and their ages have been obtained by radiometric dating based on uranium isotopes contained within them.

Fossils of living organisms do not date back as far as the oldest rocks, though we cannot be sure that life was not present at those early times. The oldest fossils, microscopic imprints left by single cells, are found in sedimentary rocks dating to about 3.5 billion years ago. Sedimentary rocks older than 3.5 billion years have generally been reshaped so much (by chemical alteration, heat, and pressure) that we would not expect to find intact fossil cells even if life

FIGURE 4.9 This false-color image shows a tiny zircon crystal—about 0.2 millimeters across—that was found embedded in a rock formation in Western Australia. Radiometric dating, using uranium isotopes within it, shows it to be about 4.4 billion years old. The image was made with a technique called cathodoluminescence, in which an electron beam is focused on the crystal, causing it to emit visible light; the colors in the image correspond to differing intensities in the emitted light.

had been present earlier. Indeed, as we'll discuss in more detail in Chapter 5, scientists have found strong evidence for life in a rock layer at least 3.85 billion years old.

In summary, while the rock and fossil records allow us to study much of the Earth's history, they do not take us all the way back to the Earth's origin. To go all the way back to the beginning, we must examine the processes that led to our planet's formation.

4.2 The Formation of the Earth

We discussed the general processes that lead to the formation of stars and planets in Chapter 1. We saw that these processes are expected to be common throughout the universe, which is one reason why we believe planets must also be common [Section 1.3]. In this section, we turn our attention more specifically to the processes as they played out in our own solar system, leading to the formation of the Earth.

Accretion

Recall that, as summarized in Figure 1.23, our solar system formed from a cloud of gas and dust that collapsed under gravity (perhaps triggered by some event such as a nearby stellar explosion); this cloud is often called the **solar nebula.** As the solar nebula shrank in size, it flattened and began to spin more rapidly, creating what we call a **protoplanetary disk** (because the planets later formed within it). The Sun formed at the center of the protoplanetary disk. Elsewhere in this disk, tiny solid particles began to condense from the gas.

Different types of particles condensed in regions of the protoplanetary disk with different temperatures, which depended on distance from the Sun. At the Earth's distance, temperatures were high enough that only metals and rock could condense. Condensation of ices, such as water ice, methane, and ammonia, required colder temperatures that were present only at greater distances from the Sun.

The first particles to condense were microscopic solid grains. These "seeds" for planet formation gradually stuck together and grew bigger. This process, in which larger objects are built from smaller ones, is called **accretion.** Accretion probably began slowly but sped up as some objects grew large enough (about 1 kilometer across) for their gravity to attract even more particles. Within a few million years, accretion had built billions of **planetesimals**—solid chunks of metal, rock, or ice that orbited the young Sun like very small planets.

As planetesimals crashed together, some grew to much more substantial sizes. Those that reached a few hundred kilometers or so across are usually called **protoplanets** to distinguish them from the more numerous but smaller planetesimals. The late stages of accretion were probably chaotic and violent.

Hundreds of protoplanets may have orbited the Sun between the present-day orbits of Mercury and Mars. Gravitational close encounters between these protoplanets would have altered their orbits, causing them to collide and grow into a smaller number of larger objects. In the final stages of accretion, the collisions probably involved the smashing together of Moon- and Mars-size objects. The accretion process would have stopped only when the few remaining objects were in orbits that didn't come close to each other. In this way, the Earth and the three other planets of our inner solar system were produced over a period of perhaps 100 million years.

If this quick overview of accretion told the entire story, we'd be left with a major mystery. The planetesimals that formed in Earth's neighborhood of the solar system should have been made only of metal and rock. Thus, if Earth had formed exclusively from these "local" planetesimals, there would have been no water to form our oceans and no gases to form our atmosphere. The simplest way to account for the presence of water and gas is to assume that at least a few water-bearing planetesimals from farther out in the solar system crashed into the young Earth. This assumption makes sense, because even today our planet occasionally is hit by asteroids or comets that formed beyond Earth's orbit. We do not yet know whether most of the water and gas was brought to Earth by asteroids that formed far enough from the Sun to include at least some water or by icy comets that formed in the neighborhood of the jovian planets. Either way, the water we drink and the air we

breathe likely were once part of objects orbiting beyond the orbit of Mars.

Differentiation

During the late stages of accretion, a second important process shaped our planet. The interior of the Earth underwent **differentiation,** in which materials separated according to their composition. In general, higher-density material sank toward the center while lower-density material rose toward the surface. (There can be exceptions to the general rule depending on how particular atoms and ions mix in the liquids.)

THINK ABOUT IT . . . *Give at least one example in which you can see differentiation occur in your daily life.* (Hint: *It's easiest to observe with liquids.*)

For differentiation to occur, our planet had to be molten or nearly molten in at least part of its interior. The heat that caused rock to melt came from three main sources. First, the impacts of accretion created heat that melted the outer layers of the young Earth. Second, as denser materials sank through the molten outer layers, their sinking generated friction that added further heat to the interior. Third, heat is continually released by radioactive decay of elements within the Earth. This heat source is still important today, but it was even more important in early times, when there was more radioactive material to decay. Note that, once the outer layers began to melt due to the first heat source (accretion), the second and third heat sources (sinking of dense materials and radioactive decay) ensured that our planet would completely melt and then differentiate fully.

Differentiation separated the Earth's interior into three major layers. Dense metals such as iron and nickel sank to the center to form the Earth's **core.** Rocky material composed primarily of *silicates*—minerals rich in silicon and oxygen—formed a thick **mantle** surrounding the core. The lowest-density rock rose to the surface, forming a thin **crust.** The crust has since changed dramatically due to ongoing geological activity (which we'll discuss shortly), but the core and mantle may be much the same today as when they formed.

Given that our deepest drills barely prick the Earth's surface (descending only a few kilometers of the nearly 6,400-kilometer distance to the Earth's center), you might wonder how we can know anything about the Earth's interior structure. Fortunately, several lines of evidence allow us to learn about the Earth's present interior structure. For example, the fact that surface rocks are considerably less dense than the Earth's overall average density

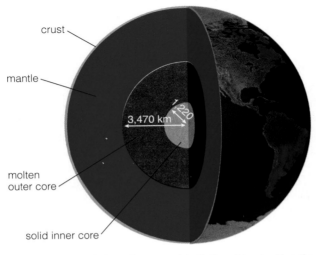

crust

mantle

3,470 km

1,220

molten outer core

solid inner core

FIGURE 4.10 The basic internal structure of the Earth, as determined from the study of seismic waves. Note that the core is divided into two regions: a solid inner core and a molten outer core.

(determined by dividing the Earth's total mass, which is found from the Earth's gravitational effects on satellites, by its total volume) tells us that much denser material must reside in the interior than on the surface. The most detailed evidence about the Earth's present internal structure comes from the study of **seismic waves**—waves that propagate both through the Earth's interior and along its surface after an earthquake. The precise speed and direction of seismic waves depend on the composition, density, pressure, temperature, and phase (solid or liquid) of the material they pass through. After an earthquake, geologists record the arrival times and the strengths of seismic waves at stations distributed all around the world. At each station, the arriving seismic waves tell us something about the average state of the material through which they have passed. By comparing data from many different stations, scientists have pieced together a fairly detailed picture of the Earth's internal structure (Figure 4.10). Note that the metallic core actually consists of two distinct regions: a solid *inner core* and a molten (liquid) *outer core*. (Because temperature rises with depth in the Earth, it may seem surprising that the inner core is solid while the outer core is molten. The explanation is that the inner core is kept solid by the higher pressure at this depth, even though the temperature is also higher.)

Formation of the Moon

The Earth is unique among the terrestrial planets in that it has a relatively large moon. The Moon's diameter is about 1/4 that of the Earth, and its mass is about 1/80 of the Earth's mass. The Moon certainly affects life in at least some ways. It is the primary cause of the tides. In addition, physiological patterns in many species appear to follow the lunar phases. So when we consider the formation of the Earth, it is natural to ask about the formation of the Moon.

Scientists were long mystified about the origin of the Moon. If the Moon had simply accreted along with the Earth, we would expect its composition to be similar to that of the Earth (because it essentially shares the Earth's orbit around the Sun and therefore would have accreted from very similar planetesimals). Instead, the Moon's composition differs from the Earth's in at least two crucial ways. First, compared to the Earth, the Moon contains a much smaller proportion of **volatile,** or easily vaporized, ingredients. In this context, volatile ingredients include not only things such as water but also elements such as lead and gold that vaporize at lower temperatures than other metals and rocks. Second, the Moon's density, which is about the same as that of the Earth's mantle, suggests that it lacks large amounts of iron like those found in Earth's core. In fact, the Moon's overall composition appears much more similar to that of the Earth's mantle and crust than to that of the Earth as a whole. Both of these differences can be explained by hypothesizing that the Moon was formed by a "giant impact" that occurred very early in the solar system's history.

The giant impact is thought to have involved a Mars-size protoplanet that slammed into the young Earth during the late stages of accretion. Computer simulations show that such a giant impact would have shattered much of the Earth, sending rock from the mantle and crust into orbit around our planet (Figure 4.11).[2] This orbiting material would have reaccreted to form the Moon, explaining why the Moon is made of material that resembles the Earth's mantle and crust. In addition, the giant impact would have vaporized volatile ingredients in the ejected rock. The vaporized volatiles would have been lost to space before the rock reaccreted to make the Moon, explaining the Moon's lack of volatiles.

We may never be certain that the Moon formed through a giant impact, because the evidence is circumstantial rather than direct. Nevertheless, our understanding of solar system formation makes it quite reasonable to suppose that the Earth suffered such a collision. Numerous large protoplanets probably roamed the early solar system, and the young planets must have been easy targets. Moreover, at least a couple of other planets (notably Mercury and

---

[2] The Moon is gradually moving farther from the Earth (a consequence of the same tidal friction [Section 8.1] that is very slowly increasing the length of the day on Earth) and was probably only about one-tenth its current distance from Earth at the time it formed. Thus, the impact would have ejected material into an orbit of this distance, not the distance of the Moon's current orbit.

FIGURE 4.11 Artist's conception of the giant impact of a Mars-size object with the Earth, as may have occurred soon after the Earth's formation. The ejected material, which came from the Earth's mantle and crust, ended up orbiting the Earth and reaccreted to form the Moon. This idea explains the Moon's overall similarity to the Earth's mantle and crust, while the heat of the impact explains the Moon's lack of volatiles.

Uranus) also have features that are best explained by giant impacts. Some questions about the giant impact hypothesis still remain; for example, why doesn't the Moon have more iron from the impacting protoplanet? However, the giant impact hypothesis is now widely accepted as our best explanation for the formation of the Moon.

The Age of the Earth

We've stated that the Earth and our solar system formed about 4.6 billion years ago. How do we know? As we've discussed, the oldest Earth rocks do not date back this far. Some Moon rocks brought back by the Apollo astronauts are older, dating to over 4.4 billion years ago but still not to the very beginning of the solar system. (The age of lunar rocks also means that the giant impact thought to have created the Moon must have occurred at least 4.4 billion years ago.) If we want to study the very oldest rocks in our solar system, we must look to meteorites that have fallen to Earth from space.

Radiometric dating tells us the amount of time that has passed since a meteorite most recently solidified. Some meteorites are younger than the solar system as a whole, because they are fragments of asteroids that formed and later shattered in collisions. However, a large number of meteorites date back to about 4.6 billion years ago, and none are older than this. We therefore infer that this age represents the time at which the meteorites and the solar system originally formed. The chemical structure of these meteorites agrees with what we expect for material that solidified in the protoplanetary disk, so we conclude that their age tells us when solid material first began to condense in our solar system. That is, the material that became the planets began to condense 4.6 billion years ago. (More precise work shows the age of the solar system to be 4.55 billion years, with an uncertainty of about 0.02 billion years.)

How old is the Earth itself? Given that the Earth formed from material containing the same ratios of lead isotopes originally present in meteorites (a fact that has been verified by detailed study of lead ores of different age found on the Earth), we can use lead isotope ratios in Earth rocks to find the Earth's age. The method is somewhat complex in its details, but it is simply based on the fact that lead isotope ratios change over time as uranium decays into lead (see Table 4.1). The results show the Earth's age to be within about 20 million years of the age of the meteorites, suggesting that the Earth accreted fairly quickly. Moreover, because much of the Earth's original lead must have sunk to the core during differentiation (while uranium tended to remain in the crust due to its atomic and ionic properties), this method can also be used to estimate when differentiation occurred. The results show that differentiation occurred within about 100 million years (0.1 billion years) after the birth of our solar system. From all the evidence, we conclude that the Earth formed and differentiated about 4.5 billion years ago.

4.3 The Hadean Earth

Even after the Earth formed and differentiated into the core–mantle–crust structure, impacts continued to pound the surface. Meanwhile, the same internal heat that had melted the interior for differentiation fueled volcanic eruptions. Together, the effects of impacts and eruptions must have dramatically transformed the surface of our planet, explaining why we find no Earth rocks dating back to the first half-billion years of Earth's history. As shown in Figure 4.6, these first half-billion years represent what we call the Hadean eon on the Earth. In this section, we'll investigate some of the crucial processes that occurred during the Hadean and paved the way for the subsequent evolution of life.

a The full moon as it appears from Earth.

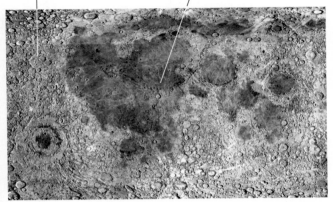

Lunar highlands are ancient and heavily cratered.

Lunar maria are huge impact basins that were flooded by lava. Only a few small craters appear on the maria.

b This flat map shows the entire surface of the Moon in the same way that a flat map of the Earth represents the entire globe. The "lunar highlands" are heavily cratered, and radiometric dating shows that these portions of the lunar surface are quite ancient—some highland rocks date back to over 4.4 billion years ago. The lunar maria are about a half billion or more years younger, and their relatively few craters tell us that the impact rate had dropped dramatically by the time they formed.

FIGURE 4.12 The Earth's Moon offers evidence about how impact rates have changed with time.

The Heavy Bombardment

The process of accretion did not end suddenly. The young planets must have shared the solar system with vast numbers of "leftover" planetesimals (some of which still remain as asteroids and comets). Over time, many of these objects must have crashed into the planets. On planets and moons with solid surfaces, such collisions leave **impact craters** as visible scars. A small telescope reveals the presence of numerous impact craters on the Moon (Figure 4.12a). The Earth must have experienced even more impacts than the Moon, both because it presents a larger target and because its stronger gravity tends to attract more objects. However, relatively few impact craters remain visible on Earth, because erosion and other geological processes (including volcanic eruptions and plate tectonics) gradually erase the scars of impacts. Still, scientists have identified more than 100 impact craters on Earth [Section 5.5].

Studies of the Moon show that the rate of impacts originally was quite high and then dropped fairly dramatically (Figure 4.12b). Regions of the Moon with rocks that date back to more than 4 billion years ago are covered by craters upon craters, proving that the Moon was subjected to a great many impacts during the first few hundred million years after its formation. In contrast, the younger **lunar maria**—so named because they look like the smooth surfaces of oceans seen from afar (*maria* means "seas")—have very few craters. The maria are huge impact craters that are smooth because they were

covered over by molten lava. The large impacts fractured the lunar crust, creating cracks through which molten lava escaped at some later time. The lava flooded the crater basins, covering the existing craters. Radiometric dating of rocks from the maria shows that the lava flows occurred between about 3.0 and 3.9 billion years ago, or about half a billion to a billion and a half years after the solar system formed. The craters we now see within the maria must be the result of impacts that occurred after the lava flows, and the relatively small number of these craters tells us that impacts have been relatively rare since the time the maria formed.

THINK ABOUT IT . . . *Crater counts can be used to estimate the ages of surfaces of solid planets and moons throughout the solar system. Consider Jupiter's moons Io and Callisto (two of its four largest moons). Callisto is heavily cratered, while Io shows no evidence of impact craters at all. What can you conclude about the ages of the surfaces of these moons? How can you explain these differences? Note that both moons are assumed to have formed at the same time in the early solar system.*

More detailed studies of the ages of lunar craters tell us more precisely when the impact rate dropped. Before about 4.0 billion years ago, the cratering rate on the Moon was so high that measuring it precisely is difficult. However, the rate dropped rapidly thereafter, and it began to level off by about 3.8 billion years ago. The cratering rate presumably followed the same basic pattern on Earth and the other planets. We therefore say that the **heavy bombardment,** that is, the period of time during which the planets were heavily bombarded by leftover planetesimals, ended some 3.8–4.0 billion years ago. On Earth, the end of the heavy bombardment coincides with the end of the Hadean eon (see Figure 4.6). Thus, we conclude that the Hadean eon was marked by frequent impacts, at a rate far higher than the Earth has experienced since.

Outgassing: The Origin of the Atmosphere and Oceans

Unlike much larger planets, the Earth was not born with its atmosphere in place. Our planet was too small and too warm to capture significant amounts of hydrogen and helium gas from the solar nebula (only the jovian planets captured such gas), and no other gases were present in large enough quantities to make a substantial atmosphere. However, as we discussed earlier, some planetesimals from farther out in the solar system brought water and other hydrogen compounds to the forming Earth. As a result, water and gases such as methane and carbon dioxide became trapped in the Earth's interior rock. When this rock melts, the water is released and the gases escape. We call this process **outgassing,** because gas trapped in the interior is released to the surface. Outgassing probably released most of the water vapor that condensed to form our oceans as well as most of the gas that formed our atmosphere.[3]

Volcanoes are a major source of outgassing, as is evident in any photo of a volcanic eruption (Figure 4.13). Some outgassing also occurs as a result of impacts—the heat of an impact can melt rock and allow gas to escape. In addition, some studies (of isotope ratios for inert gases) suggest that much of the Earth's atmospheric gas was released quite early, possibly at the time the Earth underwent differentiation. In this case, the outgassing occurred because the Earth was almost fully melted, allowing gas to percolate outward from the interior. In fact, careful studies of the old zircon grains found in some Earth rocks (see Figure 4.9) suggest that oceans were already present at the time these grains solidified, some 4.3–4.4 billion years ago. For oceans to have

FIGURE 4.13 This photo shows the eruption of Mount St.Helens (in Washington State) on May 18, 1980. Note the tremendous amount of outgassing that accompanies the eruption. (However, Mt. St. Helens' outgasses recycled rather than juvenile gas.)

been present, substantial outgassing must have occurred before the Earth was about 200 million years old.

Studies of present-day volcanoes show that the primary gases released by outgassing are water (H_2O), carbon dioxide (CO_2), nitrogen (N_2), and sulfur-bearing gases (H_2S or SO_2). Because the Earth's overall composition has not changed much since its formation, we expect that the same gases were released in early times.[4] Water vapor condensed to make rain, gradually building up the Earth's oceans. The gases that remained airborne made up the Earth's early atmosphere.

---

[3] Some water and gas may also have been supplied directly to the surface by impacts after the Earth formed. This process may have begun to create an atmosphere even before the Earth was fully formed—perhaps when it was only one-third its current radius. It is difficult to determine the relative contributions of this process and outgassing, but recent studies of comet composition suggest that outgassing was more important to the formation of the atmosphere and oceans.

[4] Some present-day volcanoes release recycled gases, while others release "juvenile" gases (gases that have not previously been outgassed). The outgassing composition is determined from those releasing juvenile gases. In addition, the crust and mantle probably began to absorb oxygen once it was produced by life, which may also have altered the outgassing composition.

The composition of the early atmosphere was very different from that of the atmosphere today. The early atmosphere was probably dominated by carbon dioxide (for reasons we will discuss shortly), whereas today's atmosphere is dominated by nitrogen (about 78% of the atmosphere) and contains only trace amounts of carbon dioxide (less than 0.1%). More important to us, the early atmosphere contained essentially no molecular oxygen, whereas the present atmosphere is about 21% oxygen. Thus, we could not have breathed the atmosphere on the early Earth. We now believe that the Earth's present oxygen atmosphere came about through the action of living organisms, as we will discuss further in Chapter 5. Life therefore must have arisen in an oxygen-free environment.

Impact Sterilization of the Hadean Earth

If the Earth had oceans as early as 4.3–4.4 billion years ago (as suggested by the evidence in the zircon grains), we might wonder whether life could have arisen in these young oceans. As we'll discuss in Chapter 5, the young Earth should have had all the ingredients needed for life, and the presence of oceans would have meant plenty of liquid water. However, if any life did arise at such an early time, it was probably extinguished by the impacts of the heavy bombardment.

The idea that early life might have been wiped out by impacts is known as **impact sterilization.** Although it would take a huge impact to make our planet uninhabitable, the heavy bombardment probably included quite a few impacts of sufficient size. Calculations based on the energy and heat release from impacts suggest that the impact of an asteroid 350–400 kilometers across would provide enough heat to completely vaporize the oceans. Besides vaporizing the oceans, such an impact would have raised the surface temperature to as high as 2,000°C (3,600°F). (The temperature rise comes both from the heat released upon impact and through the absorption of sunlight by the vaporized water.) It's highly unlikely that life could survive such effects, so such an impact would have effectively sterilized the planet. Slightly smaller impacts, by asteroids 150–190 kilometers across, would have vaporized the top few hundred meters of the oceans. These impacts probably would have been sufficient to kill off any life as well, though there's a possibility that life might have survived in the deepest parts of the early oceans.

Thus, while it is conceivable that life could have arisen on the Earth as early as 4.4 billion years ago, it is highly unlikely that any such organisms could have survived long enough to become our ancestors. Life might even have arisen and been extinguished multiple times on the early Earth, but the common ances-

tor of all life today must have arisen after the last sterilizing impact. We do not know exactly when or how many times sterilizing impacts occurred during the heavy bombardment. However, estimates of the impact rate suggest that the last sterilizing impact probably occurred between about 3.9 and 4.2 billion years ago. The timeline in Figure 4.6 shows the period during which impact sterilization was likely to have occurred.

Interestingly, there are still four asteroids in the asteroid belt that are large enough to completely vaporize the oceans if they were to hit the Earth, and another 90 or so that could vaporize the top few hundred meters of the oceans. However, these asteroids are all in stable orbits, so the Earth is safe from such catastrophic impacts. (However, some more-distant objects—such as the object Chiron that orbits the Sun approximately between the orbits of Saturn and Uranus—may be in unstable orbits, so we cannot completely rule out such a large impact in the distant future.)

Origin of the Continents

The Earth's present surface contains two fairly distinct types of crust (Figure 4.14). **Seafloor crust** is made primarily of the relatively high-density igneous rock called *basalt,* which commonly erupts from volcanoes like those along mid-ocean ridges and in Hawaii. Seafloor crust is typically only 5–10 kilometers thick, and radiometric dating shows that it is quite young—usually less than 200 million years old. **Continental crust** is made of lower-density rock, such as granite, and includes some rock that dates all the way back to the end of the Hadean, about 4.0 billion years ago. Continental crust is much thicker than seafloor crust—typically 20–70 kilometers thick —but it sticks up only slightly higher because its weight presses it down farther onto the mantle below.

FIGURE 4.14 Earth today has two distinct types of crust.

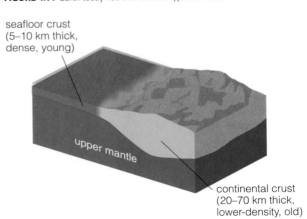

seafloor crust
(5–10 km thick,
dense, young)

upper mantle

continental crust
(20–70 km thick,
lower-density, old)

How did the Earth come to have two types of crust? The answer lies in the way crust is formed and recycled through volcanoes. Early on, volcanoes would have erupted only basalt, which is a mixture of those minerals in mantle rock that melt most easily. Thus, the Earth's early crust must have looked much like seafloor crust today. Making continental crust requires remelting the seafloor crust in a way that allows lower-density rock to separate from higher-density rock. As we'll discuss shortly, the action of *plate tectonics* recycles seafloor crust over time, explaining both the young age of modern seafloor crust and the ongoing formation of continental crust today. We do not yet know whether plate tectonics occurred during the Hadean and early Archaean eons. However, the old age of some continental crust tells us that, with or without plate tectonics, volcanoes were already recycling the Earth's crust by the end of the Hadean eon. Thus, while the late-Hadean Earth would have looked quite different from the Earth today, it already had oceans and the beginnings of the present continents.

4.4 Earth's Ongoing Active Geology

As the heavy bombardment ended, the Earth gradually settled into a stabler state. Of course, occasional impacts continue to occur, sometimes with dramatic effects on the evolution of life [Section 5.5]. Since the end of the Hadean eon, however, most of the Earth's geology has been shaped by internal processes. In this section, we discuss the internally driven geology that continues to reshape our planet today.

Internal Heat and Active Geology

The Earth is a geologically active planet, by which we mean it has features such as volcanic eruptions and earthquakes. In contrast, the Moon has no active volcanoes and no significant moonquakes, making it geologically "dead" (at least in terms of internally driven geology). Why are the Earth and the Moon so geologically different? The answer is internal heat. The Earth's internal heat drives volcanoes, and we'll soon see that it causes earthquakes and other geological processes as well. The Moon is geologically dead because it lacks sufficient internal heat to drive such geological processes.

Given that the Earth and the Moon both formed within a short time of each other, you might wonder why the Earth remains much hotter inside than the Moon. The difference in internal temperature is due mostly to size: Large worlds tend to retain their internal heat much longer than do smaller worlds. You can see why by picturing a large world as a smaller world wrapped in extra layers of rock. The extra rock acts as insulation, making it take longer for interior heat to escape into space. If you now add the fact that the larger world contains more heat in the first place, it's clear that a large world will take much longer to cool than a small world. The Moon's relatively small size probably allowed it to cool substantially within a billion years or so after its formation, while the Earth's interior still remains hot enough to keep iron molten in the outer core.

THINK ABOUT IT . . . *Give an example from everyday life of a small object cooling faster than a larger one and of a small object warming up more quickly than a larger one. How do these examples relate to the issue of geological activity on the Earth and the Moon?*

This same idea explains the geological state of all the terrestrial worlds. Mercury, the smallest of the four terrestrial planets, is probably as geologically inactive as the Moon. Venus, which is almost the same size as the Earth, should still retain plenty of internal heat and therefore should be geologically active. Mars is intermediate in size and probably has a low level of ongoing geological activity (see Chapter 7).

Mantle Convection and the Lithosphere

Today, most of the Earth's interior is solid rock. The only exceptions are the outer core (see Figure 4.10) and places closer to the surface where some rock is partially melted, providing the molten rock of volcanic eruptions. In our everyday lives, we tend to think of rock as the ultimate in strength and stability, hence the phrase "solid as a rock." However, the strength of rock depends on its temperature and the surrounding pressure, and under the conditions found in the Earth's mantle even "solid" rock can flow gradually. The idea of a solid flow may seem a bit strange, but a good analogy is the popular toy Silly Putty. The putty can feel pretty solid, especially when it is cold. You can even form it into a ball and bounce it. But if you put a pile of it on a table or inside its "egg" container, after a few weeks you'll see that it has flowed slowly outward.

At the base of the mantle, heat from the core causes the rock to expand. This expansion makes it less dense and lighter than the rocks that lie above it, creating an unstable situation in which the lighter rocks and the denser rocks will tend to change places. The flow of the rock is very slow, but over long periods of time the hotter, lower-density rock will rise. Near the top of the mantle, rock can cool as its heat flows to the surface (by conduction), causing it to contract and sink. This creates the characteristic flow pattern called **convection,** in which hot material expands and rises while cooler material contracts and falls. You are probably familiar with convection in other situations. It can occur any time a substance is strongly heated from below. For example, you can see convection in a pot of soup on a hot stove, and convection is important in weather because the warm air near the ground tends to rise while the cool air above tends to fall. The ongoing process of convection creates **convection cells** (Figure 4.15). Keep in mind that mantle convection is quite slow. Typically, any material in this flow moves only a few centimeters per year. At this rate, it would take about 100 million years for a particular piece of rock to be carried from the base to the top of the mantle.

The convection cells do not extend all the way to the surface. Near the top of the mantle, the lower temperatures make the rock more rigid, and the rock remains rigid to the surface. Thus, a layer of rigid rock essentially "floats" upon the softer, convecting rock below. We call this layer of rigid rock the **lithosphere** (*lithos* means "stone" in Greek). The lithosphere encompasses both the crust and the uppermost portion of the mantle, extending to a depth of about 100 kilometers.

The flow of the mantle fuels the movement of lithospheric rock above.[5] You can see how by studying the arrows in Figure 4.15, which indicate the flow directions resulting from convection. Where the flow is upward along the border of two adjacent convection cells, rock from the mantle rises up into the lithosphere. Where the flow is downward, rock from the lithosphere plunges down into the mantle. Along the tops of convection cells, the lithosphere may move in one direction or another along the surface. These competing motions have broken the lithosphere into a set of large **plates** that move around the surface of the Earth, somewhat like a tightly packed set of bumper cars, in very slow motion. The motion of these plates, called **plate tectonics,** is one of the defining geological features of the Earth. (The word *tectonics* comes from Greek legend, in which Tecton was a carpenter, and refers to internal forces that do "carpentry" on planetary surfaces.)

Plate Tectonics and Continental Drift

The scientific development of the idea of plate tectonics has an interesting history. In 1912, German

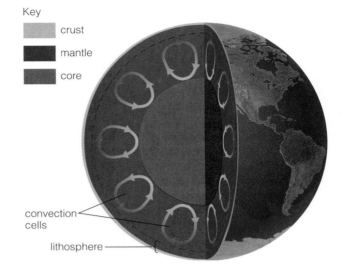

FIGURE 4.15 The lithosphere, which extends to a depth of about 100 kilometers, is a region in which rock is cooler and more rigid than the softer, warmer rock below. Note that the lithosphere encompasses all of the crust plus the uppermost portion of the mantle. Beneath the lithosphere, the mantle rock convects very gradually—typically moving at a rate of only a few centimeters per year. (The figure shows the general idea of mantle convection; however, there is some controversy as to whether the convection cells reach all the way to the base of the mantle.) The thickness of the crust and lithosphere are exaggerated, since on this scale they would appear too thin to be seen.

meteorologist and geologist Alfred Wegener proposed an idea he called **continental drift:** that continents gradually drift across the surface of the Earth. The idea came in part from the puzzlelike fit of continents such as South America and Africa (Figure 4.16). In addition, Wegener noted that similar types of distinctive rocks and rare fossils were found in eastern South America and western Africa, suggesting that these two regions had once been close together. But despite these strong hints, no one at the time knew of a mechanism that could allow the continents to push their way through the solid rock beneath them. Wegener suggested that the Earth's gravity and tidal forces from the Sun and Moon were responsible, but other scientists quickly determined that these forces were too weak to move continents around. As a result, Wegener's idea of continental drift was widely rejected by geologists for decades after he first proposed it.

In the mid-1950s, scientists began to observe geological features that suggested a mechanism for continental motion. In particular, scientists discovered *mid-ocean ridges* along which mantle material erupts onto the ocean floor, pushing apart the existing seafloor on either side. This **seafloor spreading** and other evidence led to the discovery of plates and plate tectonics. Today, we understand that continents move in concert with the underlying mantle, driven by the heat released from the Earth's interior. Because this idea is quite different from Wegener's

---

[5] Although the heat flow of mantle convection clearly drives the process of plate tectonics, it is still unclear whether the mantle simply pushes the plates apart at spreading centers or whether the leading edges of plates are pulled down at subduction zones. This is an active area of geological research.

original notion of continents plowing through the solid rock beneath them, geologists generally no longer use the term "continental drift" and instead consider continental motion within the context of plate tectonics.

Over millions of years, the movements involved in plate tectonics act like a giant conveyor belt for the Earth's lithosphere (Figure 4.17). Mid-ocean ridges occur where mantle material rises upward to the ocean bottom, creating new seafloor crust and pushing plates apart. At ocean trenches, the dense seafloor crust of one plate pushes under the less dense continental crust of another plate. This process, called **subduction,** recycles seafloor crust back into the mantle. The combination of the creation of new seafloor crust and the destruction of old seafloor crust via subduction has been occurring for at least the last couple of billion years. The rates of creation and destruction of seafloor crust are rapid enough that the seafloor must have been recycled tens of times during the history of the Earth, explaining why all of the present-day seafloor crust is young.

As the seafloor crust descends into the mantle, it is heated and may partially melt. The lowest-density material tends to melt first and then erupt from volcanoes, producing a lighter rock that becomes new continental crust. That is why volcanoes tend to be found near subduction zones and why continental crust is less dense than seafloor crust. Moreover, because continental crust does not get recycled and is continually being produced by volcanoes, the continents are slowly growing over time.

Spreading and subduction are not the only ways in which plates interact. Two continental plates crashing into each other can push each other upward, creating mountain ranges. The Himalayas are still slowly growing in this way as the plate carrying India pushes into the plate carrying most of the rest

FIGURE 4.16 The puzzlelike fit of South America and Africa shows how the continents were part of a single "supercontinent" about 200 million years ago.

FIGURE 4.17 Plate tectonics acts like a giant conveyor belt that produces and recycles seafloor crust. New seafloor crust erupts from the mantle at mid-ocean ridges, where plates spread apart. At ocean trenches, the seafloor crust pushes under the less dense continental crust, returning the seafloor crust to the mantle. The subducting seafloor crust may partially melt, leading to volcanic eruptions that produce new continental crust.

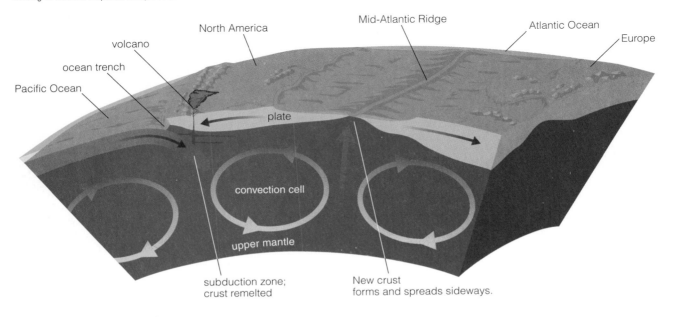

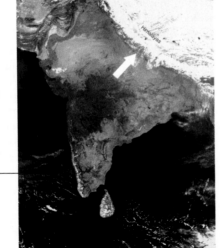

FIGURE 4.18 This satellite photo shows the Himalayas, which are still slowly growing as the Indian plate pushes into the Eurasian plate. The arrow indicates the direction of the plate motion.

of Eurasia (Figure 4.18). In other places, such as along the San Andreas Fault in California, plates are moving sideways and sliding past each other (Figure 4.19).

Overall, geologists now believe that the Earth's lithosphere is broken into about a dozen plates, as shown in Figure 4.20. On average, the plates move slowly and gradually. But the rough edges of adjacent plates often get caught on each other, resisting slippage until the tension grows too high. When the rock breaks, the plates can lurch violently, causing an earthquake. Most earthquakes therefore occur along plate boundaries.

In a few places not near plate boundaries, volcanoes occur where a plume of hot material, called a **mantle plume** (or "hot spot"), rises up from deep within the mantle. These plumes stay fixed relative to the mantle, so plates can move across them over millions of years. In some such cases, a string of volcanic islands vividly outlines the motion of the plate over the mantle plume. Hawaii offers a striking example. If you look closely at Figure 4.20, you'll see that the Hawaiian islands (including some "islands" that are not tall enough to poke above sea level) stretch across a large part of the Pacific in a long chain. Each island formed when it was over the mantle plume. Today, the plume is centered just southeast of the Big Island of Hawaii. It keeps the Big Island active and is powering an undersea volcano that will become a new island when it pokes above sea level in about a million years. Other mantle plumes occur within continental plates. Yellowstone National Park sits atop a mantle plume whose heat causes the geysers and hot springs for which the park is famous.

THINK ABOUT IT . . . *Suggest a way in which radiometric dating of rocks from the Hawaiian islands can confirm that the islands are the result of a plate moving over a mantle plume.*

Plate Tectonics Through Time

We can reconstruct the recent history of plate tectonics by taking the current motions of plates and running them backward in time (Figure 4.21). Studies of rocks and fossils confirm the results of such reconstruction. We find that about 200 million years ago all the continents were together in a single "supercontinent," sometimes called *Pangaea* (from Greek words meaning "all Earth").

Mapping the sizes and locations of continents at earlier times is more difficult, but studies of magnetized rocks (which can record the orientation of ancient magnetic fields) and comparisons of fossils found in different places around the world have allowed geologists to map the movement of the continents back to about 750 million years ago. We cannot map the motions of the plates prior to that time, but we have other evidence for the existence of plate tectonics. The oldest evidence comes from seismic studies that suggest the presence of an ancient subduction zone (found in Canada) that formed some 2.7 billion years ago. Subduction would be a sure sign of the conveyor-like action of plate

FIGURE 4.19 The San Andreas Fault marks the boundary between two plates that are moving sideways and sliding past each other, as indicated by the arrows. The map also shows a few other known faults, and the stars indicate locations of recent major earthquakes (and the years in which they occurred).

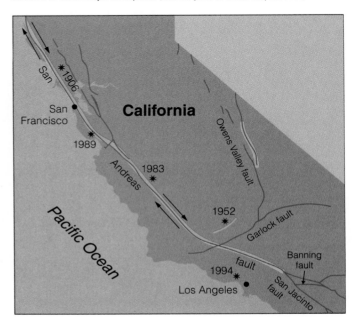

FIGURE 4.20 This relief map shows plate boundaries (solid lines), with arrows representing the directions of plate motion. Color represents elevation, progressing from blue (lowest) to red (highest).

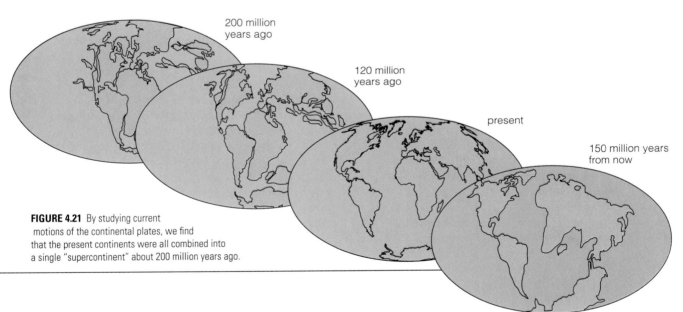

200 million years ago

120 million years ago

present

150 million years from now

FIGURE 4.21 By studying current motions of the continental plates, we find that the present continents were all combined into a single "supercontinent" about 200 million years ago.

tectonics. We do not know whether plate tectonics operated even earlier. It's possible that the lithosphere was too thin and brittle to undergo subduction before about 3 billion years ago. But the mantle must have been convecting ever since its formation during differentiation. If anything, mantle convection would have been more vigorous in early times, when there was more interior heating than there is today. (Four billion years ago, the rate of heat release from radioactive decay was about five times greater than it is today.)

Mantle convection has therefore been affecting the surface throughout the Earth's history and has been driving plate tectonics through much if not all of that history. Could plate tectonics be a crucial

The Mediterranean Sea will become mountains, Australia will merge with Antarctica, and California will slide northward to Alaska.

ingredient in making the Earth habitable? To answer this question, we must investigate the processes that regulate the Earth's climate.

4.5 Climate Regulation and Change

The key to the Earth's long-term habitability has been its relatively stable climate, which has allowed water to exist in liquid form for most of geological history. At first, a stable climate might not seem surprising. After all, the primary source of heat for the atmosphere and oceans is the Sun, and Earth's orbit of the Sun should not have changed since its formation. However, theoretical models of the Sun and observations of other Sun-like stars reveal an important fact: Stars gradually brighten with age. Models suggest that the Sun today is about 30% brighter than it was when the Earth formed, which means the young Earth received a lot less solar warmth and light than it does today. In this section, we'll explore how the Earth has nonetheless kept a well-regulated climate. We'll also investigate a few periods in which the normal regulation appears to have broken down to some extent, leading to ice ages or surprising warmth.

The Greenhouse Effect

You might be surprised to learn that if the Earth had no atmosphere, our planet would be frozen over. With a little mathematics and physics, we can calculate the expected temperature of our planet given our distance from the Sun and the amount of incoming sunlight absorbed by the Earth's surface. Such a calculation shows that the **global average temperature**—the average temperature for the entire planet —would be −17°C (1°F) in the absence of an atmosphere, well below the freezing point of water. Given that the Sun has brightened with time, an airless Earth would have been even colder in the past. But the Earth is not frozen. The actual global average temperature today is 15°C (59°F), and geological evidence shows that it has been warmer at various times in the past. How can we account for the Earth's being so much warmer than it would be without an atmosphere? The answer is that the atmosphere traps heat. The mechanism by which the atmosphere does this is called the **greenhouse effect.** You've probably heard of the greenhouse effect as an environmental problem, and we'll discuss the nature of this problem shortly. But, as we've just seen, the greenhouse effect is also critical to the existence of life on Earth. It explains why our planet is warm enough for liquid water to exist.

Figure 4.22 shows the basic mechanism of the greenhouse effect. Sunlight consists mostly of visible light, which passes easily through most atmospheric gases and reaches the Earth's surface. Some of this visible light gets absorbed by the ground. (The rest is reflected back to space by clouds or the ground.) The ground then radiates the absorbed energy back toward space—but primarily in the

form of infrared light rather than visible light. Some atmospheric gases absorb infrared light quite effectively. These infrared-absorbing gases—which include carbon dioxide (CO_2), water vapor (H_2O), and methane (CH_4)—are called **greenhouse gases.** (Note that nitrogen and oxygen do not absorb infrared light and are *not* greenhouse gases.) When a greenhouse gas molecule absorbs a photon (an individual "piece" of light) of infrared light, it begins to rotate and vibrate and then reemits a photon of infrared light. This photon is quickly absorbed by another greenhouse molecule, which does the same thing. The net result is that greenhouse gases tend to slow the escape of infrared radiation from the lower atmosphere, while their molecular motions heat the surrounding air. The greenhouse effect therefore makes the Earth's surface and lower atmosphere warmer than they otherwise would be from the absorption of sunlight alone.

THINK ABOUT IT . . . *Carbon dioxide is only a trace gas in our atmosphere today, while nitrogen and oxygen together make up some 98% of our atmosphere. Why, then, do we focus on carbon dioxide when we talk about the Earth's climate?*

The greater the concentration of greenhouse gases, the greater the degree of greenhouse warming. This leads us to a simple explanation of how the young Earth stayed warm even when the Sun was dimmer: The greenhouse gas concentration must have been much higher than it is today. In fact, theoretical studies suggest that the amount of carbon dioxide in our atmosphere may have been great enough to make the surface temperature as high as 85°C (185°F)—a temperature that only hyperthermophiles [Section 3.5]) could love. We are therefore

FIGURE 4.22 The greenhouse effect. The lower atmosphere becomes warmer than it would be if it had no greenhouse gases.

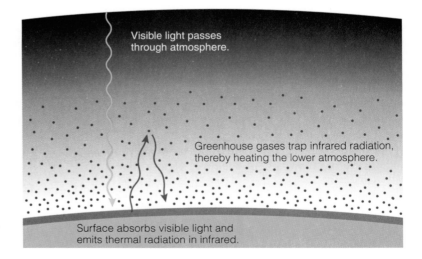

Visible light passes through atmosphere.

Greenhouse gases trap infrared radiation, thereby heating the lower atmosphere.

Surface absorbs visible light and emits thermal radiation in infrared.

Basic Characteristic of Light

In our everyday lives, we often think of light only as the illumination that makes it possible for us to see. But, in fact, most forms of light are invisible to our eyes. Let's investigate.

As discovered by James Clerk Maxwell in the 1860s, light is an **electromagnetic wave,** that is, a wave that can cause charged particles (such as electrons) to move up and down. Like other types of waves (such as water waves, sound waves, or waves on a vibrating string), light is characterized by a **wavelength** and a related **frequency** (Figure 4.23). The wavelength is the distance between adjacent peaks of the wave, and the frequency measures the number of wave peaks passing a given point each second. All light travels through space at the same speed; the speed of light, abbreviated c, is about $c = 300,000$ km/s. Thus, wavelength and frequency are inversely related. The longer the wavelength, the lower the frequency, and vice versa. This idea can be expressed with a simple formula: wavelength $\times$ frequency = speed of light.

Early in the twentieth century, scientists discovered that besides its wave properties light also has properties that make it act like lots of individual particles. These particles of light, called **photons,** can exert pressure, knock individual electrons out of atoms, or cause molecules to start rotating and vibrating. We can still characterize photons by their wavelength and frequency, but we can also add a third basic characteristic: the photon's **energy.** Figure 4.24 shows how wavelength, frequency, and energy are related for light; note that the highest frequencies and energies go with the shortest wavelengths, and vice versa.

Light can have any wavelength, frequency, or energy. The complete range of possibilities, shown in Figure 4.24, is called the **electromagnetic spectrum.** For convenience, we usually refer to different portions of the electromagnetic spectrum by different names. The **visible light** that we see with our eyes is only a tiny portion of the complete spectrum, with wavelengths from about 400 nm at the blue end of the rainbow to about 700 nm at the red end. (One nanometer [nm] is a billionth of a meter.) Light with wavelengths somewhat longer than red light is called **infrared,** because it lies beyond the red end of the rainbow. Light with very long wavelengths is called **radio** (or radio waves)—thus radio is a form of light, not a form of sound. On the other end of the spectrum, light with wavelengths somewhat shorter than blue light is called **ultraviolet,** because it lies beyond the blue (or violet) end of the rainbow; note that ultraviolet light has more energy than visible light (per individual photon), which is why it can damage skin cells, causing sunburn or skin cancer. Light with even shorter wavelengths is called **X rays;** X rays have enough energy to pass through skin and muscle but not through bones or teeth, which is why they can be used to take photographs of bone structures. The shortest-wavelength light is called **gamma rays.**

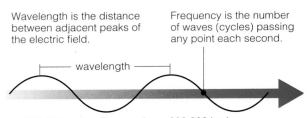

Wavelength is the distance between adjacent peaks of the electric field.

Frequency is the number of waves (cycles) passing any point each second.

wavelength

All light travels with speed $c = 300,000$ km/s.

FIGURE 4.23 Light is an electromagnetic wave characterized by a wavelength and frequency. The longer the wavelength, the lower the frequency, and vice versa.

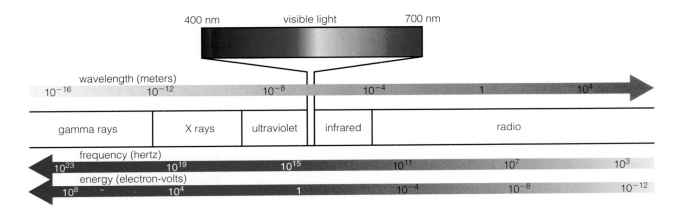

FIGURE 4.24 The electromagnetic spectrum. The unit of frequency, hertz, is equivalent to waves (or cycles) per second. For example, 10^3 hertz means that $10^3 = 1,000$ wave peaks pass by a point each second. Visible light is only a small portion of the entire electromagnetic spectrum.

led to another important question about the Earth's atmosphere and climate. If the concentration of greenhouse gases was much higher in the past, where has this gas gone?

Detailed study shows that this question goes even deeper. Our atmosphere contains only a tiny fraction of all the carbon dioxide outgassed throughout the Earth's history. About 60 times as much carbon dioxide is dissolved in the oceans as is present in the atmosphere. Far more carbon dioxide is locked up in **carbonate** rocks—sedimentary rocks made from minerals rich in carbon and oxygen—such as limestone. The total amount of carbon dioxide trapped in the oceans and in rocks is some 170,000 times the amount in our atmosphere today. If all this carbon dioxide had remained in our atmosphere, the Earth might have become so hot that the oceans would have boiled away and life could not exist. Indeed, as we'll discuss in Chapter 6, the planet Venus apparently did retain much of its carbon dioxide in its atmosphere, which is why its surface temperature is a searing 470°C (880°F). Clearly, we owe our existence to some mechanism that has removed most of the carbon dioxide from our atmosphere and locked it up in water and rock. Interestingly, this mechanism is closely tied to plate tectonics.

The Carbon Dioxide Cycle

The mechanism by which carbon dioxide has been removed from the Earth's atmosphere and by which the current small amount of atmospheric carbon dioxide remains stable is called the inorganic **carbon dioxide cycle,** or the **CO_2 cycle** for short. (It is also sometimes called the *carbonate–silicate cycle* or the *inorganic carbon cycle*.) The CO_2 cycle continually moves carbon dioxide from the atmosphere to the ocean to rock and back again, as illustrated in Figure 4.25. Because it is a cycle, there is no starting or ending point. If we view the process as starting with atmospheric carbon dioxide, it proceeds as follows:

- Atmospheric carbon dioxide dissolves in the oceans.

- At the same time, rainfall erodes silicate rocks (rocks rich in silicon and oxygen) on the Earth's continents and carries the eroded silicate minerals to the oceans.

- In the oceans, the silicate minerals react with the dissolved carbon dioxide to form carbonate minerals. These minerals fall to the ocean floor, building up layers of carbonate rock such as limestone. (During the past half-billion years or so, the carbonate minerals have been made by shell-forming sea animals, falling to the bottom in the sea shells left after the animals die. But chemical reactions would do the same thing—and apparently did for most of Earth's history—even without the presence of animals.)

- Over millions of years, the conveyor belt of plate tectonics carries the carbonates to subduction zones, and subduction carries them down into the mantle.

- As they are pushed deeper into the mantle, some of the subducted rocks melt and become magma (molten rock) for volcanic eruptions. When the volcanoes erupt, the carbon dioxide is released back into the atmosphere.

The CO_2 cycle maintains a balance between the supply of carbon dioxide released by volcanoes and the loss of carbon dioxide to the oceans and then to carbonate rocks. This balance can adjust itself slightly, which explains how the Earth maintains a fairly stable climate.

The CO_2 Cycle as a Thermostat

The CO_2 cycle is crucial to climate regulation on Earth, because the rate at which carbonate minerals form in the oceans is very sensitive to temperature: Carbonate minerals form faster at higher ocean temperatures.[6] This fact sets up feedback processes that act much like a thermostat for the planet:

- If the Earth warms up a bit, carbonate minerals form in the oceans at a more rapid rate. This, in turn, increases the rate at which the oceans dissolve CO_2 gas, thereby pulling CO_2 out of the atmosphere. The reduced CO_2 concentration leads to a weakened greenhouse effect that counteracts the initial warming and cools the planet back down.

- If the Earth cools a bit, carbonate minerals form more slowly in the oceans. This slows the rate at which the oceans dissolve CO_2 gas, allowing the CO_2 released by volcanism to build back up in the atmosphere. The increased CO_2 concentration strengthens the greenhouse effect and warms the planet back up.

Over long periods of time, these feedback mechanisms apparently have kept the Earth habitable despite changes in the Sun's brightness, changes in the rate of volcanism (which changes the output of CO_2 into the atmosphere), and other climate effects. The young Earth had liquid-water oceans despite a dim-

---

[6] More fundamentally, it is the silicate weathering rate that depends strongly on surface temperature. However, because dissolving more silicates in the oceans increases the rate of carbonate mineral formation, the latter rate also is tied directly to temperature.

FIGURE 4.25 The CO_2 cycle continually moves carbon dioxide from the atmosphere to the ocean to rock and back again.

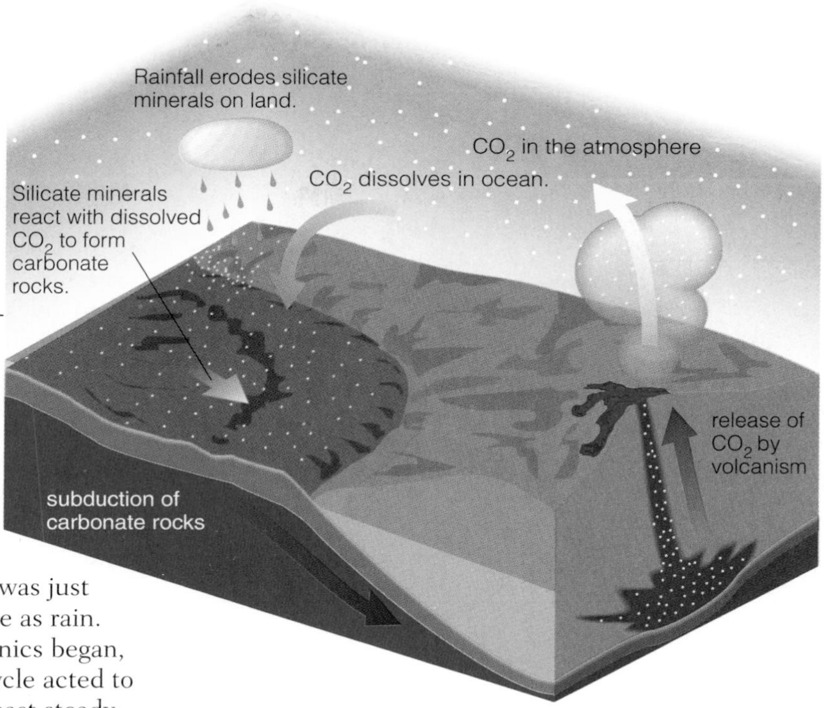

Rainfall erodes silicate minerals on land.

Silicate minerals react with dissolved CO_2 to form carbonate rocks.

CO_2 dissolves in ocean.

CO_2 in the atmosphere

release of CO_2 by volcanism

subduction of carbonate rocks

mer Sun because the greenhouse effect was just strong enough to allow water to condense as rain. Once the oceans formed and plate tectonics began, the feedback mechanisms of the CO_2 cycle acted to keep the temperature fairly steady—at least steady enough for at least some ocean water to remain liquid rather than freezing solid or vaporizing into gas.

Long-term Climate Change

While the Earth's climate has remained stable enough for liquid water to exist throughout its history, it has not been perfectly steady. There have been numerous warmer periods and numerous ice ages. Such variations are possible because the CO_2 cycle does not act instantly. When something begins to change the climate, it takes time for the feedback mechanisms of the CO_2 cycle to come into play, because these mechanisms depend on the gradual action of mineral formation in the oceans and of plate tectonics. Calculations show that the recycling time for atmospheric CO_2 through the CO_2 cycle is about 400,000 years. That is, if the amount of CO_2 in the atmosphere were to rise due to, say, increased volcanism, it would take some 400,000 years for the CO_2 cycle to restore temperatures to their current values.[7] Let's briefly investigate the Earth's climate variability.

Ice Ages Ice ages occur when the global average temperature drops by a few degrees. The slightly lower temperatures lead to increased snowfall that may cover continents with ice down to fairly low

latitudes. For example, the northern United States was entirely covered by glaciers during the peak of the most recent ice age, which ended only about 10,000 years ago. In fact, relative to temperatures over at least the past 200 million years, we are still in an ice age. This ice age has persisted for the past 35 million years or so, with periods of deeper cold interspersed with periods of warmer temperatures such as the present. Remarkably, recent evidence indicates that we can enter or leave a cold period very rapidly, within as short a time as a few decades.

The causes of ice ages are complex and not fully understood. Over periods of tens or hundreds of millions of years, the climate has at least in part been influenced by the Sun's gradual brightening and the changing distribution of continents around the globe. But during the past few million years—a period too short for solar changes or continental motion to have a significant effect—the ice ages appear to have been strongly influenced by characteristics of the Earth's rotation and orbit. For example, the Earth currently has an axis tilt of about 23.5°, but the tilt varies slightly over time (between about 22° and 25°) because of gravitational tugs from Jupiter and the other planets (Figure 4.26). Even a small change in the tilt can affect the climate. A greater tilt makes seasons more extreme, making summers warmer (and winters colder). This summer warmth tends to prevent ice from building up, which makes the whole planet warmer on average. In contrast, a smaller tilt tends to keep polar regions colder and darker, allowing ice to

---

[7] There can be (and have been) variations in the CO_2 concentration and temperature on shorter time scales, but these are affected by processes other than the inorganic CO_2 cycle alone (such as the organic carbon cycle).

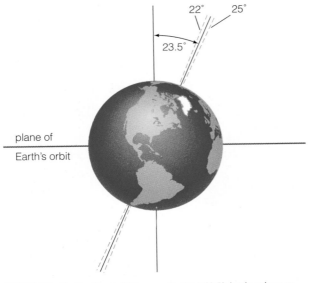

FIGURE 4.26 The Earth's axis tilt is currently about 23.5°, but it varies over time between about 22° and 25°.

build up. Changes in the tilt, along with other rotational and orbital changes, correlate well with the cycles of ice ages during recent geological history.

Snowball Earth Recently, geologists have found mounting evidence for the occurrence of several particularly long and deep ice ages between 750 and 580 million years ago. During these periods, glaciers appear to have advanced all the way to the equator. If so, the global temperature must have dropped low enough for the oceans, which cover most of the Earth's surface, to begin to freeze worldwide. This would have set up a feedback process that would have cooled the Earth even further, because ice reflects much more sunlight than does water. (About 90% of incoming sunlight is reflected by ice, compared to about 5% by water.) With less sunlight being absorbed, the surface would cool further, which in turn would allow the oceans to freeze further. Geologists suspect that in this way our planet may have entered periods now called **snowball Earth.** We do not know precisely how extreme the cold became, but some models suggest that at the peak of a snowball Earth period the global average temperature may have been as low as −50°C (−58°F) and the oceans may have been frozen to a depth of 1 kilometer or more.

How did the Earth recover from a snowball phase? Even if the Earth's surface got cold enough for the ocean surface to freeze completely, the Earth's interior still remained hot. As a result, volcanism would have continued to add CO_2 to the atmosphere. The oceans, covered by ice, would have been unable to absorb this CO_2 gas, and the CO_2 content of the atmosphere would have gradually built up and

strengthened the greenhouse effect. As long as the ocean surfaces remained frozen, the CO_2 buildup would have continued. Scientists suspect that, over a period of 10 million years or so, the CO_2 content of our atmosphere may have increased 1,000-fold. Eventually, the strengthening greenhouse effect would have warmed the Earth enough to start melting the ice on the ocean surface. Now the feedback processes that started the snowball Earth episode moved quickly in reverse. As the ocean surface melted, more sunlight was absorbed (again, because liquid water absorbs more and reflects less sunlight than does ice), which warmed the planet further. Moreover, because the CO_2 concentration was so high, the warming continued well past current temperatures—perhaps taking the global average temperature to more than 50°C (122°F). Thus, in just a few centuries, the Earth emerged from a snowball phase into a hothouse phase. Geological evidence supports the occurrence of such dramatic increases in temperature at the end of each snowball Earth episode. The Earth then slowly recovered from the hothouse phase as the CO_2 cycle removed carbon dioxide from the atmosphere.

The snowball Earth episodes must have had severe consequences for life on Earth at the time. Indeed, the end of these episodes roughly coincides with a dramatic increase in the diversity of life on Earth (the Cambrian explosion, which we'll discuss in Chapter 5). Some scientists suspect that the environmental pressures caused by the snowball Earth periods may have led to a burst of evolution. If so, we might not be here today if not for the dramatic climate changes of the snowball Earth episodes.

Human-induced Climate Change

Before we leave the topic of climate change, it's worth seeing how what we've learned applies to the important issue of human-induced climate change. You are probably aware of current concerns about **global warming,** the prospect of a warming Earth caused by human input of greenhouse gases into the atmosphere. Although some controversy remains about how serious the problem of global warming might be, as well as about what political and economic changes should be made in response to the problem, at least three facts are clear.

First, there is no doubt that human activity, especially the burning of fossil fuels, is increasing the amounts of greenhouse gases in the atmosphere. Figure 4.27 shows the rise in CO_2 concentration over about the past half-century (which is as long as these data have been collected). Data from ice cores show that the total rise since the dawn of the industrial age has been much greater still. If current trends

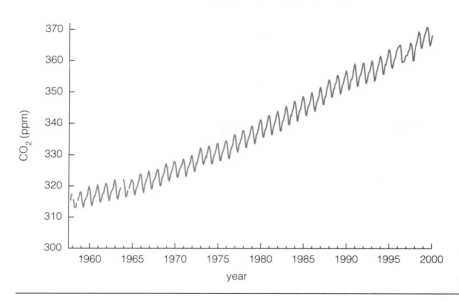

FIGURE 4.27 These data, collected for many years on Mauna Loa, Hawaii, show the increase in atmospheric carbon dioxide concentration. Yearly wiggles represent seasonal variations. The concentration is measured in parts per million (ppm). For example, the 2000 concentration of about 370 ppm means about 370 CO_2 molecules for every 1 million air molecules. (Note that 370 ppm is equivalent to $370 \div 1{,}000{,}000 = 0.00037 = 0.037\%$.)

continue, the CO_2 concentration by 2050 may be more than double what it was in the early 1800s. Other greenhouse gases (including methane and CFCs) are also being added to the atmosphere by human activity.

Second, the CO_2 cycle will not help moderate any changes due to this human activity, at least not on time scales that matter to our civilization. Remember that the recycling time for the CO_2 cycle is about 400,000 years, which is far too long to alleviate a problem that develops in just decades or centuries.

Third, there is no doubt that a continually rising concentration of greenhouse gases would eventually make our planet warm up, just as it has warmed up in the past when the greenhouse effect has strengthened. However, on time scales of decades or even centuries, some feedback mechanisms might partly counter greenhouse warming. For example, increased greenhouse gases can lead to more evaporation, which makes more clouds. Increased cloudiness, in turn, might tend to block sunlight and cool the planet, countering the warming effects of the extra greenhouse gases in our atmosphere. (This type of feedback could not operate indefinitely, because the warming would resume once the Earth became completely cloud-covered.) On the other hand, clouds can also trap heat, and how this effect balances their cooling effects depends on their altitude. As a result of scientific uncertainties about how such feedback mechanisms operate, we have corresponding uncertainties in predicting how much warming will occur in the short term and when it will

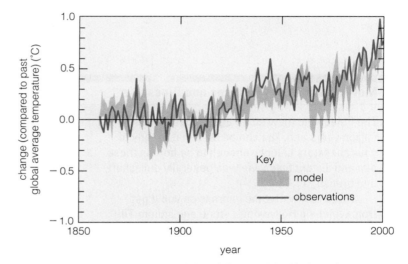

FIGURE 4.28 This graph compares the predictions of climate models with observed temperature changes (red line) since about 1860. The thickness of the green swath for the models represents the range of variation between several different models. These models take into account both natural factors (such as slight changes in the Sun's brightness and the effects of volcanic eruptions) and emission of greenhouse gases by humans. The agreement is not perfect—telling us we still have much to learn—but it is good enough to give us great confidence that greenhouse gases are, indeed, producing global warming.

occur. Nevertheless, observations indicate that temperatures are, in fact, increasing globally in a manner consistent with the rise predicted from enhanced greenhouse warming (Figure 4.28).

Not only are we unable to predict the precise amount of global warming that human activity may cause, but we are similarly unable to predict the precise effects that might arise from any global warming. Clearly, global warming would mean a higher global average surface temperature, but beyond that the process becomes very complicated.

Global weather patterns would ensure that different regions would experience different degrees of warming. Some regions would even become colder. Other regions might experience more rainfall or become deserts. And the greater overall warmth of the atmosphere would tend to mean more numerous and more intense storms, though it may be difficult or impossible to know just where such storms would strike. Secondary effects, such as those arising from ecological changes, pose an even more intractable problem. For example, it is difficult to know how forests and other ecosystems would respond to any climate changes induced by global warming, which means we cannot easily predict the impact of such changes on food production, fresh water availability, or other issues critical to the well-being of human populations.

Given the many uncertainties associated with global warming, you could spend an entire course—or an entire career—trying to understand its effects or discussing what we should do about it. Perhaps the most important lesson is that advanced civilizations, like ours, are capable of altering the way their planets function. Thus, the prosperity of advanced civilizations must surely depend on how well they manage that capability.

THINK ABOUT IT . . . *Global warming poses a great political challenge because we have no way of gauging the precise magnitude of the threat it poses. The most cautious approach would try to eliminate the threat, but that would involve major and possibly unnecessary changes to our fossil fuel–based economy. The least cautious approach would do nothing until we understand the threat better, but we might not gain such understanding until it is too late to deal with the consequences. If you were a political leader, how would you deal with the uncertain threat of global warming?*

Movie Madness: Flintstones and Dinosaurs

The Paleolithic (new stone age) began roughly two million years ago. You can be sure that for any of your ancestors unlucky enough to be born in those dim and distant times, life was generally dull, short, and brutish.

That's reality. But the impression you'd get from watching Hollywood's stone age sitcom *The Flintstones* is that our predecessors enjoyed modern lifestyles and contemporary situations. The only real difference between their existence and yours was that all the Flintstones' technology was rock-based, with domesticated dinosaurs supplying the motive power. Cars rolled on stone wheels, smiling dino's served as cranes and derricks, and air travel was facilitated by obliging pterodactyls. Fred Flintstone and his buddy Barney were smooth-skinned and beardless (try shaving with a rock razor), and regularly took showers using elephants to provide the spray, although it's a puzzle how those bathroom pachyderms avoided being preyed upon by the dinosaurs.

Needless to note, most of this doesn't make sense. *The Flintstones* is time out of joint from top to bottom, and the most serious offense is mingling our Paleolithic ancestors with dinosaurs. After all, the genus *Homo* has only been around for a few million years, and *Homo sapiens* have been strutting our stuff for only the last 100 thousand. Dinosaurs, on the other hand, disappeared about 65 million years ago.

It's all harmless entertainment, and no worse than Hollywood's tendency to portray aliens from other worlds as basically the same as contemporary humans, aside from a few gnarls and knobs on their greenish foreheads.

However, some non-Hollywood folks have seriously argued that dinosaurs and humans really were contemporaries. In the 1960s, a claim was made that fossilized human footprints could be seen alongside dinosaur prints in the Paluxy riverbed near Glen Rose, Texas. If true, this would indicate that our reading of the geological record is in serious error. But careful investigation by paleontologists has shown that the supposed "man tracks" were, in fact, just more dinosaur prints in which the small, three-pronged toe marks had been filled in by mud or eroded away.

Humans never met dinosaurs, and we could never have used them to power society. Fortunately, our predecessors eventually learned to work with metals, invent engines, and learn about the world through science. It was these developments—not chipped rocks and docile raptors—that give us a lifestyle that our stone age ancestors could never have imagined.

4.6 Geology and Life

So far in this chapter, we've focused primarily on physical processes, but these processes have deep connections to the presence of life on our planet. For example, we believe that life helped shape our atmosphere by changing its composition from its early oxygen-free state to its current oxygen-rich state. Similarly, we can infer that, by the time life arose, Earth's environment must have been compatible with the origin of life. In addition, careful study of the fossil record shows some fairly sudden and major changes that we have not yet discussed. If we hope to understand these and other connections between geology and life, we must turn our attention to how life developed and evolved. That will be the topic of our next chapter. But before we leave our discussion of geology, it's worth focusing briefly on the question we raised at the end of Section 4.4: Could plate tectonics be a crucial ingredient in making the Earth habitable?

Given the very early appearance of life on Earth, it seems unlikely that plate tectonics was necessary to the origin of life. However, plate tectonics appears to be crucial to the CO_2 cycle and hence to keeping Earth's climate stable over billions of years. Thus, while plate tectonics may not be needed for microbial life, it might be necessary if a planetary climate is to remain stable long enough to allow the development of large organisms or an advanced civilization. In that case, the prospects for finding extraterrestrial intelligence might depend in part on whether plate tectonics ought to be rare or common among planets. We'll discuss this issue further in later chapters, but the bottom line is that we do not yet know the answer. If plate tectonics can arise easily, then we might expect to find many Earth-like planets throughout the universe. On the other hand, if the onset of plate tectonics required circumstances that the Earth is "lucky" to have—such as a precise combination of size and distance from the Sun—then civilizations might be very rare.

THE BIG PICTURE

In this chapter, we surveyed the Earth's geological history and some of its implications for life. As you continue your study, keep in mind the following key ideas:

- We can read the Earth's geological history by studying the rock and fossil records. This history is not mere speculation. It is recorded in ways that we can verify independently in rocks and fossils from many places around the world and through the well verified technique of radiometric dating for determining ages. While we may never know every detail of the Earth's history, we already have a complete enough picture to understand the major processes and events that have shaped our planet.

- The Earth has been shaped by active geology driven by internal heat. This geology made life possible by causing the outgassing of the material that made our oceans and atmosphere. Moreover, through the action of plate tectonics, geology has kept our planet's climate stable enough for water to remain liquid for the past 4 billion years or more.

- Geology plays a role in making life possible, but life, once it takes hold, can also change the conditions on a planet. The oxygen in our atmosphere is a direct consequence of life. Today life affects our planet in another way: Our advanced civilization is capable of changing the way our climate functions, with consequences that we cannot fully predict.

- We do not yet know how common habitable planets may be in the universe, but geology clearly plays a crucial role in habitability. We'll return to geological considerations many more times throughout this book.

Review Questions

1. What do we mean by the *rock record* or *fossil record*? Why is it important?

2. Describe the three basic types of rock. How does recycling occur among these rock types? What is a *mineral*?

3. Briefly describe how sedimentary *strata* are made. How do they enable us to determine the relative ages of rocks and fossils buried within them?

4. Summarize the *geological time scale*. What are *eons*, *eras*, and *periods*?

5. Briefly describe how radiometric dating allows us to determine the age of a rock. Be sure to explain what we mean by a *radioactive isotope*, parent and daughter isotopes, and a *half-life*.

6. Briefly summarize how the Earth formed through *accretion*.

7. How, why, and when did *differentiation* result in the Earth's *core–mantle–crust* structure?

8. Briefly describe the giant impact hypothesis for the formation of the Moon and the evidence favoring this hypothesis.

9. How old is the Earth? How do we know?

10. What was the *heavy bombardment*? When did it occur? What do we mean by *impact sterilization,* and how might such events have occurred?

11. Briefly describe how *outgassing* led to the origin of our oceans and atmosphere. How did the Earth's early atmosphere differ from Earth's current atmosphere?

12. Briefly describe the conveyor-like action of *plate tectonics* and the evidence for this action. How does plate tectonics account for the observed differences in *seafloor crust* and *continental crust*? What evidence do we have for the operation of plate tectonics in the Earth's distant past?

13. Briefly describe the mechanism by which the *greenhouse effect* warms a planet. What are the most common *greenhouse gases*?

14. What has happened to most of the carbon dioxide outgassed through Earth's history? Describe the *carbon dioxide cycle* and how it helps regulate the Earth's climate.

15. What are *ice ages*? What do we mean by *snowball Earth* periods? Briefly describe possible causes of each.

16. Briefly describe three key facts about human-induced *global warming*.

Discussion Questions

1. *The Age of the Earth.* Some people still question whether we have a reasonable knowledge of the age of the Earth or the ability to date events in the Earth's history. Based on what you have learned about both relative and absolute ages on the geological time scale, do you think it is reasonable for scientists to be confident of ages found by radiometric dating? Is there any scientific reason to doubt the reliability of our chronology of the Earth? Explain.

2. *Plate Tectonics and Us.* Based on what you learned in this chapter, discuss whether you think the existence of plate tectonics on a planet is crucial to the development of a civilization. That is, can you imagine cases in which civilizations might arise on planets without plate tectonics? Defend your opinion.

3. *Global Warming.* Why do you think so little has been done to combat the potential threat of global warming? What more, if anything, do you think we should do at this time? Do you think it likely that we will respond to the threat of global warming before we suffer severe consequences from it? Defend your opinions.

Problems

Sense or Nonsense? In **problems 1–8,** evaluate each given statement and decide whether it makes sense. Explain your reasoning clearly.

1. We can expect that if there are paleontologists a few million years from now, they will find the fossil remains of almost every human who ever lived.

2. Nearly all the rocks I found in the lava fields of Hawaii are igneous.

3. The most common rock type in the strata of the Grand Canyon is sedimentary rock.

4. Although the Earth contains its densest material in its core, it's quite likely that terrestrial planets in other star systems would contain their lowest-density rock in their cores and their highest-density rock in their crusts.

5. If you had a time machine that dropped you off on the Earth during the Hadean eon, the first thing that would kill you would probably be getting bonked on the head by an impact.

6. If there were no plate tectonics on Earth, the continents would not move about the surface.

7. Without the greenhouse effect, there probably would be no life on Earth.

8. We don't need to worry about human-induced global warming, because the carbon dioxide cycle will restore the Earth's temperature balance before we can do any harm.

9. *Earth Without Differentiation.* Suppose the Earth had never undergone differentiation. How would the Earth be different? Write two or three paragraphs discussing likely differences. Explain your reasoning carefully.

10. *Earth Without Plate Tectonics.* Suppose plate tectonics had never begun on the Earth. How would the Earth be different? Write two or three paragraphs discussing likely differences. Explain your reasoning carefully.

11. *The Moon and Life.* Some astrobiologists have suggested that the existence of the Moon has been very important to the evolution of advanced life on Earth (see Chapter 10). What does the giant impact hypothesis for the formation of the Moon tell us about the likelihood of finding other Earth-like worlds with large moons? If the Moon was important to the development of advanced life, what would this mean for the likelihood of finding other civilizations? Explain.

12. *Geological Properties of Silly Putty.*

    a. Roll room-temperature Silly Putty into a ball and measure its diameter with a ruler. Place the ball on a table, and gently place one end of a heavy book on it. After 5 seconds, measure the height of the squashed ball.

    b. Warm the Silly Putty in hot water and repeat the experiment in part (a). Then chill the Silly Putty in ice water and repeat. How did the different temperatures affect the rate of "squashing"?

    c. Did the heating and cooling have a small or large effect on the rate of "squashing"? How does this experiment relate to planetary geology?

*13. *Dating Lunar Rocks.* You are analyzing Moon rocks that contain small amounts of uranium-238, which decays into lead with a half-life of about 4.5 billion years.

    a. In one rock from the lunar highlands, you determine that 55% of the original uranium-238 remains; the other 45% decayed into lead. How old is the rock?

    b. In a rock from the lunar maria, you find that 63% of the original uranium-238 remains; the other 37% decayed into lead. Is this rock older or younger than the highlands rock? By how much?

*14. *Carbon-14 Dating.* The half-life of carbon-14 is about 5,700 years.

    a. You find a piece of cloth painted with organic dyes. By analyzing the dye in the cloth, you find that only 77% of the carbon-14 originally in the dye remains. When was the cloth painted?

    b. A well-preserved piece of wood found at an archaeological site has 6.2% of the carbon-14 that it must have had when it was living. Estimate when the wood was cut.

    c. Suppose a fossil is 570,000 years old, which is 100 half-lives of carbon-14. What fraction of its original carbon-14 would remain? Use your answer to explain why carbon-14 generally is not useful for dating fossils of this age.

*15. *Martian Meteorite.* Some unusual meteorites thought to be chips from Mars contain small amounts of radioactive thorium-232 and its decay product lead-208. The half-life for this decay process is 14 billion years. Analysis of one such meteorite shows that 94% of the original thorium remains. How old is this meteorite?

Web Projects

1. *Local Geology.* Learn as much as you can about how geological features in or near your hometown were formed. Write a one- to three-page summary of your local geology.

2. *Global Warming.* Find recent scientific data concerning human-induced global warming. How are these new data affecting the nature of the global warming debate? Summarize your findings in a one- to two-page report.

3. *Creationist Arguments Against Radiometric Dating.* A quick Web search on "radiometric dating" (or "the age of the Earth") will reveal both scientific descriptions of the techniques and arguments by creationists seeking to claim that radiometric ages are invalid. Investigate one or more of the creationist arguments against the validity of radiometric ages. Evaluate the arguments from a scientific standpoint. Are you able to follow the arguments clearly, or do they use jargon or mathematics designed to make the technique seem too hard for the average person to understand? Contrast the creationist arguments with the scientific discussion of radiometric dating. Write a short report summarizing why few scientists accept the arguments questioning the validity of radiometric dating and whether you agree or disagree.

CHAPTER 5
The Origin and Evolution of Life on Earth

We explored the nature of life in Chapter 3, learning how evolution drives the processes that allowed simple early cells to evolve into the great diversity of life on Earth today. In Chapter 4, we saw how the Earth's physical environment has been shaped by geological processes, making possible the origin of life and keeping our planet habitable through time. But we have yet to discuss the deepest questions of all: How did life arise in the first place, and what events shaped life's evolution into its current diversity of species?

The complete answers to these questions still elude us, but scientists have put together a fairly detailed outline of what must have occurred. We know that the early Earth was a natural laboratory for organic chemistry, and we have at least some ideas about how this chemistry might have led to life. The fossil record shows us how life evolved subsequently, and it is full of important surprises. For example, plants and animals appeared only relatively recently, and we've discovered that external forces, such as the impacts of asteroids or comets, can dramatically alter the course of evolution.

We begin this chapter with a discussion of how scientists search for life's origins by studying both fossil evidence from billions of years ago and the genomes of organisms alive today. We'll then explore current ideas about the dramatic transition from pure chemistry to living organisms on the early Earth and briefly follow the story of life's subsequent evolution.

a These large mats at Shark Bay, Western Australia, are colonies of microbes known as "living stromatolites."

FIGURE 5.1 Rocks called stromatolites offer evidence of photosynthetic life as early as 3.5 billion years ago.

b The bands visible in this section of a mat are formed by layers of sediment adhering to different types of microbes.

c This section of a 3.5-billion-year-old stromatolite shows a structure nearly identical to that of a living mat. Thus, it offers strong evidence of having been made by microbes, including some photosynthetic ones, that lived 3.5 billion years ago.

5.1 Searching for Life's Origins

The fossil record details much of the history of life on Earth, but it does not tell us precisely how, when, or where life began. In this section, we'll explore how scientists are trying to learn about the origin of life.

Before we begin, it's important to note that while we may have plausible ideas about life's origins on Earth, we probably will never know what actually occurred. Rocks from the time of life's origin have been either destroyed or subsequently altered so much that we cannot fully read the information they once contained. As a result, the best hope we have of learning how chemistry becomes biology is through laboratory experiments. We have already learned much from such experiments, and we can expect to learn much more in the future. However, even if we someday create life in the laboratory, we will not know whether life arose in precisely the same manner on the early Earth.

Despite the seeming impossibility of learning the complete story of life's origins, scientific study can significantly narrow the debate. For example, the origin of life must predate the oldest fossils, and early life must have survived under the environmental conditions of the young Earth. Note that these aspects of the debate don't change if life migrated to Earth from elsewhere, a possibility we will consider in Section 5.2. In both cases, we need to understand when and where life first came to live on our planet. By combining our understanding of the Earth's geological history with what we can piece together about

early life on Earth, we can develop plausible if not precise scenarios for the origin of life.

When Did Life Begin?

Perhaps the first thing we'd like to know about the origin of life on Earth is when it occurred. We may never know exactly when the first organisms lived, but we can limit the possibilities by looking for the oldest evidence of life that we can find. Three lines of evidence point to life's arising quite early in the Earth's history.

The first line of evidence comes from rocks called **stromatolites** (from the Greek for "rock beds"), rocks characterized by a distinctive, layered structure (Figure 5.1). In size, shape, and interior structure, ancient stromatolites look virtually identical to mats formed today by colonies of microbes—sometimes called "living stromatolites" (Figure 5.1a). Living stromatolites contain layers of sediment intermixed with different types of microbes. Microbes near the top generate energy through photosynthesis, and those beneath use organic compounds left as waste products by the photosynthetic microbes. The living stromatolites grow in size as sediments are deposited over them, forcing the microbes to migrate upward in

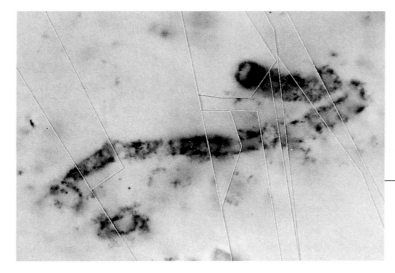

FIGURE 5.2 Microfossils of ancient living cells? This microscopic photograph shows structures that some researchers believe to be ancient fossil cells dating to 3.5 billion years ago. Other researchers, however, argue that the structures were formed by nonbiological processes.

order to remain at the depths to which they are adapted. In this way, the colonies build up a distinctive layered structure (Figure 5.1b). The structure found in a 3.5-billion-year-old stromatolite is nearly identical (Figure 5.1c). The similarity of structure suggests a similar origin, implying the existence of microbial colonies some 3.5 billion years ago. It also suggests that some of these ancient microbes produced energy by photosynthesis, which means that fairly sophisticated metabolic processes already existed. If so, then life had already evolved significantly by 3.5 billion years ago. We cannot be absolutely certain of this fact from the stromatolites alone, because geological processes of sedimentation can mimic the layering caused by life. But the evidence is strongly suggestive, especially when combined with other lines of evidence.

A second line of evidence comes from individual fossilized cells and points to the presence of life as early as 3.2–3.5 billion years ago. Finding ancient microscopic fossils is quite challenging, because rocks become increasingly rare with age, because the oldest rocks have been altered significantly be geological processes, and because microscopic fossils can be difficult to distinguish from mineral structures. Claims of ancient fossils have therefore come under close scrutiny, leading to uncertainty in the precise age of the oldest microfossils.

As recently as 2001, there appeared to be fairly strong evidence for microfossils dating to 3.5 billion years ago. Microscopic photos of sections of rock from a formation in northwestern Australia showed structures that look like individual cells (Figure 5.2), and chemical analysis showed that some of the structures contain organic carbon. However, a recent reanalysis of the rock containing these structures has called this interpretation into question. The researchers who made the original claim of microfossils had assumed that the rock had formed on the floor of a shallow sea, but it now appears the rock formed below the seafloor at the base of a hot spring. This evidence suggests that the structures may have been formed by mineral processes acting on organic molecules in the rock. It is possible that the organic molecules came from the remains of microbes living in the hot spring, but it is also possible that they are of nonbiological origin.

Despite the uncertainty surrounding these particular structures, it is quite likely that we have evidence for fossils nearly as old. Rocks from at least two other locations, dating to between 3.2–3.5 billion years old, also show what appear to be fossilized cells. Many researchers expect the biological origin of these samples to withstand further scrutiny. While ongoing debate and uncertainty about the nature of these different rocks and of the structures contained within them persist, it is through such discussions that scientists are able to figure out how to properly interpret the rock record and to understand the earliest history of life on Earth.

We are unlikely to find fossil cells older than 3.5 billion years, because the few sedimentary rocks that date back earlier than this have been reshaped too much by geological processes to show any intact fossil remains. However, a third line of evidence may push the origin of life to more than 3.85 billion years ago—or some 350 million years before the oldest stromatolites and fossilized cells. This evidence comes from the analysis of ratios of carbon isotopes in ancient metamorphic rocks (Figure 5.3).

Carbon has two stable isotopes: carbon-12, with six protons and six neutrons in its nucleus, and carbon-13, which has one extra neutron. Carbon-12 is far more common, but any inorganic carbon sample always contains a small proportion of carbon-13 atoms mixed in with the more numerous carbon-12 atoms. (Typically, carbon-13 accounts for about 1 out of every 89 carbon atoms.) When living organisms metabolize carbon, they incorporate carbon-12 atoms slightly more easily than they do carbon-13 atoms. As a result, living organisms—and fossils of living organisms—always show a slightly lower fraction of carbon-13 atoms than that found in inorganic material. Some rocks older than 3.85 billion years show the same lower fraction of carbon-13 as living organisms and fossils. Although the lower carbon-13 ratio seems reasonably likely to indicate that the rocks contain remnants of life, it may also have arisen from nonbiological processes. (We say that the rocks are *older than* 3.85 billion years because the rocks themselves have not been dated. Instead, rocks in layers just above them date to 3.85 billion years old, so the rocks below must be older.)

If life existed earlier than 3.85 billion years ago, we may never find any recognizable traces. Recall that the oldest identified Earth rocks are only slightly older than 3.85 billion years (about 4.0 billion years [Section 4.1]), and these old rocks are both very rare and heavily altered by heat and pressure from tectonic activity. Thus, even if early organisms left an imprint in such ancient rocks, the evidence may now be gone. Nevertheless, even an age of 3.85 billion years has important implications for understanding the origin of life.

As we have seen, the young Earth probably experienced numerous large impacts during the heavy bombardment, which ended between about 3.8 and 4.0 billion years ago [Section 4.3]. If any life had arisen before the heavy bombardment ended, these impacts probably would have extinguished it and sterilized our planet. We cannot know exactly when the last sterilizing impact occurred. It might have been as early as 4.1–4.2 billion years ago or as late as about 3.9 billion years ago.

We therefore have both early and late constraints on the origin of life—or at least of the organisms from which all present life evolved. On the early side, these organisms probably could not have arisen more than about 4.1–4.2 billion years ago, since earlier life would have been extinguished by a sterilizing impact. On the late side, the carbon isotope evidence in ancient sedimentary rocks suggests that life may already have existed by 3.85 billion years

FIGURE 5.3 This ancient rock formation on the island of Akilia (off the coast of southern West Greenland) may hold the oldest known evidence for life on Earth. The evidence itself comes from analysis of a metamorphic rock cut from within it, which has a carbon isotope ratio suggesting that it contains remnants of life. Because the rock is metamorphic, its origin cannot be dated precisely. However, it was found beneath what appears to be sedimentary layers dating to 3.85 billion years ago, so it is at least that old.

ago, and the fossil evidence suggests that life was widespread by 3.5 billion years ago. (Both sets of constraints are illustrated on the timeline in Figure 4.6.) Thus, it seems likely that life arose and colonized our planet in a period of no more than about 500 million years, and possibly as little as 100 million years or less. Although this is quite a long period of time by human standards, it is very short compared to the age of the Earth.

THINK ABOUT IT . . . *What does the seemingly rapid appearance of life on Earth suggest about the difficulty of life's arising on a planet with suitable conditions? Based on your answer, does life elsewhere seem more or less likely than it might have before we had such tight constraints on the time of life's origin?*

DNA Molecules as Living Fossils

The fossil record tells us a great deal about how life evolved, but it is not the only way to study evolution. Another, completely independent way of learning about early life involves comparing DNA sequences of living species in order to determine which ones evolved earlier. Let's examine how this technique allows us to use DNA molecules as "living fossils."

Consider the DNA of an organism that long ago became the common ancestor of all life today. (Recall from Chapter 3 that a great deal of biochemical evidence points to a common ancestor for all life today, including the fact that all organisms use DNA as their hereditary material.) Mutations created variations on this DNA, and each new species therefore had slightly different DNA sequences than did the older species from which it evolved.[1] Over millions and billions of years, continuing evolution led to new species with DNA molecules increasingly different from the DNA of the common ancestor. But, always, the new molecules were built by changes to the older ones so that, in principle, the changes are traceable and can be used to explore the evolutionary history of life on Earth.

Determining the sequence of bases in an organism's DNA is a difficult and time-consuming task, and to date only a few dozen organisms have had their complete genome sequences determined. In many more cases, biologists have compared smaller pieces of the DNA of many species.[2] By comparing the DNA sequences in similar genes among many different species, biologists can map the evolutionary history of the genes. For example, two species with very similar DNA sequences probably diverged relatively recently in evolutionary history, while two species with very different DNA sequences probably diverged much longer ago.

---

[1] Genomes can change through other mechanisms besides mutation. For example, bacteria can incorporate entire genes from other organisms in a process called *lateral gene transfer*. Today, this process is the primary way bacteria gain resistance to antibiotics, and it is also used in genetic engineering. It may have been common among early living organisms.

[2] A particularly common technique relies on determining the sequence of bases in molecules of ribosomal RNA (rRNA), which tells us the sequence of the DNA that coded for it; rRNA is thought to evolve more slowly than most genes, so it provides a good proxy for overall evolutionary changes.

Such comparisons of DNA sequences are what led biologists to the conclusion that life can be divided into three major domains: Bacteria, Archaea, and Eukarya (see Figure 3.12). Through more-detailed analysis, biologists have mapped out relationships among many types of organisms in what is sometimes called the *tree of life* (Figure 5.4). Each of the three domains represents one major branch of the tree, with each domain branching further in many other ways.

The tree of life does not tell us *when* different organisms evolved; rather, it shows how closely or distantly related they are. For example, the tree shows that, genetically, animals and plants are closely related to each other. In contrast, all of the Eukarya are genetically quite different from all of the Bacteria. Note that we do not expect any living organism to be much like the common ancestor of all life on Earth. All modern species have presumably evolved significantly over time, so even the "simplest" bacterium must be quite complex compared to the common ancestor. Nevertheless, organisms on branches that emerge closer to the "root" of the tree must represent the descendants of organisms that branched off from other parts of the tree at earlier times in evolutionary history.

Unfortunately, while the three domains are well established by their significant genetic differences, the precise ordering of the branches within each domain is still quite uncertain, primarily because of the relatively small number of organisms that have been studied to date. Indeed, given that mutation rates were probably much higher in early times than

---

FIGURE 5.4 The tree of life, showing evolutionary relationships determined by comparison of DNA sequences in different types of organisms.

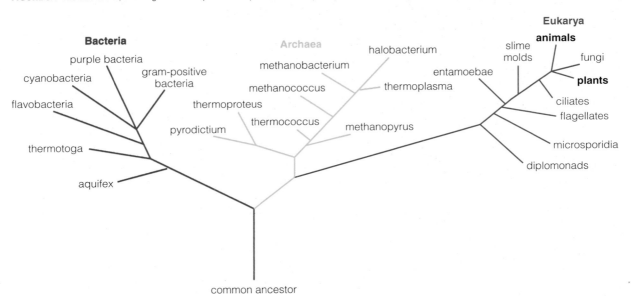

they are today (because the cellular machinery for correcting errors was less well developed), it now seems likely that all three domains arose quite early —perhaps more than 3.5 billion years ago—and that nearly all the known branches split off soon after that.

Despite the uncertainty in the branching order, it appears that many organisms on branches closest to the root are extremophiles such as those living near deep-sea volcanic vents (black smokers) or in hot springs like those in Yellowstone (see Figures 3.17 and 3.18). These organisms are adapted to life in hot water (making them *hyperthermophiles*) and use chemical energy rather than photosynthesis to fuel their metabolism. Thus, it appears that the common ancestor of all modern life on Earth resembled these extremophiles more closely than it did any other organisms living today.

Where Did Life Begin?

We can combine what we've learned about the probable nature of early organisms with geological considerations to come up with ideas about *where* life first arose on Earth. Let's begin by considering locales that we can probably rule out.

It seems unlikely that life arose on the land surface. Recall that the early atmosphere contained no molecular oxygen, which means our planet could not have had a protective layer of ozone. Ozone is a form of oxygen produced in the upper atmosphere by interactions between oxygen and ultraviolet light from the Sun. (It is O_3, whereas the oxygen we breathe is O_2.) Today, ozone shields the Earth's surface from the Sun's dangerous ultraviolet radiation. Before the ozone layer existed, any surface life would have been exposed to very high levels of this radiation. While we can't rule out the possibility that life might have arisen in such an environment—indeed, some organisms today can survive high-radiation conditions[3]— the environment would have been much more hospitable underwater or in rocks beneath the surface.

A few decades ago, many scientists favored shallow ponds as the first home for life, largely because a famous experiment of the time (the Miller–Urey experiment, which we will discuss shortly) suggested that these ponds might have been full of organic compounds. As the ponds evaporated, the compounds would have become more concentrated, making it easier for them to combine into more com-

Common Misconceptions
Ozone—Good or Bad?

Ozone often generates confusion, because in human terms it is sometimes good and sometimes bad. In the stratosphere, ozone acts as a protective shield from the Sun's ultraviolet radiation. However, ozone is poisonous to most living creatures and therefore is a bad thing when it is found near the Earth's surface. In fact, ozone is one of the main ingredients in urban air pollution, produced as a by-product of automobiles and industry. Some people wonder whether we might be able to transport this ozone to the stratosphere and thereby alleviate the effects of ozone depletion. Unfortunately, this plan won't work. Even if we could find a way to transport this ozone, all the ozone ever produced in urban pollution would barely make a dent in the amount lost from the stratosphere.

plex molecules that could lead to life. However, more recent experiments have not supported this scenario and have suggested that shallow ponds would have lacked a source of chemical energy sufficient to support the origin of life.

Given the DNA evidence suggesting that early organisms survived in conditions like those near deep-sea vents and in hot springs, it seems reasonable to favor these locations as the first home of life. The simplest scenario envisions life originating around deep-sea vents or surface hot springs, where plenty of chemical energy was available to fuel chemical reactions that might have led to life (see Appendix D). Of course, we cannot rule out the possibility that life originated elsewhere, such as within rocks, and adapted rapidly to these hot-water environments. For example, the last major impact of the heavy bombardment might have raised global temperatures so only organisms that had already adapted to hot-water environments could survive.

We have at least some reason to favor deep-sea vents over surface hot springs as the earlier home for life. Smaller impacts were more common than larger ones during the heavy bombardment, so partial vaporization of the oceans should have been more common than complete vaporization. No place on Earth was completely safe, but the deep ocean was clearly safer than springs at or near the surface. We may never know for certain where life first arose on Earth, but deep-sea vents look like a good bet.

5.2 How Did Life Begin?

Even the simplest living organisms today—and the similar organisms inferred to have lived more than 3.5 billion years ago—seem remarkably advanced. Metabolic processes involve many intricate molecules and enzymes working together. The complex chemistry of DNA and RNA is deeply intertwined with the proteins and enzymes that help in making them [Section 3.4]. Indeed, every cellular component and process depends on many other components and processes, making it difficult to imagine how one could have developed before another. Nevertheless, life is here, so it must have arisen somehow. In this section, we'll explore what science can tell us about how simple chemical processes on the early Earth might eventually have led to living organisms.

Before we begin, it's worth noting a couple of important caveats. First, we are assuming that life began under the chemical conditions present on the early Earth. If life migrated to Earth, a possibility we will discuss shortly, it's conceivable that it originally arose in a somewhat different chemical environment. Second, we have not said anything about the possibility that life arose through any kind of divine intervention, because such a possibility falls outside the realm of science. Scientifically, we can ask only whether there are natural, chemical processes that could have led to life. As we'll see, a great deal of evidence suggests that there are.

Organic Chemistry on the Early Earth

Life today is based on the chemistry of a wide variety of organic molecules. Thus, it's logical to assume that the first life was somehow assembled from organic molecules produced by chemical reactions—without biology—on the early Earth. Scientific experiments support the idea that the early Earth may have been much like a giant laboratory for organic chemistry.

Today, the Earth's oxygen-rich atmosphere prevents complex organic molecules from forming readily outside living cells. Oxygen is such a highly reactive gas that it tends to attack chemical bonds, removing electrons and destroying organic molecules. But as early as the 1920s, some scientists recognized that the oxygen in Earth's atmosphere was a product of life and inferred that the Earth's early atmosphere must have been largely oxygen-free. They further hypothesized that the chemicals in this early atmosphere, fueled by energy from sunlight, would spontaneously create organic molecules. (This idea was proposed independently by Russian biochemist A. I. Oparin and British biologist J. B. S. Haldane.) This hypothesis was put to the test in the 1950s, in a famous experiment credited to Stanley

FIGURE 5.5 A few decades after his famous experiments, Stanley Miller poses with a re-creation of the original Miller–Urey experiment. The flask to the left contains liquid water to represent the sea; the gases, representing the atmosphere, consist of water vapor, methane, ammonia, and hydrogen. The central flask was supplied with energy in the form of electrical discharges, such as the one shown here, which simulated lightning. Below this flask, the gas was cooled so it could condense and flow back into the flask with the liquid water. Chemical reactions in the experiment produced a wide variety of organic molecules. However, more recent discoveries have shown that the atmosphere used in the experiment was somewhat different from the Earth's actual early atmosphere.

Miller and Harold Urey and now known as the **Miller–Urey experiment** (Figure 5.5).

The original Miller–Urey experiment used small glass flasks to simulate chemical conditions that they thought represented those on the early Earth. One flask was partially filled with water to represent the sea and heated to produce water vapor. Gaseous methane and ammonia were added and mixed with the water vapor to represent the atmosphere. These gases flowed into a second flask, where electric sparks simulated lightning and provided energy for chemical reactions. Below this flask the gas was cooled so it could condense to represent rain and then was cycled back into the water flask. The water soon began to turn a murky brown, and a chemical analysis (performed after letting the experiment run for a week) showed that it contained many amino acids and other organic molecules. Thus, the experiment appeared to support the idea that organic compounds could form spontaneously under the conditions thought at the time to have been present on the early Earth.

Later discoveries have called into question the relevance of the original Miller–Urey experiment. In particular, we now believe that methane and ammonia were not very abundant in the early atmosphere and that carbon dioxide instead was the most common gas [Section 4.3]. Adding carbon dioxide to the Miller–Urey experiment greatly reduces the amount of organic material produced.

Many scientists have since conducted variations on the Miller–Urey experiment with different mixes of gases thought to be more representative of the Earth's true early atmosphere. The energy source has also been varied, with some experiments using ultraviolet light (to simulate the effects of sunlight) rather than electrical discharges (which simulate lightning). Although these experiments yield a much smaller total proportion of organic material than did the original Miller–Urey experiment, they still produce a great variety of organic molecules. In fact, various experiments have now produced all the amino acids found in most living organisms, several complex sugars and lipids, and all five of the chemical bases used in DNA and RNA. Nevertheless, given the low overall yield of organic molecules, it's likely that additional sources of organic molecules were necessary for life to begin.

We know of at least two other potential sources of organic molecules. First, strong evidence suggests that many organic molecules were brought to Earth by impacts. Analysis of meteorites shows that they often contain organic molecules, including complex molecules like amino acids. Similarly, study of comet nuclei in space suggests that they too contain organic molecules. Apparently, organic molecules can form under the conditions present in interplanetary space and can survive the plunge to Earth. Given the large number of impacts that must have occurred during the heavy bombardment, substantial amounts of organic material may have been brought to Earth's surface by asteroids, comets, and meteors. The heat and pressure generated by the impacts may also have facilitated the production of organic molecules in the Earth's atmosphere and oceans.

A second additional source of organic molecules may have been chemical reactions near deep-sea vents. As these undersea volcanoes heat the surrounding water, a variety of chemical reactions can occur between the water and the minerals. These chemical reactions would have occurred spontaneously in the conditions thought to have prevailed on the early Earth, and they would have resulted in the production of the same types of organic molecules thought to have been necessary for the origin of life.

It's likely that all three sources of organic molecules—chemical reactions in the atmosphere, impacts of asteroids and comets, and chemical reactions near deep-sea vents—played a role in shaping the chemistry of the early Earth. More important, given three different ways of obtaining organic molecules, it seems likely that at least parts of the early Earth were like a natural "organic soup" of molecules needed for life. The likelihood of an organic soup near deep-sea vents is particularly encouraging, given the evidence suggesting that life may have originated in such locales. Overall, laboratory studies point strongly to an early Earth containing all the building blocks needed to make life. The next question, then, is how these building blocks might have assembled themselves to make a living cell.

The Transition from Chemistry to Biology

Variations on the Miller–Urey experiment have produced the essential building blocks of life, but, to paraphrase the late Carl Sagan, these building blocks represent only the notes of the music of life, not the music itself. Viewed in terms of simple probability, the likelihood of a set of simple building blocks ramming themselves together to form a complete living organism is at least as small as that of letting monkeys loose in a roomful of musical instruments and hearing Beethoven's Ninth Symphony. It simply wouldn't happen, even if the experiment was repeated over and over again for millions of years. There must have been at least a few intermediate steps—each involving a chemical pathway with a relatively high probability of occurring—that eased the transition from chemistry to biology.

The Search for a Self-replicating Molecule One way to explore the transition is to work backward from organisms living today. Perhaps the single most important feature of any living organism is its ability to reproduce, so we might begin by looking for individual molecules that would be capable of replication. Double-stranded DNA is far too complex—and its replication far too intertwined with RNA and proteins—to be a likely candidate for the original self-replicating molecule. So we are looking for a molecule that is simpler than DNA but still capable of making fairly accurate copies of itself.

The most obvious candidate is RNA. RNA is much simpler than DNA because it has only one strand rather than two and its backbone structure requires fewer steps in its manufacture. But it still possesses hereditary information in the ordering of its bases, and in principle it can serve as a template for making copies of itself. However, for a while there seemed to be a problem with this idea. In modern organisms, neither DNA nor RNA can replicate

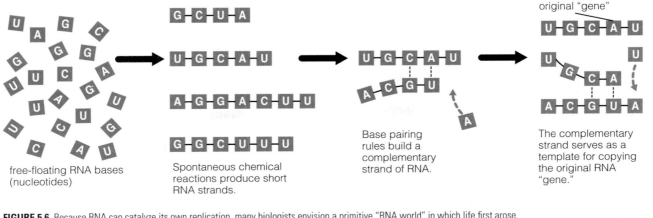

free-floating RNA bases (nucleotides)

Spontaneous chemical reactions produce short RNA strands.

Base pairing rules build a complementary strand of RNA.

The complementary strand serves as a template for copying the original RNA "gene."

FIGURE 5.6 Because RNA can catalyze its own replication, many biologists envision a primitive "RNA world" in which life first arose. Short strands of RNA may have self-assembled from components available in the environment, perhaps with the aid of clay or other minerals. Some strands may have been capable of catalyzing their own replication. Self-replicating RNA molecules may then have competed with one another and evolved through a sort of molecular natural selection. A, G, C, and U represent the four RNA bases.

itself. Both require the help of enzymes. These enzymes are proteins, which are made from genetic instructions contained in DNA and carried out with the help of RNA. This fact seemed to present a "chicken and egg" dilemma: RNA cannot replicate without enzymes, and the enzymes cannot be made without RNA.

A way around this dilemma was discovered in the early 1980s by Thomas Cech and his colleagues at the University of Colorado, Boulder. They found that RNA can catalyze biochemical reactions in much the same way as enzymes (work for which Cech shared the Nobel Prize in 1989). We now know that RNA molecules play this type of catalytic role in many cellular functions, and we call such RNA catalysts *ribozymes* (by analogy to enzymes). A plausible extension of Cech's work is that RNA or RNA-like molecules might catalyze their own replication. In laboratory experiments to date, RNA-catalyzed reactions have been able to partially (but not completely) replicate another RNA molecule. Encouraged by this result, many biologists envision that early life arose from an "RNA world" in which RNA molecules served both as genes and as chemical catalysts for copying and expressing those genes.

In this scenario, short strands of RNA-like molecules were produced spontaneously on the early Earth (by mechanisms we'll discuss shortly). Some of the molecules partially or completely catalyzed their own duplication (Figure 5.6), but they varied in their success at this replication. The result was a molecular analog to natural selection: The RNA molecules that replicated faster and more accurately soon came to dominate the population. Copying errors introduced mutations, ensuring the production of many variations of successful molecules. This allowed the molecular evolution to continue, as those RNA molecules best suited to the particular environmental conditions were more likely to continue replicating.

Eventually, the RNA world gave way to the present DNA world. DNA is a more flexible hereditary material and is less prone to copying errors. The similarities between DNA and RNA suggest that DNA molecules could plausibly have evolved from RNA molecules. Moreover, the fact that RNA today is involved in making proteins, which are the workhorses of the modern cell, suggests that RNA existed prior to DNA and retained some of its roles after DNA took over as the hereditary molecule.

Assembling Complex Organic Molecules By itself, the organic soup on the early Earth probably was too dilute to favor the creation of complex molecules like RNA from smaller building blocks. So if there really was an RNA world, how did the RNA molecules come to exist? Laboratory experiments suggest a possible answer. When we place hot sand, clay, or rock in a dilute organic solution, complex molecules self-assemble. Clay may have been especially important to the origin of life.

Clay consists essentially of silicates that have reacted with water, so it should have been common on the early Earth. (In this context, *clay* refers to a particular physical structure of the minerals, not to a particle size commonly associated with moist dirt.) Minerals in clay contain layers of molecules to which other molecules, including organic molecules, can adhere. Molecules from the dilute organic soup may have stuck to the clay, with the mineral surface structure forcing them into such close proximity that they reacted with one another to form longer chains. Indeed, strands of RNA up to nearly 100 bases in length have been produced in the laboratory in just this way. The repetitive nature of the structure of clay may even have facilitated this production by providing a repeating pattern to act as a template.

Other inorganic minerals may have played a similar role. One possibility is the mineral iron pyrite

(FeS$_2$), or "fool's gold." The surface of pyrite contains positive charges to which organic molecules can bind. Moreover, the formation of pyrite releases energy that might help fuel the creation of complex organic molecules.

We have no way to be sure that inorganic minerals really did facilitate the production of RNA and other molecules on the early Earth, and unless laboratory experiments someday show that such minerals can take simple organic molecules all the way to something resembling a true living organism, we won't know if such a scenario is really possible. Nevertheless, we have found plausible ways in which a dilute organic soup may have undergone reactions that built more complex molecules, including self-replicating RNA. The idea of an RNA world no longer seems far-fetched and perhaps was even likely under the conditions present on the young Earth.

Early Cell-like Structures The basic unit of living organisms today is the cell, but so far we have talked only about freely floating organic molecules. If these molecules could instead have been confined in some kind of cell-like structure—such as within a simple, membrane-enclosed compartment that we'll call a *pre-cell*—all the subsequent steps to life would have been much easier.

The advantages of confining organic molecules within a pre-cell are twofold. First, keeping the molecules close together should have increased the rate of reactions among them, making it more likely that cooperative relationships like those in modern cells would evolve. Second, an enclosure would have essentially isolated its contents from the outside world in a way that should have facilitated natural selection among RNA molecules. For example, suppose a particular self-replicating RNA molecule assembled amino acids into a primitive enzyme that sped up replication. If the enzyme floated freely within the organic soup, it might just as easily have helped the replication of other RNA molecules as of the one that made it. But inside a pre-cell, the enzyme would help only the RNA that made it, giving this RNA an advantage over less capable RNA molecules in other pre-cells (Figure 5.7).

Where might such pre-cells have come from? Once again, laboratory experiments suggest that they may have appeared readily on the early Earth. We know of two types of simple membranes that can spontaneously form into pre-cells. First, if we cool a warm-water solution of amino acids, they can form bonds among themselves to make an enclosed, spherical structure (Figure 5.8a). Although they are not alive, these structures exhibit many lifelike properties. For example, they can grow in size by absorbing more short chains of amino acids until they reach an unstable size at which they split to form "daugh-

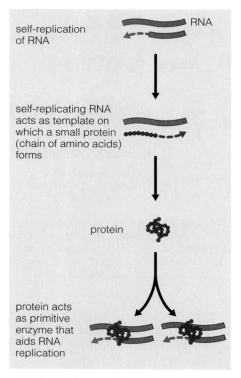

a This diagram shows a self-replicating RNA molecule that also serves as a template for making a primitive enzyme that helps its own replication.

b If the RNA and the enzyme are isolated from the outside environment inside a pre-cell, then only the molecules in this particular pre-cell will benefit from the new enzyme.

FIGURE 5.7 A possible origin of molecular cooperation.

ter" spheres. They can also selectively allow some types of molecules to cross into or out of the enclosure. Some of these spheres even store energy in the form of an electrical voltage across their surfaces, which can be discharged in a way that facilitates reactions inside them. The second type of membrane forms spontaneously when we mix lipids with water (Figure 5.8b). (Recall that lipids are one of the four major types of cellular components [Section 3.2].) Either or both types of membranes might have served as the enclosures of pre-cells, confining RNA and other organic molecules within them.

Was a membrane necessary in order to allow replication of RNA molecules, or were replicating RNA molecules a prerequisite to enclosure in a cell? It's not yet clear which came first. Life might have originated with free-floating RNA molecules that subsequently took advantage of cell membranes to enhance their ability to survive and multiply. Or cell-

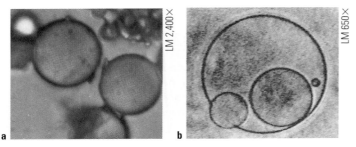

FIGURE 5.8 Laboratory versions of pre-cells. (**a**) These microscopic spheres were made by cooling a warm-water solution of amino acids. Although they are not alive, they exhibit many lifelike properties. (**b**) These microscopic membranes are made from lipids that when mixed with water spontaneously form enclosed droplets.

like membranes might have originated spontaneously and led to a sufficient concentration of molecules within their interiors to allow replication of RNA-like molecules. Either way, it seems likely that the transition from nonlife to life was gradual. There might never have been a particular moment when we would clearly say that the "first living cell" had appeared on the scene.

Handedness We've neglected one important issue about the chemistry of life. Recall that many organic molecules come in both left-handed and right-handed versions [Section 3.2]. In general, living organisms preferentially use only one of the two versions of each type of molecule. For example, living organisms use only the left-handed versions of amino acids to make proteins. Nonbiological reactions produce the left- and right-handed versions in roughly equal numbers, so both versions should have been mixed in the organic soup on the early Earth. How, then, did life end up with preferences for a particular version of each molecule?

No one knows the answer to this question, but several hypotheses have been proposed. Perhaps one version of each type of molecule (left-handed or right-handed) has some as-yet-unknown natural advantage over the other version. Or perhaps the molecules eventually became segregated in pre-cells. Some pre-cells might have contained more of one

version or the other. It's conceivable that there might have been both left- and right-handed RNA molecules in different pre-cells, and perhaps even left- and right-handed versions of early organisms. In the latter case, one version may have ultimately died out, leading to the handedness we observe in living organisms today.

Putting It All Together We began this section by noting that something as complex as a living cell could not simply have self-assembled from basic building blocks alone. However, we've found a plausible set of scenarios that, if correct, would make it much easier for life to form. In essence, we've supposed that life on the Earth originated in a series of steps, each of which seems reasonably likely. Thus, the complex structures of life were built up gradually rather than originating all at once. Let's review the sequence we have envisioned (Figure 5.9):

1. Through some combination of atmospheric chemistry, chemistry near deep-sea vents, and impacts of asteroids and comets, the early Earth developed at least localized areas in which amino acids, building blocks of nucleic acids, and other organic molecules were dissolved in a dilute "organic soup."

2. More complex molecules, including short strands of RNA, grew from the building blocks in the organic soup, perhaps with the aid of reactions using clay or other mineral surfaces as templates for their assembly. Some of the RNA molecules were capable of self-replication.

3. Membranes that formed spontaneously in the organic soup enclosed some of these complex molecules, making "pre-cells" that facilitated the development of cooperative molecular interactions.

4. Natural selection among the RNA molecules in pre-cells gradually led to an increase in complex-

FIGURE 5.9 A summary of the steps by which chemistry on the early Earth may have led to the origin of life.

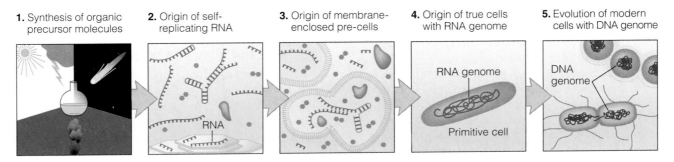

1. Synthesis of organic precursor molecules **2.** Origin of self-replicating RNA **3.** Origin of membrane-enclosed pre-cells **4.** Origin of true cells with RNA genome **5.** Evolution of modern cells with DNA genome

RNA

RNA genome

Primitive cell

DNA genome

ity, until eventually some of these structures became true living organisms.

5. Natural selection then rapidly improved and diversified life. DNA became the favored hereditary molecule, and life has continued to evolve ever since.

We may never know whether life actually originated in this way, in some similar way, or in some completely different way. Nevertheless, this scenario seems quite reasonable and perhaps even "easy" given geological time scales. It seems especially reasonable given that a number of different components of the scenario have been demonstrated in laboratory experiments. Even if life did not originate in this way, it seems that it could have—which suggests that the actual path to life must have been equally easy, or it would have followed the path we've described. In summary, we have reason to believe that the origin of life was a *likely* consequence of conditions on the early Earth, in which case it might be equally likely that life arose on many other worlds.

THINK ABOUT IT . . . *We've noted that the probability of life's arising by randomly mixing simple organic building blocks is so small as to seem impossible. Yet, in the scenario we've described, the likelihood of getting life seems quite good. In your own words, describe why these two probabilities are so different.*

Could Life Have Migrated to Earth?

We have no reason at present to think that life came to Earth from somewhere else, because it seems quite plausible that life could have arisen here on its own. Nevertheless, it's still possible that life migrated to Earth (an idea sometimes called "panspermia").

The idea that life could travel through space to land on Earth once seemed outlandish. After all, it's hard to imagine a more forbidding environment than that of space, where there's no air, no water, and constant bombardment by dangerous radiation from the Sun and stars. However, the presence of organic molecules in meteorites and comets tells us that the building blocks of life can survive in the space environment, and some microbes seem capable of surviving at least moderate periods of time in space [Section 3.5]. It therefore seems possible that life could migrate from one planet to another, perhaps within the interior of a rock blasted from a planetary surface by an impact.

Indeed, the key question probably is not whether life *could* migrate through space but whether we have any reason to suppose it originated elsewhere rather than right here on Earth. Many scientists have debated this question, with the debate taking many different twists. Today, most ideas about migrating life fall into one of two broad categories.

The first broad idea suggests that life does not form as easily as we have imagined, at least under the conditions present on the early Earth, and that the relatively short time between the last sterilizing impact and the earliest appearance of life was not long enough for life to have originated here. In this view, the only explanation for life on Earth (other than invoking the supernatural) would be migration from elsewhere. Although this idea in some sense only moves the problem of life's origin to another place, it at least allows for the possibility that more time was available for life to develop (or conditions were more conducive to rapid development) in this other place than on Earth. The primary drawback to this idea is that other worlds in our solar system probably had similar constraints on the early origin of life—such as sterilizing impacts during the heavy bombardment. The only way to get significantly more time for an origin of life is to suppose that life migrated from another star system, but interstellar migration by microbes seems highly unlikely. Such a journey would require millions or billions of years; living organisms would almost surely be killed by exposure to cosmic rays during this time. Moreover, calculations suggest that the probability of a rock from another star system hitting Earth is extremely low, which also explains why we have never found a meteorite from beyond our own solar system.

The second broad idea, in contrast, suggests that life forms so easily that we should expect to find life originating on any planet with suitable conditions. Earth might not have been the first planet in our solar system with conditions suitable for life. If life originated on a different planet first—say, on Mars—it might have migrated to Earth and taken hold on our planet as soon as conditions allowed. In essence, this idea suggests that life could have originated on Earth but never got the chance because life from another planet got here first.

Like most of what we've discussed concerning the possible origin of life, we simply do not have enough data to decide definitively whether life began on Earth or followed either migration scenario. But the mere possibility of life's migrating among worlds is interesting and will surely be investigated further as scientists continue to study the question of life in the universe.

5.3 Early Evolution and the Rise of Oxygen

Regardless of how it originated, life on Earth has flourished for some 3.5–4.0 billion years. For most of that time, life was very simple; complex, multicelled organisms arose only recently on the geological time scale. In this section, we'll briefly trace how the microbes that lived 3.5 billion years ago gradually evolved and how photosynthetic bacteria produced the oxygen we breathe today.

Early Microbial Evolution

The earliest organisms must have been quite simple, and the tree of life (see Figure 5.4) suggests that Bacteria and Archaea—which are both prokaryotes lacking a cell structure—came before Eukarya. However, we are not certain how much before. Although the earliest known fossils of cells with a clear nucleus date to a time some 1–2 billion years after the origin of life, the genetic evidence in the tree of life is consistent with the idea that simple eukaryotes appeared almost simultaneously with early Bacteria and Archaea. Because cell nuclei do not fossilize well, the fossil record might not reveal the presence of obvious eukaryotes even if they were living alongside other microbes. Nevertheless, complex, multicellular organisms—all of which are eukaryotes—did not develop until much later in the Earth's history. Microbes had the Earth to themselves for perhaps the first 1–2 billion years after the origin of life. Even today, the total biomass of microbes far exceeds that of multicellular organisms like fungi, plants, and animals.

The first microorganisms undoubtedly had at least a few enzymes and a rudimentary metabolism. Because the atmosphere was essentially oxygen-free, they must have been **anaerobic,** meaning that they did not require molecular oxygen. (By contrast, we are *aerobic* organisms—we cannot survive without molecular oxygen.) A few decades ago, when the original Miller–Urey experiment suggested that the early Earth might have been covered with a concentrated "organic soup," biologists guessed that these early organisms were chemoheterotrophs (see Table 3.1) that absorbed organic compounds from the surrounding environment. However, this idea lost favor when we learned that the original Miller–Urey experiment overestimated the concentration of the organic soup (because the experiment did not use an "atmosphere" rich in carbon dioxide).

Today, the most widely favored hypothesis holds that the first microorganisms were chemoautotrophs that obtained energy from *inorganic* chemicals. Some modern Archaea that appear to be fairly close to the root of the tree of life (see Figure 5.4) obtain their energy from chemical reactions involving hydrogen, sulfur, and iron compounds. Archaea thriving in hot sulfur springs, for example, get their energy in this way. Moreover, hydrogen and iron compounds were abundant on the early Earth, perhaps especially so in hot springs and near deep-sea vents. Because other evidence suggests that life may have originated in these locales, it seems reasonable to assume that early life used the available inorganic chemical energy. (See Appendix D for discussion of possible chemical pathways.)

Natural selection probably resulted in rapid diversification among the early life-forms. Modern DNA replication involves a variety of enzymes that help keep the mutation rate low. Early organisms, with a much more limited set of enzymes, probably experienced many more errors in DNA copying. Because more errors mean a higher mutation rate, evolution would have been rapid among early microbes. As life diversified, many new metabolic processes evolved, and the surviving organisms evolved along the different branches that we now know as Bacteria, Archaea, and Eukarya.

Photosynthesis and Oxygen Production

Perhaps the most important new metabolic process was photosynthesis, which enabled organisms to obtain energy from sunlight. Photosynthesis probably evolved gradually. At first, some organisms may have developed light-absorbing pigments that absorbed excess light energy—especially ultraviolet—that was harmful to life near the ocean surface. Over time, some of these pigments evolved to enable the cell to make use of the absorbed solar energy. Modern organisms known as purple sulfur bacteria and green sulfur bacteria may be much like the early photosynthetic microbes. These organisms use hydrogen sulfide (H_2S) rather than water (H_2O) in photosynthesis and therefore do not produce any oxygen.

Photosynthesis using water evolved perhaps as early as 3.5 billion years ago (possibly even earlier), appearing first in organisms called **cyanobacteria** (sometimes called "blue-green algae"). Indeed, the earliest fossil cells resemble modern cyanobacteria, lending credence to this idea. Because they released oxygen as a by-product of their photosynthesis, they literally changed the world. It took a billion years or more for oxygen to build up in the atmosphere, but in the end we owe the existence of our oxygen atmosphere primarily to the action of microscopic cyanobacteria (Figure 5.10).

The Rise of Oxygen

Oxygen is a highly reactive gas that would disappear from the atmosphere in just a few million years if it were not continually resupplied by life. Fire, rust, and the discoloration of freshly cut fruits and vegetables are everyday examples of **oxidation reactions** —chemical reactions that remove oxygen from the atmosphere. Many elements and molecules can participate in oxidation reactions. Today, most of the reactions that remove oxygen from the atmosphere occur in living organisms that use oxygen, including ourselves. Before oxygen-breathing organisms

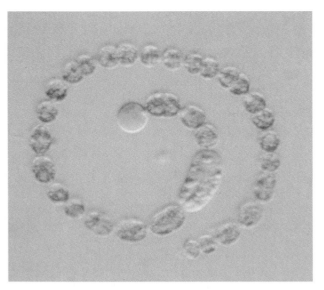

a The blue-green color of this lake (in Cape Cod, Massachusetts) is the result of a population explosion, or "bloom," of cyanobacteria.

b This micrograph shows individual cyanobacteria from the lake.

FIGURE 5.10 A bloom of cyanobacteria. Cyanobacteria split water and release oxygen in photosynthesis and are thought to have been responsible for the rise of oxygen in the Earth's atmosphere.

evolved, oxidation reactions primarily involved volcanic gases, dissolved iron in the oceans, and surface minerals (especially those containing iron) that could react with oxygen. Such reactions essentially "rust" the minerals, causing them to turn red. In the oceans, oxidation reactions with dissolved iron create minerals that precipitate to the bottom, forming "red beds" on the ocean floor. On land, the reddish color of much of the Earth's rock and clay is a direct result of oxidation reactions involving surface minerals.

When cyanobacteria first began producing oxygen, nonbiological oxidation reactions must have been common and rapid. None of the Earth's rock had yet been exposed to significant amounts of oxygen, so surface rock could easily react with oxygen, removing it from the atmosphere. Over millions of years, as some surface rock became saturated with oxygen (that is, had absorbed as much oxygen as possible), plate tectonics continually brought new rock to the surface.

Studies of rocks that are 2–3 billion years old, especially rocks of a type called *banded iron formations,* show that the atmosphere during that time contained less than 1% of the amount of oxygen it contains today. The banded iron formations were made from iron-containing minerals dissolved in the oceans, and such iron minerals cannot dissolve if there is substantial oxygen in the atmosphere and ocean. Because cyanobacteria were already producing oxygen, we conclude that oxidation reactions were removing this oxygen from the atmosphere as fast as the cyanobacteria could produce it. Thus, it took at least a billion years of activity by cyanobacte-

ria before their oxygen production began to significantly affect the atmosphere.

Rock evidence shows that the oxygen content of the atmosphere began to rise about 2.0 billion years ago, but we are unsure how rapidly it rose or precisely when it reached current levels. As we'll discuss in more detail shortly, relatively large multicellular, oxygen-using organisms appeared only about 550 million years ago. This suggests, but does not prove, that the oxygen level remained less than about 10% of its current level until that time. Clear evidence of an oxygen level near its current value first appears in the fossil record only about 200 million years ago. That is when we first find charcoal in the fossil record, implying that enough oxygen was present in the atmosphere for fires to burn.

THINK ABOUT IT . . . *Suppose oxygen built up to the point at which we could breathe the atmosphere about 550 million years ago. What fraction (percent) of Earth's history would this represent? How does your answer change if oxygen became sufficiently plentiful for our breathing only 200 million years ago? What do your answers tell you about the difference between our planet's being habitable in general and being habitable for us?*

The rise of oxygen created a crisis for life, because oxygen attacks the bonds of organic molecules. Many species of microbes probably went extinct. Among the surviving species, some managed to avoid the effects of oxygen because they lived in (or migrated to) underground locations where the oxygen did not reach them. We still find many anaerobic microbes in such locales today. Other organisms survived because, as the oxygen content of the atmosphere gradually rose, they evolved new metabolic processes that allowed them to thrive in the presence of oxygen. Plants and animals, including us, still use the metabolic processes that evolved in response to the "oxygen crisis" faced by living organisms between 1 and 2 billion years ago.

5.4 Eukaryotic Evolution: From a Slow Start to an Explosion of Diversity

The oxygen crisis must have spurred the evolution of many adaptations, but it is an enormous leap from Bacteria and Archaea to complex eukaryotic cells. Besides having a distinct cell nucleus, many modern eukaryotes have multiple chromosomes that allow them to carry much more DNA than prokaryotes, as well as complex life cycles with a variety of reproductive processes—including, in some cases, sexual reproduction. How did eukaryotes evolve from simpler prokaryotes?

Early Eukaryotes

The first known fossil evidence of eukaryotic cells dates to about 2.1 billion years ago, by which time oxygen was beginning to accumulate in the atmosphere. However, as we've discussed, comparisons of DNA sequences have led some scientists to suggest that prokaryotes and eukaryotes split from a common ancestor much earlier. In either case, the rise of oxygen clearly played a key role in the evolution of eukaryotes, because cells can produce energy (that is, make molecules of ATP [Section 3.3]) much more effectively by using oxygen than through anaerobic processes. As aerobic organisms evolved, they were able to develop adaptations that demanded much more energy than would have been available to their anaerobic ancestors.

Eukaryotes probably arose through a combination of at least two major adaptations that evolved in prokaryotes. First, some prokaryotic species may have developed specialized infoldings of their membranes that compartmentalized certain cell functions, ultimately leading to the creation of a cell nucleus (Figure 5.11a). Second, some relatively large ancestral host cell (either a large prokaryote or an early eukaryote) absorbed smaller prokaryotes with specialized functions. The absorbed prokaryotes developed a **symbiotic relationship** with the host cell, that is, a relationship in which both the invading organism and the host organism benefit from living together (Figure 5.11b).

The key evidence favoring symbiosis (the development of a symbiotic relationship) comes from the existence of structures within eukaryotic cells that look much like tiny prokaryotes trapped inside. The most important of these "cells within cells" are **mitochondria,** the cellular organs in which oxygen helps produce energy (by making molecules of ATP), and **chloroplasts,** structures in plant cells that produce energy by photosynthesis. Mitochondria and chloroplasts even have their own DNA and reproduce themselves within their eukaryotic homes. Moreover, DNA sequences in mitochondria and chloroplasts clearly group them with Bacteria rather than with Eukarya. Thus, it seems a near-certainty that mitochondria and chloroplasts *were* once individual bacteria. Assuming that these bacteria had already evolved their ability to make efficient use of oxygen (in the case of mitochondria) or to carry out photosynthesis (in the case of chloroplasts), a symbiotic relationship might have developed quite easily. The host cell would benefit from the energy produced by the absorbed bacteria, while the bacteria would benefit from the protection offered by the host cell.

Further support for this hypothesis comes from the many cases in which living organisms today form symbiotic relationships. Other evidence comes from the fact that some present-day bacteria show striking similarities to mitochondria and chloroplasts, suggesting that the latter evolved from independent organisms.

The rise of oxygen ignited an explosion of eukaryotic diversification. Generally speaking, more variations are possible on complex structures than on simple ones, so the complexity of eukaryotic cells allowed for the selection of many more adaptations than were possible in prokaryotic cells. Even single-celled eukaryotes exhibit much more diversity in structure than exists among all prokaryotes, and multicelled eukaryotes enjoy diversity far beyond that. Indeed, while some prokaryotes live in multicelled colonies in which different cells perform different functions, only eukaryotes evolved into complex multicellular organisms such as fungi, plants, and animals. Moreover, multicellularity appears to have evolved independently in several different branches of early eukaryotes, suggesting that the complex structure of eukaryotic cells opened the door for the evolution of more advanced organisms.

THINK ABOUT IT . . . *One early trend in the evolution of multicellular organisms was a trend toward larger size. Briefly discuss how larger size might have conferred an evolutionary advantage.*

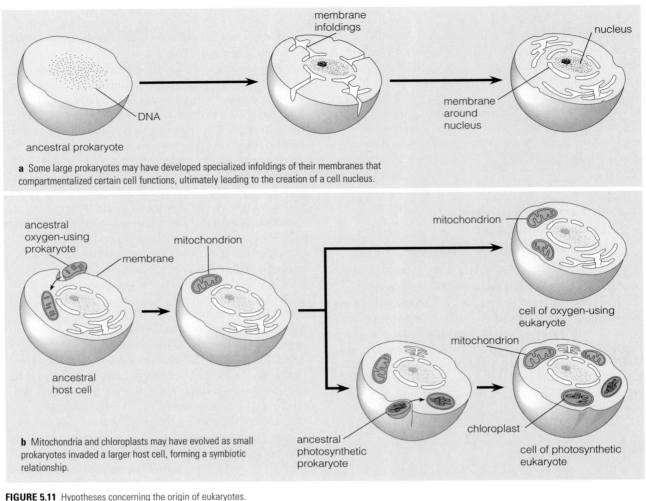

a Some large prokaryotes may have developed specialized infoldings of their membranes that compartmentalized certain cell functions, ultimately leading to the creation of a cell nucleus.

b Mitochondria and chloroplasts may have evolved as small prokaryotes invaded a larger host cell, forming a symbiotic relationship.

FIGURE 5.11 Hypotheses concerning the origin of eukaryotes.

The Cambrian Explosion

Animals may represent only one small branch on the tree of life (see Figure 5.4), but it's *our* branch. Moreover, we generally assume that extraterrestrial intelligence, if it exists, will belong to animal-like beings from other worlds. Thus, we have a special interest in understanding animal evolution.

In the broadest sense, biologists classify animals according to their basic "body plans." For example, the basic body plan shared by mammals and reptiles is fundamentally different from that of insects. Animals are grouped by body plan into what biologists call **phyla** (singular, phylum), the next level of classification below kingdoms (such as the plant kingdom and the animal kingdom [Section 3.2]). Mammals and reptiles both belong to the phylum *Chordata,* which represents animals with internal skeletons. Insects, crabs, and spiders belong to the phylum *Arthropoda,* which represents animals with body features such as jointed legs, an external skeleton, and segmented body parts. Classifying animals into phyla is an ongoing effort by biologists, but modern animals appear to comprise about 30 different phyla, each representing a different body plan.

Remarkably, nearly all of these different body plans, plus a few others that have gone extinct, make their first known appearance in the fossil record during a period spanning only about 40 million years —less than about 1% of the Earth's history. This remarkable flowering of animal diversity appears to have begun about 545 million years ago, at about the time of the start of the Cambrian period (see Figure 4.6).[4] Hence, it is called the **Cambrian explosion.**

Note that the Cambrian explosion came long after the origin of eukaryotes. Since the time of the Cambrian explosion, there has been no similar development of new animal body plans. We are therefore confronted with two important and related questions: Why did the Cambrian explosion occur so suddenly yet so long after the origin of eukaryotes, and why hasn't any similar diversification happened since?

---

[4] Although the Cambrian explosion is among the most vivid events in the fossil record, a few animals are known from the late Proterozoic (including some enigmatic forms that have not survived to the present). Evidence indicates that even the lineages we first see in the Cambrian explosion actually began evolving earlier. The Cambrian explosion didn't involve only animals; other groups, such as algae, also diversified.

No one knows the answers to these questions, but many ideas have been suggested. The timing and sudden onset of the Cambrian explosion may have been influenced by at least four important factors. First, as we discussed earlier, the oxygen level in our atmosphere may have remained well below its present level until about the time of the Cambrian explosion. Thus, some scientists suspect that the dramatic change in animal life occurred at least in part because oxygen reached a critical level for the survival of larger and more energy-intensive life-forms.

A second factor, which some scientists think may have been much more important, was the evolution of genetic complexity. As eukaryotes evolved, they developed more and more genetic variation in their DNA, which opened up ever more possibilities for further variation. Perhaps the Cambrian explosion marks a point in time when organisms had become sufficiently complex that a great diversity of forms could evolve over a very short time.

The third factor may have been climate change. Recall that evidence points to a series of snowball Earth episodes ending around the time of the Cambrian explosion [Section 4.5]. The extreme climate conditions that marked these episodes may have exerted evolutionary pressure that aided the diversification of life and then fueled the Cambrian explosion when the environmental conditions eased.

The fourth factor may have been the absence of efficient predators. Early predatory animals probably were not very sophisticated, so some adaptations that later might have been snuffed out by predation were given a chance to survive if they arose early enough. Thus, the beginning of the Cambrian period may have marked a window of opportunity for many different adaptations to gain a foothold in the environment.

This last idea may partly explain why no similar explosion of diversity has taken place since the Cambrian. Once predators were efficient and widespread, it may have been much more difficult for entirely new body forms to find an available environmental niche. In addition, the fact that certain body forms were already selected during the Cambrian explosion may have limited other options. That is, while more body plans may have been possible than actually arose, once these were in existence there may not have been clear evolutionary pathways to others. Alternatively, perhaps the various body forms that arose during the Cambrian explosion represent the full range of forms possible, at least within the constraints of the genetic variability available on Earth. In any case, we and nearly all other animals living today can trace our ancestry to species that arose during the Cambrian explosion.

The Colonization of Land

Because most early microfossils are found in sediments that were originally deposited in the oceans, it's difficult to know exactly when life first migrated onto land. However, given the wide variety of environments in which microbes survive today and the fact that many different genetic lineages seem to have appeared quite early in evolutionary history, it's likely that microbial life quickly established itself wherever it could find liquid water and protection from ultraviolet radiation. Plenty of such locations are available on land—including underground and anyplace where water can pool under a shelter of overhanging rock—so it is hard to imagine reasons why microbial life would not have taken hold on land quite early. However, the situation is quite different for complex, multicellular organisms. While microbes may have thrived on land, larger organisms, including all animals, remained confined to the oceans (and perhaps other bodies of water) even after the Cambrian explosion.

For larger organisms, surviving on land was more difficult than surviving in the oceans, primarily because it required evolving a means of drawing water and mineral nutrients from the soil rather than simply absorbing them from surroundings. In addition, photosynthetic land organisms need complex bodies with some parts specialized for energy collection above ground (where sunlight is available) and other parts specialized for collecting water and nutrients from the soil. Plants (and perhaps fungi as well) were the first large organisms to develop the means to live on the land. The colonization of land by plants appears to have begun about 475 million years ago.

Another factor that may have influenced the late colonization of land was the development of the ozone layer that offers protection from solar ultraviolet radiation. As we discussed earlier, ozone is a molecule (O_3) made from oxygen atoms, so the ozone layer could not exist until the atmosphere contained some threshold level of oxygen. Uncertainties in both the oxygen levels through time and the level needed for a substantial ozone layer make it difficult to know when the ozone layer first appeared. But until it did, life on any exposed land surface would have been difficult or impossible. Large organisms could survive in the oceans because water absorbs ultraviolet light, protecting organisms living beneath its surface.

DNA evidence suggests that plants evolved from a type of alga. Some ancient algae might have survived in shallow-water ponds or along lake edges. Because such locales occasionally dry up, natural selection would have favored adaptations that allowed the algae to survive during periods of dryness. Organisms that evolved to survive entirely on land would have had even more advantages, particularly since there were no land animals around to eat them.

FIGURE 5.12 This painting, based on fossil evidence, shows a forest of the Carboniferous period. The large trees with straight trunks are seedless plants called lycophytes. The tree on the left with feathery branches is a horsetail. The plants near the forest floor are ferns. The dragonfly was the size of a bird today.

Once plants moved onto the land, it was only a matter of time until animals would follow them out of the water. Within about 75 million years, amphibians and insects were eating land plants. By the beginning of the Carboniferous period, about 360 million years ago, vast forests and abundant insects thrived around the world (Figure 5.12). These Carboniferous forests were important not only as a major step in evolution, but also because they became an important part of our modern economy. Much of the land area of the continents was flooded by shallow seas during the Carboniferous period, hindering the decay of dead plants. Thick layers of dead organic matter piled up in the stagnant waters. Over millions of years, as these layers were buried, pressure and heat gradually converted the organic matter to coal. Nearly all the coal that has helped fuel our industrial age is, in fact, the remains of these forests of the Carboniferous period.

THINK ABOUT IT . . . *Based on the preceding discussion, explain why coal is called a "fossil fuel." What other fossil fuels do we use to generate energy?*

5.5 Impacts and Extinctions

Once animals colonized the land, the evolutionary path that led to humans becomes much clearer. Reptiles evolved from amphibians; early dinosaurs and mammals followed during the Triassic period, begin-

ning about 245 million years ago. But the fossil record shows that the path was not smooth. In particular, the fossil record shows a number of striking transitions in the nature of living organisms. The most famous transition defines the boundary between the Cretaceous and Tertiary periods, which dates to about 65 million years ago. Dinosaur fossils exist below this boundary, but not above it. Somehow, after dominating the Earth for 180 million years, the dinosaurs went extinct in a geological blink of the eye. What could have caused this sudden extinction?

The K–T Event

In 1978, while analyzing geological samples collected in Italy, the father–son team of Luis and Walter Alvarez and their colleagues made a startling discovery about a thin layer of dark sediments that marks the Cretaceous–Tertiary boundary, called the **K–T boundary** for short (the *K* comes from the German word for Cretaceous, *Kreide*). They found that these sediments were unusually rich in the element iridium, which is rare on Earth's surface but common in meteorites. Subsequent studies found the same iridium-rich sediment marking the K–T boundary at many other sites around the world (Figure 5.13). The large amount of iridium found in the layer led the Alvarezes and their team to propose a stunning hypothesis: The extinction of the dinosaurs was caused by the sudden impact of an asteroid or a comet. It would take an asteroid about 10–15 kilometers in

FIGURE 5.13 The arrow points to a layer of sediment marking the 65-million-year-old K–T boundary. Note its distinctive dark color. The layer is about an inch thick at this location.

diameter to produce as much iridium as is apparently distributed worldwide in the K–T boundary layer.

Their hypothesis was not immediately accepted and still generates some controversy, but it is now clear that a major impact coincided with the death of the dinosaurs. Key evidence comes from further analysis of the thin layer marking the K–T boundary.

Besides being unusually rich in iridium, this boundary features the following:

- Unusually high abundances of several other metals, including osmium, gold, and platinum. In each case, the abundance looks much like what we commonly find in meteorites rather than what we find elsewhere on the Earth's surface.

Movie Madness: Armageddon

In 1994, a lot of people who believed the dinosaurs were the victims of a one-off accident changed their minds. That was the year comet Shoemaker–Levy 9 smacked into Jupiter, leaving entrance wounds the size of planet Earth.

Folks realized that death by rock isn't all that improbable. If it happened to the dino's, it could happen to us. An errant asteroid a dozen miles across might someday career into our planet and raise enough dust and burn enough forests to darken the world for years. We'd all slowly starve.

Alerted to this possibility for havoc and destruction, Hollywood lost little time in showing how ingenuity and some gutsy guys (with the emphasis on the latter) could save us even if Nature hurls a large space rock our way. TV specials and theatrical films soon appeared showing Earth under mortal threat from ballistic boulders.

In the film *Armageddon,* the incoming object is as big as Texas, and a mere few weeks away. That's a real slap in the face for astronomers. Picture this: an asteroid as big as Ceres (the largest rock in the asteroid belt), and the astronomers only find the darn thing when it's as close as Mars?

The end of the world is nigh, but not to worry. NASA is in high gear to divert this king-sized clod and decides the best thing to do is to blow it into two large pieces with a nuclear bomb. The two pieces will diverge slightly and sail harmlessly by opposite sides of Earth. Needless to say, geeky NASA personnel aren't up to this kind of macho mission, so the space agency recruits a bunch of oil-rig roughnecks to plant and detonate the bomb. The NASA folk refer to this group of gritty misfits as "the wrong stuff."

In fact, it's a bad idea to try to blow up an incoming asteroid. The chances are that you'd only turn a single shell into buckshot. More practical schemes envision fastening some sort of rocket engine to the side of the rock and slowly nudging it out of the way. Another approach, at least for asteroids that we find when still far from Earth, is to paint the asteroid white and let the gentle pressure of sunlight do the job. Neither scheme involves roughnecks (or human pilots at all).

The threat, of course, is real. Astronomers estimate that rocks comparable to the one that did in the dinosaurs (together with countless other species) will slam into Earth roughly every 50 million years. But by carefully keeping tabs on those asteroids that cross Earth's orbit, we can see disaster coming. We'll have years to mount a defense.

The bottom line is that this is one kind of disaster we can probably avoid. After all, unlike the dino's, we've got a space program.

FIGURE 5.14 Meteor Crater in Arizona was created about 50,000 years ago by the impact of an asteroid about 50 meters across. Because the asteroid hit the Earth at very high speed, it left a crater more than 1 kilometer across and almost 200 meters deep. The K–T impact was much larger.

- Grains of "shocked quartz" —quartz crystals with a distinctive structure that indicates they were formed under conditions of high temperature and pressure that occur in nature only in impacts.

- Spherical rock "droplets" of a type known to form when drops of molten rock cool and solidify in the air.

- Soot (at some sites), which is evidence of burning and suggests the occurrence of widespread fires.

All of these features point to an impact. Besides the theoretical and experimental evidence telling us that shocked quartz must be produced in impacts, we have observational evidence—we find shocked quartz at Meteor Crater in Arizona (Figure 5.14). The rock "droplets" in the K–T boundary layer presumably were made from molten rock splashed into the air by the force and heat of an impact. The force of the impact would have blasted other debris so high that it rose above the atmosphere. This debris would have spread worldwide, and as it fell to Earth it would have been heated by friction, much like a spacecraft reentering the atmosphere. A global rain of this hot, glowing dust from the impact would have ignited vast forest fires as it fell.

The clearest proof that an impact occurred came in 1991 with the discovery of a large impact crater that appears to match the 65-million-year age of the K–T boundary. The crater, about 200 kilometers across, is located on the coast of Mexico's Yucatán peninsula, about half on land and half underwater (Figure 5.15). Its size indicates that it was created by

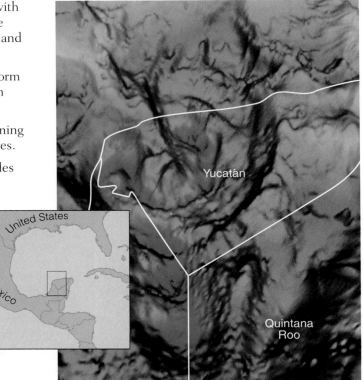

FIGURE 5.15 This false-color image, made with the aid of precision measurements of the local strength of gravity, shows an impact crater with its center near the northwest corner of the Yucatán. The white lines on the image correspond to the coastline and borders of Mexican states.

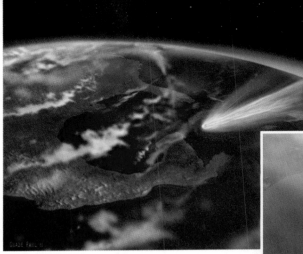

FIGURE 5.16 This sequence of paintings shows an artist's conception of the impact that caused the death of the dinosaurs some 65 million years ago.

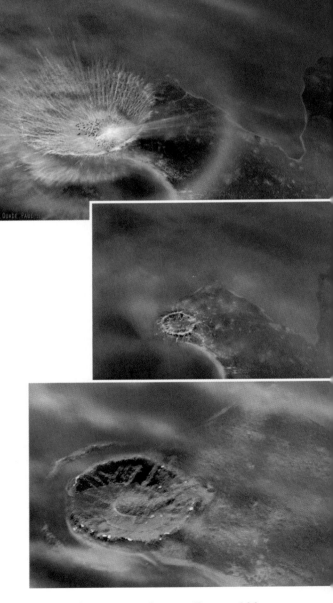

the impact of an asteroid or a comet measuring about 10 kilometers across. (It is named the *Chicxulub crater,* after a nearby fishing village.)

Figure 5.16 depicts the impact and its immediate aftermath. The asteroid or comet slammed into the Earth with the force of 100 million hydrogen bombs. It apparently hit at a slight angle, sending a shower of red-hot debris across the continent of North America. A huge tidal wave sloshed more than 1,000 kilometers inland. Much of North American life may have been wiped out almost immediately. Not long after, the hot debris raining down around the rest of the world ignited fires that killed many other living organisms. The longer-term effects were even more severe.

Dust and smoke remained in the atmosphere for weeks or months, blocking sunlight and causing temperatures to fall as if Earth were experiencing a global and extremely harsh winter. The reduced sunlight would have stopped photosynthesis for up to a year, killing large numbers of species throughout the food chain. This period of cold may have been followed by a period of unusual warmth. Some evidence suggests that the impact site was rich in carbonate rocks, so the impact may have released large amounts of carbon dioxide into the atmosphere. The added carbon dioxide would have strengthened the greenhouse effect, and the months of global winter might have been followed by decades—or longer—of global summer.

The impact probably also caused chemical reactions in the atmosphere that produced large quantities of harmful compounds, such as nitrous oxides. These compounds dissolved in the oceans, where

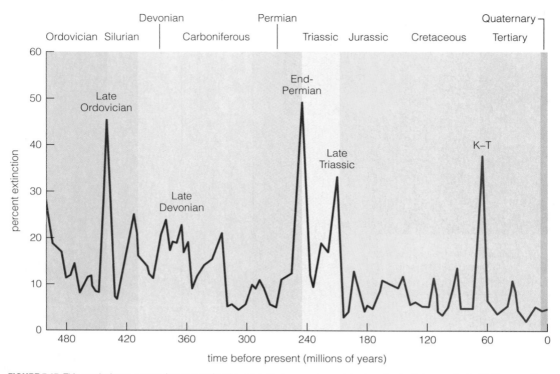

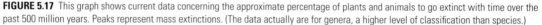

FIGURE 5.17 This graph shows current data concerning the approximate percentage of plants and animals to go extinct with time over the past 500 million years. Peaks represent mass extinctions. (The data actually are for genera, a higher level of classification than species.)

they probably were responsible for killing vast numbers of marine organisms. Acid rain may have been another by-product, killing vegetation and acidifying lakes around the world.

The fossil record suggests that the K–T impact may have ultimately killed up to 99% of all living organisms while driving up to 75% of all existing *species* to extinction—a clear example of what we call a **mass extinction,** or the rapid extinction of a large fraction of all living species. Indeed, the more remarkable fact may be that the other 25% of species survived. Among these survivors were small mammals, which may have survived in part because they lived in underground burrows and managed to store enough food to outlast the global winter that followed the impact.

The evolutionary effect of the extinctions was profound. For 180 million years, dinosaurs had diversified into a great many species large and small, while mammals remained small and rodentlike. With the dinosaurs gone, the mammals (as well as all other surviving animals) were presented with an evolutionary opening. Over the next 65 million years, evolution produced all the mammal species we see today—including us—and numerous others that have gone extinct. Had it not been for the K–T impact, dinosaurs might still rule the Earth.

Other Mass Extinctions

The fossil record shows at least several other mass extinctions besides the K–T extinction. Determining

extinction rates precisely is difficult, primarily because it requires determining the precise layer that contains the *last* occurrence of a given species. Nevertheless, we have enough data to be sure that extinction rates vary considerably with time.

Figure 5.17 shows current data on the extinction rate for plants and animals over the past 500 million years. The data reveal at least five major mass extinctions, including the K–T extinction, and numerous smaller extinction events. Could these extinctions also have been caused by impacts?

Simple probability makes impacts a reasonable hypothesis. On average, impacts the size of the K–T event should happen about every 100 million years or so. And, indeed, impacts appear to have coincided with at least some of the other mass extinctions. For example, the sedimentary strata marking the most severe mass extinction—the Permian extinction about 245 million years ago—show many of the same features as the strata marking the K–T boundary. No impact crater has been found for the Permian extinction, but that's not too surprising. Plate tectonics would probably have destroyed by now any seafloor crater more than about 200 million years old (because seafloor crust is completely recycled on that time scale), and impacts are more likely over oceans because oceans cover most of Earth's surface. Still, no other mass extinction has been tied to an impact with the same clarity as has the K–T extinction, and other causes for extinctions are possible.

Some geologists believe that episodes of unusually active volcanism may have led to climate change and

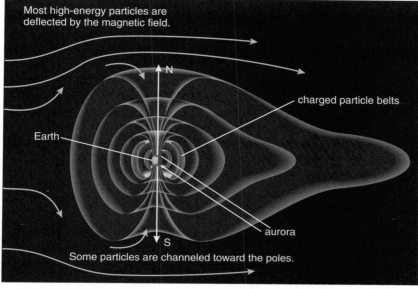

Most high-energy particles are deflected by the magnetic field.

Earth

charged particle belts

N

S

aurora

Some particles are channeled toward the poles.

a This diagram represents the Earth's magnetic field, which is invisible in reality. The magnetic field deflects most high-energy particles and channels others toward the poles. It stretches out in the direction opposite the Sun, because most of the charged particles come from the Sun.

b This photograph shows an aurora visible over the coast of Norway. Auroras are caused by light emitted as the charged particles crash into atoms and molecules in the Earth's atmosphere.

FIGURE 5.18 Effects of the Earth's magnetic field.

high extinction rates. Other types of climate change may also have caused mass extinctions. For example, if there have been other periods of cooling similar to the known snowball Earth episodes [Section 4.5], the cold climate may have driven many species to extinction. Another set of hypotheses envisions extinctions as being tied to changes in the mutation rate.

While many mutations occur simply as the result of copying "errors" within cells, others are caused by external influences. For example, ultraviolet light can cause mutations, which is why sun exposure can lead to skin cancer. If the concentration of the Earth's ozone layer varies with time, then the amount of ultraviolet light reaching the surface would also vary. Perhaps some of the extinctions occurred when the ozone layer thinned, resulting in an increase in solar ultraviolet light that caused the mutation rate to increase.

Mutations can also be caused by high-energy particles that stream continuously from the Sun (making up the *solar wind* [Section 1.3]). The Earth's magnetic field deflects most of these particles, while channeling others toward the Earth's poles, where they cause auroras (Figure 5.18). However, studies of magnetized rocks show that the Earth's magnetic

field varies in strength with time. In particular, the Earth's magnetic field apparently reverses itself—with the north magnetic pole becoming the south magnetic pole, and vice versa—every few million years on average. We are not yet sure exactly how often or why these reversals occur, but the magnetic field probably disappears altogether for thousands of years while a reversal is in progress. The mutation rate might spike upward during this time due to the absence of the normal protection from high-energy particles. Although magnetic field reversals happen much more frequently than mass extinctions, particular reversals may have occurred at times when life was more susceptible to major change and thus might have played a role in extinction events.

A related possibility ties increased mutation rates to nearby *supernovae*, or the explosions of stars. Supernovae are rare events. Out of the more than 100 billion stars in the Milky Way Galaxy, we expect only about one star per century to explode in a supernova. Most of these supernovae occur far from our solar system. Nevertheless, because the Sun and stars are all moving within the Milky Way Galaxy (see Figure 1.10), different sets of stars make up our galactic neighborhood at different times. Simple probability calculations suggest that our planet must occasionally be located within a few tens of light-years of an exploding star. A supernova generates a prodigious number of very high energy particles called *cosmic rays*. Thus, when a supernova occurs near Earth, we might expect a big upward spike in the number of cosmic rays reaching the Earth and causing mutations.

Whether or not we ultimately learn the causes of past mass extinctions, they clearly have had tremendous effects on the evolution of life. With each mass extinction, many of the dominant species on the planet have disappeared, creating changes in environmental conditions and predator–prey relationships. These changes allow new species to evolve over the millions of years that follow. Just as the K–T event apparently paved the way for the rise of mammals, other extinctions may have created similarly critical portions of the evolutionary path that made our present existence possible.

The topic of mass extinctions also holds a cautionary lesson for our species today. Human activity is driving numerous species toward extinction. The best-known cases involve relatively large and wide-ranging animals, such as the passenger pigeon (extinct since the early 1900s) and the Siberian tiger (nearing extinction). But most of the estimated 10 million or more plant and animal species on our planet live in localized habitats, and most of these species have not even been cataloged. The destruction of just a few square kilometers of forest may mean the extinction of species that live only in that area. According to some estimates, human activity is driving species to extinction so rapidly that up to half of today's species may be gone by the end of this century. On the scale of geological time, the disappearance of half the world's species in just a few hundred years would certainly qualify as another of the Earth's mass extinctions. Are we unwittingly clearing the way for a new set of dominant species?

THINK ABOUT IT . . . *The fossil record suggests that the dominant animal species are nearly always victims in a mass extinction. If we are causing a mass extinction, do you think we will be victims of it? Or will we be able to adapt to the changes so that we survive even while many other species go extinct? Defend your opinion.*

The Continuing Impact Threat

The discovery that at least one mass extinction is tied to an impact has spurred scientific concern over whether our civilization might be vulnerable to future impacts. How serious is this threat?

Small particles hit the Earth almost continuously, burning up in our atmosphere as **meteors** (some-times called "shooting stars," a misleading name that arose long before we knew their true source). Most meteors are caused by particles no bigger than a pea. The particle itself is too small for us to see, but it enters the atmosphere at such high speed (typically between about 45,000 and 250,000 km/hr) that it burns up and heats the surrounding air, producing the meteor flash. About 25 million particles enter our atmosphere and burn up as meteors each day. Interestingly, these particles add a total of about 40,000 tons to the Earth's mass each year. This sounds like a lot in human terms, but it is negligible compared to the Earth's total mass of 6 *billion trillion* (6×10^{21}) tons.

Somewhat larger objects entering our atmosphere may be heated to the point where they explode, producing an extraordinarily bright flash called a *fireball*. (Many so-called UFO sightings are actually fireballs.) In some cases, debris from the explosion hits the ground. Because this debris consists of rocks from space, we call these rocks *meteorites*. An impacting object more than a few meters across may survive the plunge through our atmosphere intact, carrying so much energy that it vaporizes solid rock and leaves a visible crater. Meteor Crater (see Figure 5.14) is a popular tourist stop because it is one of the few easily recognizable impact craters on the Earth's surface. In fact, geologists have identified more than 100 other impact craters around the world, including the one left by the K–T impact (Figure 5.19).

No large impacts have left craters in modern times, but we know of at least one smaller impact that devastated a fairly large area. In 1908, a tremendous explosion occurred over Tunguska, Siberia (Figure 5.20). The explosion, which released energy equivalent to that of several atomic bombs, is now thought to have been caused by a small asteroid or comet no more than about 30 meters across. Atmospheric friction caused it to explode completely before it hit the ground, so it left no impact crater. Nevertheless, forests were flattened and set on fire, and the shock wave knocked over people, tents, and furniture up to 200 kilometers away. If the impact had occurred over a populated area, the death toll could have been enormous.

Overall, larger impacts should be correspondingly rarer. Figure 5.21 shows how often, on average, we expect the Earth to be hit by objects of different sizes. While the chance that our civilization is in any imminent danger of a major impact seems relatively low, it is not negligible. Moreover, the question is not whether a future impact will occur, but when. Many known asteroids have orbits around the Sun that cross the Earth's orbit or pass near enough to Earth's orbit that they may someday be gravitationally perturbed into an Earth-crossing orbit. These objects have a very real chance of striking the Earth at

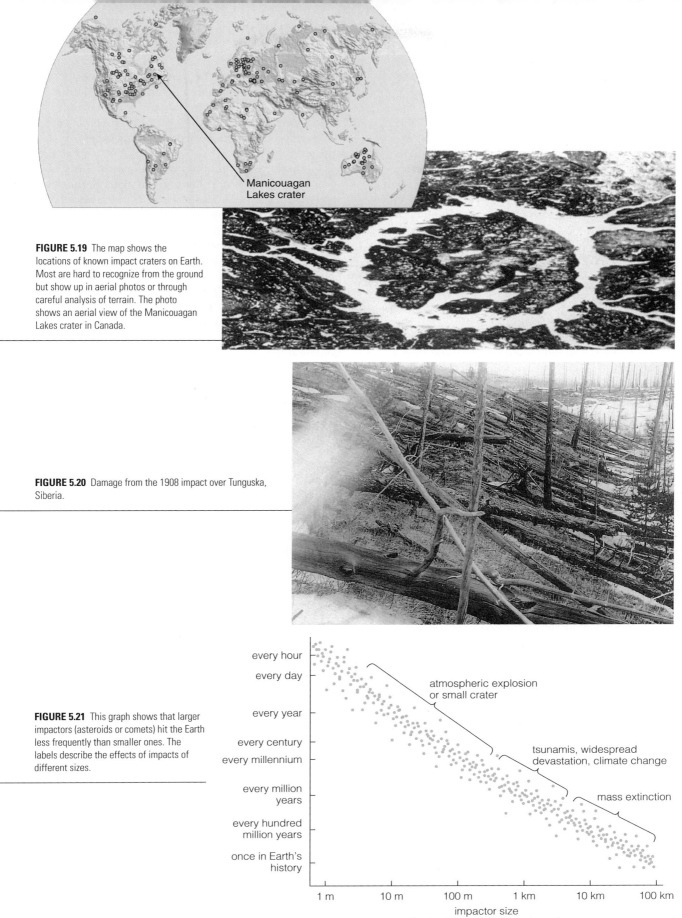

FIGURE 5.19 The map shows the locations of known impact craters on Earth. Most are hard to recognize from the ground but show up in aerial photos or through careful analysis of terrain. The photo shows an aerial view of the Manicouagan Lakes crater in Canada.

Manicouagan Lakes crater

FIGURE 5.20 Damage from the 1908 impact over Tunguska, Siberia.

FIGURE 5.21 This graph shows that larger impactors (asteroids or comets) hit the Earth less frequently than smaller ones. The labels describe the effects of impacts of different sizes.

every hour
every day
every year
every century
every millennium
every million years
every hundred million years
once in Earth's history

atmospheric explosion or small crater

tsunamis, widespread devastation, climate change

mass extinction

1 m   10 m   100 m   1 km   10 km   100 km

impactor size
(crater is about 10 times larger)

some future time. Some of these objects are in the 1-kilometer size range, and a few are as big as the 10-kilometer K–T impactor.

At present, we have no way to deflect an impacting object if we should discover one on a collision course with the Earth. Some scientists and politicians have proposed that we develop a program for identifying potential impact threats and deflecting Earth-bound asteroids or comets. The identification program is already under way with careful telescopic observations. A deflection program would be far more expensive and hence is much more controversial.

THINK ABOUT IT . . . *Study Figure 5.21. About how often should we expect an impact large enough to cause a mass extinction? About how often should we expect an impact large enough to cause a large number of human deaths (that is, "widespread devastation")? Based on your answers, do you think we should be doing anything to protect against the threat of impacts? Defend your opinion.*

5.6 Human Evolution

We've traced the course of evolution from the origin of life through the extinction of the dinosaurs. We've seen that evolution took many surprising twists and turns to that point. The subsequent evolution of mammals and humans was just as interesting. In this section, we'll briefly investigate the pathway that led to our emergence as the first species on Earth capable of learning about its own origins.

Primate Evolution

We are primates, as are all the great apes, monkeys, and prosimians (such as lemurs). The ancestor of all of today's primates lived in the trees, and many of the traits that make us so successful evolved as adaptations to tree life. For example, the limber arms that allow us to throw balls and work with tools evolved so that our ancestors could swing through trees, and our dexterous hands evolved to hang from branches and manipulate food. The eyes of primates are close together on the front of the face, providing overlapping fields of view that enhance depth perception—an obvious advantage when swinging from branch to branch. For the same reason, primates developed excellent eye–hand coordination.

Parental care is essential for young animals in trees, and primates evolved close parent–child bonds. These bonds, in turn, made it possible for primates to be born in a much more helpless state than the babies of most other types of animals. Al-

though many primate species, including us, eventually moved down from the trees, most primates continue to nurture their young for a long time. This trait reaches its extreme in humans. Human babies are nearly helpless at birth and require parental care for more years than the offspring of any other species.

Contrary to a common myth, humans did not evolve *from* gorillas or other modern apes. Rather, modern apes and humans share a common ancestor that is now extinct. Figure 5.22 shows the evolutionary history of major primate branches. Our closest living relatives, chimpanzees and gorillas, shared a common ancestor with us just a few million years ago.

The fact that modern gorillas, chimps, and humans all evolved from the same ancestor has at least two important implications for understanding our existence today. First, it shows that relatively small genetic differences can make a big difference in species success. About 98% of the DNA sequences that make up the human genome are identical to the sequences that make up the chimpanzee genome. Thus, a 2% difference in genetic material is all that separates our current success on the planet from the current predicament of chimpanzees, which survive naturally in only a few isolated locations in Africa. Second, it suggests that the evolution of intelligence is a complex process. Gorillas and chimpanzees have been evolving from our common ancestor just as long as we have, but we are the only species building cities and radio telescopes. This fact raises the question of whether advanced intelligence is an inevitable outcome of evolution. We'll save this topic for Chapter 11.

THINK ABOUT IT . . . *Briefly describe some ways in which intelligence confers an evolutionary advantage. Are there evolutionary disadvantages to increased intelligence?*

The Emergence of Humankind

Even after hominids (human ancestors) diverged from the ancestors of chimpanzees and gorillas, human evolution followed a remarkably complex path. Indeed, one of the most pervasive but incorrect myths about human evolution is that it followed a simple pathway from stooped apes to upright humans (Figure 5.23).

The reality is that there have been numerous hominid species, sometimes with two or more sharing the Earth at the same time (Figure 5.24). The earliest fossil skulls that appear to be identical to those of modern humans are about 100,000 years old. However, even then our ancestors shared the

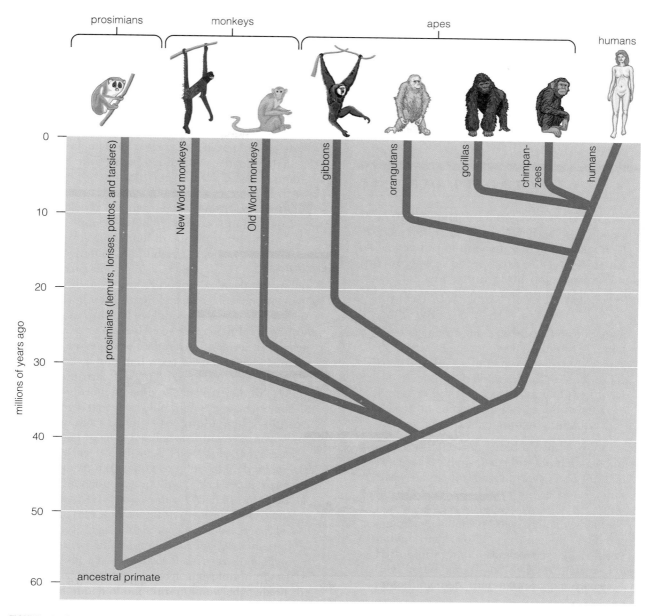

FIGURE 5.22 The evolutionary history of the major primate branches. The common ancestor of modern humans, chimpanzees, and gorillas lived between about 5 and 8 million years ago.

FIGURE 5.23 This famous type of illustration suggests that humans evolved along a simple pathway from apes. However, it is almost completely wrong.

planet with another hominid species, usually called *Neanderthals,* that was quite similar and may have had a slightly larger brain on average. The Neanderthals disappeared about 35,000 years ago, and no one yet knows why.

Deciphering the details of human ancestry is a rich field of research, and entire courses are taught on what has been learned. We will not discuss such details in this book, but before we leave the topic it's

Review Questions

1. How far back in time do we find evidence of life on Earth? Explain the evidence. What does this tell us about the ease with which life arose?

2. How do studies of DNA sequences allow us to reconstruct the evolutionary history of life? What living organisms appear to be most closely related to the common ancestor of all present life?

3. What was the *Miller–Urey experiment,* and how did it work? In what way do the original experimental conditions now seem unrealistic?

4. Describe the three ways the early Earth probably obtained the organic building blocks of life.

5. What do we mean by an "RNA world," and why do scientists suggest that such a world preceded the current "DNA world"?

6. Describe how complex molecules such as RNA might have spontaneously formed on the early Earth. How might membranes have originated?

7. Briefly summarize a plausible sequence of events that could have led to the origin of life on Earth.

8. Briefly discuss the possibility that life migrated to Earth.

9. Briefly discuss the early evolution of life, from the first organisms to the development of photosynthesis and oxygen production.

10. How did the Earth get the oxygen in its atmosphere? Why did it take so long for oxygen to become abundant?

11. Why do we think that an "oxygen crisis" might have led to greater diversity in eukaryotes? Briefly describe how eukaryotes might have evolved from simpler prokaryotes.

12. What was the *Cambrian explosion?* Briefly discuss ideas about what might have caused it and why no similar event has happened since.

13. Briefly describe when and how life colonized the land.

14. What was the K–T event, and how is it thought to have led to the demise of the dinosaurs? What evidence supports this scenario? How did this event pave the way for our existence?

15. Briefly discuss other *mass extinctions* and their possible causes.

16. Discuss the threat that future impacts may pose to us and our planet.

17. Briefly describe how adaptations that evolved so primates could live in trees have proved useful to us as humans.

18. Briefly summarize and explain a few common misconceptions about human evolution.

Discussion Questions

1. *Our Bacterial Ancestry.* Some of Darwin's early detractors complained that evolution implied we were descended from monkeys or apes. In fact, as we saw in this chapter, our evolution is built upon far more primitive organisms. The oxygen we breathe was produced by bacteria and is processed in our cells by mitochondria that probably represent bacteria living symbiotically within us. Does our relationship to bacteria affect the way we should view ourselves as a species? Defend your opinion.

2. *Mass Extinctions.* Past mass extinctions have generally caused the demise of the previously dominant species on Earth. Given the evidence that we are in the midst of causing a mass extinction today, do you believe we are endangering our own existence? If so, what should we do about it? Defend your opinions.

3. *The Missing Link.* As we discussed in this chapter, there is no longer any such thing as a critical "missing link" in human evolution, and the theory of evolution has never suggested that humans evolved *from* present-day apes. Nevertheless, huge numbers of Americans profess belief in both of these myths about evolution. Why do you think these erroneous claims continue to be popular? What can or should be done to better educate the public?

4. *Evolution by Choice.* Consider the technology we are likely to have in the near future that would enable us to genetically engineer our own species, allowing us to choose the path of our future evolution. How do you think society can or should regulate the use of this awesome power? Do you think its potential benefits outweigh its risks, or vice versa? Overall, do you think it likely that advanced civilizations, if they exist, have engineered their own evolution? Defend your opinions.

Problems

Would You Believe It? Each of **problems 1–8** describes a hypothetical future discovery. In light of our current understanding of Earth and evolution, briefly discuss whether each discovery seems plausible or surprising. Explain your reasoning clearly.

1. We discover evidence of life, in the form of a particular ratio of carbon-12 to carbon-13, in a rock that is older than 3.85 billion years.

2. We discover an intact fossil of a eukaryotic cell that is 3.0 billion years old.

3. We discover a previously unknown type of bacteria with a genome that is 95% identical to the human genome.

4. Researchers carry out an experiment in which they mix amino acids and other building blocks of life in a flask. By using a novel energy source, they cause the molecules to react in such a way that within minutes they combine to make a living cell.

5. We discover that, contrary to present belief, oxygen was abundant in the Earth's atmosphere at the time when life arose.

6. We discover a crater from the impact of a 10-kilometer asteroid that dates to about 2,500 years ago.

7. We discover an asteroid about 3 kilometers across that is on a collision course with Earth.

8. We find fossil remains of an early primate that lived about 50 million years ago and was, from all appearances, identical to a modern gorilla.

9. *Geology and Life.* In Chapter 4, we discussed the role of plate tectonics and the CO_2 cycle in climate regulation on Earth. Suppose neither of these processes had ever operated. Could life still have arisen on Earth as discussed in this chapter? If so, how far could evolution have progressed before the lack of climate regulation would have blocked further major developments? Write a one-page essay summarizing and explaining your answers.

10. *Our Accidental Existence.* The history of life on Earth suggests that many "accidents" have occurred and have shaped the path of evolution. Identify and describe three important events in Earth's history without which our current existence would have been highly unlikely. Explain your reasoning clearly.

11. *Extinction and Oxygen.* Suppose we somehow kill off a large fraction of the photosynthetic life on Earth. Could this have consequences for the oxygen content of our atmosphere? Why or why not?

12. *A Brief History of Life on Earth.* Take all the ideas about the origin and evolution of life on Earth and try to condense them into a one- to three-page essay on the history of life on Earth. Or, if you prefer, try to capture the ideas in a poem.

13. *Impact Movie Review.* View one of the recent movies concerning the threat of an impact to our civilization, such as *Deep Impact* or *Armageddon.* Based on what you have learned in this chapter, write a one- to two-page critical review in which you include discussion of whether the impact scenario is realistic.

Web Projects

1. *The Origin of Life.* NASA's Astrobiology home page frequently covers new discoveries about the origin and evolution of life. Learn about one recent important discovery, and write a short essay summarizing the discovery and how it affects our understanding of how life might have evolved on Earth.

2. *Impact Programs.* The discovery that impacts could pose a threat to our civilization has led to calls for new programs to help alleviate the threat. In a few cases, legislation has even been proposed to implement such programs. Learn about one proposal for dealing with the impact threat. Write a short essay explaining the proposal and discussing your opinion of its merits.

3. *Extinction.* Learn more about how biologists estimate the rate at which human activity is driving species to extinction. What conclusions can we draw about the present rate of extinction? Based on this rate, are we in danger of causing a mass extinction comparable to past mass extinctions on Earth? Summarize your findings in a one-page essay.

CHAPTER 6

Searching for Life in Our Solar System

Having studied Earth's habitability and life, we turn to the search for life elsewhere in our solar system. Because there are many places to look—eight other planets, dozens of moons, and thousands of known asteroids and comets—we need a strategy to help focus our efforts on the worlds most likely to be habitable.

The first step in such a strategy is to determine what we are looking for, so we begin this chapter by discussing the environmental requirements for life. Next, we'll discuss some of the tools we can use to learn about the habitability of distant worlds. Finally, we'll take a brief tour of our solar system as we know it today, discussing why some worlds seem promising as potential abodes of life and others do not. This tour will lay the groundwork for more detailed discussion of the most promising places in the chapters that follow.

In our solar system, we are looking primarily for microbes. We have learned enough to be confident that no other advanced civilization has ever existed in our solar system. Yet, the discovery of life of any kind would be profound, both to our understanding of biology and to philosophical considerations of our place in the universe. Even if we don't find life in our own solar system, the search itself will teach us much about the characteristics that can make a planet habitable and will thereby help us when we extend the search to other planetary systems.

6.1 Environmental Requirements for Life

There's no place like home—at least, not within our own solar system. No other world has an atmosphere that we could breathe or abundant surface water that we could drink. No other world has a combination of surface temperature and pressure under which we could survive without a heated (or cooled) and pressurized spacesuit. Few worlds have atmospheres that offer any protection from dangerous ultraviolet radiation from the Sun or from high-energy particles from space. Indeed, without undertaking major engineering projects to build self-contained environments, we have no hope of long-term survival on any other world in our solar system.

However, when we discuss habitability, we generally mean an environment in which life of *some* kind might survive, not necessarily human life. This greatly broadens the possibilities. After all, we could not have survived even on our own planet for much of its history [Section 5.3], yet life flourished just the same. Past and present life on Earth has managed to thrive in a far greater variety of environments than we ourselves can survive in. If we are going to identify potentially habitable worlds in our solar system, we must specify the range of environments that we can consider acceptable for life. We've touched on some of these ideas in previous chapters. Here, we'll try to tie them all together into a clear list of environmental requirements for life.

Elements of Life

Perhaps the most obvious requirement for life is a set of chemical elements with which to make the components of cells. Life on Earth uses about 25 of the 92 naturally occurring chemical elements, although just four of these elements—oxygen, carbon, hydrogen, and nitrogen—make up about 96% of the mass of living organisms [Section 3.2]. Thus, a first requirement for life might be the presence of most or all of the elements used by life.

Interestingly, this requirement probably can be met by almost any world. Recall that all chemical elements in the universe besides hydrogen and helium (and a trace amount of lithium) were produced by stars [Section 1.3]. Although all of these "heavy elements" are quite rare compared to hydrogen and helium, they are found just about everywhere. Moreover, the elements oxygen, carbon, and nitrogen (which make up most of the mass of living matter) are the third, fourth, and sixth most abundant elements in the universe, respectively.

Heavy elements are continually being manufactured by stars and released into space by stellar deaths, so their proportion compared to hydrogen and helium gradually rises with time. Heavy elements make up about 2% of the chemical content (by mass) of our solar system; the other 98% is hydrogen and helium. In some very old star systems, which formed before many heavy elements were produced, the heavy element proportion may be less than 0.1%. Nevertheless, every star system we've studied has at least some amount of all the elements used by life. Moreover, when planetesimals [Section 4.2] condense within a forming star system,[1] they are inevitably made from heavy elements, because the more common hydrogen and helium remain gaseous (except for the small proportion of hydrogen that becomes part of ices such as water, methane, and ammonia). Thus, planetesimals everywhere should contain the elements needed for life, which means that objects built from planetesimals—planets, moons, asteroids, and comets—also contain these elements. The nature of solar system formation explains why the Earth contained all the elements needed for life, and it is why we expect these elements to be present on other worlds throughout our solar system, galaxy, and universe.

Note that this argument doesn't change even if we allow for life very different from life on Earth. Life on Earth is carbon-based, and most biologists believe that life elsewhere is likely to be carbon-based as well [Section 3.2]. However, we can't absolutely rule out the possibility of life with another chemical basis, such as silicon or nitrogen. The set of elements (or their relative proportions) used by life based on some other element might be somewhat different from that used by carbon-based life on Earth. But the elements are still products of stars and would still be present in planetesimals everywhere. No matter what kind of life we are looking for, we are likely to find the necessary elements on almost every planet, moon, asteroid, and comet in the universe.

A somewhat stricter requirement is the presence of these elements in molecules that can be used as ready-made building blocks for life, just as the early Earth probably had an "organic soup" of amino acids and other complex molecules [Section 5.2]. Recall that Earth's organic molecules likely came from some combination of three sources: chemical reactions in the atmosphere, chemical reactions near deep-sea vents in the oceans, and molecules carried to Earth by asteroids and comets. The first two sources can occur only on worlds with atmospheres or oceans,

---

[1] We do not yet know whether planetesimals can condense only in star systems with relatively high proportions of heavy elements or if they were also able to condense in the old stars with heavy elements in proportions of 0.1% or less.

respectively. But the third source should have brought similar molecules to nearly all worlds in our solar system.

Studies of meteorites and comets suggest that organic molecules are widepsread among both asteroids and comets. Because every world was pelted by asteroids and comets during the heavy bombardment [Section 4.3], every world should have received at least some organic molecules. However, these molecules tend to be destroyed by solar radiation on surfaces unprotected by atmospheres. Moreover, while these molecules might stay intact beneath the surface (as they evidently do on asteroids and comets), they probably cannot react with each other unless some kind of liquid or gas is available to move them about. Thus, if we limit our search to worlds on which organic molecules are likely to be involved in chemical reactions, we can probably rule out any world that lacks both an atmosphere and a surface or subsurface liquid medium such as water.

Energy for Life

A second key requirement for life is an energy source to fuel metabolism [Section 3.3]. Life on Earth uses a wide variety of energy sources. Some organisms get energy directly from sunlight through photosynthesis. Others get energy by consuming organic molecules (for example, by eating photosynthetic organisms) or through chemical reactions with inorganic compounds of iron, sulfur, or hydrogen.

Sunlight is available everywhere in our solar system, though it becomes much weaker with increasing distance from the Sun. The energy available in sunlight decreases with the *square* of the distance from the Sun (Figure 6.1). For example, if we could put a leaf of a particular size on a world twice as far from the Sun as Earth, the leaf would receive only one-fourth as much energy as the same leaf on Earth (over the same period of time). At 10 times the Earth's distance from the Sun, it would receive only $1/10^2 = 1/100$ of the energy it would receive on Earth. Thus, photosynthetic life on worlds far from the Sun would have to be either much larger than life on Earth (giving it a larger surface area for collecting light), much more efficient at collecting solar energy, or much slower in its metabolism and reproduction.

Chemical energy sources also should be readily available on many worlds. Chemical reactions can occur under a wide variety of circumstances, but only if the potential reactants are brought into contact. This means that the ongoing reactions needed to provide energy for life can occur only on worlds where materials are being continually mixed (see Appendix D). On a practical level, this probably requires either an atmosphere to mix gases or a liquid medium to mix materials on or below a world's surface.

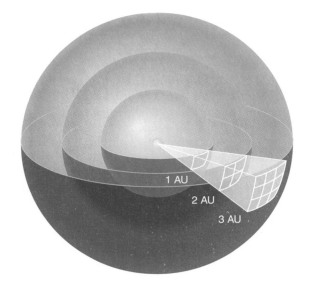

FIGURE 6.1 Any given amount of sunlight is spread over a larger area with increasing distance from the Sun. As shown in this diagram, the area over which the sunlight is spread increases with the square of the distance: At 2 AU the sunlight is spread over an area $2^2 = 4$ times as large as at 1 AU, and at 3 AU the sunlight is spread over an area $3^2 = 9$ times as large as at 1 AU. (Recall that 1 AU is the average Earth–Sun distance, or about 150 million kilometers.) Thus, the energy contained in sunlight (per unit area) *declines* with the square of the distance from the Sun.

Among the terrestrial worlds of the inner solar system, the availability of liquid water probably is key to whether ongoing chemical energy is available for life. On Earth, water is available in at least four distinct ways: in the oceans (and lakes and rivers); in the atmosphere as water vapor and as droplets or ice crystals in clouds; in subsurface groundwater; and as ice in the polar caps, glaciers, and winter ice and snow. Moreover, Earth's distance from the Sun and moderate greenhouse effect allow most of the water to be liquid rather than frozen as ice or evaporated as water vapor.

For a planet such as Mars, which is generally too cold to allow ice to melt to liquid water at the surface, internal heat holds the key to chemical energy. If enough heat is available to power active internally driven geological processes, such as volcanism, then there is probably also enough heat to allow a liquid medium to bring chemical reactants into contact. Recall that larger worlds tend to retain internal heat for longer time periods than do smaller worlds [Section 4.4]. The Moon and Mercury have both cooled significantly and therefore lack the heat needed to drive active internal geology. But Venus and Mars probably still have enough internal heat to support surface or subsurface chemical reactions. A smaller potential source of chemical energy on Venus and

Mars—or any world with an atmosphere—is lightning or other electrical discharges.

The situation is a bit more complicated in the outer solar system. The jovian planets (Jupiter, Saturn, Uranus, and Neptune) certainly remain hot inside, but their lack of a solid surface means that they cannot support geological processes in the same manner as solid planets. However, their atmospheres are in constant motion with lots of mixing and have frequent lightning or other electrical discharges. Thus, chemical energy is available in the jovian planet atmospheres, though scientists debate whether it is sufficiently concentrated to support life. (There are other reasons to doubt the presence of life on the jovian planets, which we will discuss in Section 6.3.)

The moons of the jovian planets are all smaller than Mars. If they were similar to the terrestrial planets in composition and in their heat sources, they would be too small to have significant active internal geology. However, the jovian moons tend to contain a great deal of ice, because they formed in parts of the solar system where it was cool enough for ices to condense from the solar nebula (see Figure 1.22). Because ices melt at much lower temperatures than do metal or rock, the jovian moons can have internally driven "ice geology"—such as internal convection involving gradual movement of icy material—with much less internal heat than is needed for the "rock geology" of the terrestrial planets. In addition, at least some of the jovian moons (notably Io and Europa) have an ongoing source of internal heat quite different from any heat source on the terrestrial planets. (We'll discuss this source, *tidal heating*, in Chapter 8.) The available heat can melt some of the ice, allowing liquid water, and in at least one case (Io) it can raise temperatures enough to melt silicate rock. Numerous moderate- to large-size moons in the outer solar system show evidence of active internal geology at some point in their histories and may have had enough heat to allow for some liquid medium.

Does Life Need Water?

If any one ingredient seems absolutely essential to all life on Earth, it is liquid water. Recall that water plays at least three vital roles for life on Earth [Section 3.3]: It dissolves organic molecules, making them available for chemical reactions within cells; it allows for the transport of chemicals into and out of cells; and it is involved directly in many of the metabolic reactions that occur in cells. It is difficult to imagine life in the absence of a liquid substance to play these roles. But could these roles be fulfilled by some liquid other than water?

No one knows the answer for sure, but there are a number of constraints to consider. For example, a substance that might fulfill the roles of water must, like water, be fairly common. On Earth, the only liquid besides water commonly found is molten rock, which is so hot that it's difficult to imagine life surviving within it. However, several other common substances might take liquid form on colder worlds. Table 6.1 lists the temperature range over which water and three other potential candidates—ammonia, methane, and ethane—remain liquid. The given temperature ranges are those that apply under the atmospheric pressure on Earth. At different pressures, or with the presence of dissolved minerals, the ranges can be different. For example, salt water can remain liquid at temperatures slightly below 0°C, and under sufficient pressure water can remain liquid at temperatures well above 100°C. (In some places on Earth, such as near deep-sea vents, the pressure keeps water liquid at temperatures as high as about 375°C.) Despite such variation in melting and boiling temperatures, the ranges in Table 6.1 provide a useful comparison.

THINK ABOUT IT . . . *Oil (petroleum) is also found in liquid form on Earth. Why doesn't oil seem likely as a liquid medium for the origin of life? (Hint: Oil is a fossil fuel.)*

Table 6.1 quickly reveals one advantage of water over the other substances: It remains a liquid over a wider and higher range of temperatures. (The range for ethane is almost as wide, but at much lower temperatures.) A wider range makes it more likely that the substance can stay liquid through changes in the weather or climate. The higher temperatures for water may be even more important to life. As a general rule, chemical reactions proceed more slowly at lower temperatures because the molecules themselves move more slowly. Typically, the rate of a given chemical reaction drops in half with each 10°C decrease in temperature. Thus, chemical reactions in

Table 6.1 Companion of Potential Liquids for Life. Freezing and boiling points (under 1 atmosphere of pressure) for common substances that may be found in liquid form in our solar system. The last column gives the width of the liquid range, found by subtracting the freezing point from the boiling point.

Substance	Freezing Temperature	Boiling Temperature	Width of Liquid Range
Water (H_2O)	0°C	100°C	100°C
Ammonia (NH_3)	−78°C	−33°C	45°C
Methane (CH_4)	−182°C	−164°C	18°C
Ethane (C_2H_6)	−183°C	−89°C	94°C

liquid ammonia, methane, or ethane would proceed much more slowly than similar reactions in water. Any life using these other liquids would therefore need to have a much slower metabolism than life on Earth. The slower rate of reactions may also make it less likely that life would arise in the first place in these other liquids, because the origin of life probably requires many complex chemical reactions.

Two other properties give water less obvious but perhaps crucial advantages over the other liquids. The first involves an oddity in the way water freezes. Most substances are denser as solids than as liquids, but water is a rare exception (Figure 6.2a). Ice is less dense than liquid water, which is why ice floats. This property helps life survive on Earth. In the winter, when surface temperatures are low enough for water to freeze, floating ice forms a layer on the top of lakes and seas. This layer of ice insulates the water beneath, allowing it to remain liquid—which allows life to survive within it. No other liquid in Table 6.1 shares this property with water—that is, all the others sink when frozen.

This same property of water may be even more important to long-term climate stability. During periods when the Earth cools a bit, such as the ice ages [Section 4.5], lower temperatures allow more ice to form on the surface. If this ice sank, the surface would still be covered with liquid water, which in turn would freeze and sink. This process would continue until no liquid water was left at the surface—that is, until lakes and oceans were completely frozen. Instead, because ice floats, a cool period thickens the insulating layer of ice on the surfaces of lakes and oceans, making it less likely that they will freeze completely.

The second property that distinguishes water from the other liquids in Table 6.1 is the way electrical charge is distributed within its molecules (Figure 6.2b). Within individual water molecules, the electrons tend to be distributed in a way that makes one side have a net positive charge and the other side have a net negative charge. (Molecules with such charge separation are called *polar* molecules; the term *polar* comes from having the positive and negative charges concentrated at opposite ends, or "poles," of the molecule and not from anything related to temperature.) This charge separation affects the way in which water dissolves other substances. Molecules and salts that also have charge separations dissolve in water easily. Molecules that do not have any charge separation—such as molecules of oil—do not dissolve in water.

On Earth, the charge separation property of water is critical to life. Living cells have membranes that do not dissolve in water, so the membranes effectively protect the interior contents of cells. If we place

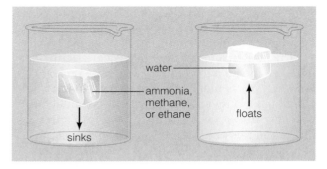

a Most substances are denser as solids than as liquids, so when solid and liquid forms exist together, the solid form sinks through the liquid. Water is a rare exception in that ice floats because it is less dense than liquid water.

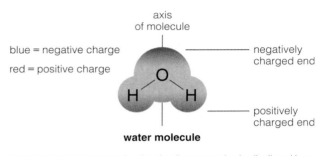

b Within individual water molecules, the electrons tend to be distributed in a way that makes one side have a net positive charge and the other side have a net negative charge.

FIGURE 6.2 Two properties of water give it advantages over other possible liquids as a basis for life.

living cells in liquid ethane, methane, or ammonia—none of which have charge separation—their membranes can dissolve. Charge separation also makes possible the formation of a special type of chemical bond, called a *hydrogen bond*, that is very important to the biochemistry of life on Earth. (The formation of hydrogen bonds as water freezes also explains why ice floats, because these bonds force the molecules into a slightly expanded structure.) Thus, among the liquids listed in Table 6.1, Earth-like living cells can survive only in liquid water.

To review, we've identified three advantages of water over other candidate liquids for life: (1) a wider and higher range of temperatures over which it remains liquid; (2) the fact that water ice floats, whereas the other substances sink when frozen; and (3) the charge separation of water molecules. These advantages make a strong case for the need for liquid water as a basis for life, but the case is not definitive. For example, it's relatively easy to imagine circumstances under which the third advantage might be turned around. If life elsewhere had cell membranes made of molecules with different charge separation properties, they might dissolve in water but not in ammonia, methane, or ethane. In that case (we do not know if it's possible), life might exist quite easily in other liquids while finding water to be poisonous.

The bottom line is that, while water clearly is an excellent medium for supporting life, we do not know if other liquids could also support life. It's difficult to tell whether the advantages we've cited for water are truly universal or specific only to terrestrial life. Given this uncertainty, along with the fact that other liquids probably exist on some of the worlds in our solar system, we should not be too quick to rule out the potential for those worlds to support life. In general, however, water is much more abundant than other liquids both in our solar system and elsewhere (largely due to the fact that oxygen is the third most abundant element in the universe, after hydrogen and helium), so it remains our prime candidate for a liquid medium.

Summary of Environmental Requirements

We began this section with the goal of making a list of environmental requirements for habitability that can help us decide which worlds to focus on as we begin the search for possible abodes of life in our solar system. In the broadest terms, we've found that life has three major requirements:

1. A source of elements and molecules from which to build living cells.

2. A source of energy for metabolism.

3. A liquid medium for transporting the molecules of life.

The first requirement is probably met by most if not all worlds. The second requirement is somewhat more limiting, but there are still plenty of worlds that should have sufficient sunlight or chemical energy for life. Moreover, any world that meets the third requirement—the need for a liquid—will probably also meet the second, because the liquid can facilitate chemical reactions with other planetary materials (especially if it is liquid water). Thus, based on our current understanding of life, we can consolidate the requirements into a single "litmus test" for habitability:

For a world to be considered potentially habitable, it must have a liquid medium, such as liquid water or one of the other liquids listed in Table 6.1.

This requirement for a liquid certainly narrows the possibilities for life in our solar system, but not as much as you might at first guess. The wide variety of habitats in which we find both liquid water and life on Earth, including the deep ocean and underground, tells us that habitability requires only the presence of a liquid *somewhere,* not necessarily on the surface. Any world with active internal geology may have liquids in localized areas of its interior. Some worlds, such as Jupiter and Saturn, have liquid droplets within their atmospheres.

There are at least two important caveats to our discussion. First, we may need to make a distinction between the ability of a world to support an origin of life and its ability to support life that might have arrived from elsewhere. For example, while most scientists have concluded that life could not have arisen indigenously in Jupiter's atmosphere, some scientists have imagined organisms that could survive there if they somehow arrived from somewhere else (see Section 6.3). Second, we must remember that everything we know about life comes from the study of life on only a single world—our own. We've already noted the difficulty of knowing whether the advantages of water as a liquid medium are truly universal or specific to life on Earth. Similarly, other apparent attributes of life may not be as crucial elsewhere as they are on Earth. It is possible that our discussions of habitability are based on a far too narrow view of life. Indeed, science fiction writers have imagined all sorts of bizarre life-forms existing under conditions far outside those we've examined here. Nature might be even more inventive. Nevertheless, a search for life must start somewhere, and it makes sense to begin by looking for the conditions that seem most likely to support life. If it turns out that our initial search is too narrow, we can always expand it in the future.

6.2 Exploring the Solar System

We've identified the types of environments we seek as potential habitats for life. Now we turn our attention to how we look for such habitats in our solar system. On Earth, we can sample potential habitats directly. We walk across the land surface, we drill into the crust, and we use submarines to explore the oceans. Unfortunately, a similarly broad strategy is far too costly to consider for any other world, at least at the present time. Therefore, we must rely on less direct ways of studying environmental conditions. In this section, we'll discuss a few of the most important ways in which we explore the solar system today.

The Challenge of Exploration

Generally speaking, we can study other worlds in our solar system in three basic ways:

1. We can make observations from Earth, using telescopes on the ground or in Earth orbit.

FIGURE 6.3 This photo shows astronaut Buzz Aldrin on the Moon in July 1969. His visor reflects an image of astronaut Neil Armstrong, the first human to set foot on another world. Six Apollo missions each took two astronauts to the Moon's surface while a third astronaut remained in lunar orbit. The Apollo astronauts brought hundreds of kilograms of Moon rocks back to Earth, and study of these rocks has helped us learn about the history of our solar system.

2. We can send robotic spacecraft to study a world closer up, in some cases sampling the surface (or atmosphere) or even returning samples to Earth.

3. We can send humans to explore another world.

To date, humans have visited only one other world: our Moon. Between 1969 and 1972, NASA's Project Apollo sent a series of spacecraft to the Moon. Six missions landed successfully—Apollo 11, 12, 14, 15, 16, and 17—giving 12 humans the opportunity to walk on another world and collect samples to bring back to Earth for study (Figure 6.3). (Apollo 13 suffered an accident in space; the astronauts returned safely to Earth but did not land on the Moon.)

Sending humans to any other world is far more challenging. To see why, recall the scale of the solar system (see Figures 1.7 and 1.8). At the 1-to-10-billion scale that we introduced in Chapter 1, the Moon is only about 4 centimeters—or roughly two thumb-widths—from the Earth. On this same scale, Mars is about 8 meters from the Earth at its nearest (when it comes closest to Earth in its orbit). That is, Mars at its nearest is about 200 times farther from Earth than is the Moon. With current technology, sending astronauts to Mars would require at least several months in each direction, plus whatever time the astronauts spent on the surface. Building and launching a spaceship that could support a human crew for such a long period would be technologically challenging and very expensive. The challenge would be even greater for other potential targets of human exploration.

Because of these challenges, virtually all solar system exploration to date has been carried out either telescopically from Earth (or Earth orbit) or with robotic spacecraft. In Section 2.4, we briefly discussed some of the ways in which telescopes and spacecraft help us learn about other worlds. Now we are ready to explore these methods in somewhat more detail.

What Can We Learn with Telescopes?

The power of telescopes comes primarily from their ability to collect far more light than the human eye and to see smaller details than human eyes can see. Today's largest telescopes, such as the 10-meter-diameter Keck telescopes in Hawaii (Figure 6.4), can collect more light than a million human eyes. Moreover, we can attach cameras and instruments to telescopes to further increase their capabilities. For example, a camera can take a time-exposure photograph that collects far more total light than can be seen by a brief peek through the telescope. In addition, we can use cameras or other instruments to study light that is invisible to the human eye, such as infrared or ultraviolet light.

Broadly speaking, we can divide telescopic studies of our solar system into three main categories of observation: imaging, spectroscopy, and long-term monitoring. Let's investigate each of these categories.

Imaging At its simplest, imaging means using a telescope to produce a picture of an object. Prior to the advent of photography in the nineteenth century, astronomers peered through telescopes and then drew pictures of what they saw. Photography enhances imaging because it provides more accurate recordings than can be drawn by hand and because it allows us to make time-exposure images.

a The two telescopes, each 10 meters in diameter, can be seen in the open areas of their twin domes.

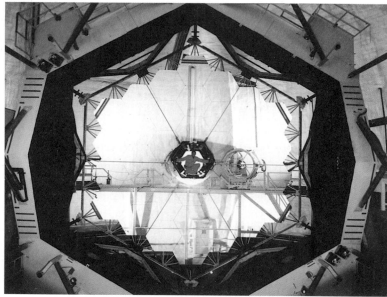

b Each telescope's mirror is actually made from 36 individual, hexagonal mirrors that work together to bring light to a single focus. These mirrors make the honeycomb pattern that surrounds the hole enclosing the man in this photo.

FIGURE 6.4 The Keck telescopes sit atop Mauna Kea in Hawaii.

Many people mistakenly assume that telescopes are designed to "magnify" images, but their purposes are actually subtler. Because a telescope collects so much light, a telescopic image allows us to see objects that are too faint to be seen otherwise. For example, many stars that are invisible to the naked eye can be spotted easily with telescopes. Note that magnification is not an issue here, because stars appear only as points of light even with powerful telescopes. Thus, while we can discover a star's existence by seeing a point of light, we generally cannot see any details on its surface.

In other cases, telescopes do reveal details, but not simply because they make objects look bigger. Rather, telescopes improve the **resolution** of an image. If you take a picture of the Moon without a telescope, greatly enlarging the picture won't help you see details in its craters. Instead, the picture will become grainy as you enlarge it. A picture taken through a large telescope will show new details— features too small to see without the telescope. Galileo used his first telescope to see details of light and shadow on the Moon, proving that the Moon had mountains and valleys. Today, telescopes can reveal details more than a thousand times smaller than what can be seen by the naked eye. For example, the resolution of the Hubble Space Telescope (Figure 6.5) would be sufficient for you to read this book from a distance of about 800 meters (one-half mile). In principle, larger telescopes (such as the Keck telescopes) offer even better resolution.

In practice, telescopic resolution is often limited by the Earth's atmosphere. The twinkling of stars that looks so beautiful to the naked eye is a headache for astronomers. Twinkling is not a property of the stars themselves but rather a result of starlight's being jiggled about by turbulence (air motions) in our atmosphere. This jiggling of light tends to blur photographs taken through telescopes. Some new technologies (most notably, *adaptive optics*) can help telescopes on Earth overcome this blurring, but the ultimate solution to atmospheric blurring is to put telescopes above the atmosphere. The Hubble Space Telescope is not nearly as big as the largest telescopes on the ground—its primary mirror is 2.4 meters in diameter, giving it only about 1/16 the light-collecting area of the 10-meter Keck telescopes. But it routinely achieves better resolution because it does not have to peer through the Earth's atmosphere.

THINK ABOUT IT . . . *Drop a coin to the bottom of a glass of water. Notice how the coin seems to move as you look through the water, even while it is actually stationary on the bottom. How is this similar to twinkling? Now compare the "twinkling" of the coin to that of a tiny pebble that you drop to the bottom. Based on what you see, can you explain why planets tend to twinkle less than stars? (Hint: While stars appear as points of light even through powerful telescopes, planets appear as disks through telescopes.)*

Resolution can also be improved by linking telescopes together through a technology called **interferometry.** An interferometer allows several smaller telescopes to work together to achieve the resolution of a much larger telescope. (The light-collecting area is not equivalent to that of the larger telescope, but only to the sum of the areas of the individual telescopes.) Interferometry is now routine with radio telescopes. For example, the Very Large Array (VLA) near Socorro, New Mexico, consists of 27 individual radio dishes that can achieve a resolution equivalent to that of a single radio dish almost 40 kilometers across (Figure 6.6). Interferometry is technologically

FIGURE 6.5 This photo shows astronauts working on the Hubble Space Telescope in the Space Shuttle cargo bay during a 2002 servicing mission. The Hubble Space Telescope orbits the Earth, which keeps it above the blurring effects of the atmosphere. Its position in space also allows it to "see" infrared and ultraviolet light that does not penetrate the Earth's atmosphere.

FIGURE 6.6 The Very Large Array (VLA) in New Mexico consists of 27 telescopes that can be moved along rails. The telescopes work together through interferometry and can achieve an angular resolution equivalent to that of a single radio telescope almost 40 kilometers in diameter.

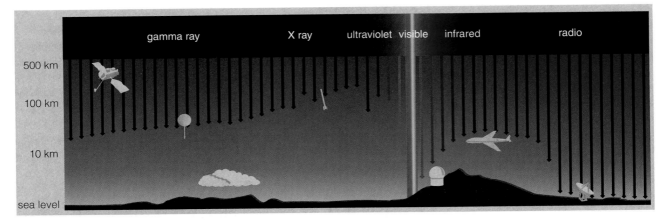

FIGURE 6.7 This diagram shows the approximate depths to which different wavelengths of light penetrate the Earth's atmosphere. Note that most of the electromagnetic spectrum—except for visible light, a small portion of the infrared, and radio—can be observed only from very high altitudes or from space.

more difficult with shorter-wavelength light. Nevertheless, astronomers have learned to link telescopes together for interferometry with infrared light and are beginning to achieve success with visible light as well.

To describe resolution quantitatively, we usually state the minimum angular distance over which we can distinguish points of light. For example, the resolution of the human eye is about 1 arcminute, or 1/60 of 1°. (If you extend your finger at arm's length, it is about 1° across, so an arcminute is 1/60 of this angular width.) This means that if you look at two stars in the sky that are separated by *more* than 1 arcminute, you'll see them as two distinct stars. But if the two stars are closer together than 1 arcminute, your eye will think there is only one star because you'll see the light from the two stars merged. Ground-based telescopes are typically limited by the atmosphere to a resolution of about 1 arcsecond, which is 1/60 of 1 arcminute (or 1/3,600 of 1°). In contrast, the Hubble Space Telescope routinely achieves a resolution of better than 0.1 arcsecond. Optical interferometers may do much better still. In the future, scientists hope to launch groups of telescopes that will work as interferometers in space, achieving far better resolution than any telescope today. Already, NASA is developing plans for an optical interferometer in space that would be capable of achieving resolution as small as 0.001 arcsecond [Section 10.3]. Such a telescope would have sufficient resolving power to give us our first crude images of planets around other stars.

Telescopes in space have another important advantage over telescopes on the ground: They can "see" light that does not penetrate the Earth's atmosphere. Most forms of light are blocked by the Earth's atmosphere; only radio waves, some infrared light, and visible light can be studied by telescopes on the ground (Figure 6.7). All other forms of light—

including most infrared, ultraviolet, X rays, and gamma rays—can be studied only with telescopes in space. (Different types of telescopes are needed for different forms of light; for example, radio telescopes require different designs from those of visible light telescopes, which are themselves different from X ray telescopes.) Although this light is invisible to our eyes, it can be recorded as images with specialized instruments. Because different forms of light are produced in different ways, images made with nonvisible light often reveal information that we could not otherwise obtain. For example, the planet Venus appears to be covered with a bland haze when photographed in visible light, but images made with ultraviolet light show clouds in its atmosphere (Figure 6.8).

Spectroscopy We can learn much from telescopic images, but for many purposes spectroscopy is an even more powerful technique. Spectroscopy involves collecting light through a telescope, then dispersing it into a *spectrum* in much the same way a prism disperses white light into a rainbow of color (Figure 6.9a). However, astronomical *spectrographs* can reveal much more detail in the spectrum than can a prism. This detail offers a wealth of information about distant objects. For example:

■ In different parts of the spectrum, an object may emit its own light, absorb incoming light, or reflect incoming light. We can learn a great deal by studying the precise patterns of emission, absorption, and reflection. For example, the planet Mars emits its own infrared light, absorbs blue light from the Sun, and reflects red light from the Sun (hence its red color). Careful study of Mars's spectrum tells us the planet's surface temperature and helps us identify minerals and ices on its surface.

b This photograph of Venus was taken with cameras sensitive to ultraviolet light on NASA's Pioneer Venus Orbiter. The ultraviolet image reveals cloud structure. The dark and light colors in this photograph correspond to different reflectivities of ultraviolet light.

FIGURE 6.8 Observing objects with nonvisible light can reveal new information.

a This photograph shows Venus as it appears in visible light. The thick clouds form a bland haze with no obvious detail. (Photo taken by the Galileo spacecraft during its Venus flyby en route to Jupiter.)

FIGURE 6.9 An astronomical spectrum is made by collecting light in a telescope and then passing the light through a spectrograph that disperses it into its component colors.

a A prism illustrates the basic idea: The prism disperses white light into a rainbow of color.

b An astronomical spectrum can show much more detail than we see in light dispersed by a typical prism. Here, we see an idealized spectrum of warm molecular hydrogen gas (H_2). The specific pattern of the many bright lines tells us that they were produced by molecular hydrogen rather than by something else.

- In many cases, spectra show patterns of bright and dark lines, called **spectral lines** (Figure 6.9b), which serve as chemical "fingerprints." Every chemical element, every ion of each element, and every molecule produces its own, unique pattern of spectral lines. (Recall that an ion is an atom that is electrically charged because the atom has either lost or gained one or more electrons.) Thus, careful study of a spectrum can allow us to determine the chemical composition of a distant world. In this way, we can learn the molecular composition of a planet's atmosphere or the chemical composition of a star. Different isotopes of an element also have slightly different spectra, so we can sometimes even determine isotopic ratios in distant worlds.

- By studying spectral lines in detail—for example, how bright or dark they are, how wide they are, and what precise set of atoms and ions is represented in a spectrum—scientists can infer even more information about distant objects. Among other things, spectral lines can tell us an object's temperature, rotation rate, and, in some cases, magnetic field strength. For objects with atmospheres, spectral lines can tell us their atmospheric density and pressure and provide information about atmospheric motions.

Perhaps more than any other astronomical technique, spectroscopy would have seemed like magic to scientists as recently as two centuries ago. Indeed, before scientists discovered and began to learn the meaning of spectral lines in the nineteenth century, many people assumed that we would never be able to discover what distant planets and stars are made of. The power of spectroscopy is limited only by how much light we can collect, not by distance. Spectroscopy has allowed us to determine the composition of planets, stars, and galaxies without ever leaving the Earth. It is an indispensable tool in the search for life in the universe, because it allows us to learn about habitability from a distance. Someday, we may even make the first conclusive detection of life outside our solar system by finding unmistakable spectral signatures in the atmosphere of some distant planet [Section 10.3].

Long-term Monitoring The third major use of telescopes is for monitoring distant objects over periods of many years or decades. This use may involve either imaging or spectroscopy (or both), repeated many times. Through long-term monitoring, we can learn about weather and climate on other planets in our solar system. In principle, we could carry out long-term monitoring even more effectively with spacecraft orbiting another planet. In practice, however, long-term monitoring is usually carried out with

FIGURE 6.10 Jupiter's Great Red Spot is clearly visible in this photograph from the Hubble Space Telescope. The Great Red Spot is a gigantic, high-pressure storm; it varies in size but typically is large enough that it could swallow two or three Earths. Records of telescopic observations tell us that this storm has persisted on Jupiter for at least three centuries, or ever since telescopes became powerful enough to allow us to see it.

telescopes on Earth. Interplanetary spacecraft tend to be fragile, and most operate for no more than a few years at best. Telescopes, on the other hand, can monitor a planet indefinitely. Moreover, telescopic data were being collected long before the era of planetary exploration began. For example, Jupiter's Great Red Spot has been seen through telescopes for more than three centuries, telling us that this high-pressure storm has persisted for at least that long (Figure 6.10). Similarly, we have learned about the seasonal patterns of dust storms on Mars through telescopic observations carried out over many decades.

Long-term monitoring with telescopes on Earth may become even more important in the future. In the past, opportunities for long-term monitoring were somewhat limited by the fact that only a few large telescopes existed. The 1990s witnessed the beginning of a revolution in telescope-making that has greatly increased our ability to observe. At the beginning of the 1990s, the world's most powerful telescope was the 5-meter telescope on Mount Palomar (near San Diego), which opened in 1947. Just over a decade later, this telescope is no longer on the top 10 list of the world's largest telescopes. With so many large telescopes available, scientists may soon be able to devote greater amounts of observing time to long-term monitoring of the planets and moons in our solar system.

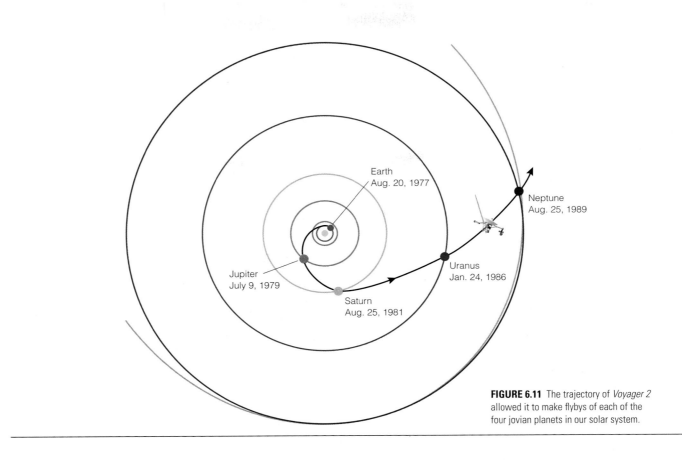

FIGURE 6.11 The trajectory of *Voyager 2* allowed it to make flybys of each of the four jovian planets in our solar system.

Robotic Spacecraft

Although Earth-based telescopes have taught us a lot about the planets, the recent revolution in our understanding of the solar system has come about largely through robotic spacecraft that actually go to the worlds we wish to study. Robotic spacecraft operate primarily with preprogrammed instructions. They carry radios for communication, which allows them to receive additional instructions from Earth and to send home the data they collect. Thus, most robotic spacecraft make one-way trips from Earth, never physically returning but sending their data back from space in the same way we send radio and television signals around the world.

The robotic missions we send to explore other worlds fall into one of four broad categories: flybys, orbiters, landers or probes, and sample return missions. The choice of spacecraft type depends both on scientific objectives and on cost. In general, the lowest-cost way to visit another planet is with a **flyby,** in which a spacecraft flies past the planet once and then continues on its way. While a flyby offers only a relatively short period of close-up study, it is an excellent way to take a first detailed look at another world. In addition, a well-planned flyby mission can visit multiple planets. For example, the *Voyager 2* spacecraft flew past Jupiter, Saturn, Uranus, and Neptune before continuing on its way out of our solar system (Figure 6.11).

To a large extent, the primary purpose of a flyby is to obtain higher-resolution telescopic images than we can obtain from Earth. Flybys generally carry small telescopes, cameras, and spectrographs. Because these instruments are brought within a few tens of thousands of kilometers of other worlds (or closer), they can obtain much higher resolution images and much clearer spectra than even the largest telescopes currently viewing these worlds from Earth. In some cases, the views afforded by flybys may give us information that would be very difficult to obtain from Earth. For example, *Voyager 2* helped us discover Jupiter's ring and learn about the rings of Saturn, Uranus, and Neptune through views in which the rings were backlit by the Sun—views possible only from beyond each planet's orbit and not from Earth. Flybys may also carry instruments to measure local magnetic field strength or to sample interplanetary dust. In addition, gravitational effects of the planets and their moons on the spacecraft provide information about object masses and densities and sometimes even about how mass is distributed within an object's interior. Like the backlit views of the rings, these data cannot be gathered from Earth. Indeed, most of what we know about the masses and compositions of moons comes from spacecraft that have flown past them.

THINK ABOUT IT . . . Look back at Figure 1.8 to see how the solar system appears on a scale of 1 to 10 billion. About how big is Jupiter on this scale, and how far is it from Earth (at its nearest)? If a spacecraft flies within 10,000 kilometers of Jupiter's cloudtops, how far is it from Jupiter on the 1-to-10-billion scale? Based on your answers, explain why spacecraft can obtain much higher resolution images of planets than we can obtain from Earth.

Flybys tend to be cheaper than other missions because they are generally less expensive to launch into space. Launch costs depend largely on weight, and onboard fuel is a significant part of the weight of a spacecraft heading to another planet. Once a spacecraft is on its way, the lack of friction or air drag in space means that it can maintain its orbital trajectory through the solar system without using any fuel at all. Fuel is needed only when the spacecraft needs to change from one trajectory (orbit) to another. Moreover, with careful planning, some trajectory changes can be made by taking advantage of the gravity of other planets. If you look closely at the *Voyager 2* trajectory in Figure 6.11, you'll see that it made significant trajectory changes as it passed by Jupiter and Saturn. In effect, it made these changes for free by using gravity to bend its path rather than by burning fuel. (This technique is known as a "gravitational slingshot.")

In contrast to the low fuel requirement for a flyby, an **orbiter** must change from an interplanetary trajectory to one that puts it in permanent orbit around another world. This maneuver requires significant additional fuel, which is one reason why orbiters are generally more expensive than flybys for equivalent weights of scientific instruments. Of course, once an orbiter reaches its destination, it can study another world for a much longer period of time than can a flyby. Orbiters are used for many types of observations. Like flybys, they often carry cameras, spectrographs, and instruments for measuring magnetic field strength. Some missions to Venus and Mars have used radar or lasers to make precise altitude measurements of surface features. In the case of Venus, radar observations give us our only clear look at the surface, because we cannot otherwise see through the planet's thick cloud cover (see Figure 6.8).

The most "up close and personal" study of other worlds comes from spacecraft that send **probes** into their atmospheres or **landers** to their surfaces. For example, in 1995 the *Galileo* spacecraft dropped a probe into Jupiter's atmosphere (Figure 6.12). The probe collected temperature, pressure, composition, and radiation measurements for about an hour as it descended before being destroyed by the high pressures and temperatures that occur deeper in Jupiter's

atmosphere. On planets with solid surfaces, a lander can offer close-up surface views and local weather monitoring. Some landers, such as the *Pathfinder* lander that arrived on Mars in 1997, carry robotic rovers that can venture across the landscape (see Figure 1.2). Landers can also conduct automated experiments on the planetary surface. For example, the Viking landers, which landed on Mars in 1976, carried out simple experiments designed to measure the composition of the surface and to look for signs of life [Section 7.3].

While probes and landers can carry out experiments on surface rock or atmospheric samples, the experiments must all be designed in advance and must fit inside the spacecraft. These limitations make scientists long for missions that will scoop up samples from other worlds and return them to Earth for more detailed study. To date, the only such **sample return missions** have been to the Moon. U.S. astronauts collected samples during the Apollo missions, and the Soviet Union sent robotic spacecraft to collect rocks from the Moon in the early 1970s. Many scientists are working toward a sample return mission to Mars, which they hope will occur within the next decade or so [Section 7.5]. (A slight variation on the sample return mission is the Stardust mission, currently en route to collect a sample of comet dust and return it to Earth in 2006.)

FIGURE 6.12 Artist's conception of the Galileo mission's probe entering Jupiter's atmosphere on December 7, 1995. This view, looking upward from beneath the clouds, shows the suitcase-size probe falling with its parachute open above it.

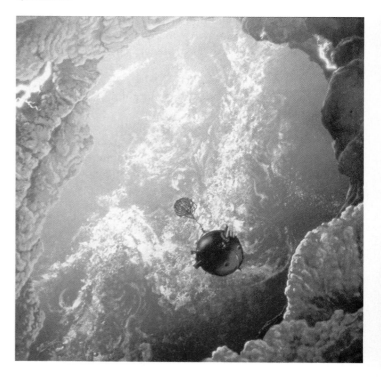

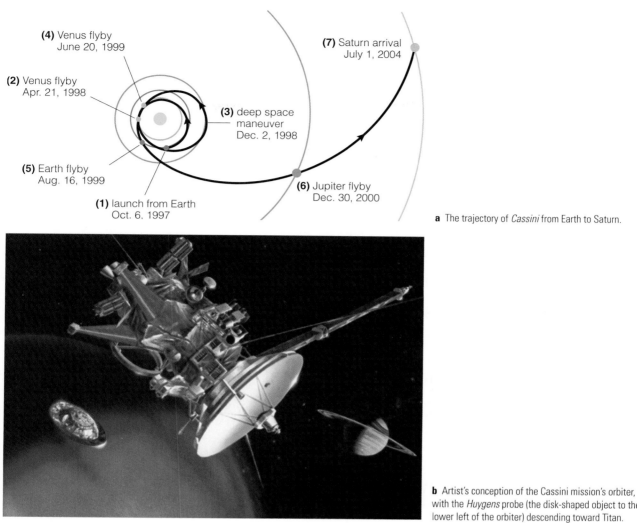

(4) Venus flyby
June 20, 1999

(2) Venus flyby
Apr. 21, 1998

(3) deep space maneuver
Dec. 2, 1998

(7) Saturn arrival
July 1, 2004

(5) Earth flyby
Aug. 16, 1999

(1) launch from Earth
Oct. 6, 1997

(6) Jupiter flyby
Dec. 30, 2000

a The trajectory of *Cassini* from Earth to Saturn.

b Artist's conception of the Cassini mission's orbiter, with the *Huygens* probe (the disk-shaped object to the lower left of the orbiter) descending toward Titan.

FIGURE 6.13 The Cassini mission to Saturn.

Many missions combine elements of more than one type of spacecraft. For example, the Galileo mission to Jupiter included an orbiter that studied Jupiter and its moons, as well as the probe that entered Jupiter's atmosphere. Each of the two Viking missions to Mars consisted of both an orbiter and a lander. The Cassini mission, which will arrive at Saturn in 2004, included flybys of Venus, Earth, and Jupiter, and it has both an orbiter that will study Saturn and its moons and a probe, called the *Huygens probe,* that will descend through the atmosphere of Saturn's moon Titan (Figure 6.13).

Exploration Past, Present, and Future

Thanks to the past several decades of study—both with telescopes on Earth and with robotic spacecraft —we now know the general characteristics of all the major planets and moons in our solar system. Tele-

scopes will continue to play an important role in future observations, particularly for long-term monitoring. But for the purpose of learning about the prospects for life and habitability in our solar system, we will probably depend primarily on spacecraft.

Table 6.2 lists some significant robotic missions of the past and present. Note that the next few years promise many new discoveries as missions arrive at their destinations. Over the longer term, all the world's major space agencies have high hopes of launching numerous and diverse future missions to study our solar system. Of course, these hopes are subject to the vagaries of politics and budgets for scientific exploration. If they come to fruition, however, the next couple of decades will see spacecraft that answer many fundamental questions about the habitability of worlds in our solar system, and perhaps even tell us whether any of them have supported—or still support—life.

Table 6.2 Selected Robotic Missions to Other Worlds

Name	Type	Destination	Lead Space Agency	Arrival Year	Mission Highlights
Cassini	Orbiter	Saturn	NASA	2004*	Includes lander (called *Huygens*) for Titan
Mars Exploration Rovers	Lander	Mars	NASA	2004*	Twin rovers to study the surface in two locations
Mars Express	Orbiter and lander	Mars	ESA**	2004*	Climate and surface studies of Mars
Nozomi	Orbiter	Mars	Japan	2004*	Japanese-led mission to study martian atmosphere
Contour	Flyby	3 comets	NASA	2003*, 2006*, 2008*	Detailed study of three comet nuclei
Mars Odyssey	Orbiter	Mars	NASA	2001	Detailed study of martian surface features and composition
Near Earth Asteroid Rendezvous (NEAR)	Orbiter	Eros (asteroid)	NASA	2000	First spacecraft dedicated to in-depth study of an asteroid; mission ended with a soft landing on the surface
Mars Global Surveyor	Orbiter	Mars	NASA	1997	Detailed imaging of surface from orbit
Mars Pathfinder	Lander	Mars	NASA	1997	Carried *Sojourner,* the first robotic rover on Mars
Galileo	Orbiter	Jupiter	NASA	1995	Dropped probe into Jupiter; close-up study of moons
Magellan	Orbiter	Venus	NASA	1990	Detailed radar mapping of surface of Venus
Voyager 1	Flyby	Jupiter, Saturn	NASA	1979, 1980	Unprecedented views of Jupiter and Saturn; continuing out of solar system
Voyager 2	Flyby	Jupiter, Saturn, Uranus, Neptune	NASA	1979, 1981, 1986, 1989	Only mission to Uranus and Neptune; continuing out of solar system
Vikings 1 and 2	Orbiter and lander	Mars	NASA	1976	First landers on Mars

*Scheduled arrival year.   **European Space Agency

6.3 A Biological Tour of the Solar System

Imagine living a century from now, when we have sent orbiters and landers to every major world in our solar system and when humans may even be living and working on many worlds. At that time, a "life in the universe" course might begin with a true biological tour of the solar system. One world at a time, we would discuss its habitability, the characteristics of its life or why it does not have its own indigenous life, and the prospects of survival (or the results of actual experiments) for any life from Earth that we might transport there.

We cannot undertake such a complete biological tour today. However, we have already learned enough to say which worlds seem most likely to be habitable —that is, to meet the environmental requirements for life—and which worlds seem least likely to be capable of harboring any life. This information will help guide the search for life in our solar system over the next century, because it will help us determine where to focus our efforts. And, as a practical matter, it will help us decide where to devote our attention in this book.

Mercury   Earth   Earth's Moon

FIGURE 6.14 Comparative global and surface views of the Moon and Mercury. The global views show the Moon and Mercury (and Earth) to scale.

Among all the worlds of our solar system, two stand out as being especially likely to be habitable: the planet Mars and Jupiter's moon Europa [Section 1.1]. Our study of each of these worlds indicates that it may have liquid water (or has had it in the past), either at the surface or beneath the surface. Each also has all of the elements necessary to support life, as well as chemical energy that can power metabolism. These worlds therefore seem to meet all of the environmental conditions that we think are necessary to support both the origin and the continued existence of life. Several other moons of the outer solar system also seem to hold at least some promise of habitability, including two other large moons of Jupiter—Ganymede and Callisto—and Saturn's moon Titan. Therefore these worlds deserve in-depth discussion. We will devote Chapter 7 to the study of Mars and Chapter 8 to the study of Europa and the other potentially habitable moons of the outer solar system. For the rest of the solar system, we'll embark on our biological tour right here, focusing on a brief survey of current understanding of the habitability or nonhabitability of the various worlds.

The Moon and Mercury Although Mercury is a planet and the Moon orbits planet Earth, the two worlds share many characteristics. Surface views show them both to be pockmarked with craters (Figure 6.14). Both are relatively small, so they have lost most of their internal heat. With no outgassing and with weak gravity that would make it difficult to hold an atmosphere in any case, both worlds are essentially airless.[2] Mercury and the Moon also share the distinction of being among the places least likely to

be habitable in the solar system, primarily because neither world is likely to have any liquids anywhere.

As we discussed in Section 4.2, the Moon has a far smaller proportion of water and other volatiles than Earth, probably as a result of vaporization that occurred in the giant impact thought to have formed the Moon. The only place the Moon may have any significant amount of water is in craters near its poles. The bottoms of these polar craters are perpetually shielded from sunlight, making them extremely cold and dark. As a result, ices supplied over the eons by impacts (or possibly by outgassing from the lunar interior) may lie frozen at the crater bottoms (Figure 6.15). However, even if such ices are present in lunar craters, they exist only because of the extreme cold and would never be exposed to enough heat to melt into liquid form. Thus, ice in lunar craters probably has no relevance to the Moon's lack of habitability, though it may be useful to *us* as a source of water for future colonies on the Moon.

We are less certain about the presence of water on Mercury. Because it formed so close to the Sun, Mercury must have accreted primarily from planetesimals that lacked any ices. But, as was the case with Earth, a few planetesimals from farther out in the solar system may have brought water and other ices to Mercury during its formation. If so, some of these materials may be chemically bound in subsurface rock on Mercury. However, Mercury's high average density suggests that, like the Earth, it was rocked by

---

[2] Technically, the Moon and Mercury have extremely thin atmospheres created by diffusion of atoms and molecules from the interior and by release of gases from impacts, but these atmospheres are far thinner than what we would consider to be an excellent vacuum in laboratories on Earth.

FIGURE 6.15 This color-coded map shows evidence for water ice in craters near the Moon's north pole. The portion of the Moon shown includes all latitudes north of 70°N, and the colors represent the concentration of hydrogen atoms measured by NASA's Lunar Prospector mission in the late 1990s. (The measurements were actually made by neutron detection techniques.) The dark blue and purple regions have the highest hydrogen concentrations, which are probably due to the presence of water ice on or just below (within a half-meter of) the surface. Note that these regions are very near the pole, where some crater bottoms are permanently shielded from the Sun. Similar results were found for the Moon's south pole.

a giant impact some time shortly after it underwent differentiation. The giant impact is thought to have ejected much of Mercury's original mantle into space, leaving behind a planet consisting predominantly of an iron-rich core. This impact likely would have ejected most or all of any water that accreted during the planet's formation.

Mercury, like the Moon, may have ices brought by impacts in crater bottoms near its poles, where temperatures are always far below freezing because no sunlight ever hits and there is no atmosphere to bring heat from other parts of the planet. But, as is also the case with the Moon, there is little chance that any of these ices ever melt into liquid form.

The Moon and Mercury may be lost causes when it comes to hope for finding life, but they are still very important to our general understanding of planetary systems. In particular, studies of the Moon and Mercury teach us a great deal about the origin and history of our solar system and help us learn about why some worlds are habitable and others are not. Thus, while the Moon and Mercury are almost certainly lifeless, they are still important targets of study for planetary scientists.

Movie Madness: *2001—A Space Odyssey*

It's the year 2001, and five clean-cut astronauts and a conniving computer named HAL are on their way to Jupiter.

Why are they going to this behemoth world? Is it to study Jupiter's complex, churning atmosphere? Umm, no that's not it. Are they on a mission to examine the giant planet's imposing magnetic field? Er, negative on that. Perhaps they're searching for signs of simple life on the moons Europa, Callisto, and Ganymede, all of which could have unseen oceans beneath their crusty exteriors? Actually, no, that's not it either.

No, these jovian rocket jockeys are headed to the king of the planets because some unknown race of aliens has been messing with our solar system. Four million years ago, some kindly extraterrestrials took a look at Earth and decided our planet's simian population had potential. They planted a dark gray monolith that, when touched by the local apes, converted them from a bunch of howling half-wits into tool-using primates.

This is an amazing tale (and one that, if true, will confound all future paleontologists), but the aliens weren't content merely to have introduced intelligence on Earth. They also buried a second monolith on the moon, figuring that a brainy species might eventually develop enough technology to leave its planet and find this weird artifact. The lunar monolith, in turn, directed us to Jupiter.

So that's why the interplanetary spacecraft *Discovery One* is on its way to the biggest world of the solar system. The hope, it seems, is that by journeying to this massive planet *Homo sapiens* can finally learn whatever profound secrets these altruistic aliens wish to share.

2001 is as much a space oddity as a space odyssey. Forget the fact that the message found at Jupiter is mostly a puzzling psychedelic light show. Ignore the fact that the ship's smooth-talking, onboard computer is more scheming than Machiavelli. No, in the end the most distressing thing about *2001* (other than the dismaying truth that in 2001 the real space program wasn't even close to sending manned missions to Jupiter) is that our past and our destiny are simply some extraterrestrials' science-fair experiment. We're no more than a bunch of choreographed puppets.

Someday it's quite likely that we will travel to the other worlds of the solar system, not just in our mind's eye or in the virtual worlds of our theaters and planetaria, but in honest-to-goodness spacecraft. However, we will make these voyages of discovery based on our own curiosity and quest for knowledge—and not because some control-freak aliens pointed to the gas-giant worlds of the outer solar system with their monolith fingers and said "Go there."

FIGURE 6.16 This photograph shows a small patch of the surface of Venus; part of the lander that took the picture is in the foreground to the lower right. The Soviet Union sent several landers to Venus during the 1970s. Because of the high temperatures and corrosive atmosphere, none of them functioned for more than an hour after landing.

Venus Venus has been called Earth's "sister planet" because it is relatively close to us in the solar system and is nearly identical in size to Earth. Venus is completely enshrouded in thick clouds, and past generations of scientists could only guess about the surface conditions. Some imagined a pleasantly warm planet. Although Venus is closer to the Sun, it is not *that* much closer—about two-thirds as far from the Sun as is Earth. Thus, if Venus had an Earth-like atmosphere, we might expect it to be only a little hotter than Earth. In fact, calculations show that with an Earth-like atmosphere, Venus's global average temperature would be about 35°C (95°F). This type of thinking led past generations of science fiction writers to speculate that Venus might be a lush, tropical paradise.

THINK ABOUT IT . . . *A good way to think about the expected temperature difference between Venus and Earth is to again consider the 1-to-10-billion scale model of our solar system. Recall that at this scale the Sun is about the size of a grapefruit, and Venus and Earth are located about 10 and 15 meters away, respectively. Imagine that you have a grapefruit that is extremely hot. If the heat feels comfortable at a distance of 15 meters, how would it feel at 10 meters? Discuss the difference qualitatively, and relate it to the expected temperature difference between Venus and Earth.*

Such dreams evaporated in 1962, when the *Mariner 2* spacecraft flew past Venus and discovered that its surface is scorching hot. (Some earlier radio observations from Earth had hinted at the high temperature, but other explanations for the data seemed possible prior to the spacecraft observations.) The surface temperature is an incredible 450°C (about 850°F)—far hotter than a pizza oven. Moreover, this temperature persists planetwide, both day and night. All the while, a thick atmosphere bears down on the surface with a pressure 90 times that on Earth's surface—equivalent to the pressure at a depth of nearly 1 kilometer (0.6 mile) in the oceans. Besides Venus's crushing pressure and searing temperature, the atmosphere of Venus contains sulfuric acid and other chemicals that are toxic to us. Indeed, landers sent to Venus by the Soviet Union provided data for only a brief period before being disabled by the high temperatures (Figure 6.16). Far from being a beautiful sister planet to Earth, Venus resembles a traditional view of hell.

What causes such extreme conditions on Venus? The answer is the *greenhouse effect*—the same effect that makes our own planet so comfortable [Section 4.5], but operating much more strongly. Recall that, in the absence of the greenhouse effect, Earth would be frozen over. Our planet is habitable because a moderate greenhouse effect traps enough heat to raise temperatures above the freezing point of water. Thus, while human contributions to the greenhouse effect are often in the news as an environmental problem, the naturally occurring greenhouse effect is a very good thing for life on Earth. Viewed from this perspective, Venus provides proof that it is possible to have too much of a good thing.

The strong greenhouse effect on Venus—often called a **runaway greenhouse effect** (for reasons we will discuss in Chapter 9)—comes primarily from carbon dioxide in its atmosphere. Earth has a modest greenhouse effect because carbon dioxide makes up only a fraction of 1% of our atmosphere. In contrast, carbon dioxide makes up more than 96% of Venus's far thicker atmosphere. Interestingly, we can explain this difference in atmospheric carbon dioxide by contrasting the fate of carbon dioxide gas on each planet. Both planets have outgassed similar total amounts of carbon dioxide over the course of their histories. However, whereas almost all of the carbon dioxide outgassed on Earth is now locked up in carbonate rocks or dissolved in the oceans (170,000 times as much as is in our atmosphere [Section 4.5]), all of Venus's outgassed carbon dioxide remains in its atmosphere. Thus, Venus's high temperature results from the strong greenhouse effect produced by its atmospheric carbon dioxide, and the high pressure results from the sheer volume of this gas.

Its high surface temperature all but rules out the possibility of life on Venus today. It is far too hot for liquid water, let alone liquid ammonia, methane, or ethane. However, Venus may have been habitable in the past. Recall that the mechanism by which carbon dioxide gets locked up in Earth's carbonate rocks involves the carbon dioxide cycle (see Figure 4.25). This cycle depends on both plate tectonics and the presence of oceans: Carbon dioxide dissolves in the oceans, where it reacts chemically to form carbonate minerals. Venus has no similar cycle because it lacks oceans (and apparently lacks plate tectonics as well) and therefore has no mechanism for removing carbon dioxide from the atmosphere. However, it is possible that Venus had oceans in the distant past, perhaps around 4 billion years ago. We'll discuss the possibility of past oceans on Venus, and what might have happened to them, when we consider the evolution of habitability in Chapter 9.

If Venus once had oceans—a big "if"—it's conceivable that life arose at that time. Unfortunately, we may never be able to find out. Crater counts suggest that Venus's entire surface is less than about a billion years old, meaning that volcanism or tectonic processes have reshaped or paved over rocks from earlier times. Thus, even if life arose and survived until the runaway greenhouse effect made Venus uninhabitable, any fossil evidence was probably destroyed long ago.

Jovian Planets Each of the four jovian planets—Jupiter, Saturn, Uranus, and Neptune—reigns over a set of moons that make their systems look much like solar systems in miniature. Together, the jovian planets have more than 80 known moons, and additional small moons may yet be discovered. A few of the large moons, including Jupiter's moon Europa and possibly Saturn's moon Titan, are likely to be habitable. But what about the jovian planets themselves? Let's consider Jupiter, since it is the best studied of the four jovian planets.

A basic examination of Jupiter shows it to be an unlikely place to find life. Jupiter's composition is nearly the same as that of the Sun—it is composed almost entirely of hydrogen and helium. It has no solid surface (though it may have a core of metal and rock), and the pressure rises enormously with depth. If you plunged into Jupiter's atmosphere, you would continue to fall until the rising pressure crushed you. Deeper inside, the gaseous atmosphere gives way to far stranger states of matter, most of which are never found on Earth except under extreme laboratory conditions.

Despite these drawbacks, a few scientists have speculated about life in Jupiter's atmosphere. The temperature is far below freezing at the cloud tops, but it rises rapidly with depth. Indeed, Jupiter has several cloud layers, each formed as different types of gases condense to make liquid droplets. For example, clouds made from droplets of liquid water form at a depth of a little over 100 kilometers into Jupiter's atmosphere. Given the comfortable temperature and the presence of liquid water, Jupiter seems potentially habitable at this depth in its atmosphere. Could there be life floating in the atmosphere at these "just right" depths?

Unfortunately, Jupiter's atmosphere appears to present a fatal problem for life: strong, vertical winds with windspeeds that would make a hurricane seem like a gentle breeze in comparison. Any complex organic molecules that might form would quickly be carried to depths at which the heat would destroy them, making it difficult to see how any life could arise indigenously. We might imagine microbes reaching Jupiter on meteorites from elsewhere, but again the vertical winds make their survival seem impossible. Such microbes would be thrown quickly onto a nonstop elevator ride between cloud layers that are unbearably cold and others that are insufferably hot.

The only way that anyone has imagined life surviving in Jupiter's atmosphere is by supposing that it might have some sort of buoyancy that allowed it to stay at the right height while the vertical winds rushed by it. However, such buoyancy would require large gas-filled sacs, making the organisms themselves enormous. Given that we cannot envision a way for microbes to survive, there seems to be no way for large, buoyant organisms to evolve in the first place. They might survive if they arrived on Jupiter from elsewhere, but we know of no way that such large organisms could survive a journey through space, even if they existed elsewhere. As a result, most scientists do not consider Jupiter to be habitable. For similar reasons, none of the other jovian planets (Saturn, Uranus, and Neptune) are likely to be habitable either.

Small Moons, Asteroids, Comets, and Pluto Our brief biological tour now brings us to the numerous smaller bodies in our solar system: small and medium-size moons, asteroids, comets, and Pluto. Before we begin, it's worth noting why we consider Pluto in this group of "small" bodies. Pluto is the smallest planet by far, with a mass significantly less than 1% of Earth's mass. Moreover, it shares much more in common with hundreds of recently discovered large cometlike objects in the outer solar system —usually called *Kuiper belt objects*—than it does with any of the other planets. Like the Kuiper belt

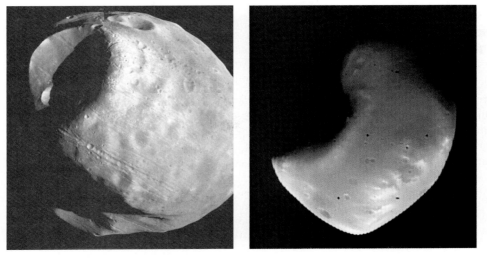

FIGURE 6.17 Mars has two small moons, Phobos (left) and Deimos (right). Phobos is only about 13 kilometers across, and Deimos is only about 8 kilometers across—making each of these moons small enough to fit within the boundaries of a typical large city. Like asteroids, these moons are too small and too cold to be habitable. Indeed, they probably once *were* asteroids, becoming moons only after they were captured into Mars's orbit. (The capture probably happened early in Mars's history, when an extended atmosphere created the friction necessary to slow the passing asteroids enough for them to end up as moons.)

Phobos   Deimos

objects, it is icy in composition and has an orbit that is fairly eccentric and significantly inclined (17°) to the plane in which all the other planets orbit. Most astronomers now consider Pluto to be the largest known Kuiper belt object and it remains possible that we will discover larger ones. Indeed, some astronomers have gone so far as to suggest demoting Pluto from its status as a planet. However, the majority currently favor keeping its planetary designation in deference to history—Pluto was named the ninth planet after its discovery in 1930.

The small size of asteroids and comets probably rules out present life, because they almost certainly lack any liquid medium today. These bodies are too small to have any leftover internal heat that might melt any ices contained within them. Moreover, the comets, including the Kuiper belt objects, and Pluto are so far from the Sun that they are in a perpetual state of deep freeze. (An exception might be the rare comets that make close passes by the Sun, during which times ices near their surfaces can melt or evaporate, but such processes occur for periods of only a few years in their multibillion-year histories.)

However, recall that we have substantial evidence of complex organic molecules in both asteroids and comets. Studies of meteorites show that many of them must have contained liquid water in the past, just after they formed. The liquid water may have persisted over time periods as long as a few tens of millions of years. Could life have arisen during these time periods on asteroids and comets? We cannot rule it out, but we have found no evidence of past life in meteorites from the asteroid belt, and most scientists consider it unlikely that life could have originated on asteroids or comets.

The many small moons in our solar system include the two tiny moons of Mars, Phobos and

Deimos (Figure 6.17); Pluto's moon Charon; and dozens of small to moderate-size moons orbiting the jovian planets. Like asteroids and comets, these moons are unlikely to harbor any life, past or present. Again, they are so small and so far from the Sun that their ices never melt, except perhaps for very short periods of time following impacts. With no liquid medium available, they probably are not habitable.

Summary If we allow for life sufficiently different from that found on Earth, we cannot truly rule out any particular place as a home to life. But if we focus our efforts on life that shares at least some characteristics with life on Earth, such as the need for liquid water or some other liquid medium, the possibilities are much more limited. In that case, we can effectively rule out finding life on most of the worlds in our solar system. Nevertheless, this process of elimination still leaves us several promising abodes for life —namely, Mars, Europa, and perhaps a few other large jovian moons. We will therefore devote the next two chapters to investigating our current understanding of the habitability of these worlds.

Although we've found most of the worlds in our solar system to be unlikely candidates for harboring life, they are still important to our general understanding of the conditions for life in the universe. In particular, study of these worlds is crucial to considerations of how planetary systems form and why some worlds are habitable and others are not. This in turn helps us understand the prospects for life not only in our own solar system, but among other star systems as well. We still have much to learn about these issues, which is why every world in our solar system will remain a target of scientific interest for decades or even centuries to come.

THE BIG PICTURE

In this chapter, we have discussed the general requirements for habitability, investigated the tools used to determine habitability, and surveyed current knowledge concerning the habitability of the many worlds in our solar system. As you continue your study, keep in mind the following "big picture" ideas:

- The general environmental requirements for life are much broader than the requirements for complex beings such as humans. In particular, when we consider our solar system, we find no place besides Earth that is likely to be home to intelligent life but several places where microbial life could potentially survive.

- Someday soon we may search directly for life on other worlds in our solar system. But for now we can assess only the habitability of worlds, using data collected by telescopes on Earth or robotic spacecraft.

- Despite the limitations of our current knowledge, we can broadly assess the prospects of habitability for all the worlds in our solar system. We find at least a few worlds that seem potentially habitable, most notably Mars and Europa. Other worlds may have been habitable in the past, including Venus and perhaps even some asteroids and comets.

Review Questions

1. Why do we expect the elements of life to be widely available on other worlds? How does the requirement of organic building blocks further constrain the prospects of habitability?

2. Discuss in general terms the availability of energy sources for life on other worlds.

3. Why is a liquid medium important for life? Why does water seem the most likely liquid medium for life? Briefly discuss a few other potential liquids for life.

4. Summarize the three major environmental requirements for life. Which seems the most constraining, and why?

5. What is imaging, and what do we mean by the *resolution* of an image? Discuss typical image resolutions for the eye, for ground-based telescopes, and for the Hubble Space Telescope.

6. What is *interferometry,* and how does it improve image resolution?

7. What are the advantages of telescopes in Earth orbit compared to ground-based telescopes?

8. Discuss the kinds of information about distant objects that we can learn through spectroscopy. What are *spectral lines,* and how are they like chemical fingerprints?

9. What can we hope to learn about other worlds in our solar system through long-term telescopic monitoring?

10. Briefly discuss the general characteristics of and differences between robotic *flybys, orbiters,* and *probes* and *landers.* What is a *sample return mission?*

11. Briefly summarize why the Moon and Mercury are unlikely to be habitable.

12. Summarize the past and present habitability of Venus. What is the *runaway greenhouse effect?*

13. Briefly discuss the habitability of the jovian planets.

14. What are the prospects for habitability of the many small bodies in the solar system? Why do we consider Pluto in this category?

Discussion Questions

1. *Bizarre Forms of Life.* Discuss some potential forms of life that have appeared in science fiction and fall outside the general types of life that we've discussed in this chapter. Which forms seem more plausible, and why? Overall, do you think the definition of life we've used in this chapter is too constraining? Defend your opinion.

2. *Artificial Life.* Imagine that future humans decide to breed new organisms tailored to as many different environments as possible. Discuss some of the places in our solar system where we could potentially plant such artificially created species, even if life probably would not arise naturally in those places. Do you think it likely that we will someday develop life-forms for other worlds? What are the philosophical ramifications of being able to custom-tailor life for worlds that don't have any natural life?

3. *Future Astrobiology Course.* Imagine that you are living a century from now and are taking a course about life in our solar system. Based on the current rate of exploration and reasonable rates for the future, how much more do you think we will know then about life in our solar system than we know now? Speculate about some of the discoveries you think may occur in the next century.

Problems

Would You Believe It? Each of **problems 1–8** gives a statement that a future explorer might someday make. In each case, decide whether the claim seems plausible in light of current knowledge. Explain your reasoning clearly.

1. On the smallest moon of Uranus, my team discovered a vast, subsurface ocean of liquid water.

2. I went ice fishing in a lake of liquid ethane on a distant moon of Saturn.

3. At the new colony on the Moon, we obtain water from a well drilled about a kilometer into the crust.

4. At the new colony on the Moon, we obtain water piped in from a crater at the Moon's north pole.

5. I was part of the first group of people to land on Venus, where we found extensive fossil evidence of past life.

6. My team visited several comets and Pluto and found that the surface material of Pluto was nearly identical to that of the comets.

7. We sent a robotic airplane into the atmosphere of Jupiter, but we could not keep it at a steady altitude and it was quickly ripped apart.

8. On a moon of Neptune, we discovered photosynthetic life with a metabolism that runs nearly a hundred times as fast as that of any photosynthetic organism on Earth.

9. *Spectroscopic Magic.* Write a short summary of the things we can learn about a distant planet through spectroscopy and why people a few hundred years ago might have thought it impossible for us ever to learn such things.

10. *Greenhouse Effect.* The text makes the statement that the greenhouse effect on Venus proves "that it is possible to have too much of a good thing." Explain this statement in two or three paragraphs.

11. *Mission Plan.* Suppose you could send a robotic mission of any type (flyby, orbiter, probe/lander) to any one of the places discussed in this chapter as being *unlikely* to be habitable. Which place and type of mission would you choose, and why? Defend your choice clearly in a one- to two-page essay.

12. *Transplanting Life.* Suppose you could genetically engineer organisms on Earth in any way that you chose. What, if any, features could you give them that would enable them to survive on (a) the Moon or Mercury; (b) Venus; (c) Jupiter; (d) Io—Jupiter's moon that is subject to almost constant volcanic eruptions so that its entire surface is almost continuously repaved by fresh lava flows; (e) a comet.

13. *Science Fiction Life.* Choose a science fiction book or movie that describes some form of alien life that falls outside the bounds of the type of life we have considered in this chapter. Write a one- to two-page critical review in which you discuss the plausibility of the life-form.

Web Projects

1. *Project Apollo.* Learn more about NASA's Apollo project, the only set of missions that has ever sent humans to another world. Describe the goals and objectives of each of the Apollo missions. Which were successful, and which were not? What lessons does Apollo offer for future attempts to send humans to the Moon and beyond? Summarize your findings and your opinions about lessons for the future in a one- to two-page essay.

2. *Planetary Missions.* Visit the Web page for one of the missions listed in Table 6.2. Write a one- to two-page summary of the mission's basic design, goals, and current status.

3. *Interferometry.* Visit the Web page for a potential future space interferometry mission, such as the Space Interferometric Mission (SIM) or the Terrestrial Planet Finder (TPF). Briefly describe how and why interferometry is involved in the mission design.

CHAPTER 7

Mars

The idea of a civilization on Mars was once taken so seriously that the term *Martians* became nearly synonymous with alien life. Spacecraft sent to Mars in the 1960s shattered this image of a world of cities and sophisticated beings, revealing a cold, dry, cratered landscape. But while we can rule out a martian civilization, the existence of past or present microbial life on Mars remains a subject of intense scientific investigation and debate.

Substantial evidence suggests that water once flowed on Mars, perhaps even in recent geological times. In particular, it seems likely that Mars once had surface or subsurface environments similar to those in which life thrived on the early Earth. If life arose on Mars (or was transplanted there from the early Earth), we may be able to find its fossil remains. It's even possible that life still survives somewhere on Mars.

Whether or not we someday discover martian life, by studying Mars we are sure to learn a great deal about planetary habitability. Based on current hypotheses concerning the origin of life on Earth, Mars appears to have met all of the requirements for life to have arisen there as well. If we find life, it will help us understand more about biology in general. If we don't find life, Mars will still teach us much about the range of conditions under which life can arise. In this chapter, we'll discuss our past and present understanding of the habitability of Mars and the nature of our search for life on the red planet.

7.1 Fantasies of Martian Civilization

Shining brightly and noticeably red in the nighttime sky, Mars has long captured the human imagination. It figures prominently in many ancient myths and legends. But the idea that Mars might harbor an advanced civilization arose only a little over 200 years ago. Before we discuss what we know about Mars today, let's go back in time and look at how Mars came to be so closely associated with the possibility of extraterrestrial life.

The story begins with the noted English astronomer William Herschel (1738–1822). Though best known for discovering the planet Uranus, Herschel made numerous other astronomical discoveries. He was joined in many of these discoveries by his sister and fellow astronomer Caroline Herschel (1750–1848), who assisted with the observing and helped grind lenses for their telescopes. The Herschels often observed Mars, noting its polar ice caps and discovering that the length of its day (24 hours 37 minutes) is very similar to an Earth day. In a talk presented to Britain's Royal Society in 1784, William Herschel also claimed that Mars possessed an atmosphere and that consequently "its inhabitants probably enjoy a situation in many respects similar to our own." With the mention of "inhabitants," the possibility of living beings on the red planet had been broached by a respected scientist in an academic setting. Martians were assumed to exist. (It should be pointed out that Herschel was not overly particular when it came to populating the cosmos. As far as he was concerned, everything in the solar system—including the Moon and the Sun—was inhabited.)

During the following century, Mars rose to the top of the astronomical charts. Herschel had spoken of permanent spots on the planet, and in the 1860s Angelo Secchi in Rome noted some peculiar streaky features as well. But it was the work of Giovanni Schiaparelli, director of the Milan Observatory, that transformed the perception of Mars for the next hundred years. In 1877, Schiaparelli claimed to see a network of 79 linear features that he called *canali*—translated as "channels," or more commonly (and incorrectly) as "canals." Coming amid the excitement that followed the 1869 opening of the Suez canal, Schiaparelli's discovery soon inspired visions of artificial waterways built by a martian civilization. Schiaparelli himself remained skeptical of such claims, and it's not clear whether he even thought the *canali* contained water. But his work caught the imagination of a young Harvard graduate, Percival Lowell.

Lowell, whose degree was in mathematics, came from a wealthy and distinguished New England family. His brother Abbott became famous as a president of Harvard, and his sister Amy gained fame as a poet. After spending a few years as a businessman and as a traveler in the Far East, Percival Lowell turned to astronomy. Impassioned by his belief in the martian canals and enabled by his wealth, Lowell commissioned the building of an observatory. He was aware that not all astronomers could make out Schiaparelli's *canali*, but rather than questioning the existence of the canals he instead guessed that they were difficult to observe because most astronomical observatories were affected by poor "seeing" (a term used to describe the clarity and stability of the air above an observatory). Seeing varies from place to place and from night to night and tends to be best in places with dry air and high altitude. After considerable research, Lowell chose Flagstaff, Arizona, for his new observatory, because he believed it had the best overall seeing conditions in the United States. The Lowell Observatory opened in 1894 (and is still operating today). Over the next two decades, Lowell meticulously mapped nearly 200 canals that he claimed to see on Mars, identifying each of them with a Latin name.

In 1895, Lowell published detailed maps of the canals, as well as the first of three books in which he explained not only his belief that the canals were constructed by a martian civilization but also his conclusions about the nature of that civilization. Lowell correctly deduced that Mars was generally arid, and he guessed that the polar caps were made of water ice (in reality, they are largely carbon dioxide ice). He therefore imagined that the canals were built to carry water from the poles to thirsty cities nearer the equator. From there it was a short step to imagine the Martians as an old civilization on a dying planet. The global network of canals convinced Lowell that the Martians were citizens of a single, global nation. Such ideas provided the "scientific" basis for H. G. Wells's *The War of the Worlds*, published in 1898.

THINK ABOUT IT . . . *Think of as many popular references to a civilization of "Martians" as you can. For example, consider science fiction stories or movies, television shows, advertisements, or music that refers to intelligent beings on Mars. What do these references tell us about the influence of science on the public imagination?*

Lowell was an effective advocate for the canals, but they do not really exist. Even in his own time, other scientists shot holes through most of Lowell's claims. Numerous astronomers, including some who had initially believed in the existence of canals, soon began to question Lowell's observations. These other astronomers could not see the canals when they looked through telescopes with their eyes, and no canals showed up in photographs taken through telescopes. Moreover, while some canal-like features may have seemed to be present when seeing was poor, they were absent when seeing was best—the

Movie Madness: Missions to Mars

There was a time, not so long ago, when the term "Martian" was just about synonymous with "space alien." You could frequently meet the Martians at the local cinema, where they were busy invading Earth and ruining everyone's whole day.

The classic example of this type of smooth move was in the film (and novel) *War of the Worlds*, in which sophisticated Martians abandoned their turf to grab ours. Mars, you see, was drying up and dying. Earth, on the other hand, was a world with abundant water: a sanctuary for our desperate neighbors.

In fact, space probes of the last four decades have shown that the martian surface is apparently lifeless, and there are no obvious indications of technologically sophisticated inhabitants, either dead or alive. So Hollywood, ever flexible, has switched gears. Now it is Earthlings that go to Mars—often to find hidden signs of habitation that would startle astrobiologists.

In the film *Red Planet*, our descendants try to rebuild Mars into a kinder, gentler world in order to escape environmental disaster on Earth. Robotic craft are sent to melt the polar caps and sow the planet with blue-green algae in a barely plausible bid to produce a warm, breathable atmosphere. In the course of this terraforming project, visiting humans stumble across some complicated life forms—indigenous Martians—which look like economy-size lice. The lice eat everything from space suits to spacemen, and frankly it's a puzzle why they haven't eaten Mars itself.

As implausible as this may be, astronauts in another space opera, *Mission to Mars,* find something even less reasonable. While checking out an odd mountain in Mars's Cydonia region, they discover that it's really a massive, alien "face," disguised by dust and rock. Venturing inside, the astronauts eventually learn that the ancient martian civilization that built the face was wiped out by an asteroid a half-billion years ago. Just before abandoning their planet, the Martians had launched a rocket to seed Earth with their DNA. This molecular emigration supposedly produced the Cambrian Explosion that began the reign of multicellular life on Earth. One has to wonder how the Martians could be confident that the jellyfish and trilobites that resulted from their seeding would eventually evolve to humans that would look pretty much like . . . the long-gone Martians!

Life on the Red Planet may, indeed, have once existed, and perhaps still does. But all that we know about conditions on Mars strongly suggests that this life would never have resembled either voracious lice or us. It's probable that any real Martians would only be visible in a microscope.

---

opposite of what we would expect if they were really there. Other scientists questioned Lowell's basic assumptions and interpretations. Writing in 1907, Alfred Russel Wallace (the codiscoverer of evolution by natural selection [Section 3.1]) used physical arguments to suggest that Mars must be too cold for liquid water to flow and that the polar caps must be dry ice (frozen carbon dioxide). He also pointed out a major flaw in Lowell's interpretation of the canals: Lowell's canals all followed straight-line paths for hundreds or thousands of miles, whereas real canals would be built to follow natural contours of topography (for example, to go around mountains). In summarizing this argument, Wallace wrote that such a network of canals, "as Mr. Lowell describes, would be the work of a body of madmen rather than of intelligent beings."[1]

What was Lowell seeing? In only a few cases do his canals correspond to real features on Mars. For example, the canal he claimed to see most often (which he called *Agathodaemon*) coincides with the location of the huge canyon network now known as Valles Marineris (the horizontal gash visible in Figure 1.2). A few other canals also roughly follow the contours of real features on Mars, but most of the canals were pure fantasy. Figure 7.1 compares a telescopic photo of Mars with one of Lowell's maps of the same regions. You can probably see how the dark and light regions match up in the photo and the drawing, but seeing any canals requires a vivid imagination.

Lowell's story illustrates both the pitfalls and the triumphs of modern science. The pitfall is the fact that individual scientists, no matter how upstanding and dedicated, may still bring personal biases to bear on their scientific work. In Lowell's case, he was so convinced of the existence of canals and Martians that he simply ignored all evidence to the contrary. But the story's ending shows why modern science ultimately is so successful. Despite Lowell's stature, other scientists did not accept his claims on faith. Instead, they sought to confirm his observations and to test his underlying assumptions. They found that Lowell's claims fell short on all counts. As a result, Lowell became an

---

[1] Excerpted in K. Zahnle, "Decline and Fall of the Martian Empire," *Nature*, vol. 412, 12 July 2001.

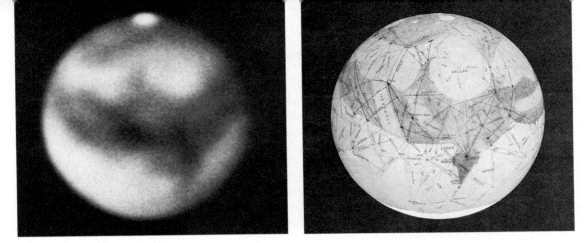

FIGURE 7.1 Can you see how the markings on Mars in the telescopic photo on the left might have resembled the geometrical features in the drawing by Percival Lowell on the right? Try blurring your eyes.

increasingly isolated voice as he continued to advocate a viewpoint that was clearly wrong.

The story also illustrates the divide that often persists between scientists and the general public. Although nearly all scientists had abandoned belief in Martians within the first few decades of the twentieth century, among the public the canal myth persisted much longer. Public belief in Martians remained widespread enough to create a famous panic during Orson Welles's 1938 radio broadcast of *The War of the Worlds*—many people thought an invasion was actually under way (Figure 7.2). Books and movies about martian civilization remained popular throughout the twentieth century. Even today, a few die-hard proponents claim to see evidence of a past martian civilization even though recent high-resolution photographs make their claims increasingly implausible.

7.2 A Current Assessment of the Habitability of Mars

The public debate about martian canals and cities was not entirely put to rest until NASA began sending spacecraft to Mars. In 1965, NASA's *Mariner 4* spacecraft flew to within 6,000 miles of the martian surface, transmitting a few dozen television-quality images of the landscape below. When the photos were given to President Lyndon Johnson, he looked at them with surprise. Mars's surface was littered with craters, not canals. The red planet looked as lifeless as the Moon. Measurements of the atmospheric pressure and temperature made from the spacecraft indicated a cold, dry planet seemingly incapable of supporting life.

Nevertheless, all was not lost when it came to the potential for life on Mars. There was no evidence of any intelligent beings, but the thin atmosphere and the polar caps left open the possibility of the exis-

FIGURE 7.2 This front-page story from the *New York Times* described the panic caused by Orson Welles's 1938 broadcast of *The War of the Worlds*.

tence of microbes or perhaps even some primitive plants or animals. On July 20, 1976, seven years to the day after Neil Armstrong's history-making walk on the Moon and nearly a century since Schiaparelli's description of *canali*, the thin skies above Mars were pierced by a NASA space probe. *Viking Lander 1* touched down without difficulty on the Chryse Planitia, a sprawling, rock-strewn plain about 1300 kilometers north of the martian equator. It was followed nearly two months later by its twin, *Viking Lander 2*, which landed almost halfway around the planet. Meanwhile, two Viking orbiters studied the planet from above.

When the Viking landers' cameras opened their eyes in the frigid martian air, they found a bleak

(*continued on p. 174*)

Among the tens of thousands of photographs snapped as part of the Viking missions to Mars in the late 1970s, one achieved special fame. The *Viking 1* orbiter snapped a picture showing a feature on Mars that looked remarkably like a human face (Figure 7.3). The "face on Mars" soon became a cottage industry, with proponents arguing that it must be the work of a lost martian civilization or of alien beings visiting our solar system. It still appears regularly in supermarket tabloids.

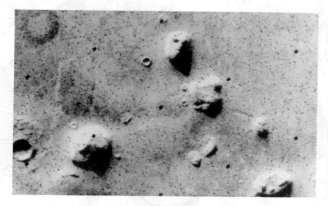

FIGURE 7.3 This Viking 1 image shows the "face on Mars" and the surrounding area. The face is on a tall mesa about 1 kilometer across. Some of the other features were claimed to represent a "fortress" and "pyramids." The black dots are instrument artifacts.

The feature in the picture certainly looks like a face. However, with many similar features existing on Mars, it seemed a near-certainty that the human likeness was a coincidence. The people promoting an artificial origin for the face on Mars ignored evidence to the contrary and claimed that NASA was conspiring to hide the truth from the public—despite the fact that it was first described as a face by a NASA scientist in a NASA press release. A few proponents even claimed at the time that a failed NASA mission to Mars (Mars Observer) had actually achieved martian orbit successfully and was imaging the face and that NASA was hiding the results from the public.

Because of the intense public interest surrounding the face on Mars, NASA made it a target of further observations with the Mars Global Surveyor mission, which achieved martian orbit in 1997. New photographs soon showed that the scientific analysis had been correct: The feature did not look like a face when viewed with higher resolution. Figure 7.4 compares the original Viking photo with a more recent photo from Mars Global Surveyor.

From a scientific standpoint, the controversy surrounding the "face on Mars" is over (if it ever really began). There is no reason to doubt that it is a natural geological feature, especially since it looks so similar to many other features in the same region of Mars. Its coincidental resemblance to a face in the Viking 1 image was a product of shadows and the human tendency to see patterns where none exist—just as we see faces in clouds, a man in the Moon, and canals on Mars.

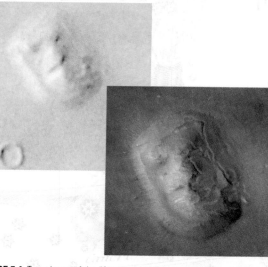

FIGURE 7.4 Two photos of the "face on Mars." The photo on the left is the original image from Viking 1 in 1976 (a close-up view of the face in Figure 7.3). The much higher resolution photo on the right is from the Mars Global Surveyor mission.

A few die-hard proponents still claim either that it was a face that has been heavily altered by natural erosion or that there was a NASA conspiracy to doctor the image before it was released and thereby hide the true evidence of a martian civilization. It's hard to blame them for their desire for compelling evidence on Mars of an extraterrestrial civilization. After all, there's a certain appeal to all the adventures that past science fiction writers imagined taking place among the inhabitants of Mars. Fortunately, for anyone heartbroken at the thought that Mars has never harbored a civilization, Mars offers other reasons to smile: In searching through hundreds of other photos from Mars missions, NASA scientists came upon a crater that shows a happy face on Mars (Figure 7.5).

FIGURE 7.5 This Viking image shows a crater on Mars that appears to show a happy face.

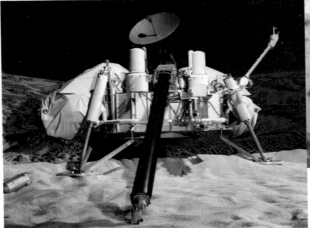

a This photograph shows a working model (actually, a spare spacecraft) of the Viking landers, identical to those that landed on Mars. It is on display at the National Air and Space Museum in Washington, D.C.

b This photo of the surface of Mars was taken by *Viking Lander 2* in 1979. It shows a thin coating of ice on the rocks and soil.

FIGURE 7.6 Two Viking landers settled on Mars in 1976.

landscape with red dust and scattered rocks stretched out against a pinkish sky. No creatures stared back at the cameras, and no plants were huddled in the weak sunlight. For months the images continued to come in, but the view scarcely changed. Nothing grew other than some occasional patches of frost, and nothing moved other than windblown dust (Figure 7.6). Though neither lander could move from the spot at which it had settled, each had a robotic arm with which it collected soil for some on-board experiments designed to look for microbes. Although some results initially looked promising, the final verdict was that no evidence of life had been found. (We will discuss these experiments in Section 7.3.)

The Viking orbiters and landers provided a wealth of scientific data about Mars. But they also left many questions unanswered, and the scientific community was itching for follow-up missions. Unfortunately, budgetary and political considerations, along with the failure of two Russian missions (Phobos 1 and 2) and one American mission (Mars Observer) to Mars, all conspired to stop spacecraft exploration of Mars for some twenty years. The long mission drought did not end until July 4, 1997, with the landing of *Pathfinder* and its little rover, *Sojourner* (Figure 7.7). Although *Sojourner* could travel only a few tens of meters from *Pathfinder*, this was enough to check the chemical composition of many nearby rocks. (*Sojourner* was named for Sojourner Truth, an African-American heroine of the Civil War era who traveled the nation advocating equal rights for women and blacks.) *Pathfinder* was followed soon after by an orbiter, *Mars Global Surveyor*, and more recently by the orbiter *Mars Odyssey.*

The Pathfinder and Mars Global Surveyor missions marked the beginning of a new era of intensive scientific study of Mars (which we'll discuss in Section 7.5). By combining the new data with earlier Viking data and telescopic studies, we are beginning to put together a realistic assessment of the habitability of Mars. In the rest of this section, we'll discuss the current state of this assessment.

Mars Today

The present-day surface of Mars may look much like some deserts or volcanic plains on Earth, but it would not be a very comfortable place to visit in person. Table 7.1 summarizes some basic Mars data. Note that the surface temperature is usually well below freezing, with a global average of about −53°C (−63°F). The atmospheric pressure is less than 1% that on the surface of the Earth, making the air so thin that no human could survive more than a few minutes without a pressurized spacesuit. The air contains only trace amounts of oxygen, so we could not breathe it. The lack of oxygen also means that Mars lacks a substantial ozone layer, so much of the Sun's damaging ultraviolet radiation passes unhindered to the surface.

FIGURE 7.7 A view of a martian floodplain at the landing site for the *Pathfinder* lander (partially visible in the foreground). The little rover, *Sojourner,* is visible in the upper right, studying a rock that scientists named Yogi.

Table 7.1 Basic Mars Data

Average Distance from the Sun (1 AU = average Earth–Sun distance)	1.52 AU (227.9 million km)
Orbital Period	1.881 Earth years
Equatorial Radius	3,397 km
Mass (Earth = 1)	0.107
Rotation Period	24 hr 37 min
Surface Gravity (Earth = 1)	0.38
Atmospheric Composition	95% CO_2, 2.7% N_2, 1.6% argon
Average Surface Temperature	–53°C
Average Surface Pressure (1 bar ≈ sea-level pressure on Earth)	0.005 bar

Nevertheless, these conditions are much less extreme than those on the Moon (mainly because of the moderating effects of the atmosphere), and it's easy to imagine future astronauts living and working in airtight research stations while occasionally donning spacesuits for outdoor excursions. Martian surface gravity is about 40% that on Earth, so everyone and everything would weigh about 40% of Earth weight. As a result, astronauts could walk around easily even while wearing spacesuits with heavy backpacks.

Visitors to Mars would find some notable similarities to Earth. The martian day is only about 40 minutes longer than an Earth day, so adapting to the patterns of day and night should be easy. The tilt of the martian axis is about 25°—only slightly greater than Earth's 23.5° tilt—so Mars has seasons much like those on Earth. However, because the martian year is nearly twice as long as an Earth year, the seasons last nearly twice as long on Mars.

The martian seasons also differ from seasons on Earth in another important way that is due to the nature of Mars's orbit. Seasons on Earth are caused by the tilt of the axis (Figure 7.8). The Earth's axis remains pointed in the same direction (toward the north star) throughout the year, so on one side of the orbit the Northern Hemisphere is tilted toward the Sun and on the other side it is tilted away from the Sun. Therefore, as shown in Figure 7.8, the Northern Hemisphere receives more direct sunlight when the Southern Hemisphere receives less direct sunlight, and vice versa. That is why the two hemispheres experience opposite seasons.

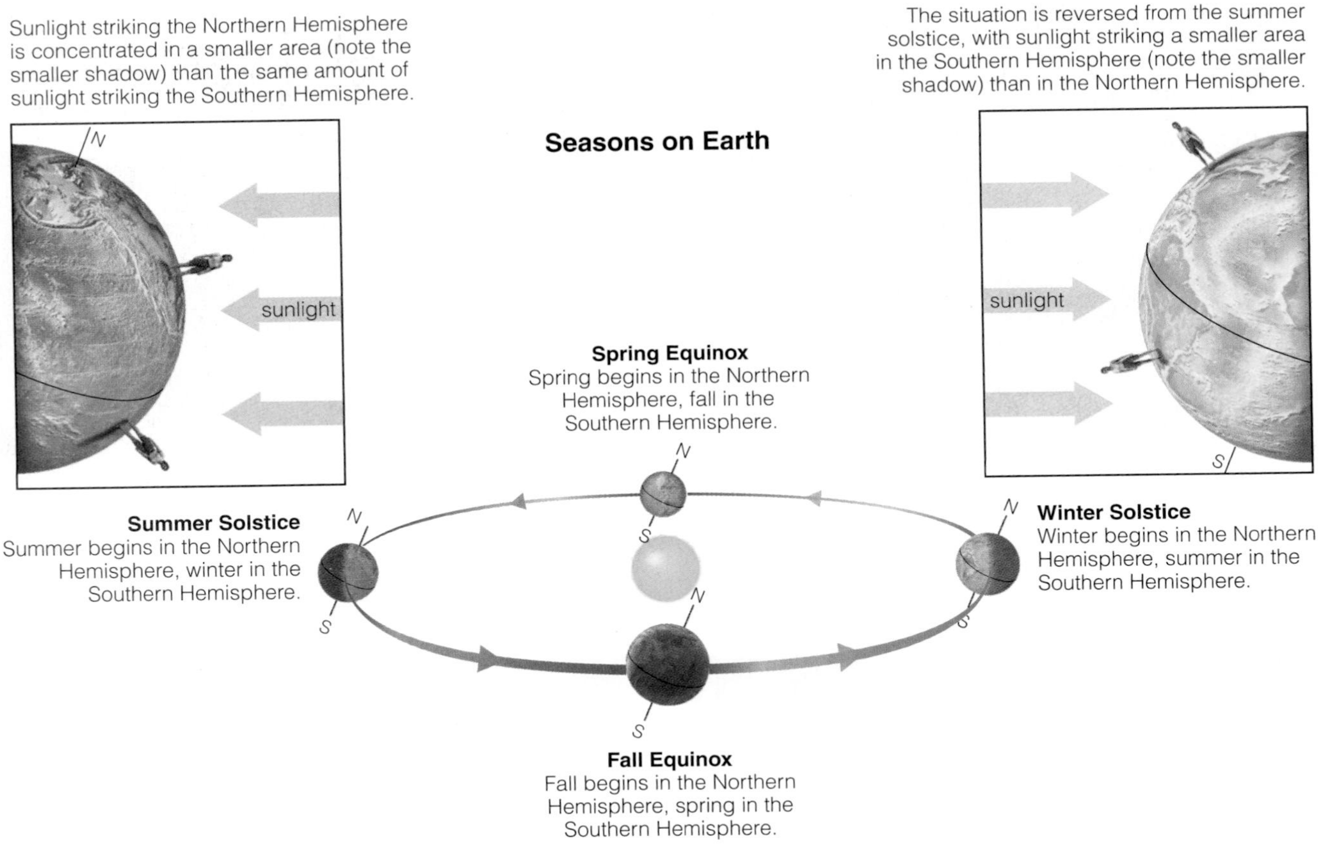

Sunlight striking the Northern Hemisphere is concentrated in a smaller area (note the smaller shadow) than the same amount of sunlight striking the Southern Hemisphere.

The situation is reversed from the summer solstice, with sunlight striking a smaller area in the Southern Hemisphere (note the smaller shadow) than in the Northern Hemisphere.

Seasons on Earth

sunlight

sunlight

Spring Equinox
Spring begins in the Northern Hemisphere, fall in the Southern Hemisphere.

Summer Solstice
Summer begins in the Northern Hemisphere, winter in the Southern Hemisphere.

Winter Solstice
Winter begins in the Northern Hemisphere, summer in the Southern Hemisphere.

Fall Equinox
Fall begins in the Northern Hemisphere, spring in the Southern Hemisphere.

FIGURE 7.8 Seasons occur on Earth because, even though the Earth's axis remains pointed toward Polaris throughout the year, the orientation of the axis *relative to the Sun* changes as the Earth orbits the Sun. Around the time of the summer solstice, the Northern Hemisphere has summer because it is tipped toward the Sun, and the Southern Hemisphere has winter because it is tipped away from the Sun. The situation is reversed around the time of the winter solstice, when the Northern Hemisphere has winter and the Southern Hemisphere has summer. At the equinoxes, both hemispheres receive equal amounts of sunlight. (The solstices and equinoxes were named by people living in the Northern Hemisphere, so the names of the four points reflect the Northern Hemisphere seasons.)

For Earth, the axis tilt is the only significant influence on the strength of sunlight reaching the surface at different times of year.[2] On Mars, however, the shape of the orbit introduces a second important effect. Mars has a more elliptical orbit that puts it significantly closer to the Sun during its southern hemisphere summer and farther from the Sun during its southern hemisphere winter (Figure 7.9). Hence, the martian seasons are much more extreme for the southern hemisphere than they are for the northern hemisphere. In addition, because planets move faster in the portions of their orbits closer to the Sun, the southern hemisphere's summer is shorter in duration and more intense than the northern hemisphere's longer, milder summer.

The seasons cause one of the major features of Mars's climate: seasonal changes in the pressure and the carbon dioxide content of the atmosphere. When it is winter in, say, the northern hemisphere, polar temperatures drop so low that carbon dioxide freezes out of the atmosphere as solid ice. This ice forms a polar cap of carbon dioxide frost that can be as much as a meter thick at the pole and extend as far south as latitude 40°. At the same time, it is summer in the southern hemisphere, where the frozen carbon dioxide in the polar cap sublimes into carbon dioxide gas. (By "sublimes," we mean that the material goes directly from the solid, or ice, phase to the gas phase without first melting into a liquid. In general, *subli-*

---

[2] The Earth's orbit is not a perfect circle, and the distance to the Sun varies about 3% between the nearest and farthest points in Earth's orbit. However, in terms of the seasons, this effect is overwhelmed by effects resulting from the distribution of the Earth's continents and oceans. In particular, the climate-moderating effects of the oceans keep temperatures more uniform throughout the year in the Southern Hemisphere (which has more ocean and less continent) than in the Northern Hemisphere, despite the fact that Southern Hemisphere summer and winter occur when the Earth is nearest and farthest from the Sun, respectively.

Seasons on Mars

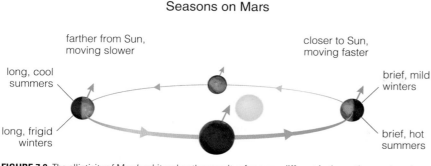

farther from Sun,
moving slower

closer to Sun,
moving faster

long, cool
summers

brief, mild
winters

long, frigid
winters

brief, hot
summers

FIGURE 7.9 The ellipticity of Mars's orbit makes the severity of seasons different in the northern and southern hemispheres.

mation refers to the phase change from solid to gas, while *evaporation* refers to the phase change from liquid to gas.) As the southern summer ends and the northern summer begins, the whole process reverses, with carbon dioxide gas subliming at the north pole and freezing at the south pole. Overall, as much as a third of the total carbon dioxide of the martian atmosphere cycles seasonally between the north and south polar caps. Figure 7.10 shows the dramatic shrinking of the north polar ice cap as the northern hemisphere summer approaches.

THINK ABOUT IT . . . *Understanding the difference between sublimation and evaporation is important to visualizing the behavior of ices and liquids on Mars. It's easy to watch evaporation on Earth, because any puddle of water soon evaporates. Sublimation, though less obvious, is also common. An easy way to see sublimation is with frozen carbon dioxide, also known as "dry ice." How can you tell that dry ice is subliming? (If you've never seen dry ice, it is readily available; try looking up sources of it in your local yellow pages.)*

FIGURE 7.10 These images from the Hubble Space Telescope show Mars in polar projection so that the north pole is in the center of the images. The image at left shows the polar cap just as spring begins in the northern hemisphere, when the cap is near its maximum size. The middle image is in mid-spring, and the image at right shows the polar cap in early summer.

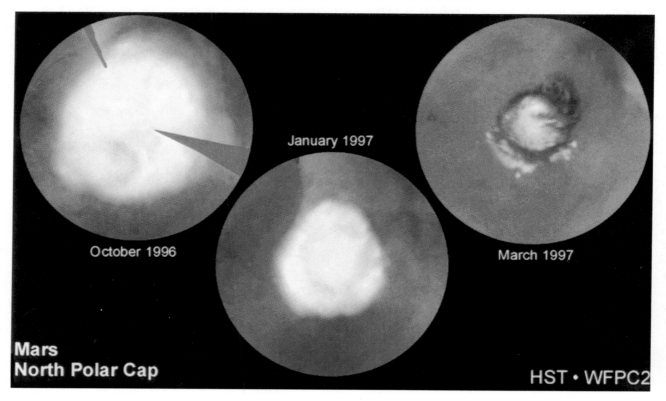

January 1997

October 1996

March 1997

Mars
North Polar Cap

HST • WFPC2

FIGURE 7.11 This image, from the Mars Global Surveyor, shows the residual north polar cap during northern hemisphere summer. The white material is primarily water ice, with dust mixed in.

Over the course of the northern hemisphere summer, all the frozen carbon dioxide at the north pole sublimes, leaving a residual polar cap that lasts through the rest of the summer season, made of water ice (H_2O) mixed with martian dust (Figure 7.11). Some of the water ice sublimes to add water vapor to the atmosphere, but the temperature does not rise high enough for the ice to melt. Given that seasonal changes are more extreme for the southern hemisphere, we might similarly expect to see all the carbon dioxide sublime from the south polar ice cap during southern hemisphere summer. However, this is not the case; instead, the south polar ice cap retains frozen carbon dioxide throughout the martian year. The reason for this is not fully understood, but it appears to involve several factors, including elevation, the way atmospheric dust affects heat transport,

and the history of the polar caps over the last few hundred thousand years.

Winds generated by the temperature differences near the edges of the polar caps sometimes initiate local dust storms. Occasionally, local dust storms in low southern latitudes expand during summer into huge dust storms that enshroud the entire planet (Figure 7.12). At times the martian surface becomes almost completely obscured by airborne dust. As the dust settles out, it can change the surface appearance over vast areas (for example, by covering dark regions with brighter dust); such changes fooled past astronomers into thinking they were seeing seasonal changes in vegetation. The dust storms also leave Mars with a perpetually dusty atmosphere that gives the martian sky its pale pink color.

Liquid water is not stable on the martian surface today—any liquid water would tend to evaporate or freeze almost immediately. Thus, we do not find liquid water on the surface of Mars, even though the midday temperature near the equator often rises high enough to melt any frost that collects on surface rock. If this frost does melt, it evaporates very quickly. More commonly, it probably sublimes directly from ice to water vapor.

While there is no liquid water on the surface today, Mars clearly has substantial amounts of water. Besides evidence of water ice in the residual summer north polar cap, we often find water vapor and ice crystals in the martian atmosphere, sometimes forming clouds. (The south polar cap probably also contains at least some water ice mixed in with its carbon dioxide ice.) In addition, it's likely that Mars has substantial amounts of subsurface ice, a possibility supported by recent observations from *Mars Odyssey*. Spacecraft instruments have detected near-surface

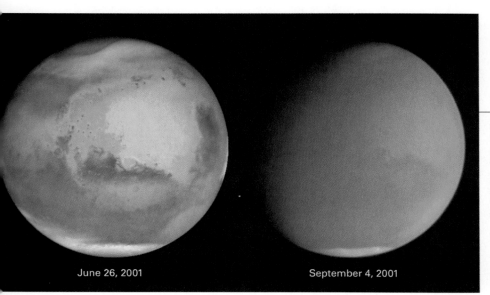

June 26, 2001   September 4, 2001

FIGURE 7.12 These two Hubble Space Telescope photos contrast the appearance of Mars in the presence and absence of a global dust storm. If you look carefully at the first image, you can see localized dust storms near both polar caps (look toward the upper-right edge of the southern cap). The second image shows how, just over two months later, these storms had grown into a planet-wide dust storm.

FIGURE 7.13 This map, made with observations from *Mars Odyssey*, represents the hydrogen content of the martian surface soil. The blue areas contain the most hydrogen, probably because they represent regions in which the top meter or so of the surface soil contains water ice.

hydrogen, probably coming from water ice frozen into the top meter or so of the surface soil (Figure 7.13). As we'll discuss shortly, Mars may also have subsurface ice at greater depths (a hundred meters or more), and in some places this ice may be melted to make underground pockets of liquid water. If any life exists on Mars today, it most likely lives in such pockets, perhaps resembling the rock-dwelling bacteria (or *lithophiles*) found deep underground on Earth in the Columbia River Basalt [Section 3.5].

The Geography of Mars

The surface of Mars may be desolate and barren today, but it was not always so. Many surface features appear to have been shaped by liquid water, leading scientists to conclude that Mars must once have had a much more hospitable climate. But before we discuss the evidence for surface water and ideas about the climate history of Mars, it's useful to get our bearings by looking at the large-scale geographic features of the planet.

Figure 7.14 shows the full surface of Mars, with the poles at the top and bottom and the equator running horizontally across the middle (much the same way similarly shaped maps show the full globe of Earth). Keep in mind that Mars is only about half as large in diameter as the Earth, so its surface area is only about one-fourth that of the Earth (because surface area is proportional to the square of the radius). Thus, if you placed this map next to a map of Earth on the same scale, the Earth map would be four times as large in area. Interestingly, the total land area of Mars is about the same as the total land area of Earth, because Earth's continents occupy only about one-fourth of the surface area (the rest is ocean).

THINK ABOUT IT . . . *Try to find a map of Earth that has about four times the area (twice as long and wide) as Figure 7.14. (If you cannot find a map of the right size, use any Earth map and try to photocopy it at the appropriately scaled size.) Compare the size of various Earth features, such as continents and oceans, to that of various Mars features.*

A number of key geographical features are clearly visible in Figure 7.14. The most striking feature is the dramatic difference in terrain in different parts of the planet. Note that much of the southern hemisphere has relatively high elevation and is scarred by numerous large impact craters. The northern plains, in contrast, show very few impact craters and tend to be below "sea level" for Mars, by which we mean the average martian surface level. Mars should have been densely cratered everywhere during the heavy bombardment that occurred in the early history of the solar system [Section 4.3], so the regions with few craters must have had their original craters erased or covered over since then. Thus, we can conclude that the southern highlands show an older surface than do the northern plains. These northern lowlands show features that are characteristic of lava flows, suggesting that eruptions of a very fluid lava covered up the older impact craters. Interestingly, we see faint "ghost" craters in some of these regions, suggesting that the lava flows were not thick enough to completely erase the underlying features and confirming that the entire planet was once densely cratered.

Perhaps the next most significant features are the large Hellas basin (to the east) and the Tharsis bulge (to the west). Hellas is an impact basin created by the impact of a large asteroid. Notice the large depression in its middle and the "doughnut" of ejected debris that surrounds it. Given the large number of impact craters visible within the basin, it must be one of the

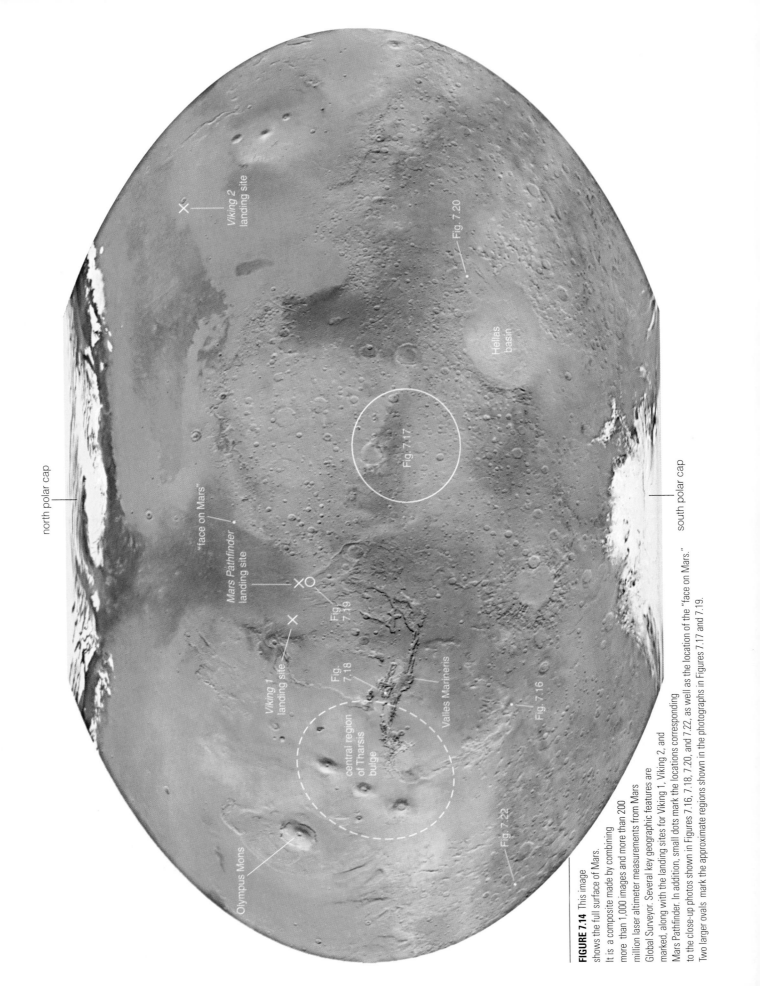

FIGURE 7.14 This image
shows the full surface of Mars.
It is a composite made by combining
more than 1,000 images and more than 200
million laser altimeter measurements from Mars
Global Surveyor. Several key geographic features are
marked, along with the landing sites for Viking 1, Viking 2, and
Mars Pathfinder. In addition, small dots mark the locations corresponding
to the close-up photos shown in Figures 7.16, 7.18, 7.20, and 7.22, as well as the location of the "face on Mars."
Two larger ovals mark the approximate regions shown in the photographs in Figures 7.17 and 7.19.

oldest features on Mars, perhaps as old as 4.4 billion years. Tharsis, in contrast, was probably created by a long series of volcanic eruptions emanating from a long-lived mantle plume. (Recall that a mantle plume, such as the one forming the Hawaiian islands on Earth, involves hot material rising up from deep in the mantle [Section 4.4].) Tharsis is some 4,000 kilometers across, with most of it at elevations about 2–10 kilometers above the martian "sea level." Volcanic eruptions from this mantle plume probably also built the giant volcanoes around Tharsis. One of these volcanoes, Olympus Mons, appears to be the largest volcano in the solar system. Its base is some 600 kilometers across, making it large enough to cover an area the size of Arizona. Its peak stands about 26 kilometers above the "sea level" for Mars, or some three times as high as Mount Everest stands above sea level on Earth. Much of Olympus Mons is rimmed by a cliff with a face as tall as 6 kilometers in places.

East of Tharsis and just south of the equator, we see the long, deep system of valleys called *Valles Marineris.* Named for the *Mariner 9* spacecraft that first imaged it, Valles Marineris is as long as the United States is wide and almost four times as deep as the Grand Canyon. Valles Marineris probably formed largely through tectonic processes that cracked the surface and left its tall cliff walls (the tectonics probably were associated with the tremendous mass of material that formed the Tharsis bulge). The cracking at the western end of Valles Marineris is typical of what we expect from such a tectonic origin. However, other features of the valleys appear to have been shaped by flowing water. In some places, the canyon walls show evidence of

layering that may have been caused by numerous lava flows or deposits of sediments. Moreover, the canyon is so deep in places that some of its walls probably were once several kilometers underground and hence may have been exposed to subsurface liquid water. For this reason, Valles Marineris may be one of the best places to look for fossil evidence of past martian life.

The varying terrain and geological features of Mars provide important clues to its geological history. In particular, detailed comparisons of crater counts on different parts of Mars allow us to divide the surface broadly into regions of three different ages (Figure 7.15). The oldest and most heavily cratered regions appear to have changed little since at least 3.7 billion years ago. Planetary scientists call the period before 3.7 billion years ago on Mars the **Noachian era** (pronounced "no-ah'-ki-an"), corresponding to the period of heavy bombardment [Section 4.3]. The period following the heavy bombardment, which we think of as the middle history of Mars, involves surfaces that appear to have formed between about 3.7 and 1.0 billion years ago, during what we call the **Hesperian era.** The crater counts indicate that the youngest surfaces on Mars are less than about 1.0 billion years old and formed during what we call the **Amazonian era.** Table 7.2 summarizes the names and time periods of the geological eras on Mars.

Table 7.2 Geological Eras on Mars. The eras are based on surface region ages determined from crater counts.

Era	Time
Noachian	4.6–3.7 billion years ago
Hesperian	3.7–1.0 billion years ago
Amazonian	1.0 billion years ago to the present

FIGURE 7.15 This simplified map of Mars shows how different surface regions have different ages based on crater counts. (You might want to compare this simplified map to the more detailed view in Figure 7.14.) The dark brown regions date to the Noachian era, which came to an end about 3.7 billion years ago. The light brown regions date to the Hesperian era, which lasted from about 3.7 to 1.0 billion years ago. The youngest surfaces, shown in tan, are less than 1.0 billion years old.

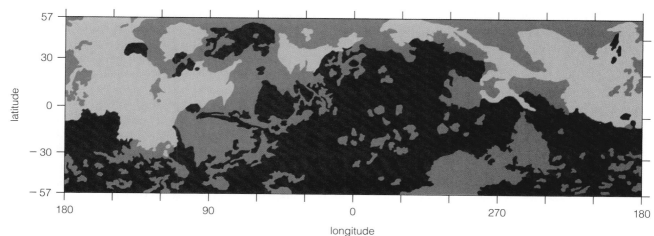

By examining the types of geological features that appear on the surfaces of different ages on Mars, we can get an idea of both what processes helped shape the surface and when they operated. For example, we can look at features that indicate volcanic eruptions, such as lava flows or volcanoes, and deduce the history of volcanism. We find that the volcanic eruption rate has decreased steadily since at least about 3.5 billion years ago and has been extremely low in recent geological times.

THINK ABOUT IT . . . *Does the Earth's surface have regions of different ages similar to what we find on Mars? Explain.*

Until recently, it seemed likely that Mars no longer had any volcanic activity. We would expect Mars to be much less volcanically active than Earth because of its smaller size, which has allowed its interior to cool much more than has Earth's interior in the $4\frac{1}{2}$ billion years since the planets formed. In addition, the presence of impact craters on volcanic slopes suggests that most of the large martian volcanoes have been inactive for the past billion years. Taken together, these facts point to a lack of recent volcanic activity. However, analysis of meteorites that appear to have come from Mars (so-called "martian meteorites" [Section 7.4]) offer a different perspective. Radiometric dating of such meteorites shows that some of them are made of volcanic rock that crystallized from molten lava as recently as 180 million years ago—a time that is within the most recent 4% of Mars's history (180 million years is about 4% of 4.6 billion years). Given this evidence for geologically recent volcanic eruptions, it is likely that Mars remains volcanically active today.

Evidence of Liquid Water on Mars

Large-scale features of Mars are important to our overall understanding of the planet, but features that might have been carved by liquid water are of particular interest for our investigation of Mars's habitability. Close-up studies of the surface show clear evidence for liquid water in the martian past. *Mariner 9* and the Viking orbiters captured the first such evidence. Figure 7.16 shows channels that look much like terrestrial river drainage systems. They appear to have been carved by running water, though the precise mechanism by which they were made remains a source of scientific controversy. Some scientists believe they formed like terrestrial river valleys, from runoff after rainfall. Others suspect that they were formed by erosion due to water-rich debris flows or by "sapping"—a process in which the surface collapses as water flows out from underground and undercuts the banks. More likely, a combination

FIGURE 7.16 This photo, taken by the Viking orbiter, shows ancient riverbeds that were probably created billions of years ago.

of these processes was involved. Regardless of the specific process by which the channels formed, they tell us that liquid water must have existed on Mars at or just below the surface. Moreover, the branching nature of the channels indicates that they must have been carved relatively gradually, meaning that liquid water must have been stabler at the surface than it is today. Because these channels are found almost exclusively in heavily cratered terrain dating to the Noachian era, we conclude that Mars must have had a warmer climate during this period, which ended about 3.7 billion years ago.

Other evidence for a warmer or wetter Mars during the Noachian era comes from more detailed analysis of the ancient, heavily cratered terrain. Small impact craters (less than about 15 kilometers in diameter) are nearly absent, and larger impact craters show evidence of substantial erosion (Figure 7.17). Apparently, water erosion was widespread during the early history of Mars, eroding the rims of large craters and erasing smaller craters altogether. Moreover, there is evidence that the craters may once have been filled with water. Figure 7.18 shows a close-up of the bottom of one crater with evidence of a past lake. Liquid water may have once filled the crater like a pond, allowing sediments to build up from material that settled to the bottom. The sculpted patterns in the crater bottom were probably created as erosion exposed layer upon layer of sedimentary rock, in much the same way the action of the Colorado River exposed the sedimentary layers visible in the walls of the Grand Canyon on Earth.

Craters in younger (Hesperian and Amazonian) regions of the surface show far less erosion than do the ancient craters of the southern highlands, suggesting that little or no rain has fallen since the end

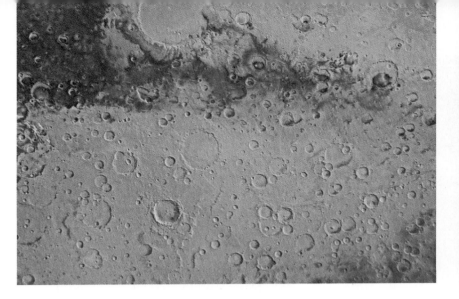

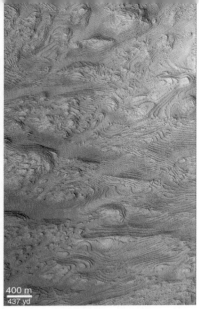

FIGURE 7.17 This photo shows a broad region of the southern highlands on Mars (see Figure 7.14 for approximate location and size). The indistinct rims of the many large craters appear to have been eroded by ancient rains. Relatively few smaller craters are present, suggesting that these have been erased by erosion.

FIGURE 7.18 Close-up view of the floor of a crater bottom, showing evidence of a past lake. The sculpted patterns apparently reveal layers of sedimentary rock. The sediments may have been deposited over thousands of years at a time when the crater was filled with water.

of the Noachian era. Detailed comparisons suggest that the erosion rate during the Noachian era was about 1,000 times greater than the erosion rate at any time since (still barely comparable to the lowest erosion rates found in the driest areas on the Earth). Nevertheless, many surface features show evidence of more recent water flows. This is not as far-fetched as it may sound. Although liquid water is *unstable* on the surface of Mars today, a large quantity of water may take a short time to freeze or evaporate completely. Thus, a catastrophic release of floodwaters— from beneath the surface, for example—could survive long enough to carve channels and other surface features.

Some features on surfaces dating to the Hesperian or early Amazonian era do indeed appear to have been shaped by catastrophic floods. For example, the *Pathfinder* landing site is in the midst of what appears from orbit to be a great floodplain. The surface view from *Pathfinder* (see Figure 7.7) supports the idea that a great flood passed through the region. The departing waters left rocks stacked against each other in the same manner as do floods on Earth. In other places, wide channels appear to have been carved by floodwaters, leaving streamlined islands in their wake (Figure 7.19). The water that raged through these channels can be traced upstream to its source region. Because no surface reservoir is visible there, the water must have emerged from beneath the surface. Although we do not yet understand how or why the water emerged from underground, the evidence of flooding suggests that liquid water has been present within the martian crust, probably kilometers below the surface where temperatures were warm enough to keep the water from freezing. Moreover, the floods

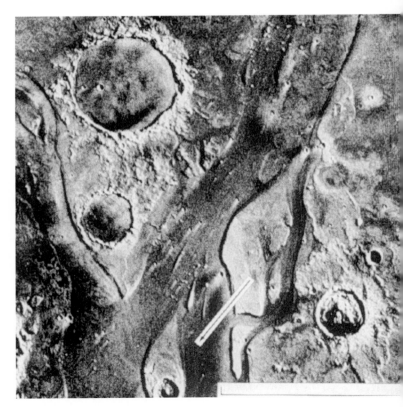

FIGURE 7.19 This channel, which is over 50 kilometers wide in places, appears to have been carved by floodwaters flowing across the surface. The arrow points to an island carved into a streamlined shape as the waters flowed around it.

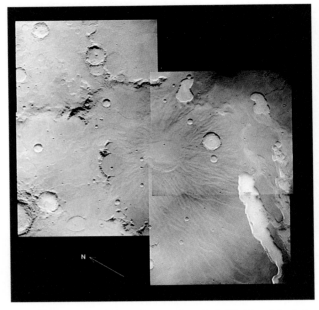

FIGURE 7.20 This set of photos shows a volcano (called Hadriaca Patera) at the center, with its caldera just to the right of the border between the photos. Note the many channels flowing downhill from the central caldera and the wide flood channel (called Dao Vallis) toward the lower right.

FIGURE 7.21 This map shows Mars color-coded by elevation. Blue areas are the farthest below "sea level," and red and brown areas are the highest above it. Note that the entire north polar region is quite low in elevation. Some scientists speculate that this low-lying region may once have held an ocean.

appear to have occurred at least sporadically throughout the Hesperian and Amazonian eras, suggesting that liquid water has been present in the crust for most or all of martian history.

Some large volcanoes show water-carved channels on their flanks, suggesting a link between volcanic heating and some of the floods. Figure 7.20 shows a volcano with numerous downhill channels going outward in all directions from its central caldera (the "crater" in the center of a volcano). Toward the lower right, we see a much wider channel that was probably carved by floodwaters released during one or more eruptions. These features suggest the past existence of underground pockets of liquid water near volcanoes—a potential habitat for life. If martian volcanoes remain active today, such pockets of water might still exist.

Most of the large flood channels appear to have drained into the northern lowlands, where the water may have collected before freezing, evaporating, or being absorbed into the surface. One of the most intriguing ideas about water on Mars is the suggestion that an ocean may once have filled the basin in the north polar region (Figure 7.21). Careful study shows the presence of features that may look like shorelines along the boundaries of this possible ocean, although the evidence is not particularly strong. Future studies should help us learn whether and when such an ocean might have existed.

Taken together, the evidence we have discussed so far makes a convincing case for liquid water on

Mars in the distant past. The evidence suggests that liquid water was present at the surface at least intermittently during the Noachian era, prior to about 3.7 billion years ago. During subsequent times, it is likely that liquid water has been present within the martian crust. The evidence for liquid water in the crust is strongest for times in the distant past, but at least some evidence points to liquid water in the crust in fairly recent geological times.

Prior to the arrival of *Mars Global Surveyor* in 1997, most scientists believed that no water had flowed at the surface during the past several hundred million years or more. However, photos taken by this orbiting spacecraft offer tantalizing hints of water flows in much more modern times. The strongest evidence for liquid water in recent times comes from photos of gullies on crater and channel walls. Figure 7.22 shows one example; note the striking similarity to the gullies we see on almost any eroded slope on Earth. The gullies probably formed when underground water broke out in small flash floods, carrying boulders and soil down the slope. Because they are relatively small (note the scale bar in Figure 7.22), the gullies should be gradually covered over by blowing sand during martian dust storms. Thus, gullies that are still clearly visible must be no more than a few million years old at most. Geologically speaking, this time is short enough to make it quite likely that water flows are still forming gullies today.

FIGURE 7.22 This photo from *Mars Global Surveyor* shows gullies on a crater wall. Scientists suspect that the gullies were formed by water released from underground in small flash floods. These gullies are no more than a few million years old, and some may still be forming today.

If water still flows on occasion, it may have great significance for the possible present-day survival of organisms that might have arisen on Mars long ago. However, Mars today is clearly drier and less hospitable than the Mars of the distant past, especially the Mars of the Noachian era, when substantial water erosion occurred. If we wish to understand the prospects of finding life on Mars, we must come up with some ideas about how and why the climate changed over time.

The Climate History of Mars

Based on the geological evidence we have discussed, Mars must have had a much warmer atmosphere early in its history than it has today. The erosion patterns that we trace to the Noachian era probably could not have formed unless liquid water was much stabler at the surface. If water was liquid on the surface, the temperature must have been much higher than it is today, because temperatures now are almost always well below freezing everywhere on the surface.

If Mars had a warmer and thicker atmosphere in the past, it must also have had a much stronger greenhouse effect than it has today. The primary greenhouse gas was probably carbon dioxide. Computer simulations show that liquid water would be stable on Mars today if Mars had a carbon dioxide atmosphere about 400 times as dense as its current atmosphere. The idea of such a dense atmosphere is not as implausible as it may sound. That atmospheric pressure would be only about three times that on Earth today, and the total amount of CO_2 needed to make such an atmosphere is only about 1/30 the amount present in Venus's atmosphere or the amount locked up in carbonate rocks on the Earth. Certainly, Mars has plenty of ancient volcanoes that could have released carbon dioxide into the atmosphere. However, there's a potential problem with this scenario. Recall that, at the time Mars had liquid water (more than 3.7 billion years ago), the young Sun was nearly 30% dimmer than it is today [Section 4.5]. When scientists factor this dimmer Sun into the calculations, they find that CO_2 alone probably could not have warmed Mars enough for liquid water to be stable. One way around this problem is to presume that methane also made a substantial contribution to the greenhouse effect, but we do not know whether enough methane was available on Mars.

Thus, while a past warm climate must have been due to the greenhouse effect in some way, we still do not know for sure what caused this greenhouse effect. Nevertheless, it's worth noting that if martian volcanoes outgassed carbon dioxide and water in the same proportions as do volcanoes on Earth, Mars would have had enough water to cover the entire planet to a depth of tens or even hundreds of meters. The bigger question is not whether Mars once had a denser atmosphere but rather how all its atmospheric gas disappeared. In particular, Mars must somehow have lost most of its carbon dioxide gas. This loss weakened the greenhouse effect until the planet essentially froze over.

The fate of the lost carbon dioxide is not completely clear. Some is locked up in the polar caps, and some in carbonate minerals that probably formed much like carbonate rocks on Earth [Section 4.5]. We do not yet know how much carbonate rock exists on Mars. We have identified carbonate minerals in martian meteorites, but studies of Mars from orbit have not found evidence of any substantial amounts of carbonate rock on the surface. It remains possible that Mars has much larger quantities of carbonate minerals underground, but overall it seems unlikely that the polar caps and carbonate minerals can account for the vast amounts of carbon dioxide that must once have been present in the martian atmosphere. A more recently proposed hypothesis suggests that much of the carbon dioxide escaped to space.

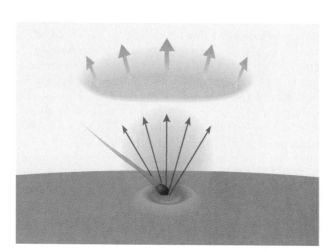

a The shock waves from an impact can blast away large amounts of atmospheric gas into space.

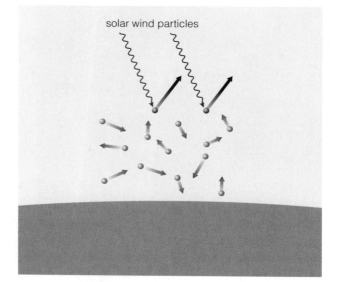

b Atoms or molecules in the upper atmosphere can be stripped from the atmosphere when they are struck by particles from the solar wind.

FIGURE 7.23 Mars probably lost most of its carbon dioxide to space through two basic mechanisms: impacts blasting away atmosphere, and molecules being stripped away by the solar wind.

The loss of a gas like carbon dioxide to space can occur in two basic ways (Figure 7.23). One way is for the gas to be blasted away by large impacts. The second is for the gas to be stripped from the atmosphere when it is hit by fast-moving particles coming from the Sun in the *solar wind* [Section 1.3]. Much of the gas in the early martian atmosphere may have been blasted away by impacts, especially during the Noachian era, which coincided with the period of heavy bombardment in the young solar system. However, based on magnetic field measurements from Mars Global Surveyor, the final blow to Mars's thick atmosphere was probably stripping by solar wind particles.

Early in its history, Mars had a global magnetic field that protected the atmosphere from the solar wind—just as Earth's magnetic field protects our atmosphere, deflecting solar wind particles or channeling them toward the poles (where they produce the auroras). But as the martian magnetic field weakened and eventually disappeared (as the core cooled), the atmosphere became vulnerable. Most of the carbon dioxide that remained in the atmosphere at the end of the Noachian era was probably stripped away. With this gas now gone, there is little hope that Mars will ever again have the warm, wet climate it had before about $3\frac{1}{2}$ billion years ago.

Much of the water once present on Mars is also probably gone for good. Like carbon dioxide, water vapor may have been blasted into space by impacts and stripped away by the solar wind. But Mars also lost water vapor in a third way. Water molecules are easily broken apart by ultraviolet light from the Sun, which penetrates the martian atmosphere because of the lack of an ultraviolet-absorbing ozone layer. Once hydrogen atoms break away from water molecules, they can escape into space fairly easily by virtue of their low mass and high speeds. Recall that temperature characterizes the average energy of atoms and molecules in a gas. All types of atoms and molecules have the same average energy at a particular temperature, but lighter-weight particles move faster than heavier particles.[3] Hydrogen is the lightest of all elements, and even at relatively low temperatures hydrogen atoms can reach speeds fast enough to escape Mars.

Once the hydrogen stripped from water molecules escaped Mars, the oxygen had nothing with which to recombine to make water again. Mars has very little oxygen remaining in its atmosphere, which means that the oxygen must also have been lost. Some was probably lost through impacts and stripping by the solar wind, some through photochemical processes in the upper atmosphere, and the rest by chemical reactions with surface rock. The oxygen literally rusted the martian rocks, giving the "red planet" its distinctive color.

---

[3] More technically, the temperature of a gas is a measure of the average kinetic energy of particles in the gas. A particle's kinetic energy is given by the formula $\frac{1}{2}mv^2$, where m is the particle's mass and v is its speed. Because all the particles in the gas have the same average value for this product, the lower-mass particles must have higher average speeds.

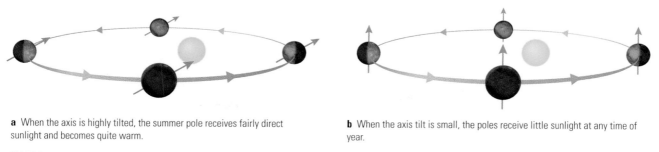

a When the axis is highly tilted, the summer pole receives fairly direct sunlight and becomes quite warm.

b When the axis tilt is small, the poles receive little sunlight at any time of year.

FIGURE 7.24 Mars's axis tilt probably varies dramatically, causing climate change because of the effect on the seasons.

The picture we have painted explains how Mars underwent the transition from a warm, wet planet to a cold, dry one. This change probably occurred late in the Noachian era or early in the Hesperian era, and it was probably complete by about 3 billion years ago. The change may not have occurred smoothly; that is, colder periods may have alternated with warmer periods depending, for example, on volcanic activity.

The martian climate probably never again underwent a change as dramatic as the one that occurred some 3–4 billion years ago, but lesser changes still occur. In particular, the tilt of Mars's axis varies over time, probably much more than does the tilt of Earth's axis. Recall that the tilt of Earth's axis varies between about 22° and 25° due to gravitational effects from Jupiter and other planets and that changes in the tilt play at least some role in the occurrence of ice ages on Earth [Section 4.5]. The effects of Jupiter's gravity on Mars are even greater, because Jupiter is closer to Mars than to Earth. In addition, our large Moon exerts a gravitational pull that helps stabilize the Earth's axis. Mars lacks a large moon, and its two tiny moons (Phobos and Deimos) are far too small to offer any stabilizing influence on its axis. Calculations show that the axis tilt of Mars may vary from as little as 0° to 60° or more, on time scales of hundreds of thousands to millions of years (Figure 7.24). When the axis is highly tilted, the summer pole will become quite warm, allowing substantial amounts of water ice to sublime into the atmosphere. When the axis tilt is small, the poles may stay in a perpetual deep freeze for tens of thousands of years. These changes must affect the amounts of carbon dioxide and water vapor in the atmosphere and hence the global temperature and pressure (but probably not enough to allow liquid water to become stable at the surface). The martian polar regions show layering that probably reflects changes in climate due to the changing axis tilt (Figure 7.25). Understanding this type of relatively recent climate change will be a major goal of future missions to Mars.

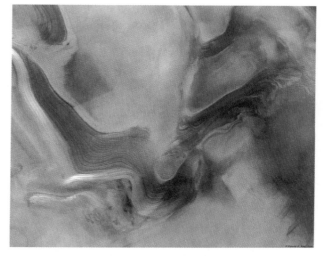

FIGURE 7.25 This photo from *Mars Global Surveyor* shows a region about 40 kilometers wide at the edge of a martian polar cap. Note the alternating layers of ice and dust, which are probably the result of climate changes caused by the changing axis tilt of Mars.

Summary: Is Mars Habitable?

Mars clearly has the elements needed for life, and energy is available for life in the form of sunlight (on the surface) and chemical energy (underground). So the question of whether Mars is habitable hinges ultimately on the availability of liquid water. If we are interpreting the evidence correctly, the Mars of some 3.5–4.0 billion years ago must have looked somewhat like the Earth of the same period, with water flowing across the surface, active volcanism, and an atmosphere rich in carbon dioxide. In short, the young planet Mars appears to have enjoyed environmental conditions very similar to those under which life arose on Earth. We therefore conclude that Mars *was* habitable in its early history, which is why we are so eager to learn whether any life actually arose.

Mars today tells a very different story. Its dry, frigid surface is probably not habitable. However, the evidence for liquid water still existing below the surface suggests that Mars *may* be habitable underground in at least some locations. It is time to turn our attention to the search for life on Mars.

7.3 Searching for Life on Mars

Back in the days of Percival Lowell, creative scientists and engineers sought ways to communicate with the imagined Martians. Several pioneers of radio broadcasting, including Nikola Tesla (1856–1943) and Guglielmo Marconi (1874–1937), suggested the possibility of interplanetary radio communication. In a way, they were right. Today we use radio to communicate with spacecraft that we send to Mars and elsewhere, and radio searches are an integral part of SETI efforts to find civilizations in other star systems (see Chapter 11). However, Tesla and Marconi assumed the radio signals would come from other intelligent beings. Both claimed to have received mysterious signals that they thought might be from Martians. Tesla wrote, "The feeling is constantly growing on me that I had been the first to hear the greeting of one planet to another."[4] At one point, even the U.S. Army joined in listening for broadcasts from Mars (Figure 7.26).

We now know that we won't be receiving any radio messages from indigenous Martians. If we want to search for life on Mars, we need to study actual martian soil to see whether it contains living microbes. This type of search was first carried out by the two Viking landers in 1976.

The Viking Experiments

Each of the Viking landers was equipped with several on-board, robotically controlled experiments. Three experiments were designed expressly to look for signs of life. A fourth was designed to analyze the content of martian soil and proved important to interpreting the results of the three biology experiments. Each lander was equipped with a robotic arm for scooping up soil samples. The robotic arms even pushed aside rocks to get at shaded soil that was less likely to have been sterilized by ultraviolet light from the Sun (Figure 7.27).

At first, the Viking experiments seemed to be giving results suggesting that they had found evidence of life on Mars. But further study led scientists to the conclusion that the experiments had detected some interesting chemistry, but no biology. In order to understand the implications of the results, let's briefly discuss each experiment and what it found.

FIGURE 7.26 Listening for martian radio signals, as pictured in *Radio Age,* October 1924.

The Carbon Assimilation Experiment The first of the three biology experiments mixed a sample of martian soil with carbon dioxide (CO_2) and carbon monoxide (CO) gas brought from Earth. In some runs of the experiment, the soil was also mixed with water. The aim was to see if any of the carbon dioxide or carbon monoxide would become incorporated into the soil, as would be the case if living organisms were using either gas as a source of carbon in their metabolism. The carbon dioxide and carbon monoxide from Earth could be distinguished from the same gases in the martian atmosphere because they had been "tagged" with radioactive carbon-14 (rather than the far more common, stable isotope carbon-12). Sure enough, the experiment found that the carbon-14 was incorporated into the soil, a result that at first seemed to suggest that metabolism was occurring. As a further test, however, the experiment was repeated with soil first heated for 3 hours to 175°C (347°F)—hot enough to break the chemical bonds between carbon and other atoms and thereby kill any Earth-like organisms that might have been present. The experiment found that the tagged carbon was still incorporated into the soil. Scientists concluded that the results were unlikely to be due to biological activity and that some sort of geochemical process was instead causing a reaction with the gases.

The Gas Exchange Experiment The second biology experiment mixed martian soil with a "broth" containing organic nutrients brought from Earth. The

---

[4] Original source: Nikola Tesla, "Talking with the Planets," *Collier's Weekly* 26, no. 19, 4 (1901); taken from S. J. Dick, *The Biological Universe,* Cambridge University Press, 1996.

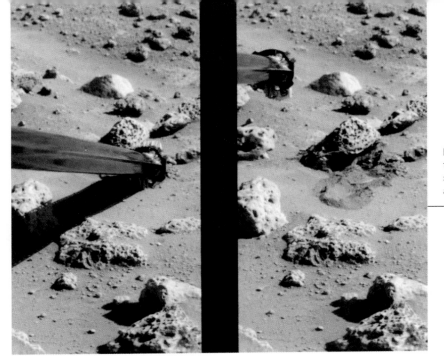

FIGURE 7.27 This pair of before-and-after photos, transmitted back to Earth by the Viking Lander 2, shows a location where the robotic arm pushed away a small rock on the martian surface.

experiment looked for a variety of gases that might be released by the respiration of martian microbes, including hydrogen (H_2), nitrogen (N_2), oxygen (O_2), methane (CH_4), and carbon dioxide (CO_2). Again, the initial results were promising for life: As soon as the soil was exposed to the nutrients, oxygen was released into the chamber. However, careful analysis showed that the oxygen was probably coming from chemical rather than biological reactions. One reason for this conclusion was that the oxygen was released in the dark, whereas Earth organisms would release oxygen only if they were undergoing photosynthesis in light. Another reason was that the reaction continued even when the soil was first heated to high temperatures that would have killed any organisms present. A third reason was that the oxygen was given off even when the soil was exposed only to water vapor, rather than to the nutrients, a result inconsistent with metabolism. Scientists concluded that the experiment had not found evidence of life but that reactions between water vapor and something nonbiological in the soil instead were giving off the detected oxygen.

The Labeled Release Experiment The third biology experiment also mixed martian soil with organic nutrients. If martian microbes were present and consumed any of the nutrients, then by-products of metabolism and respiration would show up in the form of gases released from the soil. The nutrients had been tagged with radioactive carbon-14 and sulfur-35 so that the by-products would be detectable. The experiment involved measuring the change with time in the amount of radioactivity in gases within the reaction chamber. If life were present, we would expect the level of radioactivity to rise as the organisms consumed the nutrients and released more gas and then level off as the nutrients were used up. This is precisely what happened. Moreover, repeating the experiment with prior heating of the soil sample to 50°C (122°F) substantially reduced the amount of radioactivity, and heating it to 160°C (320°F) eliminated any sign of the tagged isotopes in the chamber gas. This is just what we would expect if the heat were killing the organisms, as it would if we carried out the experiment on Earth. However, to decide whether this experiment really found evidence of life, we need to look at the results of a fourth experiment.

The Gas Chromatograph/Mass Spectrometer Experiment This experiment was not formally designed to search for life but instead was designed to measure the abundance of organic molecules in the martian soil. We might expect organic molecules to be present even in the absence of biological activity, because they would be supplied by the impact of meteorites containing organic molecules; such organic molecules are found on the Moon, for example. The experiment looked for organic molecules by heating the soil to temperatures up to 500°C (930°F), high enough to kill any organisms, break apart any organic molecules they contained, and vaporize the molecules. The vapors were passed through a *gas chromatograph,* a device that separates different gases as they pass through it. They were then measured using a *mass spectrometer* to determine the masses of molecules passing through it. If any organic life was present, the experiment should have found significant amounts of organic molecules.

However, the experiment found no measurable level of organic molecules in the martian soil. (The only organic molecules detected were solvents left over from the cleaning of the instrument before it was launched from Earth.) Given the sensitivity of the experiment, this result meant that organic molecules could not be present in the tested martian soil at a concentration of more than a few parts per billion—some 10 million times lower than the concentration we find in Earth soil. The most likely conclusion is that the soil contained no living organisms, though we cannot completely rule out the possibility that organisms were present in such low abundance that they did not show up in this experiment.

Summary of the Viking Experiments All three biology experiments gave results that initially seemed consistent with life. But the results of two of the three experiments were unchanged when the soil was heated to high temperature, making them appear inconsistent with life. The results of the labeled release experiment could be attributed to biology, but the lack of organic molecules determined by the gas chromatograph/mass spectrometer experiment was inconsistent with the level of activity that would have been required for biological processes. As a result, scientists favor the conclusion that some chemical process mimicked a biological result in this experiment. In summary, scientists concluded that the Viking experiments had not found life and that the absence of any measurable level of organic molecules made the presence of life in the tested soil highly unlikely.

Beyond Viking

Although the Viking missions did not find life on Mars, neither did they rule out the possibility that life exists there. In particular, since we expect a reasonable chance for life only where there is liquid water—which on Mars means deep underground or near volcanic sources of heat—the Viking sites were among the least likely places to find life on Mars. Why, then, did scientists choose these unpromising sites as destinations for the Viking landers?

Perhaps the main answer is that the Viking landing sites look so unpromising only with the benefit of hindsight. It was the Viking results that first led us to believe that environmental niches that could support life might exist on Mars, and the existence of life in extreme environments on Earth was just beginning to be understood at that time. Thus, these ideas on the nature of life could not be factored into decisions about the Viking landing sites. In addition, the choice of landing site for any robotic mission involves a trade-off between more interesting sites, such as the canyons of Valles Marineris or the slopes of

Olympus Mons, and safer terrain. Even when Mars is nearest Earth, it takes radio signals about 20 minutes to travel from Mars to Earth. This means that spacecraft must land by "autopilot." There is no hope of controlling the descent remotely from Earth, because by the time the cameras showed a danger (say, the danger of falling off a cliff), it would be too late to initiate a course correction. The Viking landing sites were chosen in part because they were expected to be fairly flat and safe. Despite such precautions, had one of the missions landed on top of a large rock such as the one visible in Figure 7.28 (nicknamed "Big Joe" by scientists), it probably would have been destroyed.

Future missions will eventually explore more interesting terrain, but even then robotic experiments performed aboard a spacecraft are unlikely to provide conclusive evidence for life. The problem, as the Viking experiments showed, is that even seemingly positive results may turn out to have other explanations. The experiments must be designed before any spacecraft is launched, and it's impossible to anticipate every surprising result that might arise. The only way we are likely to find definitive evidence for or against life is by being able to change the experiments as we go along, which means either sending humans to Mars to conduct the experiments or bringing samples of Mars back to Earth for study. Missions that bring Mars rock to Earth are still at least a few years off (see Section 7.5). Nevertheless, we already have a few samples of rock from Mars available for study. As we'll discuss next, one of these samples has yielded intriguing results about the possibility of life on Mars.

7.4 Martian Meteorites

Meteorites, rocks that fall to Earth from space, can be distinguished from terrestrial rocks by their composition. All meteorites contain elements and isotopes in ratios that are clearly different from the ratios found in rocks of the Earth's crust. But not all meteorites are the same, because they originated in different ways. For example, some meteorites appear to be nearly pristine pieces of rock left over from the formation of our solar system, while others appear to be the remains of asteroids that shattered in collisions. Some meteorites even come from the Moon, having been knocked off the surface by an impact. Among the more than 20,000 meteorites that have been collected, a few have unusual characteristics that lead us to believe they are pieces of rock from Mars. As of early 2002, scientists had identified 19 **martian meteorites.**

Although scientists at first were surprised to find evidence that rocks from Mars had landed on Earth,

FIGURE 7.28 The view from the Viking 1 lander. The large rock is about 2 meters across—big enough to have ended the mission if the Viking lander had landed on top of it instead of on smoother ground.

on further reflection it does not seem so strange. A large impact can hit a planet hard enough to blast some debris into space, and Mars has plenty of large impact craters. Each piece of rock launched into space by an impact will orbit the Sun like a tiny asteroid. If its orbit intersects that of Earth, the rock can come crashing down as a meteorite. Calculations show that over the history of our planet some 1 billion tons of rock from Mars should have crashed down in just this way. Thus, it's quite reasonable to expect to find a few martian meteorites, even though most of the billion or so tons of Mars rock must by now lie buried beneath the Earth's surface.

Several pieces of evidence lead scientists to believe that the claimed martian meteorites are, indeed, rocks from Mars. They all appear to be pieces of volcanic rock, which means they must have originated on a world large enough to have had volcanism at some time. Moreover, most of these meteorites are relatively young volcanic rocks—less than about a billion years old—which means they must come from a world that has had active volcanism in geologically recent times. (One known older rock among the martian meteorites is tied to the same planet of origin as these younger rocks by virtue of its similar composition.) An origin on the Earth can be ruled out by a comparison of the relative abundances of the isotopes oxygen-16, oxygen-17, and oxygen-18; these rocks have values distinctly different from those of terrestrial rocks. The nature and age of the meteorites rules out the Moon and all the asteroids as their place of origin, leaving Venus and Mars as the only likely candidates. The best evidence for martian origin comes from analysis of small amounts of gas trapped within the meteorites. This gas ap-

pears identical in its chemical and isotopic composition to that of the Mars atmosphere as measured by the Viking landers, and it is distinctly different from any other known source of gas in our solar system. The atmospheric gas appears to have become embedded in the rocks by shocks from the same impacts that knocked the rocks into space.

Meteorites from Mars would generate scientific excitement under any circumstances, because they provide unique insights into the history of the planet. Indeed, these rocks have provided us with many new and important insights into the nature of Mars. But martian meteorites became big news to the public only in August 1996, when a group of NASA scientists announced that one of them contains possible evidence of past life on Mars and perhaps even actual fossil organisms. The evidence remains a source of ongoing scientific controversy. We will address this controversy next.

ALH84001

In recent years, scientists have learned that Antarctica is one of the best places to hunt for meteorites (Figure 7.29). No more meteorites land in Antarctica than anywhere else on Earth, but it is easier to find them there. Because few terrestrial rocks end up on the Antarctic ice, a rock found on the ice has a good chance of being a meteorite. Moreover, the slow movement of glaciers tends to carry the meteorites along, concentrating them in places where the ice flow runs into mountains. In fact, most of the meteorites now in our scientific collections have been collected from the Antarctic ice in annual expeditions that have taken place since 1979.

FIGURE 7.29 This photo shows a researcher who has just found a meteorite on the Antarctic ice.

a The martian meteorite ALH84001, before it was cut open for detailed study. The small block shown for scale to the lower right is 1 cubic centimeter, about the size of a typical sugar cube.

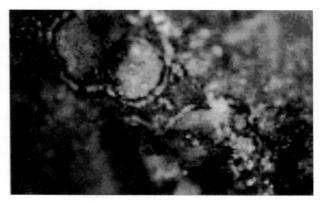

b A microscopic view of a thin slice of ALH84001 shows orange carbonate globules, indicating that the rock was once infiltrated by liquid water. The grains are about 0.1–0.2 millimeter across. The globules formed about 3.9 billion years ago.

FIGURE 7.30 The meteorite known as ALH84001 apparently comes from Mars.

In December 1984, a team of American scientists scooped up a 1.9-kilogram meteorite from the Allan Hills region of Antarctica (Figure 7.30a). It was cataloged as "ALH84001": ALH for Allan Hills, 84 for the year in which it was found, and 001 indicating that it was the first meteorite found on the expedition. It did not immediately draw special attention, but an analysis a decade later showed that it was one of the rare martian meteorites. It quickly proved itself special even among this small group of rocks.

Whereas other known martian meteorites are geologically young, radiometric dating showed ALH84001 to be a piece of volcanic rock that formed about 4.5 billion years ago. The meteorite also shows evidence of later shocks, probably due to the effects of impacts that did *not* launch it into space. In addition, it contains carbonate globules (Figure 7.30b) that formed about 3.9 billion years ago. The carbonate minerals precipitated within the rock, meaning that the rock must have been infiltrated by liquid water (as commonly occurs on Earth).

We can determine when ALH84001 was launched into space by looking for effects of exposure to cosmic rays, high-energy particles that leave telltale chemical signatures on anything unprotected by an atmosphere. Careful analysis of ALH84001's cosmic ray exposure shows that it wandered through the solar system for about 16 million years, which means it must have been blasted from the surface of Mars about 16 million years ago. We can determine when cosmic rays stopped disturbing the meteorite, and thus when it fell to Earth and gained the protection of Earth's atmosphere, by studying the decay of radioactive isotopes produced by the cosmic rays. Such analysis shows that ALH84001 landed in Antarctica about 13,000 years ago. Table 7.3 summarizes the history of ALH84001.

THINK ABOUT IT . . . *Note that the 4.5-billion-year age of ALH84001 makes it considerably older than any rocks found on Earth (besides meteorites). Yet ALH84001 apparently remained at (or near) the martian surface in recent geological times (since it was blasted from the surface only 16 million years ago). Why is it plausible for Mars to have such an old rock? If we visit Mars and collect rocks, would we expect to find other rocks of similar age?*

Evidence of Life in ALH84001? This meteorite was singled out for study for evidence of life in part because of its old age, which indicates it was present on Mars at the time when water flowed, and in part

Table 7.3 The History of Meteorite ALH84001

Time	Event
4.5 billion years ago	Solidifies from molten rock in the southern highlands of Mars
4.0–4.5 billion years ago	Affected by nearby impacts, but not launched into space
3.9 billion years ago	Infiltrated by water, leading to the formation of carbonate grains within the rock
16 million years ago	Blasted into space by an impact on Mars
13,000 years ago	Falls to Earth in Antarctica
December 27, 1984	Found by scientists
October 1993	Recognized as a martian meteorite
August 1996	Announcement that ALH84001 contains possible evidence of martian life

because of its carbonate deposits, which tell us that liquid water actually flowed through this rock. The claimed evidence of life in ALH84001 comes from detailed, state-of-the-art studies of the carbonate globules and the surrounding rock. In brief, four types of evidence were originally heralded as pointing to the existence of biology on Mars:

- The carbonate globules have a layered structure, with alternating layers of magnesium-, iron-, and calcium-rich carbonates. This layering involves minerals that would not be expected to form next to each other, suggesting some sort of process able to create minerals that are out of equilibrium with each other. On Earth, biological activity can do this.

- The globules contain complex organic molecules known as *polycyclic aromatic hydrocarbons,* or *PAHs.* These molecules can be produced by both biological and nonbiological processes (and have even been found in meteorites that are not from Mars, although in much lower abundances), but on Earth they are most commonly produced by the decay of dead organisms or by reactions between such decay products and the environment (for example, in the burning of fossil fuels).

- Within the iron-rich layers, we find crystals of the mineral magnetite similar in size and shape to magnetite grains that on Earth occur only when made by bacteria (Figure 7.31).

- Most intriguingly, highly magnified images reveal rod-shaped structures, some of which are segmented, that look much like ordinary bacteria except that they are about a hundred times smaller (Figure 7.32a). At first, this small size

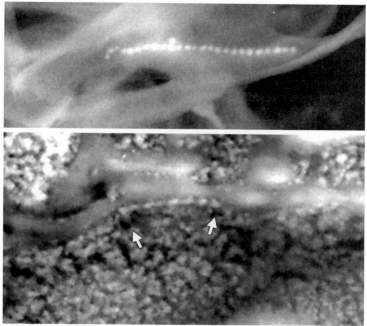

FIGURE 7.31 Top: Microscopic chains of magnetite crystals produced by bacteria on Earth. Bottom: Similar chains of magnetite crystals found in the carbonate globules of ALH84001. The similarity to the terrestrial chains has been cited as evidence of life on Mars.

seemed to rule out their being living organisms. However, biologists have recently identified similar-looking and similar-size structures on Earth that they call "nanobacteria" or "nanobes" (Figure 7.32b). While some controversy still exists as to whether the terrestrial nanobacteria are truly living organisms, they appear to contain DNA.

While none of these lines of evidence alone would prove biological activity, the original investigators argued that, on the whole, biology seemed a much more likely explanation than nonbiological processes.

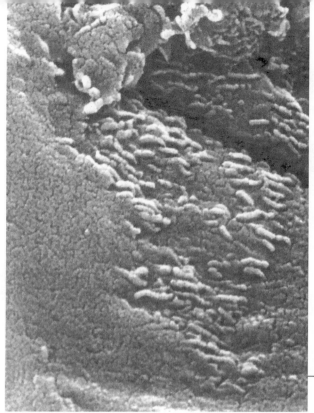

a This photo shows rod-shaped structures found in the carbonate globules of ALH84001. They measure about 100 nanometers in length and are as small as 10–20 nanometers in width.

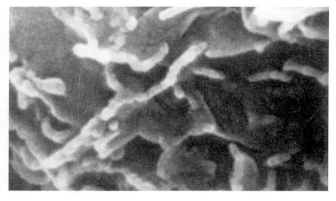

b This photo shows terrestrial "nanobacteria" in a sample of volcanic rock from Sicily. They are close in size to the structures seen in ALH84001. The scale bar at the bottom is 1 micrometer, or 1,000 nanometers.

FIGURE 7.32 Does ALH84001 contain fossils of martian organisms?

They felt that it would be a "better" scientific explanation if only a single process—biology—could account for each observation than if a different process was required to explain each result.

Are There Nonbiological Explanations for the Evidence? The four lines of evidence for fossil life in ALH84001 might seem to make a strong case for past life on Mars. However, other scientists have proposed alternative, nonbiological mechanisms that might have produced each observed phenomenon. There are potential ways to get layered carbonate without the action of life. For example, several pulses of hot water with different dissolved elements might have passed through the rock and laid down the different mineral layers. The PAHs might have been produced by chemical rather than biological processes, or, as we'll discuss shortly, they might have been introduced by terrestrial contamination before the rock was picked up from the Antarctic ice. The magnetite chains might not be as clearly indicative of life as the photo comparisons in Figure 7.31 suggest. Some scientists question whether the chains really have the shape they seem to or whether this apparent shape is an artifact of how the images were made. Moreover, even if the chains are real, the fact that they look like similar chains made by bacteria on

Earth does not prove that they were made in the same way; perhaps there are nonbiological ways of producing the chains. Finally, the rod-shaped structures don't tell us anything other than that they look a lot like nanobacteria. There could well be nonbiological ways of producing similar-looking structures. In fact, most scientists today believe that these structures are, for the most part, artifacts of how the rocks were processed in the laboratory prior to their being examined. In summary, while each of the four lines of evidence is suggestive of life, each may also have an alternative, nonbiological explanation.

Could the Evidence Be Contamination from Earth? It's also important to consider whether the claimed evidence of life in ALH84001 might actually represent contamination of the meteorite during the 13,000 years it has spent on Earth. In fact, contamination may well explain at least part of the evidence. Contamination could have occurred either on the Antarctic ice before the meteorite was found or in the subsequent handling of the meteorite. Contamination during handling seems unlikely. Like other meteorites collected on the Antarctic expeditions, ALH84001 was picked up with sterile bags and sealed in an airtight container for its trip to a laboratory at the Johnson Space Center in Houston. At the laboratory, it was handled with extreme care in clean rooms. As a further check, scientists subjected clean quartz disks to the same laboratory procedures as the meteorites. If there was laboratory contamination, the quartz disks should have been affected in the same way as the meteorites. The disks showed no contamination, so scientists are quite confident that no laboratory contamination affected the meteorites. For a while, scientists were also confident that no contamination had occurred on the Antarctic ice, because analysis showed that ALH84001 contains a higher abundance of organic molecules deep within the rock than nearer the surface. If the organic molecules represented terrestrial contamination, we would expect just the opposite. Nevertheless, a more recent analysis found modern, living terrestrial bacteria inside the rock! These organisms probably infiltrated the rock while it sat on the ice. This is not too surprising, given that some of the Antarctic meteorites are actually picked up out of little puddles of water on top of the ice. This clear contamination probably explains the presence of PAHs in the rock. Though it may not affect the other lines of evidence, it certainly weakens any case to be made about martian life in ALH84001.

Summary of the ALH84001 Controversy The extraordinary debate over whether we already have discovered evidence of life beyond Earth will undoubtedly continue as scientists subject ALH84001 and other martian meteorites to further scrutiny. The techniques that have been applied in the analysis to date are the best available, but new technologies will allow the martian meteorites to be studied in even more detail over the coming years. Still, unless a surprising and major breakthrough occurs, it is unlikely that the meteorites by themselves can provide definitive evidence of life. In the words of the late Carl Sagan, "Extraordinary claims require extraordinary evidence." The claim that meteorites from Mars hold fossil life is surely extraordinary and is unlikely to gain wide acceptance while any significant doubt remains.

It is also possible that further study will prove that ALH84001 does *not* contain fossil life. But even in that case, we still won't know whether there has ever been life on Mars. Most Earth rocks do not contain fossils. Thus, whether or not life existed on Mars, it wouldn't be surprising if the first ancient Mars rock we examine lacks fossils. In all likelihood, a final verdict on the question of life on Mars will have to wait at least until future space missions scoop up and return to Earth samples of rock from a variety of locations on Mars.

THINK ABOUT IT . . . *Given the fact that ALH84001 has apparently been contaminated by terrestrial bacteria, could we ever prove that a martian meteorite holds evidence of life on Mars? Why or why not?*

Implications for Our Understanding of the Origin of Life

The discovery of meteorites from Mars helped rekindle the debate over whether life on Earth might have been transplanted from elsewhere, especially from Mars [Section 5.2]. If there has ever been life on Mars, some organisms have almost certainly had the opportunity to hitch rides to Earth on meteorites blasted off the martian surface by impacts. The bigger question is whether any such hitchhiking organisms could survive the journey.

For a living microbe to go from Mars to Earth, it would have to survive three potentially lethal events: the impact that blasts it off the martian surface, the time it spends in the harsh environment of interplanetary space, and the fiery plunge to Earth through our atmosphere. None of these events seems to pose an insurmountable obstacle. The interiors of martian meteorites such as ALH84001 show only minimal disruption, suggesting that organisms inside these rocks could survive both the impact and the later fall to Earth.

The chance of surviving the trip between planets probably depends on how long the meteorite spends in space. Recall that, once a rock is launched into space, it orbits the Sun until its orbit carries it directly into the path of another planet, such as Earth. In the case of ALH84001, the meteorite orbited the Sun for 16 million years before crashing into Earth (see Table 7.3). It seems highly unlikely that living organisms could survive in space for millions of years. However, a few meteorites are likely to be launched into orbits that cause them to crash to Earth during one of their first few trips around the Sun. Calculations suggest that about 1 in 10,000 meteorites may travel from Mars to Earth in a decade or less. Experiments conducted in Earth orbit have already shown that some terrestrial microbes can survive at least 6 years in space, so it seems quite reasonable to imagine martian microbes arriving safely on Earth.

Based on this scenario, we can surmise that there has almost certainly been a substantial exchange of rocky material among all the terrestrial planets. That is, rocks from Venus have probably also landed on Earth (though we have not yet identified any meteorites from Venus), and rocks from Earth have probably landed on Venus and Mars. This interplanetary exchange raises at least two important philosophical questions about the origin of life.

First, it's very likely that terrestrial microbes have made and survived the journey to Mars many times. Thus, if we do find life on Mars, we'll have to wonder whether it arose there indigenously or we are finding the descendants of terrestrial organisms that migrated to the red planet. Indeed, given the evidence that Mars once had a global surface environment—as well as more recent isolated niches—in which many known Earth organisms could thrive, it would be somewhat surprising if life from Earth had not taken hold there. The only way we may ever be confident that Mars life is not transplanted Earth life will be if its biochemistry is too different from that of terrestrial life to allow for a common ancestor. Even in that case, we might wonder whether the martian life could have descended from a species that arose on Earth very early but was extinguished by a sterilizing impact before the end of the heavy bombardment [Section 4.3]. After all, if life arose and was extinguished multiple times on the early Earth, there is no reason to expect the biochemistry to have been the same with each separate origin of life.

Second, the possibility of life migrating among the planets raises the question of whether we can distinguish between an indigenous origin of life on Earth and an origin based on immigration from elsewhere. If the early environment on Mars or Venus happened to become habitable before the environment on Earth, life might have arisen on one of those planets and then migrated to and taken hold on Earth. We might never be able to determine whether the first Earth life was native or alien. We can be sure, however, that regardless of where life first arose, life took hold on Earth very early in its history.

7.5 Future Exploration of Mars

We still have much to learn about the habitability of Mars and the possibility of martian life. Our best hope for learning about these issues rests with launching more missions to Mars. The 20-year post-Viking drought in new missions to Mars was very frustrating to planetary scientists. Now, however, we appear to be in a new era of Mars exploration marked not only by frequent missions but by innovative technologies that make the new missions far less expensive than the Viking missions. This new era began with the landing of *Pathfinder* in 1997 (see Figure 7.7).

Recall that a primary factor in mission cost is the weight of the spacecraft and its onboard fuel, because heavier spacecraft cost much more to launch [Section 6.2]. Some of the recent reduction in mission cost comes from advances in miniaturization that allow scientific instruments to weigh less than they did in the past. Other cost reduction comes from new techniques for reducing the amount of onboard fuel a spacecraft must carry. For example, whereas the Viking landers used rocket power to settle gently on the martian surface (after first slowing through the atmosphere with parachutes), *Pathfinder* hit the surface at crash-landing speed but was protected by a cocoon of air bags deployed on its way down. These air bags allowed it to bounce along the surface (for more than a kilometer) until it finally came to rest (Figure 7.33). Two recent orbiting missions, Mars Global Surveyor and Mars Odyssey, saved on fuel costs by carrying only enough fuel to enter highly elliptical orbits around Mars. In each case, the spacecraft settled into the smaller, more circular orbit needed for scientific observations by skimming the martian atmosphere at the low point of every elliptical orbit. Atmospheric drag slowed the spacecraft with each orbit and, over several months, circularized the spacecraft orbit. (This technique of using the atmosphere to slow the spacecraft and change its orbit is called *aerobraking*.)

Given current technology and budgetary considerations, it makes sense to send spacecraft to Mars only when Earth and Mars have an optimal orbital alignment. Such an alignment occurs roughly every 26 months, allowing missions to Mars to be scheduled approximately every 2 years. If all goes well, these biennial missions will revolutionize our understanding of Mars over the next decade.

Mars Exploration over the Next Decade

Typical space missions today take at least 3 years to design and build, and planning and mission selection generally begin at least a few years prior to building. Thus, most missions that will visit Mars in the next decade are either already under construction or in their design phases. Beyond the next decade, the shape of Mars exploration will depend largely on what we discover in the short term.

Figure 7.34 depicts missions currently operating, under construction, or under study; keep in mind that the missions in later years are less certain than those in earlier years, because they are less far along in their planning. Note also the international character of the Mars exploration program. For example, Nozomi is a Japanese mission, *Mars Express* is being built by the European Space Agency (ESA), the *Beagle 2* lander is British, and the CNES Science Orbiter is French. (Nozomi was launched in 1998 but its arrival at Mars was delayed until 2003 due to an engine misfiring shortly after launch.) To date, about a dozen nations have made solid commitments to build instruments or spacecraft for Mars exploration.

Although we can hope that all these missions will succeed, history shows that Mars exploration is not easy. For example, both Mars missions launched by the United States in 1999 (an orbiter and a lander) failed before beginning scientific operations at Mars. These failures led to a reexamination of the exploration strategy and to new quality control safeguards. One result was that, while NASA had previously

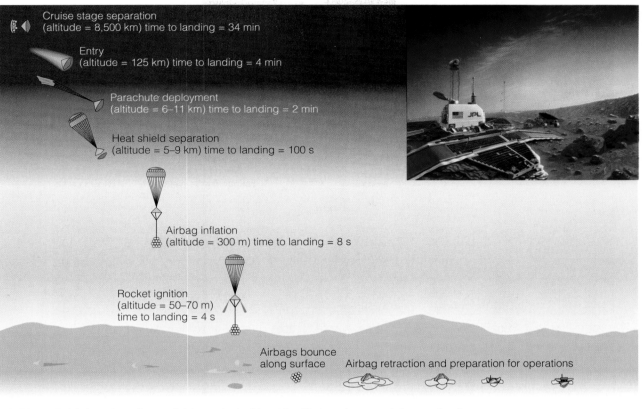

Cruise stage separation
(altitude = 8,500 km) time to landing = 34 min

Entry
(altitude = 125 km) time to landing = 4 min

Parachute deployment
(altitude = 6–11 km) time to landing = 2 min

Heat shield separation
(altitude = 5–9 km) time to landing = 100 s

Airbag inflation
(altitude = 300 m) time to landing = 8 s

Rocket ignition
(altitude = 50–70 m)
time to landing = 4 s

Airbags bounce along surface

Airbag retraction and preparation for operations

FIGURE 7.33 *Pathfinder* used parachutes and air bags to land safely on Mars. The inset shows an artist's conception of the lander on the surface.

planned to send both orbiters and landers at each biennial launch opportunity, the new plan calls for alternating orbiters and landers. Thus, for example, the orbiting *Mars Odyssey* will be followed by landers (the Mars Exploration Rovers) in 2003, with the next NASA orbiter (Mars Reconnaissance Orbiter) being launched in 2005. (However, the European Space Agency's orbiter, *Mars Express,* will be launched in 2003.) This scheme keeps subsequent orbiter and lander missions 4 years apart, which allows enough time for design changes based on any lessons—particularly lessons of failure—learned from prior missions. Another change that occurred in the wake of the 1999 mission failures concerned the timing of a sample return mission. NASA had once hoped to launch a sample return mission as early as 2005, but such a mission is now unlikely to proceed before 2011.

Despite the best efforts of scientists and engineers, it's likely that at least some of the upcoming Mars missions will fail. Nevertheless, we have already succeeded in reaching Mars many times, and we can hope that most of the missions will succeed. One way or another, a decade from now we should know far more than we do today about Mars and the possibility that it has ever harbored life.

Preventing Contamination—in Both Directions

One important consideration in Mars exploration is how to avoid contamination, in either direction. Given the likelihood that some Earth organisms could survive in at least a few locations on Mars, it's very important to make sure we don't accidentally contaminate Mars with Earth life while we are looking for Mars life. Otherwise, we might think we'd found evidence for martian life when in fact we were seeing only evidence of microbes that hitched a ride from Earth aboard a spacecraft. In addition, some people worry that terrestrial life might be able to outcompete any indigenous martian life, driving the martian life to extinction. This possibility raises an ethical question: Do we have a right to do something that could endanger native life on another planet?

Similarly, but in the other direction, the prospect of a sample return mission has caused some people to fear we might unleash dangerous martian microbes on Earth. In H. G. Wells's story *The War of the Worlds,* the martian invaders are eventually overcome not by human ingenuity but rather because they have no resistance to terrestrial disease. We now know that no intelligent Martians exist, but the prospect of martian microbes makes people wonder

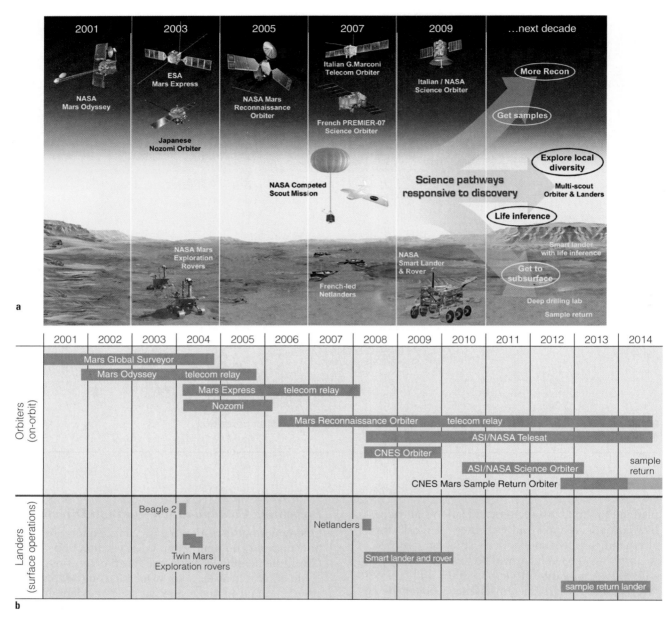

FIGURE 7.34 Plans for Mars exploration. (**a**) This painting shows the different types of spacecraft planned for Mars over the next decade. The dates shown are launch years. (**b**) The bars represent the times over which each spacecraft is expected to operate at Mars. "Telecom relay" means that a spacecraft is being used like a communications satellite for relaying data from other spacecraft but its primary scientific mission is over. CNES and ASI are the French and Italian space agencies, respectively. Note that the precise nature and timing of the missions is likely to change, particularly for those farther in the future.

whether the same thing could happen in reverse. That is, could martian microbes cause disease for which we are unprepared or outcompete terrestrial organisms on our own turf?

Clearly, the best way to solve these problems is to avoid contamination. An international treaty, signed in 1967, addresses the possibility of contaminating Mars with life from Earth. The treaty requires that any spacecraft sent to Mars must have a less than 1 in 1,000 chance of causing contamination. Today,

scientists strive for even lower contamination probabilities. Note that the specific destination and purpose of a spacecraft affects the requirements. We now know that the basic martian surface environment, such as at the Viking and Pathfinder landing sites, is not conducive to the survival of life from Earth. The surface has no liquid water or organic nutrients to sustain life, and any organisms from Earth would probably be killed by ultraviolet radiation from the Sun. Thus, even if a few microbes were to survive and reach Mars, they would have little chance of survival at most locations on Mars. On the other hand, if we send a lander to explore subsurface hot water near a volcano, we should be much more concerned about contamination, because water just might allow any immigrant microbes to survive.

The possibility of unleashing dangerous martian microbes on Earth is scarier but far less likely. Although many people don't realize it, disease-causing microbes are highly adapted to the species they infect. For example, diseases that infect plants generally do not infect animals. Indeed, "species jumping" by diseases is quite rare and generally occurs only between species that are evolutionarily close. HIV (the virus that causes AIDS), for example, is thought to have jumped from chimpanzees to humans, but on an evolutionary level this is a fairly small jump between different species of primates. Thus, even if martian microbes were accidentally released and subsequently survived on Earth, it's highly unlikely that they would cause disease. In addition, remember that martian meteorites undoubtedly land on Earth every year. If there is life on Mars, it has almost certainly reached Earth already. The fact that we do not see any harmful effects from this "natural contamination" makes it unlikely that any martian life can harm Earth life.

Nevertheless, it pays to be cautious given the high stakes involved, and samples brought back from Mars will surely be transported in sealed containers that would not break open even if they were to crash on Earth. Once here, they will be handled with extreme care in laboratories; we already know how to deal with organisms that are nearly always fatal and for which we have no cure or treatment, such as the Ebola virus. It is not a stretch to think that we would be able to deal with any harmful martian organisms.

THINK ABOUT IT . . . *In your opinion, should we allow samples from Mars to be brought to Earth, or should they be studied only in Earth orbit, such as on the Space Station? Defend your opinion.*

Human Exploration of Mars

Should we send humans to Mars? From a scientific standpoint, the advantage of sending humans to Mars is that people are far better able to make decisions while they work than any robot we can currently envision. If we sent a team of scientists to Mars with provisions to last a couple of years, vehicles for traveling around the planet, and equipment for drilling into the crust, their mission might well give us an answer to the question of whether Mars has ever harbored life.

Of course, such a mission would be extremely expensive. A spacecraft carrying humans to Mars would be far larger and heavier than any robotic spacecraft. It would have to have enough room for the people, enough provisions for both the round-trip journey and the time spent on the surface, and enough fuel to carry the ship and its crew back to Earth. There may be some creative ways to reduce the cost. For example, some people have proposed that, before sending the explorers, we could send robotic mini-factories that would manufacture fuel for the return trip from chemicals available on the martian surface. But there's no getting around the basic fact that a human mission will cost at least as much as dozens of robotic missions combined.

Sending humans to Mars also has at least one major scientific drawback: It vastly complicates the issue of avoiding contamination by terrestrial organisms. People are veritable warehouses of microbes. For example, the number of bacteria in the average person's mouth is far greater than the number of people who have ever lived. We harbor microbes on our skin, in our breath, in our food, and in our excrement. Preventing all these microbes from escaping into the martian environment during an extended stay on the planet might be nearly impossible.

All things considered, we could probably accomplish our scientific goals more cheaply and with far less risk of contamination by sticking to robotic missions. It might take longer, since we're likely to need multiple sets of robotic missions—with each new set coming 2 years after the previous set—to do the same science that humans could accomplish in a single mission. But aside from requiring greater patience, robotic missions should be able to serve our scientific purposes.

The scientific pros and cons of sending humans to Mars are fairly clear, but the history of the space program shows that human exploration has rarely been driven by science. The manned space program began for political reasons, largely as part of a "race" between the United States and the Soviet Union. For example, while the Apollo program provided valuable scientific data about the Moon, its primary purpose was to prove to the world that the Americans could get there before the Soviets. If we decide to send humans to Mars, the decision will surely be based primarily on social or political considerations rather than scientific ones. Many potential reasons for sending people to Mars have little to do with science. Some people advocate Mars exploration as a way of inspiring people, especially children, here on Earth. Others see exploration as a moral imperative. Still others imagine colonies on Mars where people could open a new frontier, much as was once the case in America.

Terraforming Mars

If we ever seek to establish colonies on Mars, we will be confronted with an entirely new set of ethical and moral dilemmas. The issue of contamination will be unavoidable with colonies. But we may be tempted to make far more significant changes to the martian environment. As we've discussed, present-day Mars is not a very pleasant place for humans to live. There has been serious discussion among some scientists about ways we might initiate wholesale changes on Mars to make it more livable. Such changes go by the name **terraforming,** because they would tend to make the planet more Earth-like.

The basic idea of terraforming Mars involves changing the environment so that the atmospheric pressure and temperature become greater. Raising the temperature might be accomplished by adding a greenhouse gas to the atmosphere, while raising the pressure would require adding a large amount of gas to the atmosphere. One suggestion involves manufacturing chlorofluorocarbons (CFCs), which are strong greenhouse gases, and releasing them into the martian atmosphere. If we could strengthen the greenhouse effect enough, the warmer temperatures might begin to release the abundant carbon dioxide that may be frozen in the martian polar caps and elsewhere beneath the surface.

The release of this carbon dioxide gas would increase the atmospheric pressure and further strengthen the greenhouse effect. The idea might just work—but putting it into practice wouldn't be easy. Because the CFCs would tend to be broken apart by sunlight, we would have to manufacture them continually and in great abundance in order to start the greenhouse warming. Calculations suggest that we would need a manufacturing capability about a million times greater than our recent CFC-manufacturing capability on Earth and would need to keep it up for a few hundred thousand years before the surface warmed enough to drive substantial quantities of carbon dioxide into the atmosphere. Thus, if it is possible at all, we have plenty of time to consider the wisdom of terraforming Mars.

Whether terraforming could ever make the outdoor martian environment fully suitable for humans is unclear. But simply raising the pressure and temperature to be more Earth-like would make Mars far more livable, even if the air did not contain oxygen for us to breathe. With Earth-like pressure and temperature, we could walk around on Mars carrying only an air tank (and some protection from ultraviolet radiation). Moreover, such conditions might allow plants to survive outdoors, making it much easier to grow food.

Terraforming poses complex ethical questions. For example, do we have the right to terraform another planet? In part, the answer may depend on whether there is life on Mars. If there is, then terraforming might well drive it extinct. But even in that case, some might argue that it is ethically permissible if the terraforming benefits humans. The ethical issues may be fewer if there is no life on Mars, but some people still argue that reshaping an entire planet to our liking would be wrong.

THINK ABOUT IT . . . *The question of our effect on life on Mars is much like questions raised about endangered species on Earth. Some people say that we have no right to drive any species to extinction—an idea that was embodied in the U.S. Endangered Species Act. Others say that potential extinctions must be weighed against the human and economic costs of preventing them. Where do you stand on this issue? How does your answer affect your opinion of whether it would be ethical to terraform Mars?*

Interestingly, some of the ethical issues involved in Mars colonization were explored by science fiction writers well before the idea of terraforming ever arose. In particular, back in the days when people believed in canals and a dead or dying martian civilization, many stories dealt with interactions between humans and Martians. For our last word on the topic of human colonization, we turn to a science fiction story written by Ray Bradbury. In "The Million-Year Picnic," he imagined a human family who escaped to Mars just as people on Earth were finishing off our civilization through hatred and war. On Mars, the family found plenty of water and the vacant cities left by extinct Martians, and they hoped to be joined by a few other families from Earth. At the end of the story, they are on the banks of a canal. One of the children asks his father about a promise made earlier:

> "I've always wanted to see a Martian," said Michael. "Where are they, Dad? You promised."
>
> "There they are," said Dad, and he shifted Michael on his shoulder and pointed straight down.
>
> The Martians were there. Timothy began to shiver.
>
> The Martians were there—in the canal—reflected in the water. Timothy and Michael and Robert and Mom and Dad.
>
> The Martians stared back up at them for a long, long silent time from the rippling water. . . .

THE BIG PICTURE

In this chapter, we have discussed past fantasies about martian civilization, our current understanding of the habitability of Mars, and the search for life on the red planet. As you continue in your study, keep in mind the following "big picture" ideas:

- Mars holds a special allure for most people not only because of legitimate scientific questions, but also because past fantasies led many people to imagine a martian civilization. Mars and Martians became deeply embedded in modern culture, helping generate great public interest in Mars exploration.

- Different regions of the martian surface appear to be almost frozen in time, representing different eras in the planet's history. As a result, we can piece together at least a partial story of Mars from its earliest times to the present. We find a planet that has gone through dramatic change. Its surface, once warm and wet, is now dry and frozen.

- Although liquid water probably has not been stable on the martian surface for the past 3 billion years or longer, we now have evidence that it still flows at or near the surface on occasion. Thus, it is likely that liquid water still exists in some locations on Mars, such as near volcanoes or deep underground.

- The actual search for life on Mars began with the Viking experiments and continues today with studies of martian meteorites. For the future, we have embarked on a new era of intensive scientific study of Mars, marked by numerous current and planned missions to the red planet.

Review Questions

1. Briefly summarize the evidence, both real and imagined, that led to widespread belief in a martian civilization about a century ago.

2. What would it be like to walk on Mars today? Briefly discuss the conditions you would experience.

3. How do martian seasons differ from Earth seasons? Briefly explain why seasonal changes on Mars are accompanied by pole-to-pole winds.

4. Why isn't liquid water stable at the martian surface, and what happens to water ice that melts on Mars?

5. Give a brief overview of the geography of Mars, explaining the nature of key features such as the Tharsis bulge, Olympus Mons, Valles Marineris, the southern highlands, and the northern lowlands.

6. What do we mean by the *Noachian, Hesperian,* and *Amazonian eras* on Mars? How does the nature of the martian surface allow us to learn about conditions during these distinct eras?

7. Summarize the evidence suggesting that Mars must have been warm and wet during the Noachian era. Why do we think the climate must have changed dramatically near the end of that era?

8. Summarize the evidence pointing to floods of liquid water in post-Noachian times.

9. What evidence suggests that water might still flow at or beneath the martian surface today? What evidence suggests that Mars might still have subsurface liquid water near volcanoes?

10. How must the martian atmosphere have been different during the Noachian era? What happened to the gas that made up the much thicker atmosphere of that time? Based on your answer, will Mars ever again have a warm and wet period?

11. Based on all the geographic and geological evidence, summarize the current view about the past and present habitability of Mars.

12. Briefly summarize the Viking experiments and their results. Explain why the experiments are not considered to have offered evidence of life, despite some seemingly positive results.

13. What are *martian meteorites,* and what makes us believe they really came from Mars?

14. Briefly summarize the possible evidence of past life discovered in studies of ALH84001 and why this evidence generates controversy.

15. Why do we now think that the terrestrial planets have exchanged large amounts of rock? What are the philosophical implications of this exchange for our investigation of the origin of life on Earth and the possibility of someday finding life on Mars?

16. Briefly summarize plans for Mars exploration over the next decade.

17. Discuss the issue of biological contamination in either direction between Earth and Mars. How serious is each issue? What steps can we take to prevent contamination in each direction?

18. Summarize the scientific pros and cons of sending humans to Mars. What other considerations are likely to play a role in decisions about such missions?

19. What do we mean by *terraforming* Mars? Why might it be tempting for future human colonists?

Discussion Questions

1. *The Role of the Martians.* Percival Lowell may have been sadly mistaken in his beliefs about Martians, but he succeeded in generating intense public interest in Mars. If he had never made his wild claims about canals and civilization, do you think we would be exploring Mars with the same fervor today? Defend your opinion.

2. *Lessons from Mars.* Discuss the nature of the climate change that occurred on Mars at the end of the Noachian era. Do you think this climate change holds any important lessons for us as we consider potential climate changes that humans are causing on Earth? Explain.

3. *Human Exploration of Mars.* Should we send humans to Mars? If so, when? How much would you be willing to see spent on such a mission? Would you volunteer to go yourself? Discuss these questions with your classmates, and try to form a class consensus regarding the desirability and nature of a human mission to Mars.

Problems

Would You Believe It? Each of **problems 1–8** describes a hypothetical future discovery about Mars. In light of our current understanding of Mars, briefly discuss whether the discovery seems plausible or surprising. Explain your reasoning clearly.

1. We discover a string of active volcanoes in the heavily cratered southern highlands.

2. We find subterranean pools of water on the slopes of one of the Tharsis volcanoes.

3. We discover that Mars was subjected to global, heavy rainfall about 1 billion years ago.

4. Photos from future orbiters show that new gullies have formed alongside some of the ones already seen by *Mars Global Surveyor.*

5. We find pools of liquid water on the surfaces of some crater bottoms in the southern highlands.

6. A reanalysis of the Viking experiments leads us to conclude that they did, in fact, detect microbes living on the martian plains.

7. A sample return mission finds fossil evidence not only of martian microbes, but also of martian plants, including tall trees.

8. We find life on Mars, but it is clearly descended from the same common ancestor as life on Earth.

9. *Martian Fossil Hunting.* On Earth, we cannot find fossil evidence of life dating to times prior to about 3.8 billion years ago. If life ever existed on Mars, is it possible that we would find older fossils than we find on Earth? Explain why or why not.

10. *Plan a Mars Mission.* Suppose you were in charge of a mission designed to land on Mars. Assume the mission carries a rover that can venture up to about 50 kilometers from the landing site. What landing site would you choose? Write a one-page summary of why you think your site is a good target for a future mission.

11. *Human Mission Requirements.* Assume that a mission will carry humans to Mars on a journey that takes a few months in each direction and allows the explorers to spend about 2 years on the martian surface. Make a list of key provisions that would be needed for the mission, explaining the purpose of each item. In addition, briefly discuss whether you think any of these provisions could be found or manufactured on Mars rather than having to be brought from Earth.

12. *Terraforming Mars.* Make a list of the pros and cons of terraforming Mars, assuming that it is possible. Overall, do you think it would be a good idea? Write a short defense of your opinion.

13. *Mars Movie Review.* Watch one of the many science fiction movies that concern trips to Mars (such as *Total Recall* or *Red Planet*). In light of what you now know about Mars, does the movie give a realistic view of the planet? Are the plot lines that concern Mars plausible? Write a critical review of the movie, focusing on these issues.

14. *Martian Literature.* Read a book of science fiction about Mars, such as H. G. Wells's *The War of the Worlds*, Ray Bradbury's *The Martian Chronicles*, or any of the Edgar Rice Burroughs books about Martians. Write a critical review of the book, being sure to consider whether it still merits interest in light of current scientific understanding of Mars.

Web Projects

1. *Upcoming Mars Missions.* Pick one of the planned Mars missions shown in Figure 7.34 and learn more about the mission and its goals. Summarize your findings in a one- to two-page report.

2. *Martian Meteorites.* Find information about the latest discoveries concerning martian meteorites and whether they hold evidence of life. Choose one recent discovery that seems important, and write a short summary of how you think it alters the debate about the habitability of Mars or life on Mars.

3. *The Mars Society.* Visit the Web site of the Mars Society, which advocates future human colonization of Mars. Learn about the society's activities and its approaches to Mars exploration. Do you agree with the society's goals and strategies? Write a short summary of your opinions.

CHAPTER 8
Life on Jovian Moons

Jupiter orbits the Sun at more than five times the Earth's distance, and the other jovian planets (Saturn, Uranus, and Neptune) lie much farther still from the Sun. Sunlight in this distant realm is faint—too weak to provide much warmth. Nevertheless, several of the moons in these frigid outer reaches of our solar system are now considered to be possible places to find life beyond Earth. In this chapter, we will investigate these exciting worlds.

We'll begin by examining how and why it might be possible for distant moons to have liquid water (or other liquids). We'll then turn our attention to the most promising potential abode of life: Jupiter's moon Europa. We'll also investigate the likelihood that the same conditions that occur on Europa might occur on two others of Jupiter's moons: Ganymede and Callisto. Finally, we will discuss Saturn's remarkable moon Titan, which has an atmosphere thicker and denser than Earth's.

Aside from helping satisfy our general curiosity about the habitability of jovian moons, our discussion in this chapter will lead to an intriguing possibility: While Earth is the only planet we know to have life and Mars seems the most likely other world to have had or still have life, it's conceivable that the jovian moons could be the most numerous homes to life in our solar system. If any of these moons do indeed prove to harbor life, the possibilities for finding biology elsewhere in the universe will be greatly broadened.

8.1 Icy Moons of the Outer Solar System

In previous chapters, we discussed life on Earth in some depth and investigated the prospects for finding life on Mars. We now turn our attention to more distant reaches of the solar system, notably the realm of the four jovian planets—Jupiter, Saturn, Uranus, and Neptune. As we discussed in Chapter 6, the jovian planets themselves seem unlikely to be habitable. However, these planets are orbited by many moons, which we call **jovian moons** because they orbit jovian planets.

The idea of finding life in the cold, outer solar system once seemed far-fetched, but several of the jovian moons now seem potentially habitable. In this section, we introduce the major moons with a brief historical review of their discovery. We will discuss why most of the bigger moons contain substantial amounts of water ice. We'll also see how tidal forces on these moons can generate heat that could in some cases lead to melting and the presence of liquid water, either intermittently or permanently, thus creating possible habitats for life.

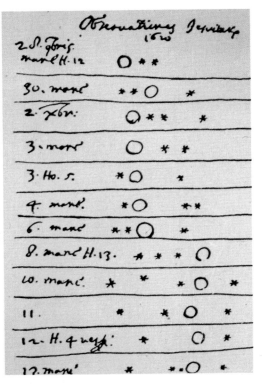

FIGURE 8.1 A page from Galileo's notebook in 1610. His sketches show four "stars" near Jupiter, which Galileo soon realized were moons orbiting the giant planet.

Discovering the Moons

Among Galileo's many discoveries [Section 2.2], one of the most notable was his finding of the four large moons of Jupiter. Having heard of the telescope's recent invention, Galileo built his own homemade versions beginning in 1609. At the time, the telescope was thought to be useful primarily for defense. It would allow you to see ships coming your way far earlier than if you had to wait until they were visible to the naked eye. Galileo, however, did something rather different with his telescopes. He turned them toward the heavens. On January 7, 1610, while gazing at Jupiter, he saw what at first seemed to be three small stars in its vicinity. What intrigued Galileo was that the three were close to one another and in a line. The following night, he looked again and was surprised to note that the three stars had moved relative to Jupiter, but not in the direction expected of background stars. A few days later he noted a fourth point of light, and within a week he realized that these four "stars" always stayed close to Jupiter and were clearly in orbit around it (Figure 8.1). In March 1610, Galileo published his results in a pamphlet he called *The Starry Messenger,* claiming to have found four bodies moving around the giant planet "as Venus and Mercury around the Sun." These four bodies are what we now call the **Galilean moons** of Jupiter; proceeding outward from Jupiter, we know them individually as Io, Europa, Ganymede, and Callisto (Figure 8.2).

Other scientists soon discovered additional moons in the outer solar system. The brilliant Dutch scientist Christiaan Huygens (1629–1695) found the largest of Saturn's moons—Titan—in 1656 (the same year he invented the pendulum clock). Before the close of the seventeenth century, Giovanni Domenico Cassini, an Italian astronomer who became director of the Paris Observatory, had discovered four more moons around the ringed planet. (Saturn's rings were first sighted by Galileo, but the resolution of his telescope was too low for him to discern what they were. Huygens was the first to realize that the rings do not touch Saturn's surface, and Cassini showed that the rings were not solid but instead were marked by a dark division, which we still call the Cassini division.) Even today, astronomers continue to make new discoveries of moons orbiting the jovian planets, though all of the larger moons have surely been discovered by now. The new discoveries involve small moons—usually no more than a few kilometers across—detected with the aid of new telescopes or spacecraft. For example, nearly two dozen small moons were discovered in photographs taken by the Voyager probes in the 1970s and 1980s, and as recently as 2002 astronomers using a ground-based telescope in Hawaii announced the discovery of 11 new moons of Jupiter.

Characteristics of the Moons

The jovian moons come in a wide range of sizes. While many small ones are not much bigger than a single mountain on Earth, others are as big as small planets. The two largest—Jupiter's moon Ganymede

FIGURE 8.2 This set of photos, taken by the Galileo spacecraft orbiting Jupiter, shows global views of the four Galilean moons as we know them today (left to right): Io, Europa, Ganymede, and Callisto. Sizes are shown to scale.

and Saturn's moon Titan—are larger than the planet Mercury. The three other Galilean moons (Io, Europa, and Callisto) and Neptune's moon Triton are larger than Pluto. Figure 8.3 shows a montage of jovian moons larger than about 100 kilometers in diameter.

Almost all the moderate- and large-size moons orbit their planet in much the same way that planets orbit the Sun: They orbit very nearly in the equatorial plane of their host world, moving in the same direction as their planet's spin. In this sense they resemble miniature solar systems within our solar system, which suggests that they were formed in a smaller-scale version of the same processes that gave birth to the planets [Sections 1.3, 4.2]. Recall that the nature of

the planets born in different parts of the solar system depended on the types of materials that could condense into solid form at various distances from the Sun. In the outer solar system, where ices were able to condense, accretion built planetesimals much larger than those that formed in the inner solar system, where only metal and rock could condense. These icy planetesimals grew large enough so that their gravity could attract significant additional amounts of gas and dust. The result was miniature versions of the solar nebula, with the jovian planets forming in the centers and their moons forming in the swirling disks that surrounded them (Figure 8.4).

More specifically, the larger moons were built up by the accretion of grains of ice and dust in the disks

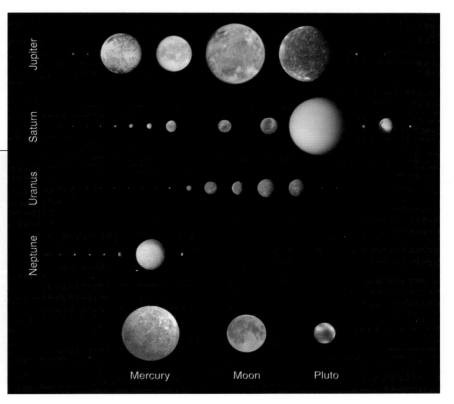

FIGURE 8.3 The larger moons of the jovian planets, with sizes (but not distances) shown to scale. Mercury, the Moon, and Pluto are included for comparison.

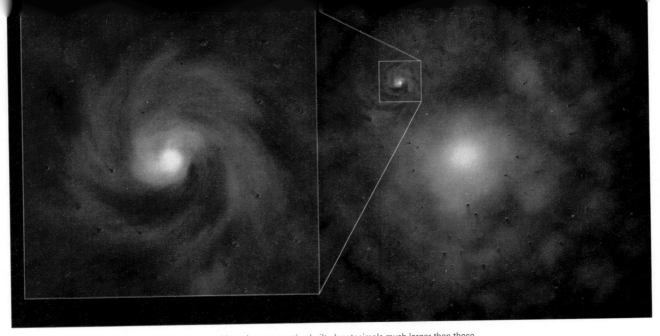

FIGURE 8.4 In the outer solar system, where ices could condense, accretion built planetesimals much larger than those that formed in the inner solar system, where only metal and rock could condense. The gravity of these large, icy planetesimals drew in more gas and dust, leading to miniature versions of the solar nebula in which the planets formed. Within each "miniature solar nebula," a jovian planet grew in the center and its moons formed in the swirling disk that surrounded it. This painting shows such a "miniature solar nebula" (left) embedded within the entire solar nebula (right).

around the forming jovian planets. Ices were abundant throughout the outer solar system, so all the jovian moons were formed from a mixture of ice and rock. The average densities of these moons are significantly lower than that of the Earth, reflecting the fact that they contain substantial quantities of ice, which is low in density. There is also variation in composition among the moons as we move from one planet to the next, because different ices condense at different temperatures. Water ice condensed easily at the temperatures near Jupiter, but methane and ammonia ice condensed only at the colder temperatures at greater distances from the Sun. As a result, Jupiter's moons contain significant quantities of water but no methane or ammonia. Moons of the more distant planets contain higher overall proportions of ice than of rock, as well as some methane and ammonia ice in addition to water ice.

The small jovian moons differ from their larger brethren both in appearance and in the properties of their orbits. Most have an irregular shape and often resemble peanuts, potatoes, or other snack foods (Figure 8.5). This is hardly surprising: The lesser gravity of these small objects is too weak to force the rigid material of which they're composed into a sphere. Many of these smaller satellites have orbits that are highly elliptical and inclined to the equator of their host planet, unlike the more nearly circular and equatorial orbits of the larger moons. In a few cases, the moons orbit backward relative to their host planet's spin. Their

FIGURE 8.5 A montage of the small moons of Saturn, shown to scale. Note the irregular shapes of these moons. (The small moons of the other jovian planets also are probably not spherical.) All of these moons are smaller than the smallest moons shown in Figure 8.3. These photographs were taken by the Voyager spacecraft.

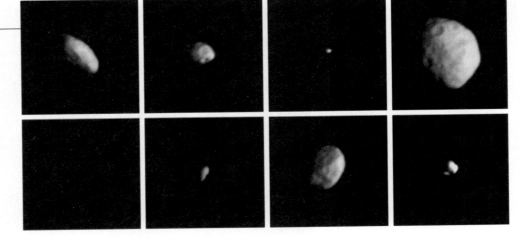

a If you do not rotate while walking around the model, you will not always face it.

b You will face the model at all times only if you rotate exactly once during each orbit.

FIGURE 8.6 The fact that we always see the same face of the Moon means that the Moon must rotate once in the same amount of time that it takes to orbit the Earth once. You can see why by imagining that you are the Moon and walking around a model of the Earth. The same idea applies to the synchronous rotation of jovian moons.

differing compositions—some are icy, and others are simply rock—suggest that many of these small, irregular satellites did not form "in place" as the larger moons did. Instead, most of the small jovian moons are probably captured comets or asteroids. While the idea of captured moons is appealing, it's not easy for a planet to directly grab a wayward asteroid or comet. Some process for reducing the object's speed as it approaches the planet is required, or else it would simply fly on by as it continued its orbit around the Sun. One possibility is that when the jovian planets were young, they had very large, "fluffy" atmospheres that might have served to slow down such small bodies when they passed nearby. Interestingly, one large moon also has orbital characteristics (notably, an orbit that is backward relative to its planet's rotation) suggesting it is a captured object: Neptune's moon Triton. No one knows how the capture of such a large object might have occurred, but it seems a near-certainty that Triton once orbited the Sun as an independent object —probably as one of the *Kuiper belt objects* of which we now consider Pluto to be a member [Section 6.3]. Indeed, the fact that Triton is larger than Pluto is one reason why some astronomers suspect that Pluto may not be the largest Kuiper belt object in the outer solar system, though no larger ones have yet been discovered.

Synchronous Rotation

Nearly all jovian moons share an uncanny trait: Like our own Moon, they always keep the same face turned toward their planet. This behavior, called **synchronous rotation,** means that each moon completes exactly one rotation around its axis while it makes one orbit around the planet. You can see how this works with a simple demonstration (Figure 8.6). Place a ball on a table to represent a planet like Earth, and walk around the ball so that your head represents an orbiting moon. If you do not rotate as you walk around the

ball, you'll be facing away from it by the time you are halfway around your orbit. The only way you can face the ball at all times is by completing exactly one rotation while you complete one orbit.

Synchronous rotation is not a coincidence; it develops quite naturally for any moon that orbits relatively close-in to its planetary parent, and it arises for the same reason that tides arise on Earth. We can therefore understand how synchronous rotation comes about by examining what causes the tides. Recall that the strength of gravity declines with distance. As a result, the Moon's gravitational pull on the near side of the Earth is slightly stronger than that on the far side. This *difference* in the strength of gravity is the cause of the tides (Figure 8.7). Note that there are two tidal bulges, which is why there are two high tides per day as the Earth rotates through the bulges. Simply put, the reason why there are two bulges is that the oceans on the near side to the Moon are being pulled out from the Earth, while the oceans on the side opposite the Moon bulge because the Earth is being pulled away from under

FIGURE 8.7 Tidal bulges face toward and away from the Moon because of the difference in the strength of the gravitational attraction in parts of the Earth at different distances from the Moon. (Arrows represent the strength and direction of the gravitational attraction toward the Moon.) There are two daily high tides as any location on Earth rotates through the two tidal bulges.

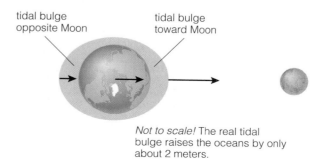

tidal bulge opposite Moon

tidal bulge toward Moon

Not to scale! The real tidal bulge raises the oceans by only about 2 meters.

them. A better way to look at tides is to recognize that the gravitational attraction toward the Moon gets progressively weaker with distance throughout the Earth and that this varying attraction tends to stretch the entire Earth—land and ocean—along the Earth–Moon line. The tidal bulges are more noticeable for the ocean than for the land only because liquids flow more readily than solids.

THINK ABOUT IT . . . *The Sun exerts a stronger gravitational force on the Earth than does the Moon—after all, the Earth orbits the Sun, not the Moon. However, the tides are caused primarily by the Moon, not by the Sun. Why is the difference in the strength of gravity across the Earth greater for the Moon, even though its overall gravity is weaker? Do you think other planets can have any significant effect on tides on Earth?*

Notice that while gravity tends to raise the tidal bulges directly along the Earth–Moon line, the Earth's rotation tends to outrun the raising and lowering of the tides so that the bulges occur ahead of this line (Figure 8.8). As a result, the Moon's gravity exerts a small amount of drag—called **tidal friction** —as it continually (and unsuccessfully) tries to pull the bulges back into line. This friction gradually slows the Earth's rotation. The effect is quite small on human time scales; tidal friction is increasing the length of Earth's day by only about 1 second every 50,000 years. (On short time scales, this effect is overwhelmed by other effects on the Earth's rotation.) But the frictional slowing adds up over billions of years. Early in the Earth's history, a day may have been only 5 or 6 hours long. A secondary effect of tidal friction is that the Moon's distance from the Earth grows larger as the Earth's rotation slows down, because the slight excess mass in the Earth's tidal bulge exerts a small gravitational tug that tends to pull the Moon slightly ahead in its orbit. This tug essentially acts to expand the Moon's orbit, and as a result the Moon gradually moves farther from the Earth. Early in its history, the Moon must have been much closer to the Earth—probably only about 10% or less of its current distance.

We can understand why the Moon is in synchronous rotation with Earth by turning the situation around. Just as the Moon raises tides on the Earth, the Earth must raise tides on the Moon—in fact, much stronger tides, because the effect of Earth's gravity on the Moon is much greater than the effect of the Moon's gravity on the Earth. If the Moon rotated rapidly, these tides would generate substantial tidal friction that would tend to slow the Moon's rotation. The tidal friction would disappear only when the Moon's rotation slowed so much that its tidal bulges always stayed on the Earth–Moon line— which can happen only if the Moon always keeps the

same face to Earth in synchronous rotation. In other words, no matter how fast the Moon may have been rotating at its birth, it was inevitable that tidal friction would slow this rotation until it was synchronous. At that point, the synchronous rotation was permanently "locked in," because there was no more tidal friction to slow the Moon's rotation further.

In much the same way, tidal friction explains the synchronous rotation of the jovian moons. If at its birth a moon was rotating faster than it orbited, tidal friction slowed its rotation until the rotation and the orbit were synchronized. Similarly, if a moon was once spinning slower than it orbited, tidal friction accelerated its rotation until it was synchronized with its orbit. Given the large sizes of the jovian planets in comparison to their moons, synchronous rotation should have set in for the close-in moons within no more than a few million years.

Tidal Heating

Just as rubbing your hands together generates heat, tidal friction can generate heat inside a planet or moon. This process probably generated substantial internal heat during the few million years during which tidal friction was bringing the young jovian moons into synchronous rotation. Because this heating arises from the effects of tides, it is one example of what we call **tidal heating.** In essence, this process draws upon the energy stored in the moon's orbit and in the rotation of the moon and the parent planet, gradually converting this energy into heat. In some cases, the combination of heat from accretion, from the decay of radioactive elements, and from tidal heating was enough to melt the interiors of jovian moons. As a result, some of the jovian moons

---

FIGURE 8.8 If the Earth always kept the same face to the Moon, the tidal bulges would stay fixed on the Earth–Moon line. But the Earth's rotation tries to pull the bulges along with it. The result is that the bulges stay nearly fixed relative to the Moon but in a position slightly ahead of the Earth–Moon line. (The effect is exaggerated here for clarity.)

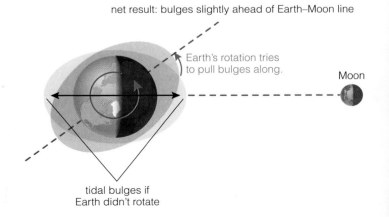

net result: bulges slightly ahead of Earth–Moon line

Earth's rotation tries to pull bulges along.

Moon

tidal bulges if Earth didn't rotate

underwent *differentiation* much as did the Earth [Section 4.2], giving them a "layered" internal structure with denser rock toward the center and less dense ice toward the surface.

Because all the jovian moons are relatively small —the largest is barely larger than Mercury—the heat generated early in their history is now long gone. We expect the moons to have some small amount of internal heat due to the ongoing decay of radioactive elements, but icy moons probably harbor much less radioactive heat than rocky moons of the same size because ice usually contains much less radioactive material mixed in than does rock. In addition, because radioactive heat input declines over time (because less radioactive material remains), it is probably not a significant heat source for the jovian moons today. However, in the case of the three inner Galilean moons (Io, Europa, and Ganymede), tidal heating provides an ongoing source of warmth.

The fact that these moons are still being tidally heated is a consequence of their orbits. Just as tidal effects can lead to synchronous rotation, they can also lead to special alignments, or **orbital resonances,** among the orbits of moons. Io, Europa, and Ganymede exhibit such resonances. For every orbit of Ganymede (7.2 days), Europa waltzes around Jupiter exactly twice (with a period of 3.6 days), and Io orbits exactly four times (1.8-day period). As shown in Figure 8.9, these three moons therefore line up about every 7 days (one Ganymede orbit), with the result that the gravitational tugs they exert on each other add up over time. The effect is much like that of pushing a child on a swing. If you time your pushes properly, a series of small pushes can add up to a *resonance* that causes the child to swing quite high. In the case of the three moons, the resonances tend to make their orbits somewhat elliptical. Like synchronous rotation, orbital resonances arise quite naturally. The orbital resonances of Io, Europa, and Ganymede came about because of a feedback with tides that the moons raise on Jupiter. Just as our Moon raises tides on Earth that are gradually causing the Moon to move outward from Earth (and slowing the Earth's rotation), the Galilean moons raise tides in Jupiter's atmosphere that cause the moons to move farther from Jupiter. The effect is greatest on the closest-in moon, so its orbit tends to move outward until it achieves a resonance with the next moon out. Then the two moons move outward in lockstep until they achieve a resonance with the third moon. This is how Io, Europa, and Ganymede came to have orbital resonances among themselves. Indeed, these moons are now moving slowly outward in lockstep toward Callisto, though it will be many billions of years before they reach a resonance with this fourth moon.

Because the resonances make them move in ellipses rather than in circles, each of the three inner

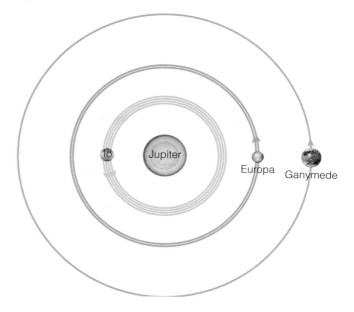

FIGURE 8.9 About every 7 Earth days (one Ganymede orbit, two Europa orbits, and four Io orbits), the three moons line up as shown. The small gravitational tugs repeat and make all three orbits slightly elliptical.

Galilean moons orbits Jupiter with a variable speed. This is a result of Kepler's second law (see Figure 2.9), which states that a body's orbital speed is greatest when it is closest to the object it orbits. Because of this change in orbital speed, the moons can't possibly keep their tidal bulges always exactly aligned with Jupiter. When they are closer-in to their planetary master, their bulge will lag behind perfect alignment; when they are moving in the slower, outer parts of their orbits, the bulge will get ahead. The result is that Jupiter's gravity is endlessly trying to move the bulge into immediate alignment. The constant change in the size and direction of the bulges, shown for Io in Figure 8.10, acts to stretch and squeeze the moon like the kneading of pastry dough. The nonstop friction from this kneading is the source of tidal heating.

FIGURE 8.10 Because Io's orbit is slightly elliptical, the size and direction of Io's tidal bulges change. The bulges and orbital eccentricity are exaggerated.

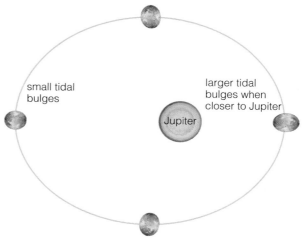

The effect is not small, as we first saw clearly in 1979 when the *Voyager 1* spacecraft glided past Jupiter and unmasked the four Galilean moons, hardly more than bright pinpoints in Earth-bound telescopes, as tortured, individual worlds. The most dramatic of these was Io, the innermost, which was found to be seething with volcanic activity (Figure 8.11). For Io, tidal heating contributes 200 times as much heat (per gram of mass) as is generated by radioactive decay on Earth—enough heat to raise Io's internal temperature to the melting point of solid rock. Indeed, it is likely that most of Io's interior is molten. The result is that Io is the most volcanically active place in the solar system. Material from vents on the sides of Io's volcanoes reaches temperatures of more than 2,000°C (3,600°F), and eruptions often spew plumes of sulfur and other gases to heights of hundreds of kilometers. Volcanic outgassing has probably expelled the water that Io once may have had, leaving it with a rocky composition and thus a higher density than the other Galilean moons.

The lack of water and the extreme volcanic activity on Io essentially rule it out as a home for life. But tidal heating has somewhat lesser effects on the more distant Galilean moons. In the case of Europa, in particular, it is possible that the level of tidal heating is "just right" for life. Without tidal heating, Europa would be wrapped in solid ice. However, its appearance offers hints that this moon has been resurfaced not with hot lava but with liquid water erupting from an ocean below. It is the possibility of a deep, subsurface ocean that encourages us to think that this small world might conceivably harbor life.

8.2 Europa: The Water World?

With its bland, nearly white exterior, Europa at first seemed the least interesting of Jupiter's four large moons. Its skin is riddled with a network of fine grooves and ridges, and very few craters are visible (see Figure 8.2). When scientists pondered the cause of its unusual appearance, they realized that there might be much more to Europa than meets the eye. In particular, its small number of craters tells us that Europa has been repaved in recent geological times —perhaps by liquid water emerging from below. However, it is extremely difficult to tell whether the

Movie Madness: *2010: The Year We Make Contact*

The monoliths are back, and they've brought plenty of buddies.

In the film *2010*, a sequel to the classic *2001*, humans are once again prompted by some enigmatic gray slabs to head for the outer solar system. It seems that an unseen race of aliens is trying to make an important point about Jupiter and its moons. So we oblige them by sending yet another batch of confused astronauts on a billion-kilometer joy ride.

However, unlike the mission in *2001*, this crotchety crew actually returns to Earth, bringing with them—like Moses coming back from the mountain—a few sage words of alien advice: "All these worlds are yours except Europa. Attempt no landing there."

What's the deal? After all, this is *our* solar system. So why are some unknown, unseen entities marking Europa off limits? That's like Mom forbidding you to go into a basement closet. Of course that's the one you'll find most interesting, and Arthur C. Clarke, the author of the novel *2010*, knew that there was, indeed, an interesting closet in the jovian system. The Voyager spacecraft had shown Europa to be an ice-covered world that could have a huge liquid ocean, and maybe even life.

In *2010*, this possibility is subtly exploited. The astronauts learn that there are chlorophyll-equipped critters somewhere below the Europan ice. Alas, life in such a deep, dark habitat is a bit of a drag. So just as the humans arrive, the sophisticated aliens who built the monoliths decide to re-engineer Jupiter, turning it into a mini-sun. They do this for the benefit of the primitive life forms on Europa. The new star eventually warms their moon and converts it to something that looks like Earth during the Mesozoic Era. The Europans, we presume, are destined to crawl out of their formerly ice-capped seas and find a monolith that will promote them to intelligent beings, the way our simian ancestors were improved in *2001*.

We lose Jupiter, but we gain a second, dimmer sun in our skies (no doubt a headache for astronomers and a source of confusion for migratory birds). But one wonders why these unseen extraterrestrials are so keen to meddle in the biological evolution of other worlds. Perhaps they just want our distant descendants to have some intelligent company right here in the solar system. It's a nice thought, but in the meantime someone needs to tell the space agencies that it's "hands off" Europa for the next few hundred millennia!

b This Galileo close-up of eruptions on Io reveals intensely hot lava, probably similar in composition to basalt from volcanoes on Earth.

FIGURE 8.11 Thanks to tidal heating, Io is the most volcanically active body in the solar system.

a Most of the black, brown, and red spots visible on the surface of Io are recently active volcanic features. The white and yellow areas are sulfur and sulfur dioxide deposits from volcanic gasses. Colors in this Galileo image are slightly enhanced.

resurfacing is caused by liquid water or by ice that is relatively warm and therefore soft enough to flow (as glaciers flow on Earth). Thus, we cannot yet be certain that an ocean of liquid water exists on Europa. In this section, we will consider the current evidence for an ocean, the implications of such an ocean for life, and future missions to Europa that may prove whether or not this moon is truly hiding vast amounts of liquid water.

The Evidence for an Ocean

Even before the Voyager spacecraft made the first detailed photos of this frosty moon in 1979, we already knew that Europa was covered with an ice shell. Spectroscopic observations made from Earth indicated the presence of water ice on the surface, although the amount was unknown. Theoretical studies, based on our ideas about the formation of moons, suggested that Europa was likely built of a mixture of rock and water. If heat had caused the moon to differentiate sometime in the past, then the less dense water would have migrated to the surface. The Voyager spacecraft confirmed at least part of this speculation: Europa's exterior was bright white and entirely ice-covered. Also, its surface was remarkably smooth, with few features rising higher than about a kilometer. Faint ridge lines, looking like a spiderweb

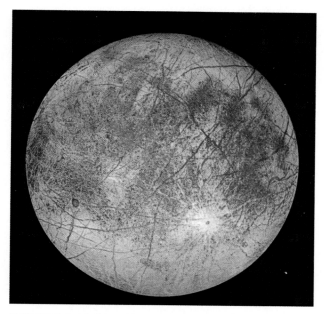

FIGURE 8.12 A global view of Europa, as seen from the Galileo spacecraft. Colors are enhanced to bring out subtle details.

of ancient highways, crisscross the surface like giant cracks in the ice. Recent images from the Galileo space probe have taught us much more (Figure 8.12).

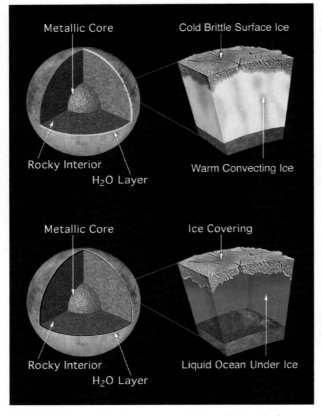

FIGURE 8.13 These artist's drawings contrast two possible models of Europa's interior. The top set shows the water layer consisting of warm ice (which can flow and convect) and a brittle crust. The bottom set shows a liquid water layer under the icy crust.

FIGURE 8.14 Chaotic terrain on Europa. Resembling icy jigsaw-puzzle pieces, this landscape suggests that liquid water has welled up from below, breaking apart the surface and then seizing it in newly frozen sea ice. The pieces can be mentally put back into place by matching up the details of the ice blocks. However, such a reassembly indicates that some of the pieces are missing, and were presumably melted during the reshuffle.

By using the Galileo spacecraft to measure subtle variations in Europa's gravitational field, scientists were able to determine its internal composition. Crudely speaking, Europa seems to consist of a central metallic (probably iron) core, overlaid by a thick sheath of silicate rock and an 80–170-kilometer-thick outer skin of water or water ice. From the gravity measurements alone, the water layer could contain just about any combination of solid ice, liquid water, or slush (partially melted ice), because all of these have about the same density (1 g/cm^3). Broadly speaking, the data lead to two possible models of Europa's interior (Figure 8.13). In one model, a brittle icy crust is underlaid by a layer of warmer ice that can flow easily and undergo convection. In the other, an ocean of liquid water lies beneath the surface. All we know for certain is that the outermost surface is frozen. Surface temperatures on Europa are brutally cold (typically −150°C), so its icy crust must be as stiff as granite.

The spacecraft pictures provide some moderately strong evidence suggesting that liquid water, not just more ice, underlies Europa's icy skin. One key piece of evidence comes from the surprising lack of impact craters. Most solar system bodies without an atmosphere are pockmarked by the occasional incoming meteorite or asteroid. Europa should be similarly blemished, especially given its proximity to Jupiter, whose hefty gravitational field pulls such cannonballs in its direction. Instead, Europa has only about 50 large craters (10 kilometers or more across), a number that should accumulate in only a few tens of millions of years, not in the billions of years since the solar system's formation. Clearly, something is erasing the craters on Europa. An obvious candidate is resurfacing by an occasional breakthrough of subsurface water.

Another suggestive piece of evidence for an underground ocean is the puzzle-piece appearance of Europa's so-called *chaotic terrain,* which resembles photos of an arctic ice pack. The surface is clogged with iceberglike blocks—some as small as football fields and others as large as a city—crisscrossed by ridge lines and suspended in what appears to have been a slushy ocean that froze. In Galileo's high-resolution photos, these blocks are often seen to be separated, as if they have rafted away from their original positions (Figure 8.14). Imagine putting your palms down on an assembled jigsaw puzzle and then spreading them slightly. Gaps will form between the pieces, and the picture printed on them will become disjointed. Europa's ridge lines similarly form a telltale picture on the ice, a picture that in places has fractured into small, jostled pieces. Individual ridges can be traced by the eye from one block to another, clearly showing the motion that must have occurred. Further evidence for an ocean comes from features that look like frozen floods, suggesting that liquid water has gushed from below and filled in some of the ridges (Figure 8.15).

While all of this photographic evidence suggests that Europa's ice-pack exterior has been churned by an underlying ocean, it still leaves a shadow of a doubt. Could it be that the ocean that produced the tortured and broken surface existed millions of years ago and is now frozen solid? Could Europa's intriguing face be merely the frozen remnant of an earlier time? Perhaps the icy crust sits atop relatively warm soft or slushy ice, and the surface features are the result of this warm ice convecting upward from below. In that case, Europa's frigid skin might extend a hundred kilometers down to the moon's rocky interior, without oceans or the possibility of life. But there's another line of evidence, arguably even more convincing than the photos, that makes the case for a liquid ocean.

In 1996, a magnetometer aboard the Galileo spacecraft detected a magnetic field near Europa. Because moons seldom have a magnetic field (though, as we'll discuss shortly, Ganymede does), researchers asked themselves what was causing magnetism on Europa. Jupiter has a strong magnetic field (which causes it to be a relatively powerful source of natural radio static), and Europa orbits within this field. Because Jupiter's magnetic equator is tilted with respect to Europa's orbit, its field at Europa's position is constantly changing as the giant planet spins on its axis. Just as a moving magnet produces an electric current in a coil of wire, so too could Jupiter's changing field induce currents in an electrical conductor on Europa. Of course, there's no giant coil of wire on Europa, but a salty ocean would conduct electricity in much the same way. The currents in such an electrically conducting ocean would act to set up a magnetic field that opposes Jupiter's—and thus would change as Jupiter's field changes. Additional measurements with Galileo's magnetometer have shown that Europa's magnetic field does indeed change as Jupiter spins. Most researchers consider this the best evidence yet that a liquid, salty ocean exists under Europa's icy surface. The magnetometer data require that Europa's subsurface water be global in extent, not limited to just a few isolated liquid pockets. They also imply that this ocean is about as salty as Earth's seas. Some of Galileo's instruments have found evidence for what appear to be salts on Europa's surface—possible seepage from a briny deep.

If we tally up the evidence for an ocean on Europa, we can list the following:

- The relatively small number of craters implies that the moon's surface is young, perhaps only a few tens of millions of years old, and has been recently repaved.

FIGURE 8.15 The smooth, dark terrain just to the right of the center of this photo mosaic probably represents an area where a liquid water geyser broke through the surface and flooded some of the ridged landscape.

- Various features on the surface (chaotic and flooded terrain) can be best explained if Europa has an icy skin atop a liquid, unseen ocean.

- Magnetometer results show a magnetic field that is likely caused by currents produced in something that conducts electricity—a briny ocean—as Jupiter's magnetic field changes.

- We know that tidal heating can supply enough heat to keep most of Europa's ice melted.

While no single one of these pieces of evidence would make an overwhelmingly strong case by itself, together they give us good reason to suspect that a liquid ocean exists on Europa. Still, the case is not definitive. If warm ice convects beneath the crust, the convection could carry the heat generated by tidal heating to the surface before it causes any ice to melt into liquid.

If there is an ocean on Europa, how big might it be? Theoretical calculations, based on the amount of tidal heating, suggest that the solidly frozen crust is about 5–25 kilometers thick. Observational evidence supports this estimate. One of the relatively few large craters on Europa seems to have had its bottom flooded by flowing water (or possibly by warm flowing ice), suggesting that the crater floor reached deep enough to pierce the icy crust and allow water from below to well up into it (Figure 8.16). Given that our models of Europa's interior suggest that the total thickness of the water/ice layer is between about 80 and 170 kilometers, a 5–25-kilometer-thick crust leaves plenty of room for a deep ocean. In fact, if a

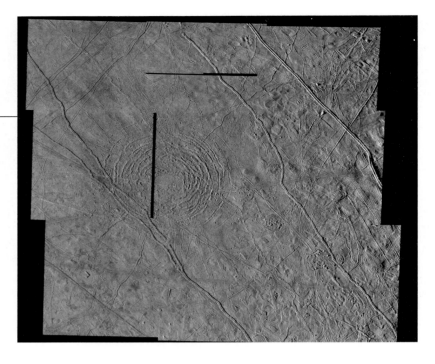

FIGURE 8.16 The concentric circles mark *Tyre Macula*, one of the few large craters on Europa. Note its flat-bottomed appearance, suggesting that it was filled in by liquid water that seeped from below. The horizontal and vertical lines represent gaps in the data.

global, subsurface ocean really exists, it could easily be 100 kilometers or more deep—some 10 times deeper than the deepest ocean trenches on Earth. A 100-kilometer-deep Europan ocean would contain roughly twice as much water as all of Earth's oceans combined.

Could Life Exist on Europa?

The oceans on Earth teem with life, and in Chapter 5 we discussed a number of reasons why many scientists suspect that life on Earth first arose in the oceans, possibly near deep-sea volcanic vents. If Europa really does have a deep liquid ocean, tidal heating has probably kept it liquid for billions of years. Might life have arisen in the Europan ocean as it arose on Earth, and could Europa be home to diverse life today?

In Section 6.1, we identified the three key environmental requirements for life: (1) a source of elements and molecules from which to build living organisms, (2) a source of energy for metabolism and growth, and (3) a liquid medium. If Europa has an ocean, then it clearly meets the third requirement. As we discussed in Chapter 6, nearly all worlds probably contain at least some quantity of the 25 elements that make up life on Earth. Europa should be no exception; its rock/water composition should have all these elements. Moreover, they should certainly be present at the rock/water boundary of the ocean floor, which is where we might expect any life to have arisen. That leaves the question of whether Europa has an energy source for life.

At first, the answer might seem obvious. After all, if there is a liquid ocean, then there must be enough heat from tidal heating and radioactive decay to keep the water temperature above freezing. Couldn't biology simply extract energy from the warm water? Unfortunately, gaining energy from a large reservoir of heat isn't possible in the absence of an adjacent "sink" of much colder water and a substantial difference in temperature between the two.[1] (For this reason, fish don't live off ocean heat, nor can ships be powered by the warmth of the seas they sail on.) Thus, in order to decide whether there might be life in a Europan ocean, we must ask whether there is enough energy in a useful form to support biology. We can separate this question into two parts. First, is there enough energy to support an origin of life? Second, if so, is there enough energy to support a reasonable total biomass? Let's begin by considering the question of energy for an origin of life.

Given the evidence that life on Earth began near deep-sea vents, we might wonder whether the same type of vents operate on Europa. Tidal heating and the decay of radioactive elements could provide the energy for such vents, and possibly even for undersea volcanoes as well (Figure 8.17). At the vents we would have a source of heat, a cooler sink (the rest of

---

[1] These conditions allow energy to be extracted by machines; for life, an additional consideration is having chemical species present that would allow reactions that can generate energy for biological use. In practice, this additional requirement is almost certainly met when we have the temperature differences noted at a rock/water boundary (see Appendix D).

FIGURE 8.17 This painting imagines a region where Europa's icy crust has been temporarily melted by heat from the eruption of an undersea volcano. We have never seen such surface melting, but if Europa really has an ocean and undersea volcanoes, it might occur on occasion.

Why does Europa look as it does? The large blocks seen in the chaotic terrain are most readily interpreted as icebergs locked in a frozen sea of material that occasionally wells up from below. But what about the complex network of fractures and ridges that cover this moon like a crazed paint job on a billiard ball? The fractures are probably due to the cracking of Europa's ice cover by the tidal forces discussed in Section 8.1. They even align with the directions of greatest stress that result from the tides.

The ridges are more intriguing. They are often double—twin mounds snaking across the surface with a narrow valley between (Figure 8.18a). It's possible that these are simply fractures where water or warm ice has oozed through from below, either bending the ice on either side of the crack into twin ridges or forming the ridges as the oozing material froze on either side of the fracture. There are also dark bands on Europa, "superhighways" 10–20 kilometers wide. These bands could be caused by tectonic processes that resulted in warm ice or water rising to the surface and freezing in the places where pieces of ice crust are spreading apart. One problem with this idea is that we don't know where the "subduction" occurs: Where on this moon does ice crust plunge back down?

Another interesting part of Europa's landscape are the *lenticulae*, roundish features that look much like freckles (Figure 8.18b). These features may be simply the result of blobs of warm ice that rise from underneath and partially or completely melt the surface.

While an undersea ocean of liquid water could explain all the intriguing details on Europa's face, it's also possible that they have been formed with nothing more than slightly warmer ice underneath the surface crust. As often happens in science, competing theories will have to be sorted out by additional observations. Only when new spacecraft can probe beneath the crust, using either radar or seismic waves, will we know for certain whether Europa has liquid oceans.

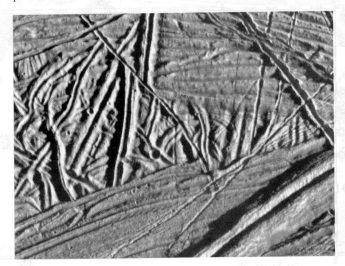

a Close-up photos show that Europa's surface cracks have a double-ridged pattern.

FIGURE 8.18 Surface features on Europa.

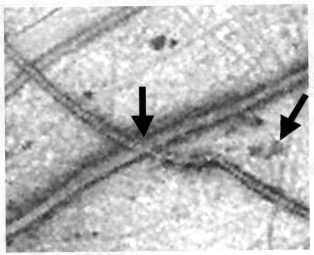

b The small, dark spots in this photo of Europa, such as the one indicated by the right arrow, are known as lenticulae. They may be places where warm ice has bubbled up just underneath the thicker, outer ice skin.

the ocean), and a large temperature difference. These conditions would facilitate chemical reactions between the water and the rock erupted from the vents, leading to the formation of complex organic molecules and the possibility of life. Indeed, on Earth chemical reactions between water and rock provide energy to many organisms, even at fairly low temperatures. All in all, if Europa has an ocean and deep-sea vents, then it would seem to have conditions quite similar to those under which life is thought to have arisen on Earth. In that case, it certainly seems plausible to imagine an independent origin of life on Europa.

If life did arise on Europa, how widespread might it be today? While deep-sea vents might lead to enough energy for an origin of life, they could not by themselves support a great biomass because they simply don't make enough energy available to organisms. Most life on Earth today gets energy from other sources, and even the great communities of organisms living near deep-sea vents get much of their energy from materials that filter down from above (such as oxygen that was ultimately produced by photosynthesis at the surface). Thus, for life to be widespread or abundant on Europa, it would need some other energy source in addition to the chemical reactions near deep-sea vents.

On Earth, the best-known source of energy for life (and the one that ultimately powers humans via the food chain) is photosynthesis [Section 3.3]. Sunlight cannot penetrate through more than a few tens of meters of ice, so it cannot directly provide energy for life in a Europan ocean. However, if photosynthesis could support life near Europa's surface, then it is possible that the energy could filter downward as organisms (dead or alive) cycle between the crust and the ocean. Although it seems unlikely, we cannot rule out photosynthetic life on Europa. If pockets of liquid water exist near the top of the ice crust, where sunlight can penetrate, then photosynthesis might provide energy to organisms living in these pockets.

Before the origin of photosynthesis on Earth, biochemical reactions may have been facilitated by energy sources such as lightning, ultraviolet radiation from the Sun, and heat released by impacts of asteroids and comets. Unfortunately, none of these energy sources are likely to be useful to life in Europan seas. Europa has no atmosphere for lightning, ultraviolet light does not penetrate the ice, and the time during which impacts were frequent enough to provide significant energy ended some 4 billion years ago. There is, however, another possible scheme for producing energy on Europa's icy surface. High-energy particles accelerated and trapped in Jupiter's magnetic field, along with some ultraviolet light from the Sun, regularly slam into the surface ice. This breaks up the ice molecules to produce small amounts of hydrogen peroxide (H_2O_2), molecular oxygen (O_2), and hydrogen (which quickly escapes). The detection of a very thin oxygen atmosphere above Europa's surface indicates that this process occurs.

These molecules, which can facilitate energy-producing oxidation reactions, [Section 5.3, Appendix D], would be mixed into the top meter of the Europan surface due to churning by small meteorites. If all of these molecules were ultimately to end up in the ocean below, they could provide energy to support life there. Unfortunately, we don't know how much of the outer ice actually gets cycled into the water below, or how often. In addition, the total amount of energy that might be available in this way is at least ten thousand times less than the amount of energy photosynthesis generates on Earth, so an ocean on Europa could not support anywhere near as much life as do the oceans on our planet.

One other known process might yield some energy for life on Europa. Some of the potassium contained in the rocky material that makes up this moon would dissolve in the ocean. The natural decay of one of potassium's radioactive isotopes would produce both hydrogen and oxygen molecules. Rough estimates suggest that these molecules could then facilitate chemical reactions that might support a small biomass.

Summarizing the case for life on Europa, we can say:

- There is good, indirect evidence that a liquid-water ocean exists.

- We expect the elements needed for life to be present in that ocean and on its floor.

- There are possible energy sources to support life, but the total available energy is quite small compared to the energy available for life on Earth.

Taken together, the evidence makes Europa seem a very good candidate for the possibility of life. That said, we should caution that it seems unlikely that sufficient sources of energy could exist in the Europan ocean to support macro-fauna—the complex sea creatures of our aquariums, for example. If life exists in Europa's oceans, it is probably quite simple and small, perhaps analogous to the most primitive single-celled organisms that have existed on the Earth.

THINK ABOUT IT . . . *Given the various uncertainties about a liquid ocean and available energy on Europa, do you consider it likely or unlikely that we will find life there? Why?*

Future Exploration of Europa

While it's tempting to envision sending a robot probe to Europa's surface that would melt or drill its way to the ocean below, the logical next step in our study of this moon is to find out if the ocean really exists. The *Europa Orbiter,* a proposed NASA spacecraft with a hoped-for launch within a decade, will attempt this. It won't land on Europa but will simply observe it from orbit with instruments that include a laser altimeter and long-wavelength radar.

The altimeter would be used to provide proof that an ocean lurks under the ice. The gravitational tug-of-war that Europa undergoes every 3.6 days in its elliptical path around Jupiter causes distortions in its shape. If a deep, liquid ocean exists under the ice, Europa's surface should regularly bulge in and out by up to 30 meters. If there is solid ice all the way down, however, then Europa won't stretch so easily and the altitude change in the bulge will be only about 1 meter. The difference can be reliably measured by the altimeter.

If altimeter measurements tell us that the ocean exists, then radar can be used to probe the surface, looking for radio reflections from an ice/water interface. This technique was used in Antarctica to

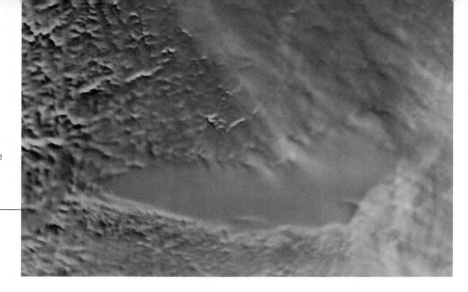

FIGURE 8.19 Lake Vostok, five kilometers beneath the surface near the center of the Antarctic continent, is comparable in size to Lake Ontario. It cannot be seen from the surface, but was discovered by ice-penetrating radar of the kind that made this image.

discover a lake hidden beneath a 5-kilometer-thick deck of ice (Lake Vostok; Figure 8.19). The radar aboard *Europa Orbiter* would be used similarly to hunt for thin spots in Europa's icy skin.

These thin spots might be preferred landing spots for any future probes, since water from below could have seeped out there most recently, possibly bringing evidence of life onto the surface. If we found any recycled water, we would analyze it for various organic molecules, such as amino acids. Because the density of life on Europa would probably be low (given the limited energy sources), we would want to melt and then filter a large quantity of the ice before examining it. And, of course, we would need to be careful to sterilize our spacecraft so that we don't accidentally transport biology from Earth to Europa and then "discover" it there.

If these experiments are encouraging, we could then dream and scheme about probes that would, indeed, manage to work their way under the ice and explore the eternally dark ocean depths where alien life might swim. In Arthur C. Clarke's story "2010," aliens from a distant star system speak to Earthlings about our own solar system. "All these worlds are yours," declare the aliens, "except Europa. Attempt no landings there." In the next few decades, we hope to ignore this fictional warning and possibly find the nearest extraterrestrial life.

8.3 Ganymede and Callisto

Europa is clearly the most likely of the Galilean moons to be habitable. However, both Ganymede and Callisto are also composed of significant amounts of water ice. Could they, too, have underground oceans, and hence the possibility of life? In this section, we investigate these other two moons of Jupiter.

Ganymede

Ganymede, the third from Jupiter of the four Galilean moons, is the largest moon in the solar system. Like Europa, Ganymede has a surface of hard, brittle ice. Craters, once sculpted, will persist for billions of years unless other geological processes erase them. However, whereas Europa's entire surface appears to have been recently repaved, Ganymede appears to have both young and old surface regions (Figure 8.20). Dark areas are often covered with craters upon craters, suggesting that these surface areas are billions of years old. Brighter regions show far fewer craters, indicating that they are much younger, and they also exhibit strange grooves. In some cases, fairly sharp boundaries separate the two types of terrain (Figure 8.21). The most likely explanation for the younger regions of ice is that they were created by "water eruptions" that occurred when internal heat caused ice below the surface to melt, erupt as watery "lava," and then freeze. The grooves may have been made either by tectonic stresses or by the eruption of water along a crack, creating the groove when refreezing caused the water to expand.

If liquid water occasionally erupts onto the surface, could it be an indication of a subsurface ocean? More intriguing evidence comes from magnetometer readings taken by the Galileo spacecraft. Ganymede apparently has its own intrinsic magnetic field—one generated within the moon—which may indicate that it has a molten, convecting core somewhat like the Earth's outer core. In addition, the magnetometer data show that part of Ganymede's magnetic field varies with Jupiter's 10-hour rotation. Just as for Europa, this is interpreted as indicating a field that is produced in an electrically conducting material under the surface—most likely a salty ocean. Another bit of evidence for an underground ocean is the

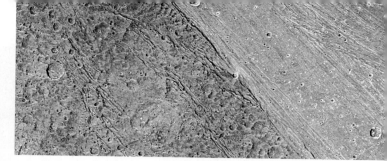

FIGURE 8.21 This close-up of Ganymede's surface shows a sharp boundary between heavily cratered ancient terrain on the left and younger "grooved terrain" on the right.

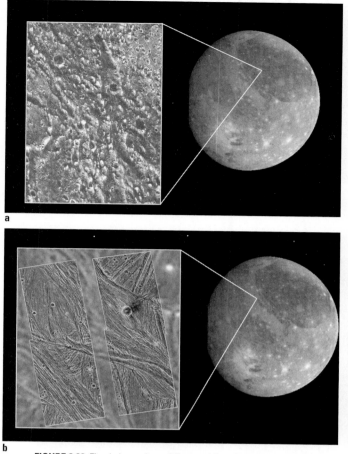

FIGURE 8.20 The darker regions of Ganymede's surface (**a**) are heavily cratered and must be billions of years old. The lighter regions (**b**) are younger-looking landscapes where eruptions of melted water have covered over the more ancient craters. These regions also show long grooves in the surface, probably formed either by tectonic stresses or when water erupts along a surface crack and expands as it refreezes.

FIGURE 8.22 Callisto is heavily cratered, indicating an old surface that nonetheless may hide a deeply buried ocean.

presence of salts on Ganymede's surface, which could conceivably be brine brought up from below.

What source of heat could be keeping water melted under Ganymede's surface? Ganymede has much less tidal heating than Europa. On the other hand, Ganymede's larger size means it should retain heat better than Europa and should have more internal heat from radioactive decay. Perhaps the combination of tidal heating and heat from radioactive decay is just enough to sustain a layer of subsurface liquid water.

While the possibility of a subsurface ocean is encouraging from the standpoint of habitability, the lesser heating on Ganymede means that the ice cover would be much thicker than on Europa—probably at least 150 kilometers thick. This would make finding life in a subsurface ocean far more difficult and the transport of possible nutrients from the surface considerably less efficient. As a result, less energy would be available to life. Life, if it exists

at all, would probably be less abundant and less evolved than seems possible on Europa. It might also be so deep below the surface that we could never gain access to it.

Callisto

Callisto is the farthest out of the four Galilean moons. Its entire surface appears to be pockmarked by countless craters that must date all the way back to the early days of the solar system (Figure 8.22). Gravity measurements suggest that this moon is mostly an undifferentiated ball of mixed ice and rock, overlaid by several hundred kilometers of water ice. Because its interior doesn't seem to be differentiated, we conclude that Callisto was never very warm inside and that neither radioactive decay nor tidal heating ever warmed this moon enough to melt it through. However, the Galileo spacecraft found an induced magnetic field for this moon, too, suggesting

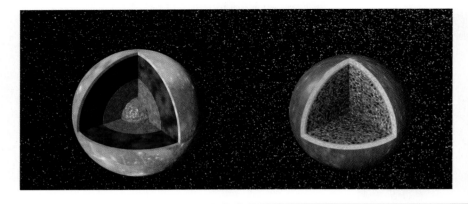

FIGURE 8.23 An artist's rendition of the internal structure of Ganymede (left) and Callisto (right). Ganymede is differentiated, with an iron and nickel metal core, surrounded by rock. The outer shell may include a liquid ocean. Callisto is a mixture of rock and ice, but it too might conceivably have liquid water under its outer skin.

—as for Europa and Ganymede—the presence of a salty, subsurface ocean (Figure 8.23).

If a subsurface ocean exists on Callisto, what heat source could keep it liquid? Unlike the other three Galilean moons, Callisto doesn't participate in the orbital resonances that cause tidal heating, meaning that the warmth required to maintain a liquid ocean would almost surely have to come from radioactive decay. This meager heat source might be sufficient due to the insulating properties of a thick, icy skin, so Callisto might indeed have a deep, unseen ocean that lies trapped between a warm radioactive inner region and an insulating outer ice layer. Support for this idea comes from observation of a large impact crater on Callisto: a 1,500-kilometer-across scar called Valhalla (Figure 8.24). On the planet Mercury, the impact that caused a similarly large feature (the Caloris Basin) apparently rearranged Mercury's mass enough to make a substantial bulge on the opposite side. No similar bulge exists on Callisto's opposite side, which may suggest that it has a shock-absorbing interior layer of liquid water. Of course, with no tidal heating, we'd expect even less energy for life to be available on Callisto than on Ganymede. Nevertheless, if an ocean exists, there is at least the possibility of life.

THINK ABOUT IT . . . *Consider all the evidence for liquid water on Jupiter's moons. On Europa, the evidence consists both of surface images and the induced magnetic field. Which piece of evidence do you find more convincing? On Ganymede and Callisto, the imaging evidence for liquid water is weak or nonexistent, but both moons*

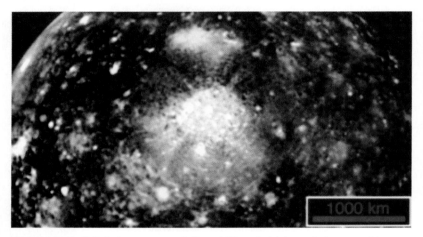

FIGURE 8.24 The Valhalla impact crater on Callisto, visible as the bright spot in the center of the photo. There is no bulge on Callisto's "back side," suggesting that the shock of this impact was largely absorbed by liquid water in the moon's interior.

show an induced magnetic field that could be produced by an invisible, salty ocean lying beneath the icy crust. Is this enough to convince you that a deep ocean exists on Ganymede and on Callisto? Why or why not?

8.4 Titan and Beyond

No other moons in the solar system have tidal heating like that experienced by the Galilean moons of Jupiter, so the prospects for finding subsurface oceans dim as we move outward from the Sun. Nevertheless, the outer solar system offers some intriguing prospects for life—or at least for interesting organic chemistry that could be a precursor to life.

By far the most promising candidate for habitability beyond Jupiter is Saturn's moon Titan. The second largest moon in the solar system, Titan is slightly larger than Mercury. It is also the only solar system moon to have a substantial atmosphere. In fact, Titan's atmosphere is even thicker than Earth's. The surface pressure is about 1.5 times that on Earth, which means that if you could visit Titan the pressure would feel fairly comfortable even without a spacesuit. The temperature, however, would not. Here, where sunlight is nearly 100 times weaker than on Earth, the surface temperature is a frigid −180°C (−290°F). Moreover, while the atmosphere is similar to Earth's in its nitrogen content (about 90% nitrogen on Titan versus 77% nitrogen on Earth), there is no appreciable oxygen to breathe. You'd need to bring along your own air and a very warm spacesuit for a visit to Titan.

While Titan's low temperatures make life there seem unlikely, some interesting chemistry is taking place on this frigid, smoggy world. In this section, we will focus primarily on the conditions on Titan and the possibility that some kind of biology might exist despite the cold. We will also briefly discuss the prospects for life on other large, cold moons of the outer solar system.

FIGURE 8.25 Titan, as photographed by *Voyager 2*. It is enshrouded by a reddish smog.

A Smoggy Moon

Titan's atmosphere was first discovered in 1944, when spectroscopic observations from Earth showed the presence of methane. However, the extent of the atmosphere was not immediately known, in part because methane is not the dominant gas (nitrogen is). Because scientists originally measured Titan's size including its haze, it was once thought to be the largest moon in the solar system. The two Voyager spacecraft, which passed by Saturn and its moons in 1980 and 1981, found that Titan's size had been overestimated because it is puffed out by a 200-kilometer-thick atmosphere. The solid part of Titan has a radius 60 kilometers smaller than that of Ganymede.

Titan's gravitational pull on the Voyager spacecraft allowed researchers to accurately determine its mass, which turns out to be nearly twice that of our Moon. Its mass, combined with its size, permits us to compute its average density, which is nearly the same as Callisto's (about 1.9 gm/cm^3). This suggests that Titan is made up of roughly equal volumes of rock and ice.

It would be interesting to know whether early heating might have differentiated Titan's interior and brought the ice to the surface. While we believe this to be the case, direct observation hasn't been possible. The Voyager cameras were capable of photographing fine detail on Titan, but they saw nothing of the surface because the moon is completely enshrouded by a reddish haze—in essence, smog (Figure 8.25). Most of the visible smog is due to chemical by-products formed when ultraviolet light from the Sun breaks apart molecules of methane.

Given that we usually think of moons as airless worlds, like our own Moon, how is it that Titan can have such a thick atmosphere? In fact, the main reason why moons generally have little or no atmosphere is their small size, which results in weak gravity unable to hold onto substantial amounts of gas and in low internal heat that does not lead to much outgassing. Titan can hold its atmosphere because of its relatively large size. A subtler question is why Titan has a thick atmosphere while Ganymede, which is even larger, does not. Two explanations for the difference have been suggested. First, recall that, at Saturn's distance from the Sun, ices of methane and ammonia should have been able to condense in the early solar system, whereas we expect only water ice at Jupiter's distance. Thus, while the outer crusts of Europa, Ganymede, and Callisto are composed largely of water ice, Titan's outer layer could be expected to have substantial amounts of methane and ammonia ice. These compounds sublime into gas (that is, go straight from the solid state to gas) at lower temperatures than does water (see Table 6.1). Thus, if internal temperatures rose on Titan during differentiation, methane and ammonia ice might have turned to gas, bubbled out of the crust, and built up an atmosphere. A second possible reason for Titan's atmosphere is that comets and asteroids hitting a moon of Saturn are traveling at lower speeds than are those that fall onto the moons of Jupiter (both because Jupiter's stronger gravity accelerates incoming objects more and because the Sun's gravity accelerates objects more at Jupiter's distance than at Saturn's). When such bodies slam into an atmosphere, they can blast much of the atmosphere away. If Ganymede once had an atmosphere, it would have been more likely than Titan's atmosphere to have been blown away by impacts.

Conditions on Titan

Despite the fact that the Voyager cameras were frustrated by smog, these spacecraft managed to tell us much about conditions within Titan's atmosphere. *Voyager 1* was given a trajectory that allowed it to sail behind Titan, so that its radio signal would pass through the smog on its way back to Earth. This ingenious experiment provided data that were used to determine both the temperature of the atmosphere and its general composition.

The atmosphere is overwhelmingly (roughly 90%) made up of molecular nitrogen (N_2), followed next by argon (about 6%), methane (CH_4), and ethane (C_2H_6). There are lesser quantities of propane (C_3H_8), acetylene (C_2H_2), hydrogen cyanide (HCN), and carbon dioxide (CO_2). What accounts for this mixture of hydrocarbons, which reads like an oil company's product line?

To begin with, once ammonia (NH_3) from Titan's icy interior made it into the atmosphere, energetic ultraviolet light from the Sun would have broken it apart into nitrogen (N_2) and hydrogen (H_2). The nitrogen molecules were heavy enough to stay put and became the principal ingredient of Titan's air. The much lighter hydrogen escaped into space. When methane (CH_4) entered the atmosphere, it too was broken apart by ultraviolet light into hydrogen (which, again, escaped) and the simpler compounds CH and CH_2. Products of the methane breakdown then reassembled themselves into more complex hydrocarbons, especially ethane (C_2H_6). Eventually, this should have led to an atmosphere so saturated with ethane that a nonstop drizzle of ethane rain began to fall.

So far, so good. But the fact that measurements from both Voyager and Earth-bound telescopes show that there's still methane gas in the air is somewhat surprising; we might expect that it would all have been converted into other molecules long ago. This puzzle can be solved if we assume that a replenishment source—a reservoir of methane—slowly feeds new gas into the atmosphere. The most reasonable source is large pools of slowly evaporating methane on the surface, where the −180°C temperature should be just warm enough to allow methane to be in a liquid state (see Table 6.1). Some of the evaporating methane would remain in the atmosphere, some would be converted to ethane by ultraviolet light and chemical processes, and some might even rain down; thus, the drizzle on Titan could contain methane along with the more predominant ethane droplets.

The idea of a rain of cold drops of liquid natural gas (which is largely made up of methane and ethane) may sound bizarre, but Titan is a moon governed by cold temperatures. Thanks to Titan's relatively weak gravity and thick atmosphere, the drops would be fat and slow—about 1 cm in size and falling at one-fifth the speed of water rain on Earth.

Because we expect that the conversion of methane into ethane and other compounds has been going on for billions of years, it's not unreasonable to imagine that Titan's surface—if we could see it—might be awash in a kilometers-deep ocean of ethane and methane. Indeed, this idea has long been popular among those who study this strange moon. However, if a global ocean exists, it should have tides resulting from Titan's elliptical orbit around Saturn. Frictional energy losses from this sloshing ocean would have converted the orbit to a more circular path. The fact that this hasn't happened suggests that a deep, worldwide hydrocarbon ocean probably doesn't exist. Radar waves bounced off Titan support this view, since the amount of reflected energy is more typical of water ice than of methane-ethane seas. Other evidence comes from recent images from Earth taken with infrared light reflected from Titan's surface, in essence allowing us to see through the clouds. These images show variations in surface brightness that probably rule out the existence of an ocean covering the entire world (Figure 8.26). However, they may suggest localized lakes of methane and ethane, or perhaps a methane-ethane equivalent of groundwater.

Life in the Cold Zone?

The temperature on Titan is so low that any surface water would be solid ice, but, as we've discussed, there might be liquid ethane or methane. In Section 6.1 we noted that water might not be the only common liquid that could support life. However, it has enormous advantages over its competitors. In particular, because methane and ethane liquids are colder than liquid water by some 200°C, the chemical reaction rates supported in these liquids would be enormously slower in Titan's seas than in our own. (These rates apply to reactions of the same type as on Earth; if life could be based on weaker chemical bonding than we find for life on Earth, the reaction rates might not be so limiting.) In addition, methane and ethane are far less able than water to dissolve other compounds or to facilitate the type of chemistry that might lead to life. Consequently, the outlook for biology on Titan is bleak.

It is not, however, completely hopeless. The ultraviolet light that hits Titan's atmosphere produces a wide range of organic molecules (the main contributors to the observed smog). Over billions of years, some of these

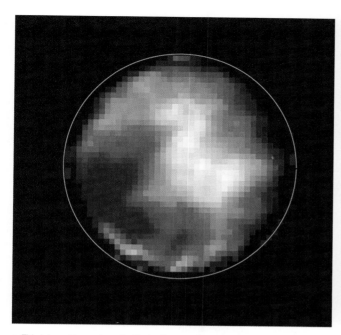

a This infrared image from the Keck telescope shows brightness variations. The dark areas could be oceans.

b Artist's conception of the surface of Titan, showing a possible ethane and methane sea.

FIGURE 8.26 Surface liquids on Titan?

compounds should have accumulated as a deep layer of organic sediment in the ethane-methane lakes or seas. Occasional impacts by comets or asteroids would provide enough heat to locally melt any water ice and create pockets of warm water that might persist for thousands of years. While it is not clear that life could form in such a short time period, some interesting chemistry would certainly have occurred along the way. In addition, the deep interior of Titan might be warm enough to melt a mixture of water and ammonia. As with Europa, Ganymede, and Callisto, there could be a hidden ocean on Titan, albeit buried under a couple of hundred kilometers of ice. What might exist within that ocean or whether it might conceivably include life isn't clear, but some minimal biology seems barely possible.

The Cassini–Huygens Mission

The tantalizing prospects of a smoggy moon with frigid seas of ethane and methane make us wish for much more information about Titan. Fortunately, it should soon be forthcoming. The 2-ton *Cassini–Huygens* spacecraft—a joint U.S. and European mission—is now on its way to Saturn (see Figure 6.13). In November 2004, the *Huygens* probe will detach from the *Cassini* orbiter and, after a 22-day cruise, parachute into Titan's atmosphere. During its $2\frac{1}{2}$-hour descent, it will explore the composition and structure of the atmosphere and should discover either the presence or the absence of hydrocarbon rain. Moreover, *Huygens* is expected to make it intact

onto Titan's surface, whether solid or liquid. Once it is on the surface, its battery power should be sufficient for perhaps half an hour of data transmission—long enough to show us a surface that we can barely imagine. The pictures could be spectacular, since the on-board camera is capable of seeing centimeter-size details. Meanwhile, the *Cassini* orbiter will use radar to "see" through the smog and map large sections of Titan, enabling us to learn more about its surface and the geological processes that have occurred there.

Even if Titan shows no signs of biology, studying this moon will give us valuable insights into chemical processes that might have been important to the beginnings of life on the early Earth. We often give lip service to the fact that our ideas about the conditions that would produce life might be too conservative. The Cassini–Huygens mission will provide our first view of another world on which the liquid environment and energy sources are quite different from those on Earth today. If life exists on Titan, it will most assuredly be life as we don't know it.

Beyond Titan

As we have progressed through this chapter, we have gradually moved down the scale of potential habitability. Europa seems reasonably likely to be habitable; Ganymede and Callisto seem somewhat less likely; and Titan, with its cold temperatures, seems to stretch to the limit the prospects for habitability.

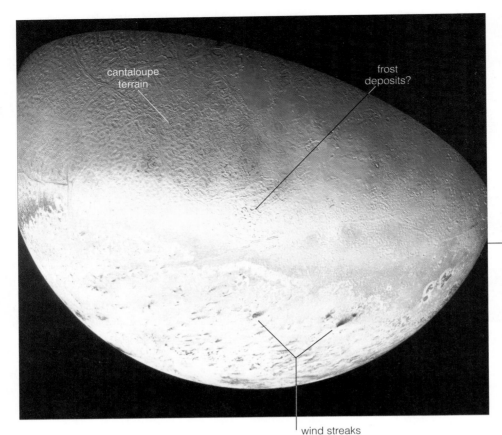

cantaloupe terrain

frost deposits?

FIGURE 8.27 The southern hemisphere of Neptune's moon Triton, photographed by *Voyager 2* in 1989.

wind streaks

Nevertheless, our surprising findings make us ask whether there could be still other habitable places in our solar system.

After the Galilean moons and Titan, the next largest jovian moon is Triton. As we discussed earlier, Triton orbits Neptune "backward," suggesting that it was somehow captured from what was once an independent orbit around the Sun. The question of how such a capture might have occurred is quite interesting, but here we are more concerned with the issue of potential habitability. Triton was photographed close-up only once, by *Voyager 2* in 1989. Its icy surface is an enigmatic mix of terrain that in some places is smooth and in others is crinkled into patterns resembling the skin of a cantaloupe (Figure 8.27). There are few impact craters. On the basis of the crater count on Triton today, its last resurfacing must have been no more than 10–100 million years ago. The only likely heat source for this resurfacing is radioactive decay, and some researchers think this may be sufficient to occasionally cause water volcanoes to erupt from a liquid ocean 150 kilometers under the surface. The water might be mixed with ammonia or methane, which would serve as antifreeze to lower its melting point.

Perhaps the most important idea emerging from our inspection of Triton is that even a moon in Nep-

tune's distant realms, where surface temperatures are horrifically cold (colder than $-230°C$), might be geologically active and could possibly hide a liquid ocean and a refuge for life. If so, and if large moons are common around the planets of other stars, then moons of jovian planets might be one of the most common habitats for life in the universe.

For our own solar system, we have now probably exhausted the possibilities for habitability. All the remaining moons, as well as Pluto (and its moon Charon) and the myriad comets, are probably too small to have ever warmed enough to melt ices to produce Titan-like seas or Europa-like underground oceans. Nonetheless, recent laboratory work has shown that amino acids and other organic compounds can form even on dry dust grains floating in space. Clearly, we cannot absolutely rule out life even in these difficult environments. But it seems a far better bet to concentrate our searches on those worlds—such as Mars and Europa—that seem most likely to fulfill the minimal requirements for biology.

THE BIG PICTURE

We have considered the moons of the outer solar system and found that several of them might offer conditions suitable for life. As you continue your studies, keep in mind the following "big picture" ideas:

■ Our own moon, a dismally dead relic of an early collision, may have misled us into thinking that only planets can harbor life. In fact, moons exhibit enormous variety. Some of the moons of the outer solar system are as large as small planets, a few might have liquid oceans, and one even has a substantial atmosphere. Life might well be possible in such places.

■ The solar system moon most likely to be habitable, Europa, is kept warm inside by a mechanism quite different from that which warms the Earth's interior. Tidal heating, the result of orbital resonances that occur among three large moons of Jupiter, can provide a continuous source of heat for billions of years. Because orbital resonances can arise quite naturally, tidal heating may be common among moons of jovian planets throughout the universe.

■ The icy moons of the outer solar system force us to rethink our basic concept of "habitability." If any of these moons are indeed homes to life, then the range of habitability is much broader than we might have guessed from studies of terrestrial worlds.

Review Questions

1. Briefly explain how the larger *jovian moons* tend to differ in general from the smaller ones. How does the formation process of the moons explain these differences?

2. What is *synchronous rotation?*

3. Briefly describe the cause of the tides on Earth, why they lead to *tidal friction,* and how tidal friction affects the Earth and the orbit of the Moon.

4. Briefly explain how the Moon's synchronous rotation arose and why synchronous rotation is common among jovian moons.

5. What is *tidal heating?* Briefly explain how it can arise and persist as a result of *orbital resonances.* How does tidal heating affect Io?

6. Discuss and summarize the evidence for an ocean on Europa.

7. Briefly discuss the prospects for life in a Europan ocean, focusing on the issue of energy for life.

8. Could Ganymede or Callisto have a subsurface ocean? Explain.

9. Briefly discuss Titan and the evidence suggesting it has ethane or methane seas.

10. Could there be life on Titan? Describe the pros and cons of the argument for life.

11. Summarize the habitability possibilities of worlds in the outer solar system. What do these possibilities tell us about the likelihood of life elsewhere in the universe?

Discussion Questions

1. *Exploring Europa.* Although Europa is a promising place to look for life, penetrating its icy crust will be difficult. How much effort and expense do you think it is worth to search for life in a Europan ocean? Should we attempt such a search soon, or should we be content to wait decades or longer? Defend your opinion.

2. *Migrating Life.* As we discussed in Chapter 7, there is a decent likelihood that life might at some point have traveled from Earth to Mars (or vice versa) on meteorites. Discuss the likelihood that life from Earth or Mars could have made its way to, and taken root on, Europa or other jovian moons.

Problems

Would You Believe It? Each of **problems 1–8** describes a hypothetical future discovery. Briefly discuss whether the discovery seems plausible or surprising. Explain your reasoning clearly.

1. We find a crater on Io that formed as the result of an impact during the heavy bombardment in the early history of the solar system.

2. The surface of Titan turns out to have very few impact craters.

3. We find life in the Europan ocean, but it consists only of microbes that live near deep-sea vents.

4. We discover that Europa, Ganymede, and Callisto all have life, but by far the most abundant and most diverse life is on Callisto.

5. We discover life on Triton that uses liquid ammonia, rather than liquid water, as its transport medium.

6. We discover a previously unknown moon of Uranus that, like Europa, probably has a subsurface ocean of liquid water.

7. In another solar system, we find a Jupiter-like planet that has only one large moon, but this moon is likely to have a subsurface ocean because of tidal heating.

8. In another solar system, we find a Jupiter-like planet with four moons that experience significant tidal heating.

9. *Europan Fish.* On Earth, fish breathe oxygen that is dissolved in the ocean. Do you expect that we will find dissolved oxygen in Europa's ocean? Why or why not? Based on your answer, if we could somehow transport fish to Europa, is it possible that they could survive in the Europan ocean?

10. *Most Likely to Have Life.* Suppose you were asked to vote in a contest to name the world in our solar system (besides Earth) "most likely to have life." Which world would you cast your vote for? Explain and defend your choice in a one-page essay.

11. *Movie Review.* A number of science fiction movies have concerned jovian moons, including *2010, Outland,* and *Gattaca.* Choose one, watch it, and write a critical review. Be sure to comment on the accuracy of any scientific content in the movie.

12. *The Sirens of Titan.* The moon Titan plays the title role in Kurt Vonnegut's novel *The Sirens of Titan.* Although the book never intended to give a realistic portrayal of Titan, it helped popularize the moon. (It even inspired a hit song of the same title, by Al Stewart.) Read the novel, and write a short critical review.

*13. *Core Radius.* Imagine a moon 1,000 kilometers in radius with an average density of 2 g/cm^3, halfway between that of water ice and rock. If this moon were fully differentiated, what would be the radius of its rock core?

Web Projects

1. *Galileo Results.* Visit the Web site for NASA's Galileo mission to Jupiter, and study some of the images of Europa, Ganymede, and Callisto. For one of the moons, create your own photo journal in which you include at least 10 photos along with a paragraph or two for each photo that explains how it relates to the question of habitability on that moon.

2. *Europa Orbiter.* Learn more about plans for the Europa Orbiter mission, including its current status and science goals. Write a one- to two-page summary of your findings.

3. *Cassini–Huygens.* Where is the *Cassini–Huygens* spacecraft right now? Learn more about the current status of the mission, and write a one- to two-page summary of your findings.

CHAPTER 9

The Nature and Evolution of Habitability

In past chapters, we have discussed life on Earth in some depth, as well as the prospects for finding life elsewhere in our solar system. In subsequent chapters, we will turn our attention to the search for life on planets and moons around other stars. A key question in that search is whether habitable planets are likely to be common or rare. Thus, before investigating other star systems, we need to discuss how the habitability of planets in our own solar system depends on factors such as their size and distance from the Sun.

We will focus particular attention on the characteristics that might make planetary *surfaces* habitable—most notably on what it takes for planets to have surface liquid water—primarily because surface life should be easier to find than subsurface life. By the criterion of liquid water on the surface, Earth is the only planet orbiting our Sun that currently has a habitable surface. However, we have good reason to believe that the surface of Mars was habitable in the distant past and that the surface of Venus also may once have been habitable.

In this chapter, we will explore the nature and evolution of surface habitability. We will contrast the climate histories of Venus, Earth, and Mars in order to develop a brief list of the key factors that influence surface habitability. We will use these factors to locate the region in our solar system within which it seems possible for a planet to have surface habitability, at least in principle, and to explore how this region evolves over time. Finally, we will consider how the habitability of the Earth is likely to change in the future and what these changes will mean for life on our planet.

9.1 The Concept of a Habitable Zone

We have considered a number of places in our solar system that might either have life now or have had life in the past, including Mars, Europa, and Titan. All of these places might be habitable in terms of being able to harbor life of some kind. However, no place in our solar system besides Earth is likely to have *surface* life at present. If there is life on Europa, it must reside in an ocean that lies beneath a thick, icy crust. If there is life on Mars today, it is almost certainly located underground, where there may be pockets of liquid water, and not on the red planet's surface.

From the standpoint of finding life in our solar system, the difference between surface habitability and habitability in general may not be that important. Future space missions will eventually allow us to explore beneath the surface of worlds like Mars and Europa, so we will find life if any exists. But finding life among other star systems poses a very different set of challenges.

We are unlikely to be able to travel to distant stars anytime soon [Section 12.1], so telescopic images and spectra offer our only realistic hope of finding life on extrasolar planets or moons. Within a couple of decades, new telescopes may be sufficiently powerful for us to identify life on worlds orbiting nearby stars [Section 10.4]. However, these technologies will enable us to discover only surface life, not underground life. For this reason, surface habitability holds special interest for our search for life beyond our solar system. In addition, surface habitability seems a prerequisite for the existence of a civilization, so the search for extraterrestrial intelligence should also be focused on places where surface habitability is likely.

If we focus on life that uses water as a liquid medium [Section 6.1], the search for surface habitability involves a search for surface liquid water. Clearly, it is the abundance of liquid water that makes it possible for our planet to be home to such diversity of surface life. When we turn our attention to the search for life around other stars, we are therefore led to ask about the conditions that make surface liquid water possible. More specifically, we can begin by asking, Why is liquid water present on Earth's surface and not on any other planetary surface in our solar system?

The simple answer to this question is that our planet happens to be located at a distance from our Sun that allows the surface to be comfortably warm, instead of so hot that liquid water would all evaporate or be lost to space or so cold that it would all be frozen. We therefore say that Earth is located within the Sun's **habitable zone.** More generally, we define the habitable zone to be the region of our solar system in which temperatures would in principle

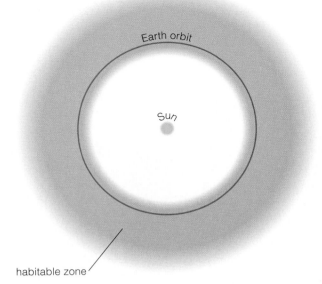

FIGURE 9.1 The habitable zone is the region of the solar system in which liquid water is possible on planetary surfaces. The Earth is clearly within the habitable zone. Later in this chapter, we will discuss the precise boundaries of our solar system's habitable zone.

allow for liquid water to exist on the surface if a planet of the right type happened to be located there (Figure 9.1).

Given that Venus today is too hot for liquid water at its surface and Mars is too cold, it's tempting to conclude that our Sun's habitable zone is narrow enough such that Earth is the only planet that falls within it. If that were the case, then our planet would not be habitable if it had formed just a little closer to or farther from the Sun. However, we already know that the reality must be much more complex. For example, the Moon is located at essentially the same distance from the Sun but is not habitable. Similarly, the evidence of past water flows on Mars suggests that the habitable zone must once have included the orbit of Mars.

In fact, the situation for Mars poses an even more difficult problem when we consider how the Sun has changed through time. Recall that the Sun is gradually brightening and therefore was dimmer and cooler in the past. Thus, we might expect the habitable zone to move outward over time as the Sun grows brighter. But the Earth has remained habitable for some 4 billion years despite the brightening Sun, while Mars, which we might expect to have become warmer as the Sun brightened, instead has become cooler. Moreover, as we discussed briefly in Chapter 6, Venus may also once have had liquid water, in which case the habitable zone at one point included its orbit as well.

Our primary goal in this chapter is to explore the boundaries of the present-day habitable zone and the evolution of the habitable zone over time. However, as we've just seen, understanding these issues requires that we first understand how and why Venus, Earth, and Mars have evolved along such different paths.

9.2 Comparative Planetary Evolution: The Case of Venus

Venus, Earth, and Mars are very different from one another today. Venus is hotter than a self-cleaning oven, with a crushing atmospheric pressure and sulfuric acid clouds. The surface of Mars is too cold for water to be liquid except near noon and near the equator, and, even when temperatures rise above freezing, its atmosphere is so thin that any liquid water would almost immediately evaporate. Earth, meanwhile, has an atmosphere seemingly in almost perfect balance for the existence of liquid water over much of its surface.

Before we can understand the factors that make a planetary surface habitable, we must understand the causes of the differences among Venus, Earth, and Mars. We have already studied the differences between Earth and Mars in some depth. In this section, we seek to understand why Venus is so unbearably hot.

Greenhouse Warming and Climate Regulation

The most immediate cause of the vast surface differences among Venus, Earth, and Mars is the differing strength of the greenhouse effect on the three planets. Recall that the greenhouse effect warms a planet's surface by slowing the escape of infrared radiation [Section 4.5]. Table 9.1 contrasts the "no greenhouse" average surface temperatures of Venus, Earth, and Mars—that is, the surface temperatures these planets would have without any greenhouse

effect—with their actual surface temperatures. Note that all three planets would be frozen over in the absence of the greenhouse effect, but greenhouse warming makes a significant contribution to the surface temperature on Earth and an extreme contribution on Venus.

The deeper question, of course, is why the greenhouse effect is so different on the three planets. In Chapter 7, we saw that the weak greenhouse effect on Mars is a result of the red planet's thin atmosphere, which we explained as the result of processes that caused Mars to lose much of its early atmosphere to space. Mars's small size and weak gravity played a major role in allowing it to lose so much atmospheric gas. Venus and Earth are nearly the same size, so we might expect any atmospheric loss processes to have affected both planets similarly. Why, then, does Venus have so much more greenhouse warming than Earth?

THINK ABOUT IT . . . *Suppose we could magically replace Venus's actual atmosphere with an atmosphere identical to Earth's. Would you expect the surface temperature to allow liquid water? Why or why not?*

As we discussed in Chapter 6, Venus has roughly the same amount of carbon dioxide as Earth. However, whereas almost all of the Earth's carbon dioxide is locked up in carbonate rocks or dissolved in the oceans, on Venus nearly all of the carbon dioxide is in the atmosphere. Atmospheric carbon dioxide is the primary source of the intense greenhouse effect on Venus.[1]

---

[1] Calculations show that carbon dioxide alone accounts for most but not all of Venus's high temperature. Other greenhouse gases—notably sulfur dioxide (SO_2), carbon monoxide (CO), and hydrochloric acid (HCl)—also make significant contributions.

Table 9.1 Comparisons of Actual Temperatures and "No Greenhouse" Temperatures for Venus, Earth, and Mars

Planet	Actual Average Surface Temperature	"No Greenhouse" Average Surface Temperature*	Greenhouse Warming (actual temperature minus "no greenhouse" temperature)
Venus	470°C	−43°	513°C
Earth	15°C	−17°	32°C
Mars	−50°C	−55°	5°C

* The "no greenhouse" temperature is calculated by assuming no change in the atmosphere other than lack of greenhouse warming. The "no greenhouse" temperature for Venus is colder than that for Earth, even though Venus is closer to the Sun, because Venus is completely wrapped in bright clouds that reflect much more sunlight than is reflected by the Earth. Because the light at Venus's surface is thus very dim, the surface would be chilly if there were no greenhouse effect.

On Earth, most of the carbon dioxide ended up in the oceans and rocks because of the carbon dioxide cycle (CO_2 cycle) [Section 4.5]. Recall that the CO_2 cycle on Earth begins with carbon dioxide dissolving in the oceans, where it reacts with sediments eroded from the continents and carried to the oceans to form carbonate minerals that fall to the ocean floor (see Figure 4.25). The carbonate rock that forms on the ocean floor is gradually carried toward continental boundaries by plate tectonics, where the rock is subducted under the continental crust. The carbonate rock is then heated as it descends into the mantle, releasing the carbon dioxide and allowing it to escape and return to the atmosphere through volcanic eruptions. This ongoing cycle regulates the carbon dioxide content of the Earth's atmosphere and acts as a thermostat to keep the Earth's climate fairly stable. The cycle also ensures that most of Earth's CO_2 resides in carbonate minerals on the ocean floor, rather than in the atmosphere as CO_2 gas.

Venus lacks a CO_2 cycle because it has no oceans in which carbon dioxide can dissolve. This explains why nearly all of Venus's carbon dioxide remains in its atmosphere, leading to its intense greenhouse warming. Again, the issue becomes more complex when we look deeper: Why doesn't Venus have oceans and a CO_2 cycle, like Earth?

Water on Venus

In fact, the question gets even more interesting when we look for water of any type on Venus today. The surface is too hot for either ice or liquid water. It is even too hot for water to be chemically bound in surface rock. The only place where Venus could conceivably have much water is in its atmosphere, but there's very little there, either. Indeed, when we add up all types of water—ice, liquid, and water vapor—the total amount of water on Earth is about 10,000 times greater than the total amount on Venus.

Broadly speaking, there are two possible explanations for the lack of water on Venus. Either Venus never had enough water to form oceans in the first place, or Venus once had more water but somehow lost it. We cannot absolutely rule out the first explanation, but it seems unlikely. At the time the planets were forming, the temperature would not have been very different at the orbits of Venus and Earth, so both planets should have accreted from planetesimals of similar composition. In fact, as we discussed in Chapter 4, these planetesimals probably contained little or no water, because temperatures in the inner solar system were too high for water vapor to condense into solid particles of ice. The water on Earth must have come from planetesimals that originated

in more distant parts of the solar system and had their orbits perturbed in such a way that they ended up crashing into our planet. Simple physical arguments suggest that such collisions must have been equally common on Venus, so Venus should have obtained water in this same way. Thus, the more likely scenario is that Venus started out with nearly as much water locked into its crust and mantle as Earth.

Any such water on Venus would have been released into the atmosphere through outgassing [Section 4.3]. We know from crater counts that the surface of Venus has been "repaved" over time, evidence of ongoing geological activity such as tectonics and volcanism. We also see numerous volcanoes on Venus, and there is evidence that some remain active[2] (Figure 9.2). Volcanoes and other geological processes would have outgassed the water. Thus, unless some mysterious process prevented Venus from getting water in the first place, there should have been plenty of water entering the atmosphere of the young planet Venus. Where did it all go?

The most likely answer is that the water was lost to space. As we'll discuss shortly, Venus may well have had an ocean in its youth. If so, the greater solar heat on Venus compared to Earth would have meant more evaporation than on Earth, and hence more water vapor in the atmosphere. Alternatively, Venus may never have been cool enough for outgassed water vapor to condense into liquid form. Either way, we expect that Venus had plenty of water vapor in its atmosphere, and some of the water molecules would have made their way into the upper atmosphere. Recall that in the upper atmosphere ultraviolet light from the Sun can break apart water molecules [Section 7.2]. Once hydrogen atoms are freed from water molecules, they escape fairly easily to space because of their low mass and high speeds. The remaining oxygen can then be locked up in surface rock by chemical reactions. Operating over billions of years, this process can easily explain the loss of an ocean's worth of water from Venus. Indeed, because Venus ultimately lost any surface water it once had, outgassed water could not be recycled back into the mantle the way it is recycled by plate tectonics on Earth. Venus may have lost water from its interior continuously throughout its history, in which case its

---

[2] The evidence for active volcanism is based on the sulfur dioxide (SO_2) content of the atmosphere. Chemical reactions with surface minerals could probably remove the SO_2 within a few million years, in which case the presence of SO_2 in the atmosphere indicates that it has been outgassed by volcanoes within the past few million years. Geologically speaking, this would be recent enough to imply that volcanoes remain active today.

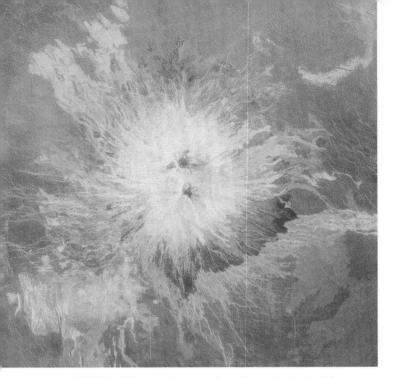

FIGURE 9.2 This image shows a region on Venus a few hundred kilometers across. The central peaks mark a volcano, and we see the paths of past lava flows radiating away from it. Because we cannot see through Venus's atmosphere, this image was created by radar, which involves bouncing radio waves off the surface. The differences in brightness represent how strongly radio waves are reflected by different regions of the surface. The yellowish color has no meaning; it was chosen arbitrarily to make the brightness differences easy to distinguish.

crust and mantle must by now be extremely dry—a situation very different from that on Earth, where significant amounts of water are chemically bound to crust and mantle rock.

Proving that Venus really did lose so much water is difficult, but evidence comes from the gases that didn't escape. Recall that most hydrogen nuclei contain just a single proton. But a tiny fraction of all hydrogen atoms (about 1 in 50,000 on Earth) contain a neutron in addition to the proton, making the isotope of hydrogen that we call *deuterium*. Water molecules that contain an atom of deuterium (called "heavy water") behave chemically just like ordinary water, and they can be broken apart by ultraviolet light just as easily. However, because a deuterium atom is twice as heavy as an ordinary hydrogen atom, it does not escape to space as easily when the water molecule is broken apart. If Venus lost a huge amount of hydrogen to space as the result of water molecules being broken apart, the rare deuterium atoms would have been more likely to remain behind than the ordinary hydrogen atoms. Measurements show that this is the case; deuterium is about 135 times more abundant (relative to ordinary hydrogen) on Venus than on Earth. Thus, a substantial amount

of water must have been broken apart and the hydrogen lost to space.

We cannot determine exactly how much water was lost and thus cannot be certain that Venus once had enough water for a true ocean. Venus must be subjected to occasional comet impacts, and comets bring a lot of water that then enters the atmosphere. Because comet water contains deuterium in a ratio similar to that found on Earth (measurements indicate that comets have about twice the deuterium-to-hydrogen ratio as does Earth), the addition of comet water tends to lower the atmospheric ratio of deuterium to hydrogen on Venus. As a result, the water loss that we can infer from the current ratio of deuterium to hydrogen in the Venusian atmosphere is only a *lower limit* on the actual water loss. The measured deuterium-to-hydrogen ratio implies that Venus has lost at least the equivalent of a global layer of water several meters deep—less than 1% of the average global depth of water on Earth. However, because this value is a lower limit, it is likely that Venus had much more water that was lost to space.

The Runaway Greenhouse Effect

We have explained how Venus might have lost its water, but why didn't this same process cause Earth to lose its water as well? The answer is the ocean itself. On Earth, most of the outgassed water vapor condensed into rain, forming the oceans, long before ultraviolet radiation could break apart much of the water vapor. The short-wavelength ultraviolet light that tends to break apart water molecules does not penetrate far into the atmosphere, let alone penetrate the ocean surface, so water in the oceans is protected. (Today, ultraviolet light is also absorbed by the ozone layer, but the ozone layer didn't exist when our planet was young.) To understand why Venus was unable to protect its water in a similar way, let's consider what would happen if we moved the Earth to Venus's distance from the Sun.

Sunlight is more intense at Venus's distance because Venus is closer to the Sun than is Earth. If we moved the Earth to Venus's distance, the greater intensity of sunlight would almost immediately raise the global average temperature by about 30°C, from its current 15°C to about 45°C (113°F). Although this is still well below the boiling point of water, the higher temperature would lead to increased evaporation of water from the oceans. It would also allow the atmosphere to hold more water vapor before the vapor condensed to make rain. The combination of more evaporation and greater atmospheric capacity for water vapor would substantially increase the total amount of water vapor in the atmosphere. Recall that water vapor, like carbon dioxide, is a greenhouse gas. Thus, the increased amount of water vapor in the

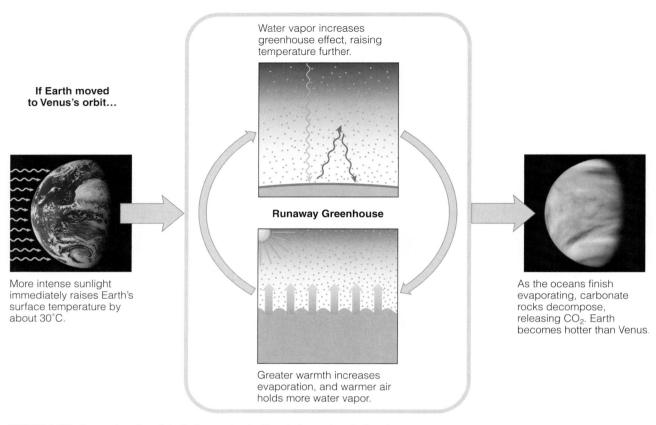

If Earth moved to Venus's orbit...

More intense sunlight immediately raises Earth's surface temperature by about 30°C.

Water vapor increases greenhouse effect, raising temperature further.

Runaway Greenhouse

Greater warmth increases evaporation, and warmer air holds more water vapor.

As the oceans finish evaporating, carbonate rocks decompose, releasing CO_2. Earth becomes hotter than Venus.

FIGURE 9.3 This diagram shows how, if the Earth were placed at Venus's distance from the Sun, the runaway greenhouse effect would cause the oceans to evaporate completely.

atmosphere would strengthen the greenhouse effect, driving temperatures a little higher. The higher temperatures, in turn, would lead to even more ocean evaporation and more water vapor in the atmosphere —which would strengthen the greenhouse effect even further. In other words, we'd have a "positive feedback" loop, in which each little bit of additional water vapor would mean a higher temperature and even more water vapor. The process would spin rapidly out of control, resulting in a **runaway greenhouse effect.** The planet would heat up until the oceans were completely evaporated and the carbonate rocks released all their carbon dioxide back into the atmosphere (Figure 9.3). By the time the runaway process was complete, temperatures on our "moved Earth" would be even higher than they are on Venus today.

This scenario suggests a simple explanation for the difference between Earth and Venus. Even though the two planets are not that different in their distance from the Sun, the difference was apparently critical. On Earth, the water rained down to make oceans, the oceans dissolved carbon dioxide that was ultimately locked in carbonate rocks by the CO_2 cycle, and the greenhouse effect was stabilized. On Venus, the temperature was just enough warmer that

oceans either never formed or soon evaporated, leading to a runaway greenhouse effect. Moreover, when the atmospheric concentration of greenhouse gases increases on Earth—as it has during past warm periods in our planet's history and during the current period of human-induced global warming—our greater distance from the Sun prevents the temperature increase from reaching the point at which a runaway process would occur. Thus, even though the Earth's climate has had warmer periods in the past, the warming has never spiraled out of control as it did on Venus.

THINK ABOUT IT . . . *Based on what you have learned about human-induced global warming, do we seem to be in any imminent danger of causing a runaway greenhouse effect on Earth? If not, could such a danger arise in the future? Explain.*

We can now return to the question of whether Venus might once have had oceans. In truth, we probably never will know, because it is unlikely that any evidence of a past ocean on Venus could have survived to the present day. The global "repaving" of Venus's surface by tectonics and volcanism would have long ago covered up surface features that could have indicated past liquid water, such as lakebed sediments or ocean shorelines. Nor are the rocks on Venus likely to offer any evidence of a past ocean. Any sedimentary rocks that might have formed in the early oceans also must have been covered up, and the high surface temperatures would have caused them to release much of their gas content back into the atmosphere, eliminating evidence that could tell us about their formation. However, recall that the Sun was probably about 30% dimmer when the planets were young, so that the intensity of sunlight at Venus at the time was not much greater than it is at Earth today. It might well have rained on the young planet Venus, and oceans could have formed. But, as the Sun gradually brightened, any oceans on Venus were doomed. The runaway greenhouse effect raised the temperature so high that all the water evaporated. In the upper atmosphere, the water molecules, unprotected from ultraviolet light, were broken apart and the hydrogen escaped to space. It is conceivable that Venus 4 billion years ago was a relatively pleasant, habitable planet with oceans of liquid water. But the water is long gone, along with any habitability that Venus might once have had.

9.3 Surface Habitability Factors

We can now combine what we have learned about Venus with our studies of Earth and Mars in past chapters to try to understand the key factors affecting surface habitability. If Venus once had liquid water, then Venus, Earth, and Mars may all have looked quite similar in their early histories. Indeed, if life can arise easily—or if it migrated among the three planets on meteorites [Section 7.4]—all three planets might have been home to life some 4 billion years ago. But Mars ultimately froze, and any remaining areas of habitability lie beneath its surface in pockets where liquid water may still reside. On Venus, the runaway greenhouse effect baked the surface far beyond the point at which water can remain liquid, and it is difficult to imagine that Venus could have even tiny pockets of habitability today. In contrast, the formation of oceans and the onset of the carbon dioxide cycle on Earth have together kept our planet habitable for some 4 billion years or more.

The stark contrast in the fates of Venus, Earth, and Mars holds key lessons about factors affecting surface habitability. The case of Venus, for example, tells us that distance from the Sun is a critical factor, since Venus seems to have been doomed to suffer a

---

Movie Madness: The Time Machine

This film, whose resemblance to H. G. Wells's original story is slightly less than marginal, features a brilliant, young New York professor from the early years of the twentieth century who travels back and forth in time, hoping to retrieve a lost girlfriend. To do so, he uses a time machine that looks as if it were kit-built from a discarded lighthouse lens, a laboratory vacuum tank, and a surplus adding machine from the Spanish-American War.

Forget the questionable physics; this story is actually about human evolution. The bulk of the film takes place in the far future, 800 thousand years hence. Thanks to an industrial accident involving some retirement home developers and the moon, all civilization on the Earth has been destroyed. Humans have evolved into two distinct races: the loveable (if somewhat simple) Eloi, and the disagreeable, ugly Morlocks. The Eloi live the life of carefree forest folk, building modest structures from bamboo and hemp. The Morlocks are slightly more sophisticated (they can smelt metal) but for some reason live underground. They only venture to the surface to hunt down the Eloi, drag them to their subterranean digs, and eat them for dinner.

Does it make sense that the most technically advanced beings on the planet will be underground heavies? In our consideration of habitable zones, we've looked at surface environments, where the raw materials of a planet are exposed to the abundant energy radiated by its sun. As we know, the energy for metabolism is tough to find below the surface, which explains why the Morlocks rely on the Eloi to eat the products of photosynthesis (veggies, for instance), and then they simply eat *them*. But this is a dangerously fragile lifestyle for the Morlocks, as they are dependent on a single surface species.

Aside from such dietary considerations, it's discouraging to think that, after a length of time several times as far into the future as we've come since the first *Homo sapiens,* the best our descendants can do is erect bamboo huts. What about expansion to other planets and travel to the stars? If *The Time Machine* is an accurate vision of what happens to thinking species, then we needn't bother with our SETI experiments (Chapter 11). After all, it's difficult to build radio transmitters out of bamboo.

runaway greenhouse effect no matter how habitable it may have been in its youth. The case of Mars tells us that other factors must also play a role, including planetary size and the way loss processes affect a planet's atmosphere. In this section, we will probe a little more deeply these three surface habitability factors: distance from the Sun, planetary size, and atmospheric loss processes. Table 9.2 shows the distance from the Sun and the radius for each planet.

Table 9.2 Distance from the Sun and Radius for Venus, Earth, and Mars

Planet	Actual Values		Values Relative to Earth	
	Average Distance from Sun	Radius (at equator)	Distance from Sun (AU*)	Radius (Earth = 1)
Venus	108.2 million km	6,051 km	0.72	0.95
Earth	149.6 million km	6,378 km	1	1
Mars	227.9 million km	3,397 km	1.52	0.53

*1 AU is the Earth's average distance from the Sun, 149.6 million kilometers.

Distance from the Sun

The first factor that affects surface habitability is a planet's distance from the Sun. After all, Mercury never had a chance at habitability because it is too hot for liquid water to exist at its surface, while planets and moons in the outer solar system are too cold to have liquid water on their surfaces.

The effects of distance can be fairly subtle, as the comparison between Venus and Earth shows. As we've seen, a planet at Venus's distance with an Earth-like atmosphere would be only a little warmer than Earth—at least until the runaway greenhouse process ran out of control. Thus, Venus's proximity to the Sun caused it to be too hot for surface habitability in a somewhat indirect way, in having made the onset of a runaway greenhouse effect inevitable. When we consider the minimum distance at which surface habitability is possible (as we will in the next section), we must therefore account not simply for solar heating, but also for processes that can lead to the evaporation of surface water. Similarly, when we consider the maximum distance at which surface habitability is possible, we must consider processes that might weaken greenhouse warming, causing all surface water to freeze.

Planetary Size

Clearly, distance from the Sun is not the only factor affecting surface habitability. If it were, then the Moon would be habitable because it is the same distance from the Sun as is the Earth. More to the point, if distance were all that mattered, then Mars should have become more habitable as the Sun grew brighter with time; instead, it froze over. Mars's lack of surface habitability appears to be less the result of its distance from the Sun than of its small size.

Recall that smaller worlds tend to lose their interior heat much more quickly than larger worlds [Section 4.4]. As shown in Table 9.2, Mars is considerably

smaller than the Earth, and its interior is probably now substantially cooler than that of Earth. Martian volcanoes, if they remain active at all, must erupt very rarely. Thus, Mars today lacks sufficient volcanism and outgassing to replace atmospheric gases that are lost either to space or to chemical reactions that incorporate them into surface minerals.

Is there a minimum or maximum size that makes it possible for a planetary surface to remain habitable over long time periods? We have traced the Earth's long-term habitability directly to the climate regulation provided by the carbon dioxide cycle, which in turn depends on the cycle of plate tectonics (see Figure 4.25). Thus, we have at least some reason to think that plate tectonics is necessary for long-term habitability on any world [Section 4.6], in which case the size requirement for surface habitability is a size that allows plate tectonics to exist.

Unfortunately, we do not really understand how planetary size is related to plate tectonics. Given the closeness in size of Venus and Earth (see Table 9.2), we might expect to find plate tectonics operating on both planets. However, we see little evidence of plate tectonics on Venus, and Venus certainly lacks the ongoing cycle of plate tectonics we see on Earth. We are not yet sure why Venus differs from Earth in this way, but one possible explanation traces the difference to the drying out of Venus caused by the runaway greenhouse effect. On Earth, volcanoes release water to the atmosphere, but this water is eventually recycled back into the crust and mantle, so Earth's interior does not dry out. On Venus, too, volcanoes would have released water to the atmosphere, but this water was not recycled into the interior because it was lost steadily to space. That is why, as we discussed earlier, we expect that Venus's interior is very dry. Because the presence of water tends to soften rock, a dry crust and mantle would tend to have thickened and strengthened Venus's lithosphere (the layer of rigid rock that "floats" on the mantle). In that case, the lack of plate tectonics may simply reflect the fact that Venus's strong lithosphere resisted fracturing into plates, which may ultimately be a result of the runaway greenhouse effect. (On the other hand, the high surface temperature on Venus would tend to weaken and decrease the thickness of the lithosphere, and we are unsure how much this effect would counter the effects due to a dry crust and mantle.)

The absence of plate tectonics on Mars probably results from some combination of its colder surface (which also makes its interior colder and stiffer) and its smaller size (which has allowed its interior to cool substantially over time). Indeed, the martian lithosphere is probably twice as thick as Earth's—perhaps too thick to break up into plates.

At this point, all we can say with confidence is that the Earth is of sufficient size for long-term surface habitability and Mars is not. Beyond that, we really don't know how large a planet must be to keep its surface suitably warm for the presence of liquid water over long periods of time. Factors of which we are unaware may well contribute to the occurrence or absence of plate tectonics and the ability to sustain an appropriate climate.

Atmospheric Loss Processes

A third factor in surface habitability is atmospheric loss processes. When it was a young planet, Mars must have had a much thicker atmosphere and a much stronger greenhouse effect [Section 7.2]. Because the Sun has brightened since that time, Mars would still be warm if it had kept its original atmosphere. But it didn't, so the greenhouse effect weakened and Mars grew colder.

Mars lost its atmosphere for two major reasons: (1) Its lack of a magnetic field allowed the solar wind to strip gases away, and (2) its low level of volcanism prevents outgassing from replenishing lost atmospheric gases. Both reasons are at least partially related to Mars's small size: The lack of a magnetic field (allowing solar-wind stripping) and the low level of volcanism are both due to the fact that the interior has cooled.[3] However, loss processes can affect larger planets as well, as the loss of water from Venus shows. In any event, atmospheric loss was clearly crucial to the dramatic climate cooling that occurred on Mars. Once most of the original atmospheric gas was lost, even a warming Sun could not keep Mars from freezing.

---

[3] Venus also lacks a magnetic field, probably at least in part due to its slow rotation. Despite its lack of a magnetic field, solar-wind stripping has not been significant on Venus, at least in comparison to the thickness of its atmosphere. For example, if Venus has in fact lost atmosphere in amounts comparable to Mars, this loss would be so small compared to the overall amount of gas in Venus's atmosphere that it would not be noticed.

Summary of Habitability Factors

We have seen how the surface habitability of Venus, Earth, and Mars has depended on each planet's distance from the Sun, its size (which presumably influences plate tectonics), and the way its atmosphere has been affected by loss processes. It is relatively easy to calculate the effects of distance on solar heating, and we also understand the physics of atmospheric loss processes fairly well. However, great uncertainty still exists in understanding the range of planetary sizes that would allow for surface habitability. Thus, when we look for a star's habitable zone, we must consider a zone in which a *planet of suitable size* could support life, although we are not yet able to define just what we mean by *suitable* in this case.

THINK ABOUT IT . . . *Using the three habitability factors, explain why the Moon is not habitable.*

9.4 The Sun's Habitable Zone

We can now apply what we have learned about surface habitability factors to make some general statements about the Sun's habitable zone. We'll begin this section by considering the habitable zone today, then examine how and why it probably changes over time, and finally consider the possibilities for life outside the habitable zone.

The Sun's Habitable Zone Today

The boundaries of the present-day habitable zone depend on the range of distances from the Sun at which a planet of suitable size could have liquid water at the surface today. The inner boundary marks the place where, if a planet were any closer to the Sun, it would be too hot for liquid water. The outer boundary marks the place where, if a planet were any farther from the Sun, surface water would freeze. Let's investigate the two boundaries separately.

The Inner Boundary For the inner boundary of the habitable zone, the limiting factor must be the ability of a planet to avoid a runaway greenhouse effect. Since Venus has suffered this fate, it must be too close to the Sun to be within the present-day habitable zone. Thus, the inner boundary of the Sun's present habitable zone must lie somewhere between the orbits of Venus and Earth.

To locate the inner boundary of the habitable zone more precisely, we can consider theoretical models of what would happen to the Earth if we moved it to various distances closer to the Sun. These calculations suggest that a runaway greenhouse effect would occur if we placed the Earth anywhere within 0.84 AU of the Sun, or about halfway between the orbits of Venus and Earth (see Table 9.2). (One AU, or astronomical unit, is the Earth's average distance from the Sun; thus, 0.84 AU simply means 0.84 times the Earth's average distance from the Sun.)

However, another factor might cause temperatures to spin out of control even beyond 0.84 AU from the Sun. According to some models, a moderate additional warming of the Earth would allow water vapor to circulate to much higher altitudes in the Earth's atmosphere (that is, a slight warming of the middle atmosphere would allow water to circulate into the stratosphere instead of remaining in the troposphere). At high altitudes, these water molecules could be broken apart by ultraviolet light from the Sun. The hydrogen would then escape to space, causing the Earth to lose this water. Then more water would rise into the upper atmosphere and would be lost in turn. This **moist greenhouse** (so-called because the upper atmosphere would become moist with water, at least until all the water was lost) would cause the Earth to lose its oceans over time. Note that the moist greenhouse leads to water loss not because the temperature is outside the range in which liquid water could exist, but rather because water that evaporates from the surface can rise into the upper atmosphere, where it can be lost to space. Eventually, this process can cause a planet to lose all its surface water. If no other factors interfere, an Earth-like planet could avoid this moist greenhouse effect only at distances greater than 0.95 AU. How-ever, the models that describe the onset of the moist greenhouse have at least a few known uncertainties (such as the effects of clouds) that might push this distance closer to the Sun.

In summary, the inner boundary of the present-day habitable zone in our solar system may be at 0.84 AU if we allow only for a simple runaway greenhouse effect, but as far out as 0.95 AU if we allow for water loss by a moist greenhouse effect.

The Outer Boundary The outer boundary of the present-day habitable zone is the distance from the Sun at which even a strong greenhouse effect would not allow a planet to stay warm enough to keep liquid water from freezing. At first, we might guess that Mars is beyond the outer boundary, since the temperature is too cold for liquid water at its surface today. However, if Mars were larger and had retained a thick atmo-sphere, it might have enough greenhouse warming to still have a habitable surface today. Thus, if it is possi-ble in principle for a planet to keep a thick atmo-sphere at such distances, the outer boundary of the habitable zone could lie beyond the orbit of Mars.

Calculations suggest that this is indeed the case. If we allow for a thick atmosphere with a strong greenhouse effect, the outer boundary of the present-day habitable zone lies at about 1.7 AU, well outside Mars's orbit (of 1.52 AU). However, there is at least one potential problem that could bring the outer boundary in closer. If the middle atmosphere of a planet is too cold, the atmospheric carbon dioxide that produces greenhouse warming will condense and rain out onto the surface. This carbon dioxide rain would limit how much carbon dioxide could reside in the atmosphere, thus preventing the atmosphere from staying thick enough for a strong greenhouse effect. This scenario might occur if the atmosphere lacked dust or any other greenhouse gas to help keep the middle atmosphere warm. In that case, the outer boundary of the habitable zone might lie at only about 1.4 AU, or just inside the orbit of Mars.

The Habitable Zone We have found two estimates each for the distances of the inner and outer bound-aries of the Sun's present habitable zone. Using the more optimistic estimates, the present-day habitable zone extends from about 0.84 to 1.7 AU. Using the more conservative estimates, it extends only from about 0.95 to 1.4 AU. Both sets of boundaries are shown in Figure 9.4. Note that, even in the more conservative case, the Sun's present-day habitable zone represents a fairly wide region in the inner solar system.

Keep in mind that we don't yet know enough about how planetary atmospheres work to know which estimate is correct—or if the truth lies some-where in between. In addition, there might be processes that can affect the atmospheric tempera-ture that we haven't thought of yet and therefore have not accounted for in our calculations of the habitable zone boundaries. Thus, these estimates of the boundaries of the habitable zone should be con-sidered just that—estimates. They might be signifi-cantly refined as we learn more in the future.

THINK ABOUT IT . . . *About how much wider is the more optimistic view of the present-day habitable zone than the more conservative view? If planets form at random locations in the inner por-tions of other stars' solar systems, how would these two different views affect the likely number of plan-ets within habitable zones around Sun-like stars in the Milky Way Galaxy?*

The Evolving Habitable Zone

Because the boundaries of the habitable zone are calculated under the assumption that we have a planet of suitable size, they depend only on the amount of heat and light put out by the Sun. Thus, when the Sun was dimmer in the past, the habitable zone must have been closer to our Sun. In the fu-ture, when the Sun is brighter than it is today, the habitable zone will lie farther from the Sun.

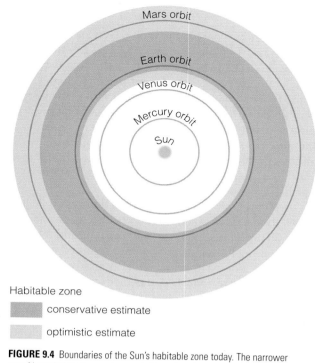

FIGURE 9.4 Boundaries of the Sun's habitable zone today. The narrower set of boundaries represents a model based on the more conservative assumptions, while the wider set represents the most optimistic scenarios.

consists of two protons and two neutrons). In a hot gas (plasma) like that in the Sun's core, the pressure depends primarily on the total number of freely moving particles, not on the types of particles. Because fusion converts four hydrogen nuclei into only one helium nucleus, the number of particles in the Sun's core gradually falls over time. Because the smaller number of particles tends to reduce the pressure, the weight of the overlying matter squeezes the deep interior and compresses the Sun's core. The compression raises the core temperature, much as a bicycle tire pump gets warm when you compress the air in it by pushing on its piston. The heating of the Sun's core in turn raises the rate of nuclear fusion. This ongoing process causes the fusion rate in the Sun to rise gradually over time, which is why the Sun slowly brightens over time.

Theoretical calculations of this gradual change in the fusion rate allow scientists to determine how the Sun's brightness changes over time. These calculations tell us that the Sun was about 30% dimmer when it was born; they also allow us to predict its future brightening. Observations of other stars of various ages show this increase in brightness with age, giving us confidence that our theoretical models are correct.

Figure 9.5 shows the results of calculations for the boundaries of the habitable zone from the Sun's birth until its death some 5 billion years from now. Notice how the habitable zone gradually moves outward from the Sun, just as we would expect.

The horizontal swaths in Figure 9.5 show the distances from the Sun at which conditions have

To determine how the habitable zone moves over time, we need to know how the Sun brightens over time. Fortunately, the process that causes the Sun to brighten is well understood. Recall that the Sun shines by fusing hydrogen into helium. It takes a total of four hydrogen nuclei (each consisting of a single proton) to make one helium nucleus (which

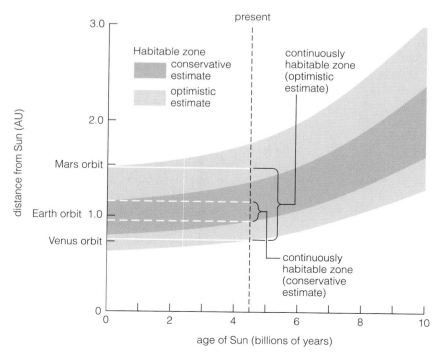

The Sun's Habitable Zone Over Time

FIGURE 9.5 This graph shows the Sun's habitable zone through time. The narrower region represents the habitable zone based on the more conservative assumptions, and the wider region represents the habitable zone based on the more optimistic assumptions. The horizontal swaths represent the zone that has been continuously habitable from 4 billion years ago to the present, again under the conservative (narrower) and optimistic (wider) assumptions.

remained habitable from 4 billion years ago to the present. This zone is often called the **continuously habitable zone,** because it has been habitable at all times since the end of the heavy bombardment about 4 billion years ago. Again, the figure shows a narrow continuously habitable zone based on more conservative assumptions and a wider continuously habitable zone based on more optimistic assumptions. Note that, under the more optimistic assumptions, both Earth and Mars are in the continuously habitable zone. Note also that the continuously habitable zone is defined for habitability only up to the present. If we look billions of years into the future, the habitable zone continues to move outward and the continuously habitable zone becomes narrower. In the next section, we will discuss how this will affect the future Earth.

THINK ABOUT IT . . . *Was Venus ever in the habitable zone? Is it in the continuously habitable zone? Under the more conservative assumptions, about when will the continuously habitable zone move beyond Earth's orbit? What will that mean?*

Habitability Outside the Zone

The concept of the habitable zone was originally developed at a time when it was assumed that life would be possible only within this zone. However, our discussions of Europa and other jovian moons have shown that this original idea was too restrictive. The habitable zone is the only place where we might expect to find liquid water on the *surface* of a planet or moon, but the case of Europa tells us that subsurface oceans may be possible elsewhere.

In fact, there are several ways habitability might present itself outside the habitable zone. One is in subsurface groundwater of any type, such as that which may still exist on Mars today. Recall that one interpretation of the gullies seen on Mars (see Figure 7.22) is that liquid water still resides just a few hundred meters beneath the martian surface. Because this water would be kept liquid due more to geological conditions than to solar heat, a Mars-like planet could have subsurface habitability even if it lies beyond the outer edge of the habitable zone. It's not too far-fetched to imagine that Mars-like planets could be more common than Earth-like planets.

The possibility of life in a subsurface ocean, such as that thought to exist on Europa, could make habitability even more common. We expect moons in the outer regions of any solar system to contain large amounts of water ice, because water is the most common of the ices that can condense in regions far from a forming star [Section 1.3]. Subsurface oceans therefore could exist on any outer solar system moon that

has enough internal heat to melt a layer of ice into liquid water. In Chapter 8, we saw how an orbital resonance leads to the tidal heating that keeps Europa's interior warm enough for the ice to melt. The same process might also help melt ice in Ganymede, and similar orbital resonances could occur in any case where a large planet has multiple large moons. In other cases, such as that of Callisto, heat from radioactive decay might be enough to melt interior ice. Thus, Jupiter-like planets in any star system might have moons similar to Europa.

Moreover, because tidal heating depends only on gravitational effects and not on distance from the Sun, tidally heated moons could exist around many stars that might not otherwise have significant habitable zones. For example, as we'll discuss in Chapter 10, stars that are smaller and cooler than the Sun have narrower habitable zones, making it less likely that planets of suitable size will exist within those zones. But these solar systems might have Jupiter-like planets with moons like Europa and thus might harbor life anyway.

It is even possible that life could exist on worlds orbiting objects that are not large enough to be stars. Recent astronomical discoveries suggest that such "substellar" objects—often called **brown dwarfs**—may be quite common (Figure 9.6). A brown dwarf is thought to be an object that forms from a spinning cloud gas much like a star but is not large enough to sustain nuclear fusion like a star. Typical brown dwarfs have masses several tens of times larger than that of Jupiter but still less than about 8% of the mass of the Sun. Brown dwarfs have no habitable zones around them at all because they are too cool and dim to provide enough energy to heat a planet or a moon. However, they may well be orbited by planets, and thus they could have Jupiter-like planets and Europa-like moons.

An even more intriguing suggestion holds out the possibility of *surface* habitability on Earth-size planets outside habitable zones. Theoretical calculations suggest that an Earth-size planet's own internal heat could keep the surface warm enough for liquid water *if* the surface were protected against heat loss by a thick hydrogen atmosphere. In our solar system, such a hydrogen atmosphere is not possible on an Earth-size planet, because solar heat would cause the hydrogen to escape into space fairly rapidly. But, as we will discuss further in Chapter 10, it is quite possible that Earth-size planets are often ejected from forming solar systems and sent into interstellar space. When an Earth-size planet is first forming, it might have a hydrogen atmosphere for a short time, particularly if it forms at a greater distance from its star or around a star cooler than our Sun. If such a planet is ejected into interstellar space before it loses this atmosphere, its thick hydrogen atmosphere might

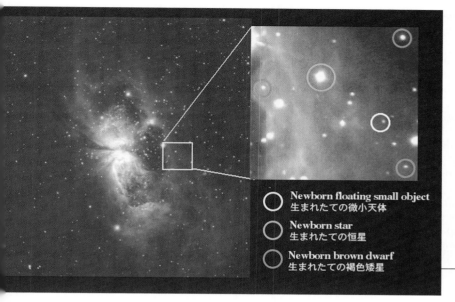

FIGURE 9.6 This beautiful photograph, taken by Japan's Subaru telescope located on the summit of Mauna Kea in Hawaii, shows a star-forming region called S106. The butterfly shape is thought to be produced by gas streaming out from a massive central star; the light from this star makes the gas glow throughout the star-forming cloud. The inset shows numerous other small objects forming in the cloud. The green circles mark brown dwarfs—objects not quite big enough to sustain nuclear fusion like an ordinary star. The entire star-forming region contains hundreds of similar young brown dwarfs. The white circle marks an object even smaller than a brown dwarf—so small that we would consider it a planet if it were orbiting a star (but it's not).

Newborn floating small object
生まれたての微小天体

Newborn star
生まれたての恒星

Newborn brown dwarf
生まれたての褐色矮星

remain for many billions of years. Indeed, such "free-floating Earths" could conceivably be quite common in interstellar space, though their small size would make them extremely difficult to detect. If they exist, such planets could have surface oceans, as well as geothermal and chemical energy much like that sustaining life underground and around deep-sea vents on Earth. Although the total available energy probably could not sustain a huge total amount of life, at least some life would be possible.

Finally, we could conceive of habitability outside the habitable zone if life can use a liquid medium other than water, such as liquid ethane. Any liquid will evaporate rapidly when atmospheric pressure is low, so the presence of surface liquids of any type requires an atmosphere. In general, this means we can hope to find surface liquids only on worlds large enough to hold significant atmospheres, but not so large that they become giants like the jovian planets, where strong vertical winds probably preclude life. In our solar system, Titan is the only world that meets this criterion. Nevertheless, moons like Titan might be relatively common among other star systems. Thus, if life based on other liquids is possible, there could be many habitable worlds similar to Titan.

In summary, it's quite possible that the majority of habitable worlds in the universe are not located within stellar "habitable zones." From this standpoint, the concept of a habitable zone might seem obsolete, because it doesn't account for the potential habitability of Europa-like or Titan-like moons or of Mars-like subsurface water or of "free-floating Earths." Nevertheless, while life might be

common on such worlds, it would be extremely difficult to detect. Also, it seems almost impossible that such worlds could give rise to advanced civilizations. Thus, the concept of a habitable zone is still quite useful when we are searching for life beyond Earth, and it is critical when we are searching for intelligent life.

9.5 The Future of Life on Earth

The Earth has remained habitable for some 4 billion years, allowing plenty of time for life to evolve and diversify and ultimately making our own human existence possible. However, the continuing evolution of the habitable zone means that the Earth's days of habitability must eventually come to an end. In this section, we will briefly investigate the future habitability of the Earth and what it may mean for our descendants.

Before we begin, it's important to note that the demise of Earth's habitability is not something worth losing any sleep over. Even under the most pessimistic scenarios, our planet will remain habitable for at least many hundreds of millions of years to come. Under more optimistic scenarios, the Earth has billions of years of remaining habitability. Compared to the length of time our civilization has existed so far, these are incredibly long time scales. If our species survives for this long, we will have had plenty of time to find ways to move to other planets in our solar system or in other star systems, or to otherwise prevent our perishing with the Earth. In

The Sun's demise in about 5 billion years might at first seem worrisome, but 5 billion years is a very long time. It is longer than Earth has yet existed, and human time scales pale by comparison. A single human lifetime, if we take it to be about 100 years, is only 2×10^{-8}, or two hundred-millionths, of 5 billion years. Because 2×10^{-8} of a human lifetime is about 1 minute, we can say that a human lifetime compared to the life expectancy of the Sun is roughly the same as 60 heartbeats in comparison to a human lifetime.

What about human creations? The Egyptian pyramids have often been described as "eternal." But they are slowly eroding due to wind, rain, air pollution, and the impact of tourists, and all traces of them will have vanished within a few hundred thousand years. While a few hundred thousand years may seem like a long time, the Sun's remaining lifetime is more than 1,000 times longer.

On a more somber note, we can gain perspective on billions of years by considering evolutionary time scales. During the past century, our species has acquired sufficient technology and power to destroy human life totally, if we so choose. However, even if we make that unfortunate choice, some species (including many insects) are likely to survive. Would another intelligent species ever emerge on Earth? There is no way to know, but we can look to the past for guidance. Many species of dinosaurs were biologically quite advanced, if not truly intelligent, when they were suddenly wiped out about 65 million years ago. Some small rodentlike mammals survived, and here we are 65 million years later. We therefore might guess that another intelligent species could evolve some 65 million years after a human extinction. If these beings also destroyed themselves, another species could evolve 65 million years after that, and so on. But even at 65 million years per shot, the Earth would have some 15 more chances for an intelligent species to evolve in 1 billion years—the length of time our planet will remain habitable under fairly conservative scenarios. Under more optimistic estimates for long-term habitability, there could be 60 or more periods—each as long as the period separating us from the dinosaurs—still to come before our planet dies. That is a lot of potential opportunities for the evolution of a species wise enough to avoid its self-destruction and to move on to other star systems by the time the Sun finally dies. Perhaps we ourselves will prove to be so wise.

the meantime, if you are seeking reasons to lose sleep, many much more immediate threats to human survival exist, including war, poverty, overpopulation, species extinction, and global warming.

THINK ABOUT IT . . . *Given the more immediate threats to our civilization, is it even worth thinking about what will happen to our planet billions of years from now? Defend your opinion.*

The End of Habitability on Earth

Thanks to the climate regulation provided by the carbon dioxide cycle, the Earth has remained habitable even as our Sun has brightened by some 30% over the past 4 billion years. As the Sun continues to brighten, the carbon dioxide cycle should continue to keep the climate relatively pleasant for at least hundreds of millions of years to come. Eventually, however, the warming Sun will cause the cycle to break down.

If you look back at Figure 9.5, you'll see that under the more conservative assumptions about the boundaries of the habitable zone, this zone will move beyond Earth's orbit in about a billion years. Thus, if these assumptions are correct, the end of habitability on Earth will come about a billion years from now. Recall that these conservative assumptions are based on the idea that a *moist greenhouse* will cause the oceans to evaporate away. That is, about a billion years from now water vapor will begin to circulate into the Earth's upper atmosphere, where ultraviolet light will break apart water molecules and allow the hydrogen to escape into space. As water is lost from the upper atmosphere, more water will evaporate to take its place, until the oceans are completely gone. At that point, it seems unlikely that any life could continue to survive on Earth. Of course, no one is yet sure whether the moist greenhouse problem will really arise in about a billion years. Many feedback mechanisms in Earth's climate are not yet well understood. For example, increased cloud cover may reduce the amount of sunlight reaching Earth's surface, preventing the onset of the moist greenhouse. That is why Figure 9.5 also shows more optimistic scenarios, under which the Earth will remain habitable well beyond a billion years into the future.

Even under these more optimistic assumptions, however, some 3–4 billion years from now the Sun will have grown so warm that the Earth will finally suffer the fate of Venus—a runaway greenhouse effect. The rising temperature on Earth will cause increased ocean evaporation, and the increased water vapor in the atmosphere will increase the greenhouse effect further (see Figure 9.3). The positive feedback won't stop until the oceans have evaporated away and all the carbon dioxide has been released from carbonate rocks. Our planet will become a Venus-like hothouse, with temperatures far too high for liquid water to exist.

We know of no natural phenomena that can prevent this runaway greenhouse effect from occurring. However, if we imagine that our descendants have become quite technologically advanced, there are several possible ways they could survive. For example, perhaps they could build a giant sunshade to reduce the amount of sunlight striking the Earth, thus preventing the runaway greenhouse effect. Moreover, the end of habitability on Earth does not necessarily mark the end of habitability in our solar system. When the habitable zone moves past the orbit of Earth, Mars will still be within it. Our descendants might survive the demise of Earth by moving to Mars and doing some well-planned terraforming [Section 7.5]. Or, if they are truly powerful, they might find a way to move our planet gradually outward from the Sun to keep it within the habitable zone. If they could gradually move Earth to the orbit of Mars over the next 3–4 billion years, then humans could stay home and our planet would stay habitable. But even this solution can be only temporary, because the Sun itself eventually will die.

Death of the Sun

The Sun's gradual "death" will begin when it runs out of hydrogen in its central core, which will happen about 5 billion years from now. Then, over the following billion years or so, the Sun will undergo a dramatic transformation. During the first few hundred million years of this period, the Sun will gradually swell to about 100 times (or more) its present radius, becoming what we call a *red giant* star.[4] At its peak, the red giant will be about 1,000 times as luminous as the Sun is today, and Earth's surface temperature will rise to 700°C or higher. Even underground life is likely to be baked to death during this period.

---

[4] You might wonder why a star should swell up and grow more luminous as it exhausts the fuel in its core. This counterintuitive process is a result of the immense pressure exerted on the core by the weight of the star's overlying gas. During the time that the central core burns hydrogen, the energy generation helps the core resist the pressure and maintain its size. Once the hydrogen in the central core runs out, the pressure causes the core to shrink and therefore rise in temperature. As this occurs, a layer surrounding the central core—a layer that still contains unburned hydrogen—ignites with nuclear fusion. In fact, this layer becomes so hot that the total rate of fusion is higher than it was while the central core fused hydrogen. The higher rate of fusion means a greater release of energy, which causes the star to puff up in size and emit more light.

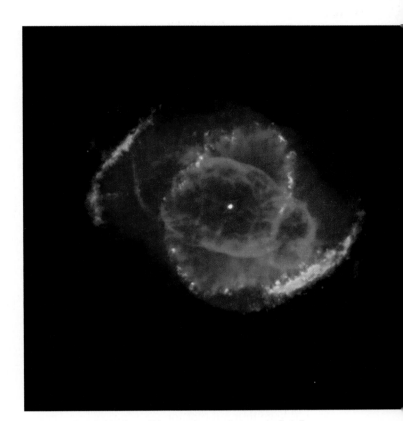

FIGURE 9.7 This Hubble Space Telescope photograph shows the Cat's Eye Nebula, one of many examples of a planetary nebula. The shells of gas were expelled from a Sun-like star that had reached the end of its life. The central white dot is the white dwarf star that remains, which is essentially just the hot core of the now-dead star. The shells of gas are moving away from the central white dwarf and will dissipate into space within a few tens of thousands of years. Some 5 billion years from now, our Sun will suffer the same fate as the star that created this nebula.

In its final death throes, the Sun will expel its outer layers into space, creating a *planetary nebula* (Figure 9.7). (Despite the name, these structures have nothing to do with planets; the term comes from the planetlike appearance of some of these nebulae when seen through a small telescope.) All that will remain of the Sun will be its hot central core; no more nuclear fusion will occur. This remaining core, known as a *white dwarf* star, will then gradually cool over time. The violence that accompanies the planetary nebula ejection will probably destroy the Earth. Even if our planet somehow survives this event and continues to orbit the white dwarf Sun, the white dwarf's light will be so feeble that the Earth's charred surface will be left in perpetual cold and dark.

Could We Still Survive?

For those who are undaunted by the thought of humans or other intelligent Earth beings surviving some 5 billion years from now, it's natural to wonder whether we could also survive the death of the Sun. The obvious solution to the Sun's death is to move to another star system. Stars that are being born today might offer great homes to us in 5 billion years. When these stars die, we can move on to still others born later. As long as there are new stars with habitable planets, we could potentially survive by migrating to new homes.

However, such long-term migration has its limits. The recycling of stellar material cannot continue forever, because dying stars return to space only part of the gas from which they were made. Over time, the galaxy will contain less and less interstellar gas. About 50 billion years from now, the Milky Way Galaxy will contain so little gas that new stars will no longer be born. What then?

Ever since Edwin Hubble first discovered that our universe is expanding, we have wondered whether the expansion will continue forever or someday stop, causing the universe to collapse back in on itself. In the past few years, astronomers have been surprised to learn that the expansion appears to be accelerating, in which case it seems that the fate of the universe is to expand forever. However, you should keep in mind that forever is a very long time. It remains possible that we will someday discover other surprises that will change our view of the fate of the universe.

If the universe continues to expand after all star formation has ceased, then life will be able to continue only around those long-lived stars that still have habitable planets. But even the longest-lived stars will eventually run out of hydrogen to fuse. Within a few hundred billion years, all existing stars will burn out and die, and the now-brilliant galaxies will fade into darkness. On much longer time scales, interactions among the burned-out stars will send many of them flying into the vastness of intergalactic space, while the rest will converge toward their galactic centers, merging into gigantic black holes. At this point, the story becomes even more speculative. If the so-called "grand unified theories" of physics are correct, the stellar corpses will eventually disintegrate into subatomic particles. Meanwhile, the giant black holes will slowly evaporate into energy and

subatomic particles in a process first described by the noted British physicist Stephen Hawking. At some point in the far distant future, the universe will consist of nothing but isolated subatomic particles and isolated photons of light, each separated from every other one by immense distances that will forever grow larger as the universe forever expands. Our current epoch of a universe filled with stars and galaxies will have been just a fleeting moment in an eternity of darkness.

This end in darkness may seem a bit depressing, but it is, after all, inconceivably far in the future. Nevertheless, it is fair to ask whether it is truly the end or instead could be followed by something else. Remarkably, some serious scientists already argue that there might be ways by which an intelligent civilization could survive even as the universe dies. But for a more humorous viewpoint, we turn to a science fiction story. Isaac Asimov, in his story "The Last Question," begins with a couple of people asking a supercomputer whether it will ever be possible to reverse the decline of the universe, averting a cold and dark end. The computer responds that the available data are insufficient to answer the question. Over billions of years, computers advance and humankind survives, making the question ever more important as the universe approaches its cold, dark fate. By the end of the story, the computer exists solely in hyperspace, outside the time and space of our universe. The universe has reached its state of ultimate darkness, with nothing left alive. Meanwhile, the computer, which Asimov calls AC, whirs on in the timelessness of hyperspace until finally it learns how to reverse the decay of the universe:

For another timeless interval, AC thought how best to do this. Carefully, AC organized the program.

The consciousness of AC encompassed all of what had once been a Universe and brooded over what was now Chaos. Step by step, it must be done.

And AC said, "LET THERE BE LIGHT!"
And there was light—

THE BIG PICTURE

In this chapter, we have tied together much of what we learned in other chapters in order to examine the concept of a habitable zone and its evolution over time. As you continue in your study, keep in mind the following "big picture" ideas:

- Despite its name, the habitable zone is not the only region around a star where habitability is possible. Instead, it refers to the region in which a planet of suitable size could have liquid water on its surface.

- Venus, Earth, and Mars may all have been habitable in the early history of the solar system. However, only Earth has remained habitable to this day. Understanding why the climates of Venus and Mars have changed and why Earth's climate has remained stable can help us understand our climate and how we can protect it.

- Estimates of the precise boundaries of the Sun's habitable zone depend on the assumptions we make about how planets are affected by solar warmth. Even with conservative assumptions, the habitable zone is significant in size, suggesting that planets in habitable zones might be relatively common throughout the universe.

- The habitable zone is gradually migrating outward from the Sun. Eventually, it will lie beyond the Earth's orbit, rendering our planet uninhabitable. Still, with sufficient technology, we can imagine our descendants finding a way to remain within the habitable zone until the Sun dies and then surviving by moving to other star systems.

- Given that an advanced civilization could find a way to survive as its star dies and could certainly find ways to overcome other natural threats such as asteroid impacts, it seems that Nature will not impose intractable problems on us for billions of years to come. If we do not survive, it is far more likely to be the result of our own failings than the result of any natural catastrophe.

Review Questions

1. What do we mean by the *habitable zone?*

2. Compare how the greenhouse effect influences climate on Venus, Earth, and Mars. Would any of these planets be habitable in the absence of greenhouse warming?

3. Briefly discuss why Venus lacks water today and the evidence suggesting that water that might once have filled oceans on Venus was lost forever.

4. What is a *runaway greenhouse effect,* and how would it occur if we put the Earth at Venus's distance from the Sun?

5. Why did a runaway greenhouse effect occur on Venus but not on Earth? Could there have been oceans on Venus before the runaway greenhouse effect set in? Explain.

6. Describe how distance from the Sun, planetary size, and atmospheric loss processes have affected the habitability of Venus, Earth, and Mars.

7. What factors affect the location of the inner boundary of the habitable zone? Be sure to explain and consider the role of a possible *moist greenhouse* in such calculations.

8. What factors affect the location of the outer boundary of the habitable zone? Briefly summarize the current boundaries of our Sun's habitable zone under both the more optimistic and the more conservative scenarios.

9. Why does the Sun gradually brighten, and how does this brightening affect the location of the habitable zone over time? What do we mean by a *continuously habitable zone?*

10. Describe several ways it is possible to have habitability outside the habitable zone. Is it possible that habitable worlds are more common outside this zone than within it? Explain.

11. Briefly discuss the fate of the Earth. When will the Earth become uninhabitable? Why?

12. Briefly discuss the fates of the Sun and of the universe and what these fates might mean to our descendants (if anyone survives that long).

Discussion Questions

1. *Are Earth-like Planets Common?* We have not yet discussed all the factors that might contribute to the question of whether Earth-like planets are common. We will discuss more of these factors in Chapter 10. However, based on what you have learned so far about solar system formation and the habitable zone, do you think Earth-like planets should be common or rare among Sun-like stars? Explain.

2. *Greenhouse Lessons.* While it seems unlikely that human activity could cause a runaway greenhouse effect on Earth, we could still cause the climate to warm substantially. Do you think we can learn anything valuable about our potential effects on Earth's climate by studying the climate histories of Venus and Mars? If so, what? Defend your opinion.

3. *A Billion Years.* At minimum, it appears that our planet will remain habitable for at least the next billion years, give or take a couple hundred million years. How long is a billion years? Think of some ways to put this time period into perspective.

4. *The Fate of Life in the Universe.* Consider the evidence suggesting that life is just a fleeting phase in the long-term history of the universe. Assuming this to be the case, how do you think it should influence our perspective on our own place in the universe? Why?

Problems

Does It Make Sense? In **problems 1–8,** evaluate each given statement and decide whether it is likely to be true or false. Explain your reasoning clearly.

1. If Venus were just a little bit smaller, its climate would be Earth-like.

2. If the Sun were twice as bright as it actually is, life on Earth would not be possible.

3. If the Sun were several times brighter than it actually is, the habitable zone would include Jupiter's orbit.

4. While the habitable zone of the Sun migrates outward over time, the habitable zones of other Sun-like stars might instead migrate inward over time.

5. Mars will someday undergo a runaway greenhouse effect and become extremely hot.

6. If the Earth someday becomes a moist greenhouse, it will mean a climate that is humid but still quite comfortable.

7. If we could somehow start plate tectonics on Venus, its surface would cool and it could have oceans of liquid water.

8. We are not yet certain, but it is quite likely that the Earth has suffered through a runaway greenhouse effect at least once in the past 4 billion years.

9. *No Plate Tectonics.* Suppose plate tectonics magically stopped on Earth, but other geological processes (such as volcanism) continued. Would the Earth's *surface* get warmer or cooler? Explain.

10. *Continuously Habitable Zone.* Is the Earth in a zone that remains continuously habitable from the Sun's birth to its death? Is any planet? Explain.

11. *Planetary Changes.* For each situation described, write two or three paragraphs explaining why the planet would or would not be habitable today.

   a. A planet the size of Mars located at the distance of Venus.

   b. A planet the size of Mars located at the distance of Earth.

   c. A planet the size of Venus located at the distance of Earth.

   d. A planet the size of Earth located at the distance of Mars.

12. *Habitable Moons.* As we'll discuss in Chapter 10, many of the newly discovered extrasolar planets are Jupiter-like in size but are located at Earth-like distances from Sun-like stars. These planets are unlikely to be habitable themselves. Could they have habitable moons? Explain the conditions under which habitable moons might be possible.

Web Projects

1. *Runaway Greenhouse Effect.* Learn more about the runaway greenhouse effect and whether it is possible that human-induced global warming could cause this to occur on Earth. How great a climate change could we potentially cause if we did nothing to slow our emissions of greenhouse gases? Write a short summary of your findings.

2. *Fate of the Universe.* Learn about astronomers who have discovered what appears to be an acceleration in the universal expansion. Why is this acceleration surprising? Why does it seem to imply that the universe will expand forever? Write a short summary of your findings.

3. *Long-Term Survival.* Read about some exotic ideas concerning how advanced civilizations might survive as the universe grows cold and dark. Do you think any of these ideas make sense, or are they just wishful thinking? Write a short essay describing one of these ideas and your opinion of it.

Then felt I like some watcher of the skies, when a new planet swims into his ken.
John Keats (1795–1821)

CHAPTER 10

The Search for Habitable Worlds

Up to this point in the book, we have focused primarily on life on Earth and the prospect of life existing elsewhere in our solar system. We have found that while Earth is the only planet known to harbor life, several other worlds in our solar system are—or at some time in the past have been—potentially habitable. If this situation is similar for the billions of other star systems in the Milky Way Galaxy and, by extension, the billions of other galaxies that pepper the universe, then the total number of habitable worlds must be enormous. The existence of a vast number of habitable worlds offers hope that among them we would find not only microbes, but also complex creatures and intelligent beings.

Unfortunately, our current data are not sufficient for us to be confident that such an immense number of habitable worlds really exists. The large planets we have detected so far around other stars are probably not of the type that would harbor life, and there is considerable debate about whether the existence of these worlds makes it more or less likely that habitable planets are common.

In this chapter, we will discuss how we search for extrasolar planets, what we have learned to date from this search, and how we might learn if distant worlds are, indeed, abodes of life. Finally, we'll discuss the controversial hypothesis that our planet is so special that it might be the only world in the galaxy to harbor intelligent beings. Although we won't be able to draw definitive conclusions, understanding the debate will help us frame the question of whether other civilizations exist that we might someday contact and how we might try to do so.

10.1 Are Habitable Planets Common?

In Chapter 1, we discussed both theoretical and observational reasons why we expect that planets are common. On the theoretical side, our current ideas about how stars and planets form suggest that the same general processes that gave birth to the planets of our solar system ought to occur around most other young stars. For example, it seems almost inevitable that a collapsing cloud of gas and dust will form a star surrounded by a spinning disk of debris (see Figure 1.14).

On the observational side, two types of evidence point to planets' being common. First, we have seen flat disks of material orbiting many other newborn stars (see Figure 1.15). The disks of young stars often contain a few percent of the mass of their stars, which is more than sufficient for constructing planets. (The combined mass of all the worlds in our solar system amounts to less than 0.2% of the Sun's mass.) Based on the observations made to date, it appears that at least a quarter to half of all young stars are surrounded by such disks. Although this fact does not tell us directly whether planets are likely to form in such disks, it certainly lends support to our theory of how stars and planets form. Second, we have detected actual planets around several dozen stars, and new planet discoveries are being made at a breathtaking rate. The first discovery of an extrasolar planet around a normal star was announced only in 1995,[1] and by the beginning of 2002 the number of known extrasolar planets was nearly 80. In this short interval, we found almost 10 times

---

[1] We are not including an earlier discovery of planets orbiting a neutron star.

THE DEVELOPMENT OF SOLAR SYSTEM FORMATION THEORY

Our current theory of how the Earth and the rest of our solar system formed, described in some detail in Chapters 1 and 4, has been remarkably successful at explaining a great many observations not only of our own system but also of other star systems. However, like any scientific theory, it is subject to constant reexamination and modification. While the strong observational evidence in its favor makes it likely that the theory is correct in its broad outlines, the recent discoveries of extrasolar planets have forced us to reconsider some of the details. One of the best ways to understand these changes is to consider them in the context of how our current theory developed.

When, at the end of the seventeenth century, the layout and motions of the planets and moons of our solar system were understood, it was only natural that scientists would speculate about how this system came to be. The eighteenth-century French mathematician Pierre-Simon Laplace (1749–1827) and his contemporary the German philosopher Immanuel Kant (1724–1804) independently came up with the idea that the Sun and the planets were born in a rotating disk of gas and dust. Laplace and Kant were inspired by obvious but still remarkable properties of the solar system. For example, all the planetary orbits lie in nearly the same plane (Pluto is an exception but had not yet been discovered), and all the planets proceed along their orbits in the same direction. In addition, most of the planets (and the Sun) spin on their axes in the same sense, and nearly all the larger moons orbit their host planets in the same direction as well. This behavior is a natural consequence of formation in a rotating disk.

The ideas of Kant and Laplace are close to what we currently believe, but other ideas have been put forth. Early in the twentieth century, it was proposed that the planets represent the debris from a near-collision between the Sun and another star. According to this "close encounter" hypothesis, at some distant time a star passed close to our Sun, gravitationally pulling blobs of solar gas into space that then cooled to become the planets. Despite its dramatic appeal, calculations showed that this hypothesis couldn't account for the orbital motions of the planets. Astronomers fell back on the disk ideas of Kant and Laplace.

The disk theory has been greatly refined as we have learned more. We now know that interstellar space contains cold, dark, and dusty clouds that can collapse under the influence of their own gravity to produce the disks. Various events could trigger the collapse, such as the shock of a nearby stellar explosion (supernova) or the slow cooling of a dense chunk of cloud as its molecules radiate away energy in the form of infrared light. In any case, the disk theory implies that virtually every time a star is born there's a chance that planets will be formed as well, so planets should be common. In contrast, if the close encounter hypothesis had been correct, we would have had good reason to think that our solar system was the only one in the galaxy. Recall that, on the scale we considered in Chapter 1, the typical separation between stars in our part of the Milky Way is equivalent to that of grapefruits located on opposite coasts of the United States. Given this wide separation and the relatively slow speeds with which stars move relative to one another, the chance that two stars would pass near enough to each other to cause a substantial gravitational disruption is extremely remote.

The outlook for planets has swung back and forth with the changing fortunes of the Kant–Laplace disk theory. Recent discoveries of disks around other stars and of extrasolar planets have boosted our confidence that the disk idea, with its prediction of plentiful planets, is at least approximately correct. But the fact that the newly discovered solar systems are quite different from our own suggests that we should expect continued modification of our theory of how planetary systems are born.

as many planets in other solar systems as exist in our own.

THINK ABOUT IT . . . *By the time you read this book, the number of known, extrasolar planets will have grown larger still. Do a quick check on the Web (using the text Web site or a search on "extrasolar planets") to find the current number of known planets. How many more have been discovered since early 2002?*

These distant solar systems are not arranged like ours. All the extrasolar planets discovered thus far are quite massive, making it likely that they are made largely of hydrogen and helium, like the jovian planets (Jupiter, Saturn, Uranus, and Neptune) of our own solar system. Much to our surprise, however, whereas the jovian planets orbit far from the Sun and follow nearly circular orbits, most of the newly discovered extrasolar planets orbit quite close to their parent star, and many have highly elliptical orbits. The large size of these planets makes them unlikely to be habitable for the same reasons that Jupiter is not habitable [Section 6.3]. Moreover, as we'll discuss in greater detail shortly, their surprising orbits may have significant implications for the possible existence of Earth-like planets within the habitable zones of stars.

We will devote most of the rest of this chapter to exploring the question of whether habitable planets could be common or rare and how we might someday observe—or even identify life on—such planets if they exist. We'll begin, however, by examining the question of what types of stars would be suitable suns for habitable planets.

10.2 Distant Suns

Perhaps the first and most obvious prerequisite to finding a planet on which life might arise is selecting a "sun" that can provide sufficient light and heat to support habitable worlds. Not all stars are potential "suns" in this way. To understand what kinds of stars might be orbited by potentially habitable planets, we must first investigate the nature of stars in more depth.

In ancient times, almost any light in the sky was considered to be a star, and in some cases we still use this historical language. For example, we often refer to meteors as "shooting stars," even though they really are just bits of interplanetary dust entering our atmosphere [Section 5.5]. Asteroids got their name, which means "starlike," because that is how they appear when first seen through a telescope, even though they are actually chunks of rock in our own solar system. Our modern definition of a star is a large ball of gas that generates energy by nuclear fusion in its hot central core (see definitions on p. 6). Thus, our Sun shines as a star because it fuses hydrogen into helium in its core.

Recall that a star goes through a "life cycle" that begins with its formation in a giant cloud of gas. Before the center gets hot enough to ignite nuclear fusion, we refer to the unfinished star as a *protostar* (*proto* comes from the Greek for "earliest form of"). A star is "born" when nuclear fusion begins in its central core. A star "dies" when it finally ceases to produce energy by any kind of fusion.

All stars spend most of their lives (about 90% of the time from star birth to star death) fusing hydrogen into helium. Stars shine fairly steadily during this period, brightening gradually as they age (for the same reason the Sun is slowly brightening [Section 9.4]). Core hydrogen fusion may continue for millions or billions of years (depending on the star's mass, as we'll discuss shortly), but eventually the central core will be so depleted of hydrogen that the fusion cannot continue. At that point, a star's life becomes more complicated. At first, it bloats up in size, becoming a *giant* or *supergiant* star. For example, when our own Sun becomes a red giant some 4–5 billion years from now, it will ultimately swell to engulf Mercury, and possibly Venus and Earth [Section 9.5]. Moreover, a star near the end of its life can produce elements besides helium. Our own Sun will someday produce carbon by fusing the helium in its core, and more massive stars can ultimately create all the other elements (through a combination of core fusion reactions and reactions that occur in their final death throes). As we discussed in Chapter 1, this stellar manufacture explains the existence of all the elements in the universe except the original hydrogen and helium that were spawned by the Big Bang, which is why we say we are "star stuff." When a star runs out of fuel for nuclear fusion and dies, it expels its outer layers into space, while the remaining core shrinks to a small, "dead" cinder. The specifics of this process vary among stars of different mass. Relatively low mass stars like our Sun end their lives comparatively gently, blowing the outer layers into space and leaving behind the type of dead star that we call a *white dwarf*. Higher-mass stars die in titanic explosions called supernovae and collapse to form bizarre types of stellar remnants known as *neutron stars* and *black holes*.

From the point of view of an astronomer, all of the various stages of stellar life are interesting. But in terms of the search for habitable planets, we are generally concerned only with stars that are in the long-lasting, hydrogen-fusing stage of their lives. Giants and supergiants may still have planets, but the rapidly changing nature of these stars (for example, bloating up in just a few million years or less)

It's "a long time ago" in someone's far-off galaxy, and the political situation is turning ugly.

The premise of *Star Wars*, a cinematic space opera that, like relatives, just keeps on returning, is that a galaxy-wide republic has been hijacked and converted to an autocratic, evil empire by despots in gray flannel suits. This may sound vaguely reminiscent of the story of Rome, but unlike what happened to that ancient civilization this political shift has encouraged a serious rebellion, a war among the stars. The rebels are led by Princess Leia (you can tell she's a princess 'cause her hair is done up like twin danish pastries), and her strategy is to take out the empire headquarters—an enormous, spherical spacecraft known to its friends as "the Death Star." The Death Star packs weaponry that can explode a planet in seconds (calculate, if you wish, the energy required to do that!). On the other hand, the rebels have "the force" on their side—a mystical ability to change the odds of every situation based on moral merit and self-discipline.

Most of *Star Wars* is battle of the sort that's familiar to any movie fan, except that the bad guys wear brittle, white plastic suits and fly spacecraft that look like box kites. But *Star Wars* offers some interesting peeks into life as it might be elsewhere in the cosmos. The rebels have their base of operations on a large planet's moon, a not impossible scenario since hefty moons could be habitable. Luke Skywalker, the young hero, hails from a world

circling a close double star. Not a problem—research has shown that planetary orbits around double stars could be stable.

There are peculiar anachronisms in *Star Wars*, however. The Death Star is obviously extremely hi-tech, and yet the principals occasionally face off using souped-up swords. Everyone jets around in spacecraft that are somehow capable of exceeding speed-of-light travel by jumping into hyperspace, and yet we often see aliens saddled up onto giant, dinosaur-like creatures—much as in *The Flintstones*.

All that can be forgiven. But *Star Wars's* biggest leap is the fact that dozens of alien races are all living contemporaneously (although it's clear that the human types are in charge). In the movie's famous cantina scene, which takes place in the wretched port city of Mos Eisely, aliens of all shapes and colors get together to do business and get drunk. As we discuss in Chapter 13, the chances that any two intelligent species (let alone dozens) appear on the galactic scene within 100,000 years of one another is quite small. If there are other societies out there, they will either be far behind us or enormously beyond our level. We won't be sharing dance music and booze with them in a seedy extraterrestrial dive.

And besides that, why does a republic have a princess, anyhow?

---

makes it unlikely that any planets could remain habitable for long. Dead stars may also have planets—indeed, the first confirmed discovery of extrasolar planets involved three small worlds orbiting a neutron star. But because a neutron star is no longer generating continuous sunshine of the type produced by ordinary stars, it would be an unlikely host for habitable planets. Consequently, in the rest of this chapter we will focus only on stars in the main, hydrogen-fusing stage of their lives.[2]

Properties of Stars

Stars vary significantly in their properties, such as mass, temperature, luminosity, and composition. The brightest stellar beacons shine with a brilliance equal to a million Suns, while other stars are so dim that

they are invisible to the naked eye even when relatively nearby. Once astronomers recognized the substantial differences among stars, their first task (like that of botanists and zoologists) was to find a useful scheme for categorizing this diverse collection of objects.

Finding such a categorization was not trivial. Stars are so far away that even when viewed with large telescopes they appear as no more than points of light. The most obvious characteristic of a star is how bright it appears in the sky, and indeed a system for describing the brightness of stars visible to the naked eye was developed by the ancient Greeks (and is still used today). Unfortunately, ranking a star by how bright it appears in the sky doesn't tell us how bright it *really* is, because brightness also depends on distance. A very bright star that is very far away will look dim, while a modestly luminous star that is nearby may shine quite brightly. For example, the brightest star in our sky, *Sirius*, is only a moderately bright star that happens to be relatively close. (Sirius is about 8.6 light-years distant, making it the seventh-nearest star other than the Sun.)

---

[2] Readers who have studied astronomy will recognize that we are talking about stars that lie on the *main sequence* of the Hertzsprung–Russell (H–R) diagram.

Refinement of the telescope eventually led to techniques for measuring stellar distances, which enabled astronomers to classify stars according to their true brightnesses. But to learn more about their properties required yet another technological step: the invention of spectroscopy [Section 6.2]. In the late 1800s, astronomers at Harvard College Observatory, under the directorship of Edward Pickering (1846–1919), began a massive effort to study stellar spectra and thereby determine other characteristics of the stars.

Making a detailed stellar spectrum is a tedious process, but at the end of the nineteenth century astronomers took an important step forward when they invented a method to record the spectra of many stars at once. To do this, astronomers mounted a glass prism in front of a telescope's objective lens and then photographed a patch of sky containing a large number of stars. On the resulting photo, the image of each star was spread out into a rainbowlike streak, and the most obvious spectral lines could be seen (Figure 10.1).

Shortly, thousands of stellar spectra were in hand. The next step was to study this wealth of data and try to make sense of what it was telling us. Pickering needed help with these tasks, so he began to hire assistants whom he called "computers." The job clearly required people well trained in physics and astronomy, but, in part because the task was seen as somewhat tedious, Pickering found few takers among the men graduating from what was then all-male Harvard College. He therefore recruited women who had studied physics and astronomy at colleges such as Wellesley and Radcliffe. Because at that time women were barred from most good positions in science, a job with Pickering represented a rare chance for career advancement. Although the work was indeed tedious, it was also cutting-edge research, and the women astronomers of Harvard College Observatory were to make many great discoveries (Figure 10.2).

At first, the astronomers found it difficult to make sense of the spectra. Pickering suggested a scheme in which stars were classified by the visibility of hydrogen lines (that is, spectral lines caused by the element hydrogen) in their spectra, using type A to designate stars with the strongest hydrogen lines, type B for those with slightly fainter lines, and so on down the alphabet to type O for stars showing the weakest lines. Following his suggestion, one of Pickering's first hired "computers," Williamina Fleming (1857–1911), classified more than 10,000 stellar spectra by 1890. The work was just beginning.

In 1896, Pickering hired Annie Jump Cannon (1863–1941), who in the course of her career would personally classify the spectra of over 400,000 stars. Within a few years of being hired, she realized that

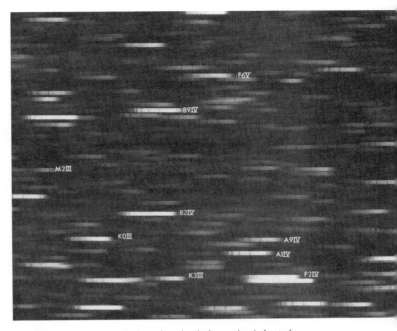

FIGURE 10.1 This photograph shows how, by placing a prism in front of a telescope's lens, astronomers can record the spectra of many stars at once. The individual spectra are somewhat crude, but they are adequate to identify the stronger hydrogen and helium lines that are the basis of the spectral type classification.

FIGURE 10.2 Women astronomers pose with Edward Pickering at Harvard College Observatory in 1913. Annie Jump Cannon is fifth from the left in the back row.

Pickering's sequence of spectral types A to O included some redundancies and, more important, fell into a much more natural order than he had supposed. She concluded that there were only seven major spectral types, which could be logically ordered as OBAFGKM, a sequence that legions of astronomy majors have memorized using the politically incorrect mnemonic "Oh, Be A Fine Girl, Kiss

Me." Cannon also subdivided each type by number; for example, stars of spectral type G could be subclassified as G0, G1, G2, and so on to G9. Our Sun is now classified as spectral type G2. The astronomical community adopted Cannon's system of stellar classification in 1910.

The stellar classifications clearly were telling us something important about the nature of stars, but no one yet knew just what that was. The answer finally came in 1925, in the dissertation of another woman working at Harvard College Observatory, Cecilia Payne-Gaposchkin (1900–1979). Relying on insights from what was then the newly developing science of quantum mechanics, Payne-Gaposchkin showed that the differences in the spectral types reflected differences in the surface temperatures of the stars. A later review of twentieth-century astronomy called her work "undoubtedly the most brilliant Ph.D. thesis ever written in astronomy."

This research quickly led to another interesting discovery. One might naively expect that stars could have many combinations of size, temperature, mass, and composition. The truth is much simpler. All stars have basically the same composition—they consist almost entirely (98% or more) of hydrogen and helium. For that reason, the physics that determines a star's characteristics is relatively simple. During the hydrogen-burning phase of its life, a star's surface temperature (and hence its spectral type) and total brightness are determined almost entirely by one thing—the star's mass. Table 10.1 lists typical properties for stars of each of the seven major spectral types.

Table 10.1 also shows that the lifetimes of stars vary considerably, which we can understand by studying the mass and luminosity columns. A star's mass essentially tells us how much hydrogen fuel it has available for fusion, while its luminosity tells us how brightly the star shines and hence how rapidly it is fusing its hydrogen. Note that there is an enormous range in luminosity: The brightest stars are nearly a billion times more luminous than the dimmest. The range in mass is much smaller. Thus, for example, an O star has some 60 times as much hydrogen fuel as the Sun (from the mass column) but burns it at a rate of a million times faster than the Sun (from the luminosity column). Therefore, it will go through all of its available fuel much faster than does the Sun. The rule is a general one: The more massive the star, the shorter its lifetime. As shown in the last column of the table, the range of lifetimes extends from just a few hundred thousand years to hundreds of billions of years.

THINK ABOUT IT . . . *Many generations of massive O and B stars have lived and died in the history of the universe. Is it also true that many generations of G, K, and M stars have lived and died? Explain.* (Hint: How old is the universe?)

Which Stars Would Make Good Suns?

We can use the properties listed in Table 10.1 to investigate the question of which types of stars might make suitable suns for habitable planets. We can immediately rule out stars of type O from their short lifetimes alone. Recall that, in our solar system, it took tens of millions of years for the planets to form by accretion [Section 4.2]. With typical lifetimes of less than a million years, O stars simply don't have

Table 10.1 Typical Properties for Hydrogen-Burning Stars of the Seven Major Spectral Types
Numbers given in "solar units" are values in comparison to the Sun; for example, a mass of 60 solar units means 60 times the mass of the Sun. Note that the Sun is a G star. (More specifically, the Sun's spectral type is G2.)

Spectral Type	Approximate Percentage of Stars in This Class	Surface Temperature (°C)	Luminosity (solar units)	Mass (solar units)	Lifetime (years)
O	0.001%	50,000	1,000,000	60	500 thousand
B	0.1%	15,000	1,000	6	50 million
A	1%	8,000	20	2	1 billion
F	2%	6,500	7	1.5	2 billion
G	7%	5,500	1	1	10 billion
K	15%	4,000	0.3	0.7	20 billion
M	75%	3,000	0.003	0.2	600 billion

enough time to make real planets. We can probably also rule out stars of type B. Although the lifetime of a B star may be just long enough for planets to form, the star's death would probably occur before the process of accretion had settled down enough for life to take hold.

Stars of types A and F, with lifetimes of 1–2 billion years, would seem to offer enough time for both the formation of planets and the beginnings of biology. After all, we have fossil evidence of life from the time our planet was just over a billion years old and carbon isotope evidence suggesting that life was widespread a few hundred million years before that [Section 5.1]. Stellar types A and F are hotter than the Sun, so their habitable zones would lie at greater distances from the central star, but there's no reason to think that planets wouldn't form in these regions.

One potential problem is that, because of their higher temperatures, A and F stars emit many times as much ultraviolet light as does the Sun. Biology on Earth, and perhaps biology in general, is vulnerable to high-energy ultraviolet light, which easily breaks chemical bonds in complex organic molecules. The intense radiation might well keep planetary surfaces sterile. However, there are at least two ways around this problem. First, ultraviolet radiation does not penetrate far into the ground, oceans, or ice. If life on Earth arose near volcanic vents in the deep oceans, as some evidence suggests, the same thing might happen on a watery planet around an A or F star. Similarly, worlds with a subsurface ocean, such as Europa may have, would offer life plenty of protection. Second, even though our Sun emits far less ultraviolet light than A or F stars, it still emits enough to make the Earth's surface sterile if not for the shielding provided by the ozone layer in the atmosphere. Planets around an A or F star might enjoy similar shielding if they had either a sufficiently thick atmosphere or an atmosphere containing sufficient oxygen. In the latter case, the additional ultraviolet light would split more atmospheric O_2 molecules, producing single oxygen atoms that would then combine to form ozone. In other words, a higher dosage of ultraviolet radiation could result in more ozone shielding. Thus, A and F stars, despite their energetic nature, seem quite capable of hosting habitable planets. However, given the fact that complex plants and animals on Earth did not arise until our planet was some 4 billion years old, the 1- to 2-billion year lifetime of A and F stars suggests that life on these planets would most likely be much simpler than life on Earth.

THINK ABOUT IT . . . *In light of current evidence concerning the past oxygen content of Earth's atmosphere (see Section 5.3), does it seem likely that planets around A or F stars could have ozone layers? Why or why not?*

Although the possibility of habitable planets around A and F stars is intriguing, Table 10.1 suggests that statistically it's not too important: Together, these types make up only about 3% of stars. G stars, like our Sun, make up another 7%—and we're well aware that G stars can have habitable planets. But if we want to know whether the *majority* of stars could have worlds capable of supporting life, we need to consider the smaller K and M types, which make up some 90% of all stars in the universe.

K and M stars have very long lifetimes that impose no limit on their ability to have planets on which life could evolve for many billions of years. Instead, the primary issue for life around these stars is the size of their habitable zones. Because these stars are dimmer than the Sun, temperatures on orbiting planets will be lower in these systems than on planets the same distances from the Sun in our solar system. More specifically, recall that the energy in starlight falls off with the square of the distance from the star (see Figure 6.1). Thus, for a planet to receive as much radiant energy as the Earth from a K star having 1/4 the Sun's luminosity, the planet would have to orbit its star at 1/2 Earth's distance from the Sun (because $(1/2)^2 = 1/4$). For a star with a luminosity 1% of the Sun's, a similar planet would need to orbit at a distance 1/10 of Earth's distance from the Sun, which means it would follow an orbit about 1/4 the size of Mercury's. The same reasoning tells us that the boundaries of a dim star's habitable zone would also be much closer to the star than are the boundaries of the habitable zone in our solar system.[3] In general, the habitable zone becomes increasingly smaller and closer-in for stars of lower luminosity (Figure 10.3). Around an M star, for example, the habitable zone would extend only over a region roughly equivalent to from 1/4 to 1/2 Mercury's distance from the Sun in our solar system. The small size of the habitable zone certainly decreases the probability that planets would be found within this region.

On the other hand, the overwhelming majority of stars are M stars, due both to their long lives—every M star ever born is still shining—and to the fact that small stars are born in larger numbers than are more massive stars. Even if their individual habitable zones are modest, there are enough M stars to host a large number of inhabited worlds. For years, this optimistic thought was moderated by two objections. First, planets orbiting close to a star get locked into synchronous rotation [Section 8.1], with one side of the

---

[3] The boundaries of the habitable zone are also affected to some extent by the star's surface temperature, which determines the wavelength at which it puts out most of its light.

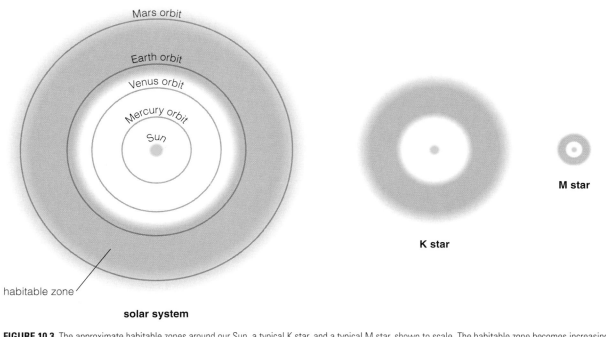

solar system

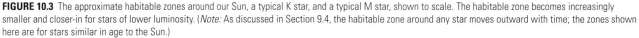

FIGURE 10.3 The approximate habitable zones around our Sun, a typical K star, and a typical M star, shown to scale. The habitable zone becomes increasingly smaller and closer-in for stars of lower luminosity. (*Note:* As discussed in Section 9.4, the habitable zone around any star moves outward with time; the zones shown here are for stars similar in age to the Sun.)

planet perpetually facing the star (much as the Moon always keeps one face turned to the Earth). Once this happens, the side away from the star becomes perpetually dark, and the atmosphere might be expected to freeze out. Second, small stars have frequent and energetic flares—sudden bursts of intense light and radiation—that might cook any complex life.

Recent research, however, suggests that neither of these objections is necessarily fatal. Even a modestly thick atmosphere containing carbon dioxide would circulate heat from the bright to the dark side on a synchronously locked world, keeping temperatures relatively uniform and allowing liquid water to exist over much of the planet's surface. An interesting thought is that the central star, viewed from such a synchronously rotating planet, would always be in the same position in the sky. This might be advantageous for plants, which could align themselves to keep their "leaves" always perpendicular to the incoming rays.

As for flares, their dangerous ultraviolet light might actually cause the production of a protective layer of atmospheric ozone, in much the way we have described for worlds orbiting stars of types A and F. And, of course, underwater or underground life would be protected in any case. The bottom line is that M stars probably can support life. If they do, given the fact that they live so long, these stars might possibly house the galaxy's oldest biology.

In summary, we have good reason to believe that the vast majority of stars are, at least in principle, capable of harboring habitable planets. However, the most common stars—small, dim stars of spectral type M—may have such small habitable zones that planets capable of supporting life could be quite rare among them.

Multiple Stars

There's another important complicating factor that we have not yet considered: Roughly half of all stars are not lone stars like the Sun but rather are members of **multiple star systems,** in which two or more stars orbit each other closely. For example, the nearest star system to our own, Alpha Centauri, is actually a triple star system. The largest and brightest stellar component is a type G star like our Sun, and the second-largest is a K star that is separated from the G star by roughly the distance between our Sun and the planet Uranus. The third star orbits at a far greater distance—about one-fifth of a light-year—swinging around the two central stars in an orbit that takes many thousands of years. Can multiple star systems have habitable planets? The answer to this question hinges on whether a planet in a multiple star system can have a stable orbit that keeps it within a habitable zone.

Let's consider planetary orbits in **binary systems,** systems with just two stars. (Systems with three or more stars are much less common.) Broadly

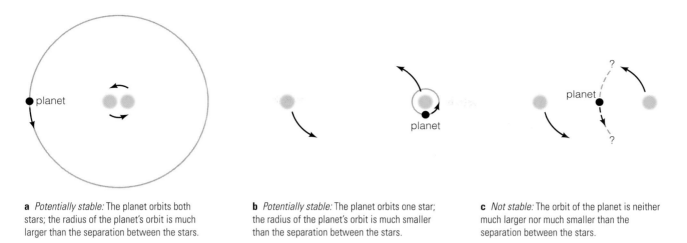

a *Potentially stable:* The planet orbits both stars; the radius of the planet's orbit is much larger than the separation between the stars.

b *Potentially stable:* The planet orbits one star; the radius of the planet's orbit is much smaller than the separation between the stars.

c *Not stable:* The orbit of the planet is neither much larger nor much smaller than the separation between the stars.

FIGURE 10.4 Three orbital possibilities in a binary star system.

speaking, we can consider three types of orbital situations in a binary star system, of which two are potentially stable:

- *Potentially stable:* A planet could orbit around the two stars together (Figure 10.4a). If the planet orbits the stellar pair at a distance considerably greater than the separation between the two stars, then gravitationally the two stars act much as one, and the planet can orbit without disruption around its distant twin masters. Computer simulations indicate that a stable orbit is possible for planets that orbit at a distance of more than about five times the separation of the stars. (The exact distance depends on the details of the system, such as the relative masses of the two stars and the eccentricity of their orbits.)

- *Potentially stable:* A planet could orbit one star or the other (Figure 10.4b). If the two stars are themselves widely separated, then a planet near either one can orbit steadily because it will feel little disturbing effect from the second star. Computer simulations indicate that stability becomes possible when the two stars are separated by more than about five times the planet's orbital distance.

- *Not stable:* If the distance between a planet and a star is not sufficiently different from the distance between the two stars, its orbit will not be stable (Figure 10.4c). The planet will experience competing gravitational tugs from both stars that will ultimately fling it out of the system (or send it crashing into one of the stars).

As an example of the first stable case, imagine that our Sun had a companion star at Jupiter's distance (which is 5.2 times that of Earth's distance). In that case, our planet's orbit (as well as those of Mercury and Venus) would be stable. We would also have a second sun to observe in the sky, providing extra research opportunities for astronomers. As an example of the second case, imagine that our Sun had a companion star at the Earth's position. The orbits of the inner planets would not be stable, but Jupiter's orbit could be.

Given these orbital possibilities, what can we say about habitability in binary star systems? Interestingly, both stable cases seem to allow for habitable planets. In the first case, the habitable zone would be some region surrounding the two stars together. In the second case, the more distant star would probably have little influence on the planet's climate, so the habitable zone would be defined solely by the star that the planet orbits. Similar conclusions probably apply to other multiple star systems.

Summary of Stellar Habitability

We began this section with the goal of determining what types of stars might potentially have habitable planets. Perhaps surprisingly, we have found relatively few limits. The only stars that we can rule out completely—the O and B stars—are very rare. G stars like our Sun clearly can have habitable planets. Even though only 7% of the stars in our galaxy are G stars, 7% of a few hundred billion stars is still an impressive tally. The small size of the habitable zone around the most common types of stars—dim, low-mass K and M stars—suggests that habitable planets may be rare around these stars, but the great abundance of such stars might compensate. Even multiple star systems seem reasonably capable of having habitable planets.

All in all, our studies of stars suggest that many or most stars could *potentially* have orbiting, habitable worlds. Whether they *do* may depend on a number of other factors. To understand these factors, we must investigate the discoveries of extrasolar planets and what these discoveries have taught us about the formation and nature of other planetary systems.

10.3 Extrasolar Planets: Discoveries and Implications

The discovery of extrasolar planets has dramatically improved our understanding of the issues surrounding the possibility of finding habitable planets around other stars. In this section, we will investigate what we have learned. We begin by discussing the techniques we can use to find extrasolar planets. We'll then move on to discuss the somewhat surprising nature of many of the planets we have found, as well as the implications of these findings.

Detecting Extrasolar Planets

The first clear-cut discovery of a planet around another Sun-like star—a star called 51 Pegasi—came in 1995. The discovery was made by Swiss astronomers Michel Mayor and Didier Queloz and was soon confirmed by a team led by Geoffrey W. Marcy and R. Paul Butler. Many more extrasolar planets have been discovered since that time, most by the team led by Marcy and Butler. Several different planet-hunting strategies have been used or proposed. If we strip away the details, however, there are really only two basic ways to search for extrasolar planets:

1. In principle, we could search for *direct* evidence of extrasolar planets by attempting to make pictures or spectra of the planets themselves. However, current technology is not capable of such direct searches.

2. Until technology makes direct searches possible, we can rely on indirect searches. Such searches look only at stars, seeking effects that can be attributed to orbiting planets.

Because all extrasolar planet discoveries to date have been indirect, we'll investigate these techniques first and then consider direct techniques.

The most common indirect techniques look for the gravitational tug that a planet exerts on its star. Although we usually think of a star as remaining stationary while planets orbit around it, this is only approximately correct. In reality, all the objects in a star system, including the star itself, orbit the system's **center of mass.** To understand this concept, think of a waiter carrying a tray of drinks. To carry the tray, he places his hand under its balance point—its center of mass. If the tray has a heavy glass of water off to one side, he will place his hand a little to that side of the tray's center. Because the Sun is far

more massive than all the planets combined, the center of mass of our solar system lies very close to the Sun—but not exactly at the Sun's center. In fact, it is located just outside the Sun's visible surface.

If you could watch from far away, you would see the Sun orbit the solar system's center of mass. You would find that the Sun orbits approximately once every 12 years but that its orbit is not regular—it loops around the center of mass in a complex way (Figure 10.5). What causes this behavior? The answer is that the Sun responds to the gravitational tugs of the planets. Jupiter, being more massive than all the other planets combined, has by far the largest effect. The Sun's basic 12-year orbit around the center of mass mirrors Jupiter's own 12-year orbit around the Sun. Thus, an extraterrestrial observer could infer the existence of Jupiter by noticing how the center of the Sun moves around the center of mass of the solar system. Saturn is the next most massive planet, so the Sun's complex path also shows the effects of Saturn's 27-year orbit. With long and careful measurement, a distant observer could discover the existence of both Saturn and Jupiter. Indeed, with patience and sufficient measurement precision, extraterrestrial astronomers could discover all of the planets in our solar system. But, as we've seen, it's far

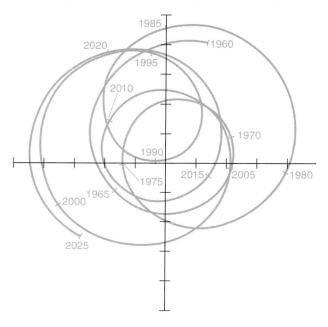

FIGURE 10.5 This diagram shows the orbital path of the Sun—that is, its orbit around the center of mass of the solar system—for the years 1960 to 2025 as it would appear from a distance of 30 light-years away if we were looking face-on at the plane in which the planets lie. The Sun's motion is caused by the gravitational tugs of its planets, and most of the motion is due to the two most massive planets, Jupiter and Saturn. The tick marks are angular distances in units of 0.0002 arcseconds. For comparison, the Hubble Space Telescope can see detail no smaller than about 0.1 arcsecond.

Planets were recognized in ancient times by their motions relative to the fixed stars of the constellations. As "wanderers" in the sky, they were clearly different from most other objects. After the Copernican revolution, they were understood to be large, spherical worlds orbiting the Sun. The idea of a planet then seemed so obvious that no one had to spend time explaining it.

Not so today. New discoveries of surprisingly small stars, massive extrasolar planets, and lonely worlds apparently drifting through interstellar space have complicated things. These new findings have prodded astronomers to think hard about what they mean when they call something a planet.

The major difficulty in defining planets is in deciding how big or how small they can be. At the small end, we have Pluto, but its status as a planet is dicey. This frigid, icy world, with its highly eccentric tilted orbit, seems strangely out of place among the planets of the outer solar system, where its planetary neighbors are the giant jovian planets. As we discussed in Chapter 6, Pluto is most likely an icy remnant from that distant time when the solar system was formed—perhaps the largest of many similar objects (though larger ones may yet be discovered) orbiting the Sun in the so-called Kuiper belt that lies beyond Neptune's orbit. None of these other Kuiper belt objects are considered planets, and it's doubtful that Pluto would have been considered a planet if it had been discovered recently rather than back in 1930.

How big can a planet be? There are gas balls larger than Jupiter but much smaller than the Sun that shine weakly in visible light. Sometimes called "failed stars" or "intermediate stars," these objects are more popularly known as *brown dwarfs* (see Figure 9.6). Because they are so dim, the first one was discovered only recently, in 1995. Unlike Jupiter, brown dwarfs are massive enough to produce some limited energy by fusing deuterium (hydrogen with a neutron in addition to its proton) in their cores. Theoretical calculations suggest that any object in the range of about 13–75 times Jupiter's mass will have the properties of a brown dwarf. Objects more massive than this will crush their interiors to the point of starting the hydrogen fusion reactions we find in the Sun and other stars. Some of the extrasolar "planets" discovered to date may actually be brown dwarfs. So should we consider them "planets" at all? Most astronomers answer no, not only because of the objects' large size but also because they are thought to form more like stars than like planets. Indeed, most brown dwarfs probably float freely in interstellar space, having been formed from collapsing clouds of gas that did not have enough mass to create a "real" star. Those found orbiting another star may have formed in much the same way as the second star in a binary star system, but with too little mass to shine as a star.

Perhaps even more challenging to our definition of a planet are objects that are clearly planetary in size but are found drifting in interstellar space, unattached to *any* star. Many or most of these objects may at one time have been planets orbiting a star and may now occupy interstellar space because they were kicked out of their stellar nests by gravitational interactions with sister worlds. Does their exile mean they are no longer planets? It seems a bit unfair, but most astronomers today are inclined to answer yes, preferring to reserve the term "planet" for bodies that orbit stars. Nevertheless, these objects have sometimes been called "free-floating planets" both by the astronomers who discovered them and in news accounts of their discovery.

The bottom line is that, for the moment at least, there is no official definition of the term "planet." In astronomy, official names and definitions are generally decided by the members of the International Astronomical Union (IAU). At a recent IAU meeting, the experts decided to hedge their bets and await further discoveries and new understandings. Meanwhile, a good working definition seems to be that a planet is a round object that orbits a star and is at least as big as Pluto and smaller than a brown dwarf. The IAU committee figures that should hold us for now.

easier to detect large planets like Jupiter and Saturn than small planets like Earth.

In a similar way, we can search for planets in other star systems by carefully watching for periodic wobbles in a star's position in the sky. There are two techniques for doing this. The first is straightforward: We make very precise measurements of a star's position to see if it is, in fact, moving slightly back and forth in the sky. The second technique is to measure a star's speed toward or away from us by looking at its spectrum for something called the *Doppler effect* (which we'll discuss shortly); variations in the star's speed may indicate motion caused by the effects of orbiting planets. Let's investigate these and other techniques in a bit more detail.

The Astrometric Technique (Indirect) The *astrometric technique* involves making very precise measurements of stellar positions in the sky (*astrometric* means "measurement of the stars"), that is, it essentially looks for motion like that shown in Figure 10.5. This technique has, in fact, been used for many decades to identify binary star systems, because two orbiting stars will move periodically around their center of mass. The technique works well for many binary systems, especially those in which the two stars are not too close together, because the stellar motions can be substantial. In the case of planet searches, however, the expected stellar motion is much more difficult to detect.

As a (sobering) example, from a distance of 10 light-years a Jupiter-size planet would cause a

Sun-like star to dance slowly over a side-to-side angular distance of only about 0.003 arcsecond—approximately the width of a hair seen from a distance of 5 kilometers. Remarkably, with careful telescope calibration astronomers can now measure displacements this small, and instruments now under development will be 5–10 times better. However, two complications add to the difficulty of utilizing the astrometric technique.

First, the farther away a star is, the smaller its side-to-side movement will appear. For example, while Jupiter causes the Sun to move about 0.003 arcsecond as seen from 10 light-years away, the observed motion is only half as large when seen from 20 light-years away. Thus, the astrometric technique works best for massive planets around relatively nearby stars.

The second complication involves the time required to detect a star's motion. Obviously, it is easier to detect larger side-to-side movements than smaller ones, and a planet with a larger orbit has a larger effect on its star. To understand why, consider what would happen if Jupiter moved farther out from the Sun. Because the center of mass of the solar system is very nearly at the balance point between the Sun and Jupiter, moving Jupiter outward would also cause the center of mass to move farther from the Sun. With the center of mass located farther from the Sun, the Sun's orbit around the center of mass—and hence its side-to-side motion as seen from a distance—would be larger. Thus, the astrometric technique works best not only for massive planets, but for massive planets that are far from their stars. However, a more distant planet takes longer to complete its orbit (in accord with Kepler's third law), which means the stellar motion it causes also takes longer. For example, whereas Jupiter causes the Sun to move around the center of mass with a 12-year period, Neptune's effects on the Sun show up with the 165-year period of Neptune's orbit. It would probably take a century or more of patient observation to be confident that stellar motion was occurring in a 165-year cycle.

Because of these complications, the astrometric technique has produced only one of the nearly 80 discoveries of extrasolar planets made by the beginning of 2002. All the others were found using the Doppler technique (which we'll describe next). But we expect the astrometric technique to prove much more valuable in the future. One of the principal goals of a NASA mission called the *Space Interferometry Mission* (SIM), which is under development for a launch around 2009, is to use the astrometric technique to search for planets around nearby stars. If it works as currently planned, SIM will be able to measure stellar positions in the sky to a precision of 1 *micro*arcsecond (one-millionth of an arcsecond), thousands of times better than our current techniques and equivalent to being able to look from Earth to measure the thickness of a dime on the Moon. With this precision, SIM should be able to detect the stellar wobbles caused by Earth-size or just slightly larger planets around the nearest dozen stars and those of Jupiter-size planets orbiting stars as far as 3,000 light-years away.

The Doppler Technique (Indirect) The *Doppler technique* offers another way to find the small movement of a star caused by orbiting planets. Rather than looking for the side-to-side motion of a star against the background of the sky, the Doppler technique involves studying a star's spectrum to look for telltale signs that the star is moving around a center of mass. The technique is based on searching for small shifts in the wavelengths of lines in the spectrum caused by what we call the **Doppler effect.**

You've probably noticed the Doppler effect on sound; it is especially easy to recognize when you stand near train tracks and listen to the whistle of a train (Figure 10.6). (You can also notice the Doppler effect with emergency sirens or even just with the "buzz" of a fast car as it goes past you.) As the train approaches, its whistle is relatively high pitched; as it recedes, the sound is relatively low pitched. Just as the train passes by, you hear the dramatic change from high to low pitch—a sort of "weeeeeeeee-oooooooooh" sound. You can visualize the Doppler effect by imagining that the train's sound waves are bunched up ahead of it, resulting in shorter wavelengths and thus the high pitch you hear as the train approaches. Behind the train, the sound waves are stretched out to longer wavelengths, resulting in the lower pitch heard as the train recedes.

The Doppler effect causes similar shifts in the wavelengths of light emitted by a star. If a star is moving toward us, then its entire spectrum—including the spectral lines within it—is shifted to shorter wavelengths (the spectral lines serve as reference lines for measuring the shift). Because shorter wavelengths of visible light are bluer, the Doppler shift of a star coming toward us is called a **blueshift.** If a star is moving away from us, its light is shifted to longer wavelengths; we call this a **redshift,** because longer wavelengths of visible light are redder. The faster the star is moving, the greater the amount of its blueshift or redshift.

The Doppler effect can be used to find planets in the following manner: If an orbiting planet causes its star to move alternately slightly toward and away from us, the starlight should shift alternately toward the blue and toward the red (Figure 10.7a). The 1995 discovery of a planet orbiting 51 Pegasi came when this star was found to be moving with a rhythmic wobble corresponding to an orbital speed of 53 meters per second (Figure 10.7b). The main diffi-

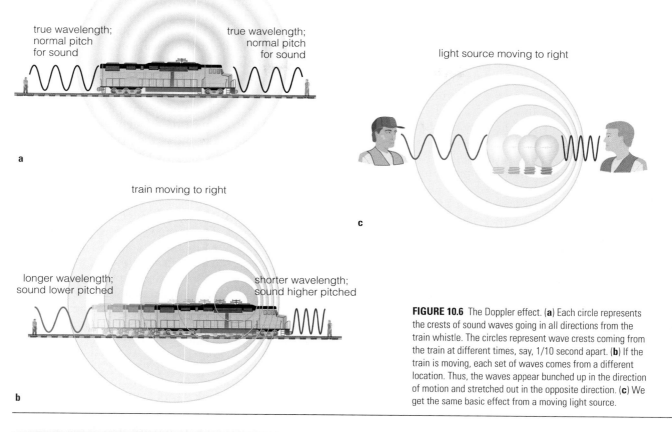

train stationary

true wavelength; normal pitch for sound

true wavelength; normal pitch for sound

a

light source moving to right

c

train moving to right

longer wavelength; sound lower pitched

shorter wavelength; sound higher pitched

b

FIGURE 10.6 The Doppler effect. (**a**) Each circle represents the crests of sound waves going in all directions from the train whistle. The circles represent wave crests coming from the train at different times, say, 1/10 second apart. (**b**) If the train is moving, each set of waves comes from a different location. Thus, the waves appear bunched up in the direction of motion and stretched out in the opposite direction. (**c**) We get the same basic effect from a moving light source.

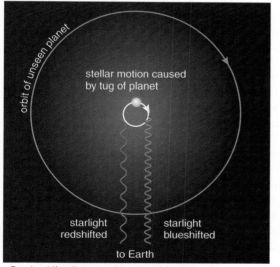

orbit of unseen planet

stellar motion caused by tug of planet

starlight redshifted

starlight blueshifted

to Earth

a Doppler shifts allow us to detect the slight motion of a star caused by an orbiting planet.

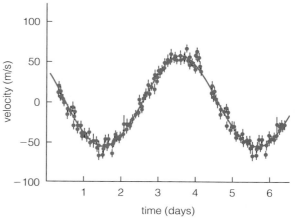

b A periodic Doppler shift in the spectrum of 51 Pegasi shows the presence of a large planet with an orbital period of about 4 days. Dots are actual data points; bars through dots represent measurement uncertainty.

FIGURE 10.7 The Doppler technique for discovering extrasolar planets.

culty with this method is that, in most cases, the shifts are very slight—much less than that found for 51 Pegasi. For example, Jupiter's gravitational pull causes the Sun to orbit around their common center of mass at a speed of only about 12 meters per second, about the speed of an automobile driving through a residential neighborhood. This represents a wavelength shift of only one part in 25 million (determined by the ratio of the Sun's orbital speed to the speed of light)—a very difficult shift to measure.

But such measurement is no longer impossible. Finding very slight Doppler shifts became feasible when astronomers started attaching special "iodine cells" to their telescopes. The cells are glass tubes filled with a small amount of iodine gas through which starlight passes on its way to a spectroscope. The result is a spectrum showing both the spectral lines from the star itself and the spectral lines from the iodine gas. Because the iodine lines will always be at the same position in the spectrum (since the cell is not moving with respect to the telescope), they make extremely precise reference points against which we can measure Doppler shifts of the star's spectral lines. It is rather like standing still in your backyard and looking through a picket fence at a car in the distance. Since you know the fence is stationary, any shift in the car's position relative to the fence must mean that the car is moving. Of course, astronomers looking at other stars are not standing still —the Earth is rotating, orbiting around the Sun, and so on. But Earth's motions are well known and can be removed from any observed Doppler shift, allowing researchers to decide when they've detected real stellar motion. By the late-1990s, astronomers were using these iodine cells to hunt for back-and-forth motions as slow as 10–15 meters per second.

Unlike the case with the stellar shifts measured through the astrometric technique, the observed orbital speeds of stars do not depend on the star's distance. Thus, as long as we can get clear spectra—which is to say as long as the star is relatively bright or we use a large enough telescope—we can discover extrasolar planets as easily around distant stars as around nearby stars. However, the Doppler technique has some limitations. In particular, whereas the astrometric technique works best for massive planets that orbit far from their stars (assuming you have enough patience to wait for the star to complete an orbit of many years), the Doppler technique works best for massive planets that are *close* to their stars. This is because gravity weakens with distance, so a planet of a given size pulls harder on its star—making the star move faster—if it is closer. Moreover, because a close-in planet has a much shorter orbital period than a farther out planet, it takes much less time to discover close-in planets even when both situations are measurable in principle. For example, the planet around 51 Pegasi has an orbital period of just 4 days, so observation over a few weeks allows us to record many orbital cycles. In contrast, even though the Doppler effect of a planet in a 12-year orbit like that of Jupiter would be measurable, we would not yet have had time since 1995 to observe even a single complete orbit.

A second limitation is that the Doppler technique measures only part of a star's wobble—the component of a star's motion that is directly toward or away from us. Suppose a planet is causing a star to move in a circular orbit. If the orbit is viewed pole-on, then the star will trace out a circle in the sky but will have no motion toward or away from us and therefore no cyclical Doppler shift (Figure 10.8a). We see a Doppler shift only if the orbit is tilted from our point of view, in which case the Doppler shift tells us the speed of the star directed toward or away from us but not its speed across our line of sight (Figure 10.8b). The Doppler shift tells us the star's full orbital velocity only if we happen to be viewing the orbit edge-on. Thus, the size of a measured Doppler shift depends on the inclination, or tilt, of the planetary system. Because we usually do not know the orbital inclination of an extrasolar planet we discover, the speed of stellar motion that we measure is usually a *lower limit* on the star's actual speed. This has an important effect on our measurement of the masses of orbiting planets. Because more massive planets cause greater stellar motion (for a given orbital distance), the speed of a star's movement can be used to calculate the mass of the orbiting planet responsible for that movement. Thus, because the Doppler technique gives us a lower limit on the star's actual speed, the mass we compute for the orbiting planet is also a minimum value. The real mass could be considerably higher (though rarely by more than a factor of about 2).

Another thing we can learn from the Doppler technique is the shape of the planet's orbit. For a circular orbit, the Doppler shift will increase and decrease in a smooth, symmetric way. If the planet is in a highly elliptical orbit, then it will speed up when it nears its sun and slow down when it is farther out, producing a more irregular change in the observed Doppler shifts (Figure 10.9).

The Doppler technique has been remarkably fruitful. Nearly all the extrasolar planet discoveries to date have been made with this technique. In addition, analysis of the best Doppler shift data can sometimes reveal motions caused by more than one planet. In 1999, such analysis was used to infer the existence of three planets around the star Upsilon Andromeda, making this the first bona fide multiple-planet solar system known beyond our own. (Others have since been discovered.)

*Transits (**Indirect**)* A third indirect way to detect distant planets is, in principle, even simpler than the two discussed so far. If we were to scrutinize a large sample of stars with planets, a small number of them —typically one in several hundred—will by chance be aligned in such a way that one or more of the orbiting planets passes directly between us and the star. The result is a **transit** in which the planet appears to move across the face of the star. We witness this effect in our own solar system when Mercury or Venus crosses in front of the Sun. The result, seen through a telescope, is a small black dot transiting

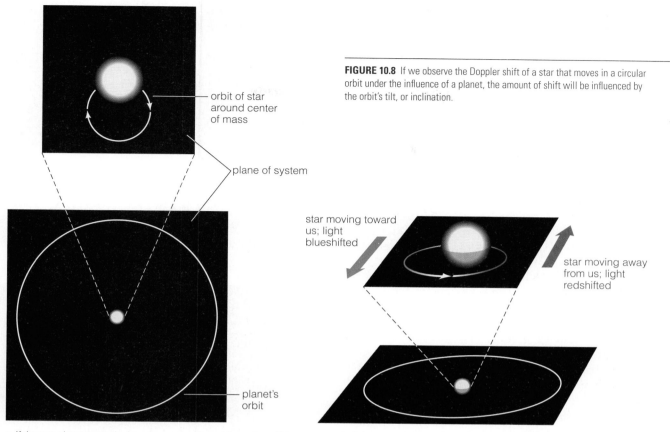

— orbit of star around center of mass

— plane of system

— planet's orbit

a If the star–planet system is viewed pole-on, no to-and-fro motion will be observed, and therefore no changing Doppler shift.

FIGURE 10.8 If we observe the Doppler shift of a star that moves in a circular orbit under the influence of a planet, the amount of shift will be influenced by the orbit's tilt, or inclination.

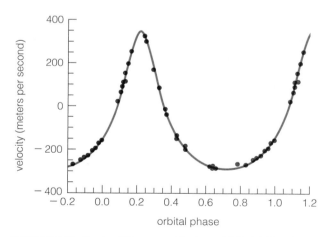

star moving toward us; light blueshifted

star moving away from us; light redshifted

b If the system is tilted, the Doppler shift tells us the portion of the star's speed directed toward or away from us. Thus, the Doppler shift for a tilted orbit gives us an underestimate of the star's full orbital speed. (We measure the full orbital speed only when we see the orbit edge-on.)

the Sun's bright disk (Figure 10.10). This effect is impossible for us to witness for other stars, which appear in our telescopes only as points of light. However, as the planet crosses the star's face, it blocks a little of the star's light. The resulting tiny decrease in the star's brightness may be measurable.[4]

THINK ABOUT IT . . . *Explain why transits are possible only for systems whose planetary orbits happen to be aligned edge-on as viewed from Earth.*

For example, if we examined another planetary system that is edge-on, we could see the brightness of its star decrease by about 1% during a few hours as a Jupiter-like world transited. Indeed, we have already observed transits caused by one large extrasolar planet, although the planet was first discovered

[4] Other processes besides planetary transits can cause periodic dimming in a star's light (for example, a long-lived star spot). Thus, transit observations can be ambiguous in some cases and must be conducted with great care and be repeated over long periods to give us confidence in their validity.

FIGURE 10.9 The Doppler shifts measured for the star 70 Virginis (from the work of Marcy and Butler). Note the uneven nature of the change in velocity, telling us that the planet causing the Doppler shift is in a highly eccentric (elliptical) orbit.

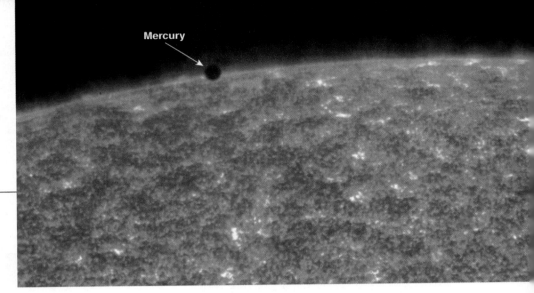

FIGURE 10.10 NASA's *TRACE* satellite captured this image of a Mercury transit on November 15, 1999. The photograph was taken with ultraviolet light, so the colors are not real; the structures seen with ultraviolet light are patches of hot gas just above the Sun's visible surface. Because it blocks a little bit of the Sun from view, the transit causes a very small temporary drop in the Sun's brightness. In principle, we should be able to detect such drops in brightness when extrasolar planets cross in front of their stars.

using the Doppler technique. The planet, which orbits a star called HD209548, transits its star every 3½ days, causing a 1.7% drop in the star's brightness for about 2 hours (Figure 10.11). This star is bright enough to be seen easily with a small telescope, and a 1.7% change in brightness is large enough to measure with an inexpensive "photometer." Thus, with equipment costing no more than a few hundred dollars, you can see for yourself clear evidence of a planet around another star.

For an Earth-size planet, the dimming caused by a transit would be only about 0.01% of a star's normal brightness, an amount that is difficult to measure with ground-based telescopes affected by atmospheric turbulence. Space-based telescopes, however, could uncover such subtle dimming. NASA already has a mission in development, called Kepler, that is designed to look for transits caused by Earth-size (or even smaller) planets through careful monitoring of the brightness of 100,000 stars during the course of 4 years. With its launch planned for 2007, Kepler will tell us whether or not solar systems in which small planets exist are plentiful. It will provide our first good inventory of extrasolar planetary systems.

Direct Techniques While upcoming missions like Kepler or the Space Interferometry Mission (SIM) may find a few Earth-*size* planets by using indirect techniques, they still won't tell us whether these planets are Earth-*like* in terms of habitability. For that we need to study the planets directly.

To see why direct detection of extrasolar planets is so difficult, recall the scale of our solar system. As we discussed in Chapter 1, seeing the Earth from the nearest star beyond the Sun (Alpha Centauri) would be like looking from San Francisco to see a pinhead that orbits a grapefruit in Washington, D.C. Moreover, it would be a really dim pinhead only about 15 meters away from a really bright grapefruit! Seeing

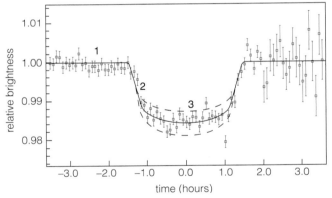

FIGURE 10.11 Careful measurements of the brightness of the star called HD209548 revealed that an orbiting planet passes directly in front of it as seen from Earth, which means that the planet's orbit must be edge-on as seen from Earth. (**a**) Artist's conception of the planet as it passes directly in front of its star as seen from Earth. (**b**) Data showing the 1.7% drop in the star's brightness that proved the planet is passing in front of the star as seen from Earth.

Jupiter would not be much easier; on the same scale, it would be a marble about 80 meters from the Sun. To put this idea into a more technological context, consider trying to see Jupiter from a distance of 10 light-years. In theory, at this distance Jupiter would

FIGURE 10.12 This artist's rendering shows one possible configuration for NASA's planned Terrestrial Planet Finder (TPF). It consists of four separate telescopes flying freely in space. The central spacecraft performs the interferometry by combining the light from all four telescopes.

appear as a point of light one-billionth as bright as the Sun and separated from the Sun by only 2 arcseconds. Unavoidable optical effects (diffraction) in the telescope would wash much of the Sun's light over the dim pinpoint that is Jupiter, and even without this effect microscopic roughness in the telescope mirror would cause random brightness variations that would render the planet invisible.

One way to improve on this discouraging situation is to observe at infrared wavelengths. Planets are too cool to emit any visible light of their own, so they shine only by the light they reflect from their stars, but they emit reasonable amounts of infrared light. They are still far dimmer than their stars, but whereas the Sun is a billion times brighter than Jupiter in visible light, it is only about 100,000 times brighter in the (far) infrared. This substantial improvement makes direct detection much more practical with infrared light than with visible light. In addition, scientists are developing new technologies such as *nulling,* which requires using two or more telescopes together as an interferometer [Section 6.2]. Nulling can be used to cancel out most of the bright light from the central star of a distant solar system. With the star blotted out, seeing the faint dots that mark its accompanying planets will be far easier. NASA plans to test nulling technology on the Space Interferometry Mission (SIM), though this mission is not expected to be able to image Earth-size planets.

NASA's planned follow-up mission to SIM, the Terrestrial Planet Finder (TPF), is tentatively envisioned as four 3.5-meter telescopes working together as an infrared interferometer with nulling capabilities (Figure 10.12). With luck, TPF may be launched as early as 2012. The European Space Agency (ESA) is considering a similar mission, called Darwin, on roughly the same time scale. (NASA and ESA are discussing ways to combine the two missions, possibly with other international partners as well.) In principle, either TPF or Darwin should have the power to conduct a conclusive search for Earth-size planets around 150 nearby, Sun-like stars. If the planets are there, these missions should *see* them (unlike, for example, the Kepler mission, which will simply infer their presence from the short-lived dimming they cause during transits of their stars). The images will be very low resolution—perhaps no more than three or four dots (pixels) for a typical planet— but the collected light should be sufficient to allow us to obtain reasonable spectra of the planets. As we'll discuss in the next section, spectroscopy may allow us to determine whether a planet is habitable, and perhaps even whether it has life.

Later in this century, astronomers hope for even more powerful interferometers either in space or on the surface of the Moon. It's certainly possible that, within the lifetimes of today's college students, we'll have optical interferometers with dozens of telescopes spread across hundreds of kilometers. Such telescopes will not only be able to detect Earth-like planets around nearby stars but also to obtain fairly clear images of those planets. We'll see whether they have continents and oceans like the Earth and perhaps be able to watch seasonal changes. Just a few centuries ago, we knew so little of the heavens that our ancestors could still imagine planets simply as "wandering stars." A mere decade ago, we had good images of the planets in our solar system but still had no conclusive evidence regarding the existence of planets elsewhere in the universe. Before the current century is out, we may have arguably better knowledge of planets around distant stars than the ancient Greeks had of our own Earth.

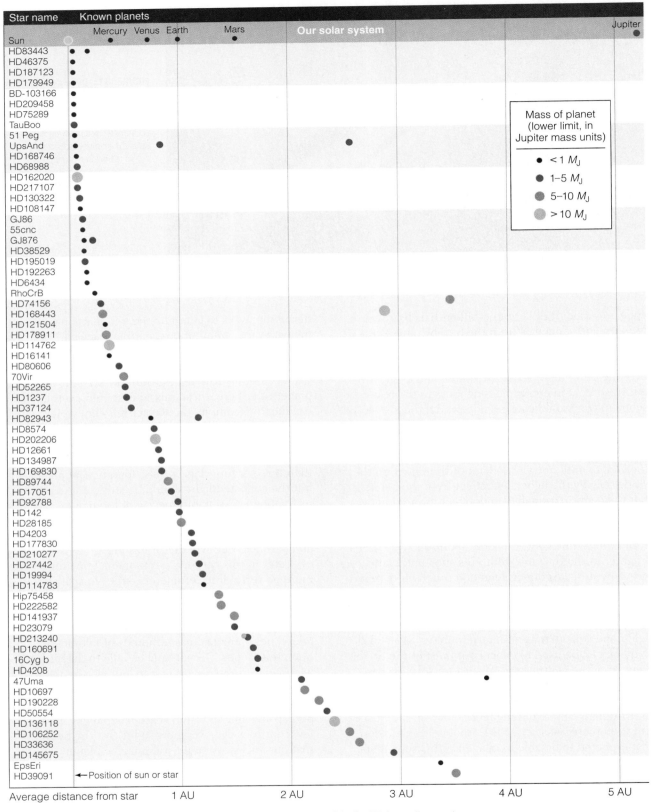

FIGURE 10.13 This diagram shows the orbital distances and approximate masses of the first 77 planets discovered around other stars. Most of the planets found so far are closer to their stars and more massive than the planets in our solar system. (Planet sizes are not to scale.)

The Nature of Extrasolar Planets

Given our discussion of planet detection techniques, you shouldn't be surprised by the fact that all the planets discovered to date around hydrogen-burning stars are far larger than the Earth. However, scientists have been very surprised by the orbital characteristics of many of these planets.

Figure 10.13 shows the approximate masses and average orbital distances for the first 77 planets discovered around such stars, with planets of our own solar system shown for comparison in the top line. Figure 10.14 compares the eccentricities (a technical measure of how much the orbit differs from a circle) of the planetary orbits to their distances, with the giant planets of our own solar system shown for comparison. If you study these figures carefully, you'll notice the following important facts:

- While all the discovered planets have masses similar to or greater than that of Saturn or Jupiter, many orbit very close to their central stars—some are closer to their star than Mercury is to the Sun (and none yet found are as far from their star as Jupiter is from the Sun).

- Most of the planets that are not close-in to their stars are in highly elliptical orbits, quite unlike the nearly circular paths followed by all the planets of our own solar system except Pluto. This means that the planets do not maintain nearly constant distances from their stars but instead are much closer to them on one side of each orbit than on the other.

- Several stars have more than one planet, proving that systems of planets exist.

Recall that our theory of solar system formation suggests that massive, jovian planets should form only in the cooler, more distant regions of the swirling disk around a young star (see Figure 1.23). In the warm, inner regions of the disk, we expect to find rocky, terrestrial worlds. Thus, the presence of massive planets so close to their stars came as a big surprise to astronomers.

Could it be that these heavyweights are very large rocky worlds? It's difficult to imagine how terrestrial planets could become so big, though we can't absolutely rule this out. However, in the case of the transiting planet around HD209548 (see Figure 10.11), we can be quite certain that the planet is jovian in nature. Because a transit can occur only if

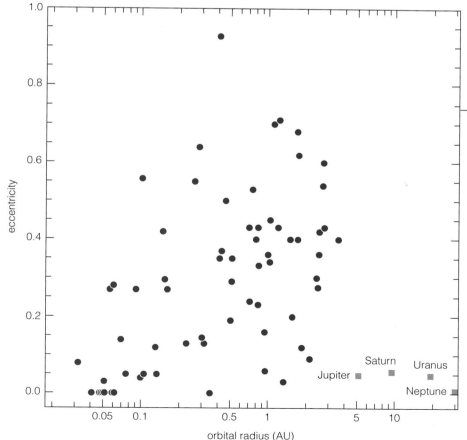

FIGURE 10.14 This graph shows the orbital radii (average distance from star) and eccentricities for the extrasolar planets discovered as of 2001. (Eccentricity is a measure of how much an orbit differs from a circle; it is 0 for a perfect circle and approaches 1 for highly elliptical orbits.) For comparison, also plotted (as squares) are the giant planets of our own solar system—Jupiter, Saturn, Uranus, and Neptune. Note that the planets in our own solar system have nearly circular orbits, but extrasolar planets tend to have much more elliptical orbits—especially those planets that are not very close to their parent star.

FIGURE 10.15 Artist's impression of the massive planet orbiting the star 51 Pegasi. At an orbital radius eight times closer to its star than Mercury is to the Sun, the temperatures on this planet are expected to be 1,000°C or greater. Planets like this are called "hot Jupiters."

we are looking at the orbit edge-on, in this case we know the planet's precise mass, not just a lower limit. In addition, because the star dims by 1.7% during a transit, we know that the planet must be just large enough to block 1.7% of the star's visible area from our view—which means that we can reliably estimate the planet's radius. From the mass and radius we can calculate the planet's average density, and it turns out to be roughly the same as the average density of Saturn. Consequently, we are confident that at least one of the close-in, but massive, planets is indeed a gas giant, and we have every reason to suspect that the same is true for the other known extrasolar planets. Because many of these planets are Jupiter-like in size but Mercury-like in orbital distance, they are often referred to as "hot Jupiters" (Figure 10.15).

Could "hot Jupiters" be habitable? It seems extremely unlikely. Their atmospheres would probably have the same strong vertical turbulence that led us to decide that Jupiter probably could not give rise to life [Section 6.3], and their proximities to their stars and elliptical orbits might only exacerbate their problems. Nevertheless, it's possible that these planets have numerous large moons, as do the jovian planets in our solar system. Some of these moons could be habitable, especially around the extrasolar planets that orbit within the habitable zones of their stars.

New Ideas About Solar System Formation

If our theory of solar system formation tells us that jovian planets can form only in the outer regions of the disks that surround newborn stars, then how can we explain the existence of "hot Jupiters" or the eccentric orbits followed by so many of these worlds? After all, the orbits in any planetary disk should tend to become circular due to an "averaging out" of the numerous collisions involved in planetary accretion.

One possibility that scientists must always consider is that something is fundamentally wrong with our model of solar system formation. That is, perhaps jovian planets *can* form in the inner solar system. Upon discovering the "hot Jupiters," scientists indeed began to consider this possibility. However, they failed to find any fundamental flaws in our basic understanding of how planets form. So they considered other ideas and now believe that the unusual orbits followed by most of the known extrasolar planets are the result of changes that occurred in these systems after the planets were born. These planets most likely did form in the outer regions of their solar systems but later migrated inward or were deflected into eccentric orbits.

How might planetary migration occur? Calculations show that friction with the abundant gas and dust in the disks of newly formed solar systems can exert a drag on young planets, causing them to spiral slowly toward their sun. In our own solar system, this drag is not thought to have played much of a role because the gas was cleared out by the young Sun's strong solar wind before it could have much effect [Section 1.3]. But it may be that the wind kicks in later in at least some other solar systems, allowing sufficient time for jovian planets to migrate inward. If they reach a spot close to the star where the disk material has thinned out, then the migration stops, and they settle into star-hugging orbits to become "hot Jupiters." Other planets may not be able to halt their inward migration and ultimately crash into their stars. Indeed, astronomers have noted that some stars have an unusual assortment of elements in their outer layers, suggesting that they may have swallowed planets.

Another possible way to account for some of the observed extrasolar planets invokes close encounters between young jovian planets in the outer regions of a disk. Such an encounter might kick one planet out of the star system entirely while sending the other inward in a highly elliptical orbit. Other possibilities include the idea that a jovian planet could migrate inward as a result of multiple close encounters with much smaller planetesimals (there is evidence that the jovian planets in our solar system have migrated a little bit by this mechanism) or that waves propagating through a gaseous disk could lead to inward migration.

The bottom line is that our discoveries of extrasolar planets have shown us that our prior theory of solar system formation was incomplete. It explained the formation of planets well, and also the simple layout of a solar system like ours. However, it needed modification in order to explain the differing layouts of other solar systems. While we are not yet sure of the precise mechanism, it now seems clear that planets can move from their birthplaces, making possible a much wider range of solar system arrangements than we had guessed before the discovery of extrasolar planets.

THINK ABOUT IT . . . *Look back at the discussion of the nature of science in Section 2.3, especially the definition of a scientific theory. Why does our theory of solar system formation qualify as a scientific theory even though we recently learned that it needed modification? Does this mean that the theory was "wrong" before the new modifications were made? Explain.*

Implications for Habitable Worlds

Planetary migration may have important implications for the question of whether solar systems of the type we have recently found can harbor habitable planets. As we have discussed, it's unlikely that the large, jovian planets themselves could be life-supporting. But couldn't there also be smaller and potentially habitable terrestrial planets in these systems? Observationally we cannot say, because our current telescopes aren't able to detect Earth-size worlds. However, theoretical work suggests that even if terrestrial-size planets were born in these systems, they could face severe problems.

The migration of a large planet is likely to cause significant disruption to the inner solar system. If the migration occurs before terrestrial planets have finished forming, the material that would have accreted onto the terrestrial worlds might be swallowed instead by the larger world. Even if the formation process is essentially complete, gravitational encounters between big planets and small ones nearly always send the smaller ones scattering. As a large planet migrated inward, it would tend to fling less massive planets into its star or into interstellar space. Thus, it seems doubtful that terrestrial planets could remain within the habitable zones of stars orbited by "hot Jupiters" or other massive planets in highly elliptical orbits. Indeed, these same gravitational encounters might also disrupt and disperse any potentially habitable moons.

All in all, the prospects for habitable planets in the extrasolar systems found so far look bleak. On the other hand, the prospects look quite good for habitable planets around stars where planetary migration did not disrupt the inner solar system. Without such disruption, we expect rocky planets to form and to remain on stable orbits, and many of these orbits should be within the habitable zones of their stars. So if we want to know whether habitable planets are common or rare, we need to know which type of solar system is more common: those with substantial migration or those without.

The evidence might seem to point to the systems with substantial migration as being far more common, since so many of the systems discovered to date appear to be of this type. However, there's an important *selection effect* that we must consider. Remember that nearly all of the planets discovered have been found with the Doppler technique and that this technique works best for massive planets in close orbits. Therefore, it is possible that the reason we are finding so many systems with "hot Jupiters" is because they are the easiest ones to find, not necessarily because they are the most common. Of the several hundred Sun-like stars that have been examined so far, approximately one in ten show evidence of one planet or more—which means that we have not detected planets around the majority of Sun-like stars. In many cases, the reason may be that these stars have planetary systems more like our own. As the number of known extrasolar planets continues to grow, we will gain more insight as to whether our solar system's layout is typical. But until we are able to make searches capable of finding small planets with upcoming missions like Kepler, SIM, and TPF, the question remains wide open.

10.4 Signatures of Habitability and Life

We do not yet know for sure whether Earth-size planets even exist around other stars. But if they do, we should begin to find them within the next several decades and get crude images and spectra, thanks to improved technology and new space missions. Let's explore how these capabilities might allow us to determine whether the planets are habitable, and perhaps whether they have life.

The first question for habitability is whether a newly discovered planet is within its star's habitable zone. (A planet outside its star's habitable zone could still harbor life, but probably not on its surface [Section 9.4]. Thus, we would not be able to detect life on such worlds.) This part will be easy, because we know roughly the extent of habitable zones around stars of different types and measuring a planet's orbital properties will tell us whether it is within this zone. But we could learn much more from images and spectra.

Suppose a telescope actually shows us a crude image—only a few pixels—of a distant world. What might we learn? To begin with, by simply watching its changing brightness, we could gain some information about the ratio of ocean to land, because seas are darker than continents. We might also find that the light from such a world changes from day to day, due to clouds, or from season to season, because of snow or ice.

A spectrum would allow us to measure many other properties. For example, even a fairly crude spectrum should allow us to gauge the surface temperature of the planet. If the image of a distant planet is bright enough, we could make a more detailed spectral analysis, one that could suggest far more convincingly that a planet is home to life. For instance, we might gain information on the types of minerals or ices on its surface. If we find that the planet has characteristics like those of Earth or Mars, we would know that it is habitable in principle though perhaps not whether it actually harbors life.

With spectra from infrared telescopes (such as TPF or Darwin), we could search for the absorption or emission features of gases in the atmosphere of a planet. Several of these gases will be easy to detect if they are present, including carbon dioxide, ozone, methane, and water vapor (Figure 10.16). While the mere presence of such gases would not necessarily point to life, their precise abundances and the combinations in which they occur could provide stronger evidence of whether life is present.

For example, Earth's atmosphere has large amounts of oxygen, the result of photosynthesis. If we found abundant oxygen in the air of another world, we would have reason to suspect the presence of life, particularly if the ratio of oxygen to the other detected gases seemed incompatible with nonbiological chemistry. To some extent, the same is true of methane, which is present in Earth's atmosphere today largely thanks to the exhaust gases produced by pigs and cows. In early times, before photosynthesis raised the oxygen level, another metabolic process known as *methanogenesis* may have been widespread on Earth, allowing microbes to expel methane rather than oxygen. The first billion years or so of Earth's biological history might have been marked by the presence of atmospheric methane, and the last 2.5 billion years by oxygen.

The bottom line is that for billions of years life on Earth has been making its presence known to anyone with a telescope large enough to find our world and make a spectrum of its reflected light. In principle, we could identify life on other worlds in the same way. Perhaps in the next few decades we will discover abundant oxygen or methane on a distant world visible to us as no more than a dot in a telescope, providing an exciting and encouraging clue that it harbors life.

10.5 Are Earth-like Planets Rare or Common?

We have discussed the current status of our search for extrasolar planets, along with the question of whether potentially habitable planets should be common or rare. We have found that our current knowledge is still quite incomplete. It may be that a very large fraction of stars have habitable planets, or it may be that the fraction of stars with planets is fairly low. Nevertheless, the example of our own solar system tells us that the fraction is greater than zero. Given the enormous number of stars in the galaxy, it seems highly likely that, one way or another, there are large numbers of Earth-size planets within the habitable zones of stars. But does Earth-*size* necessarily mean Earth-*like* in terms of habitability?

Most scientists would probably answer that, while not all Earth-size planets are likely to be Earth-like, a significant percentage probably will be. However, some scientists have questioned this idea, suggesting that the Earth may be the fortunate beneficiary of several kinds of planetary "luck." According to this recent idea, sometimes called the **rare Earth hypothesis,** the specific circumstances that have made it possible for evolution to progress beyond microbes to complex creatures (such as oak trees or humans) might be so rare that ours might be the only inhabited planet in the galaxy that harbors anything but the simplest life. This suggestion would have profound implications if true, particularly for the

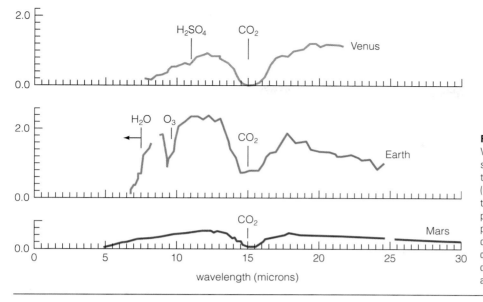

FIGURE 10.16 The infrared spectra of Venus, Earth, and Mars, as they might be seen from afar, showing absorption features that point to the presence of carbon dioxide (CO_2), ozone (O_3), and sulfuric acid (H_2SO_4) in their atmospheres. While carbon dioxide is present in all three spectra, only our own planet has appreciable oxygen (and hence ozone)—a product of photosynthesis. If we could make similar spectral analyses of distant planets, we might possibly detect atmospheric gases that would indicate life.

efforts to search for extraterrestrial intelligence that we will discuss in the next chapter. In the rest of this section, we will explore the issues underlying the question of whether Earth-like planets are common or rare.

Are Earth-size Planets Rarer Than We Think?

The first issue concerns whether we are right in guessing that Earth-size planets should be relatively numerous. We have assumed that, in general, planets can form around any star. However, the fraction of heavy elements (that is, elements other than hydrogen and helium) varies among different stars, from less than 0.1% among the old stars in globular clusters to about 2% among stars like our Sun. Proponents of the rare Earth hypothesis argue that these differences are important, because it is the heavy elements that are essential to the formation of rocky worlds. They argue that if the interstellar clouds that produce stars and planets are deficient in heavy elements, then the resulting solar systems are unlikely to include Earth-size worlds.

Two questions must be answered if we wish to judge the importance of this issue: (1) How much of the galaxy is poorly supplied with heavy elements? (2) What fraction of heavy elements is required to make Earth-size planets?

In assessing the galaxy's heavy element supply, we need to recall where these elements come from. While the Big Bang produced enormous quantities of hydrogen and helium, the remaining elements were produced by nuclear reactions in stars (and during supernova explosions) and were returned to space when the stars died. The enriched material was then recycled into new stars. Thus, stars that were born early in the history of the galaxy should have fewer heavy materials from which rocky planets could form. When we measure the composition of stars in different regions of our galaxy, we indeed find that the oldest stars—those in globular clusters—have a substantially smaller fraction of heavy elements. The same is true for stars in the galaxy's outermost regions, where the density of stars—particularly of the massive stars that would cook up heavy elements—is less and there has been less recycling of newly created heavy elements into young stars. But globular clusters and other old stars comprise only a minority of the Milky Way's stellar population. Consequently, their scarcity of heavy elements doesn't seem to affect our belief that Earth-size worlds could be common, unless a relatively minor deficit of heavy elements could forestall the birth of Earth-size worlds.

How much enriched material is required to make rocky planets? It is not simply a question of having enough raw material. Even globular cluster stars have more than enough heavy elements to construct Earth-size planets (although we don't know whether they've done so). It's a question of the formation process. Because we don't know the fine details of how rocky planets are made, the jury is still out on the importance of a given level of heavy elements in their construction. However, even if it turns out that a relatively high abundance of heavy elements is necessary, a large fraction of Sun-like systems still will make the grade. The composition of our Sun is similar to that of billions of other galactic stars.

Are Earth-size worlds common or rare? It's likely that we'll learn the answer in the coming decades as new instruments search for small planets. But even if Earth-size planets prove to be common, rare Earth advocates have other reasons to suggest that Earth-*like* worlds might not be.

Do Impacts Pummel Most Earth-size Planets?

Another issue raised by rare Earth proponents concerns the rates of impacts by asteroids and comets on planets in other solar systems. Recall that the Earth was probably subjected to numerous large impacts—some large enough to vaporize the oceans and sterilize the planet—during the heavy bombardment that went on during the first half-billion years after our planet was born [Section 4.3]. In our solar system, the impact rate lessened dramatically after that. Might the impact rate remain high much longer in other solar systems?

Although the scars of impact craters on numerous planets and moons bear witness to huge numbers of impacts in our solar system's history, the number of *potential* impacts is far higher. Thousands of asteroids still roam the region between Mars and Jupiter. Of even greater significance, by studying the orbits of comets that enter the inner solar system astronomers have concluded that as many as a trillion comets must orbit the Sun at distances far beyond the orbit of Pluto (making up what is called the *Oort cloud*). Fortunately for us, these myriad objects are essentially out of reach, posing no threat to our planet. However, we have good reason to believe that the reason they are out of reach can be traced directly to Jupiter.

This vast supply of comets is thought to have formed originally in the region of the solar system where the jovian planets were born. The comets that remain are the "leftovers" from accretion in the outer solar system (see Figure 1.23). During the early history of our solar system, the vast majority of these comets would have experienced close encounters with a jovian planet—most of them with Jupiter (Figure 10.17). As we discussed earlier, a close encounter between a big planet and a small object like a comet will tend to send the small object flying off in a new direction. Many comets must have ultimately crashed into the Sun or into the planets of the inner solar system during the heavy bombardment. The rest were flung out to the great distances at which they now reside or out of the solar system entirely.

Thus, if Jupiter had never existed, the comets might have remained in the part of the solar system where they could pose a danger to Earth. In that case, the heavy bombardment might never have ended, and huge impacts would continue to this day. From this viewpoint, our existence on Earth has been possible only because of the "luck" of having Jupiter as a planetary neighbor.

The primary question is just how "lucky" this situation might be. Our discoveries of extrasolar planets show that Jupiter-size planets are in fact quite common, though if most of them migrate inward then giants in outer orbits might be rare. Until we learn much more about the layout of other planetary systems, we have no reason to conclude that we are particularly "lucky" to have Jupiter as a planetary companion.

THINK ABOUT IT . . . *The story of life on Earth is replete with disasters that served to stress terrestrial species, resulting in the rapid evolution of new, more complex organisms. For example, the K–T impact* [Section 5.5] *apparently led to the demise of the dinosaurs and the rise of mammals. More recently, ice ages are thought to have played a major role in the evolution of modern humans. Do you think it's possible that a higher rate of impacts could be good rather than bad for life on another planet? Explain.*

Are Stable Climates Rare?

Another issue affecting the rarity of Earth-like planets concerns climate stability. Recall that, in comparison to Venus and Mars, the Earth has had a remarkably stable climate. This climate stability has almost certainly played a major role in allowing complex life to evolve on our planet. If our planet had frozen over like Mars or overheated like Venus, we would not be here today. Advocates of the rare Earth hypothesis point to at least two pieces of "luck" with regard to the Earth's stable climate.

The first piece of luck concerns the existence of plate tectonics. As we discussed in Chapter 4, plate tectonics plays a major role in the carbon dioxide cycle and hence in regulating the Earth's climate. The existence of this climate-regulating mechanism is especially important given that the Sun, like all stars, brightens as it ages. The Sun is about 30% brighter now than when the Earth formed. Yet, thanks to plate tectonics and the carbon dioxide cycle, the Earth has remained habitable. Plate tectonics probably was not necessary to the origin of life, but it seems to have been very important in keeping the climate stable long enough for the evolution of plants and animals.

FIGURE 10.17 This photograph from the Hubble Space Telescope shows the blemishes left on Jupiter in 1994 by the impacts of the fragmented comet Shoemaker–Levy 9. This impact testifies to the effect that large planets can have on small comets and asteroids. As the orbiting comet entered the inner solar system, it first passed by Jupiter. Jupiter's gravity pulled Shoemaker–Levy 9 apart and then changed its orbit so that, after rounding the Sun, it collided with the planet. The impact sites are the dark spots on Jupiter near the bottom of the photo. The largest of the blemishes is comparable in size to the Earth.

Are we really "lucky" to have plate tectonics, or are such processes inevitable on Earth-size planets in Earth-like orbits? We cannot say for sure, but as far as we know there is nothing particularly unusual about the Earth's size, composition, or orbit. Therefore, we have no reason to believe that planets with similar characteristics should be rare and no reason to think that geological processes should be any different in other cases. On the other hand, Venus is quite similar in size to Earth but apparently lacks plate tectonics. This may be the result of Venus's runaway greenhouse effect, or it may be due to factors that we do not yet understand [Section 9.3]. Nevertheless, unless some unknown, very special circumstance has encouraged long-lasting plate tectonics on Earth, it seems likely that such processes will be found on other, similar planets.

The second piece of luck regarding climate stability concerns the existence of the Moon. As we discussed in Chapter 7, Mars undergoes dramatic climate changes because the tilt of its axis varies over a significant range (see Figure 7.24). The Earth's tilt varies much less, contributing to climate stability. According to calculations that consider the causes of axis tilt variation (which are gravitational tugs from the Moon and other planets, primarily Jupiter), the reason we have a fairly stable axis and Mars does not is that we have a large Moon. In other words, if the Moon did not exist, Earth's spin axis would be sub-ject to the same large swings in its tilt as occur on Mars. Given that the Moon likely formed as a result of a random, giant impact [Section 4.2], it might seem that we are very "lucky" to have our Moon and the climate stability it brings.

Again, however, there are other ways to look at this issue. A radically different axis tilt would not necessarily make our planet unsuitable for life. For example, with sufficient atmosphere to move heat away from its sunlit hemisphere, a planet spinning on its side might enjoy a perfectly tolerable climate. Moreover, changes in axis tilt (or even having the axis flip over entirely) would probably unfold over tens or hundreds of thousands of years. As different regions of the planet became warmer or cooler, life would have plenty of time to migrate to suitable regions. Finally, the fact that our Moon might have formed in a random giant impact does not necessarily mean that large moons will be rare. At least a few giant impacts should be expected in any solar system. Indeed, Earth might not be the only planet in our own solar system to have ended up with a large moon through a giant impact: Pluto's moon, Charon, may have formed in the same way. Thus, while "luck" was certainly involved in the Earth's having a large moon, it might not be a very rare kind of luck.

Is Earth Rare or Not?

When we tot up a laundry list of ways in which luck might have made our presence on Earth possible, we may appear to be beneficiaries of remarkable or possibly even unique circumstances. That could be true, of course, but we should always be leery of probabilities calculated after the fact. For each potential argument that the Earth has been lucky, we've presented counterarguments suggesting otherwise. There's no doubt that our solar system and our world have "personality"—they exhibit properties that might be found only occasionally in other star systems. But were such properties truly essential for our existence, or were they merely a help? Our solar system has no properties obviously essential to complex or even intelligent life that other star systems would never have. Indeed, it might be that we have been cheated out of some helpful phenomena that could have sped evolution on Earth. We might be less lucky than we recognize, and creatures on other worlds might regard the nature of our planet with disappointment. Of course, speculation will not provide answers to these interesting questions, only experiments that look for intelligent life elsewhere can. It is time to turn our attention to issues surrounding the search for extraterrestrial intelligence, the subject of the next three chapters.

THE BIG PICTURE

In this chapter, we have considered the search for habitable planets beyond our own solar system. As you continue your study, keep the following "big picture" ideas in mind:

- Thanks to rapidly accumulating discoveries of planets around other stars, there's no longer any doubt that massive planets, at least, are common. However, we still do not know whether habitable planets are common or rare.

- Our discoveries of extrasolar planets have provided valuable new information about the processes involved in solar system formation, forcing us to modify our existing theory to account for planetary migration.

- Although we do not yet have an answer regarding the existence of Earth-like planets around other stars, several planned missions will address this question. Within the next few decades, we will know whether other Earth-like worlds exist. Within this century, we may have good images of such worlds.

- A recent and controversial hypothesis suggests that planets with the conditions necessary to support complex life may be very rare. However, there are counterarguments to every point raised in this hypothesis, so at present we cannot consider it to be more than speculation.

Review Questions

1. Briefly review why we have good evidence that planets of some type are common but do not yet know whether habitable planets are common.

2. How do astronomers classify stars? For each of the seven major spectral types, discuss whether stars of this type are or are not likely to have habitable planets.

3. What are *multiple star systems?* How common are they? Briefly discuss the prospects of finding habitable planets in multiple star systems.

4. Describe how we can find indirect evidence of extrasolar planets by observing their gravitational effects on their stars. Describe the two techniques by which we can measure such effects—the astrometric and Doppler techniques—and contrast their advantages and limitations.

5. What is a *transit?* How can we use transits to find extrasolar planets?

6. Why is direct detection of extrasolar planets so difficult? How might we accomplish it in the not-too-distant future?

7. Briefly discuss the general properties of the known extrasolar planets. What evidence suggests they are jovian rather than terrestrial in nature? What do we mean by "hot Jupiters"?

8. Why are the "hot Jupiters" surprising? What have they taught us about the formation of planetary systems? Discuss the prospects of finding habitable worlds in systems with "hot Jupiters."

9. Briefly discuss how images and spectroscopy might allow us to determine whether distant planets are habitable and perhaps even whether they have life.

10. What is the *rare Earth hypothesis?* Briefly summarize the arguments used to advance it and the counterarguments against each of these.

11. Briefly describe the goals of future missions such as Kepler, SIM, and TPF. How will these or similar missions help us answer the question of whether habitable planets are rare or common?

Discussion Questions

1. *The Theory of Solar System Formation.* The discussion in this chapter cites the recent modifications to our theory of solar system formation as an example of how science progresses. Do you agree that this example illustrates the hallmarks of science that we discussed in Chapter 2? Why or why not?

2. *The Copernican Principle and Rare Earth.* The Copernican revolution taught us that our planet was not the center of the universe, as had been generally believed before that time. Taking this lesson to heart, we have since assumed that our planet is not "central" or "special" in any way but rather that we are on a fairly typical planet in a fairly typical place in the universe. This principle, often called the *Copernican Principle* or the *Principle of Mediocrity,* has been borne out many times since. For example, we have learned that we are not near the center of our galaxy and that the universe has no center at all. Do you consider this principle to be in conflict with the rare Earth hypothesis? If so, does this make the rare Earth hypothesis any less scientific? Defend your opinions.

3. *What Is a Planet?* When asked what the difference is between an asteroid and a planet, one astronomer answered with "If it's round, it's a planet." Why are the larger members of the solar system—all the planets, many of the moons, and some large asteroids—round? Should "roundness" be included in the criteria for defining a planet? Explain.

Problems

Would You Believe This Headline? **Problems 1–10** give hypothetical headlines, and the dates of their appearance, concerning supposed discoveries about extrasolar planets. In each case, decide whether the headline is believable in light of what we currently know about extrasolar planets and our technological capabilities. Explain your reasoning clearly.

1. Dateline: February 16, 2009. Headline: Astronomers Conclude That Earth-size Planets Don't Exist.

2. Dateline: January 9, 2010. Headline: Astronomers Discover Earth-like World Orbiting Massive Star of Type B.

3. Dateline: June 19, 2006. Headline: Spectrum Reveals Unmistakable Evidence of Life on a "Hot Jupiter."

4. Dateline: November 7, 2005. Headline: New Images Show Oceans on Extrasolar Planet.

5. Dateline: July 20, 2006. Headline: Giant Planet Found in Our Solar System Just Beyond Pluto.

6. Dateline: October 1, 2008. Headline: Astronomers Announce First Detection of an Earth-size Planet Around Another Star.

7. Dateline: May 11, 2005. Headline: New Extrasolar Planet Has Orbit and Size Just Like Jupiter.

8. Dateline: September 15, 2035. Headline: Sun-like Star Has Three Planets with Life.

9. Dateline: March 30, 2027. Headline: More Than One-third of All Stars Now Known to Have Habitable Planets.

10. Dateline: April 2, 2007. Headline: New Observations Prove Earth Unique in Universe.

11. *Are Earth-like Planets Common?* Based on what you have learned in this chapter, form an opinion as to whether you think Earth-like planets will ultimately prove to be rare, common, or something in between. Write a one- to two-page essay explaining and defending your opinion.

12. *Nightfall.* Read the short story "Nightfall," by Isaac Asimov. If such a planet really exists, do you think the scenario described is realistic? Why or why not? Summarize and defend your opinions in a one- to two-page essay.

*13. *Transit of HD209548.* The star HD209548, which has a transiting planet, is roughly the same size as our Sun, which has a radius of about 700,000 kilometers. The planetary transits block 1.7% of the star's light.

a. Calculate the radius of the transiting planet. (*Hint:* The brightness drop tells us that the planet blocks 1.7% of the *area* of the star's visible disk; the formula for the area of a circle is $\pi \times (\text{radius})^2$.)

b. The mass of the planet is approximately 0.6 times the mass of Jupiter, and Jupiter's mass is about 1.9×10^{27} kilograms. Calculate the average density of the planet. Give your answer in grams per cubic centimeter. Compare this density to the average densities of Saturn (0.7 g/cm^3) and Earth (5.5 g/cm^3). (*Hint:* To find the volume of the planet, use the formula for the volume of a sphere: $V = 4/3 \times \pi \times (\text{radius})^3$; be very careful with unit conversions.)

*14. *51 Pegasi.* The star 51 Pegasi has about the same mass as our Sun, which means we can determine orbital properties of its planets by applying Kepler's third law. Mathematically, for a planet orbiting a star with the same mass as our Sun, this law states:

$$p^2 = a^3$$

where p is the orbital period of the planet in years and a is its average distance from the Sun in AU (where 1 AU ≈ 150 million kilometers, which is the Earth's average distance from the Sun). Through the Doppler technique, astronomers found a planet orbiting 51 Pegasi with an orbital period of 4.23 days.

a. Use Kepler's third law to calculate the planet's average distance from its star. Compare this to Mercury's average distance from the Sun, which is 0.39 AU or about 58 million kilometers. (*Hint:* Don't forget to convert the planet's period into years before applying Kepler's third law.)

b. Combine the average distance you found in part (a) with the orbital period to calculate the speed (in km/sec) at which the planet orbits its star. Why is this number so much larger than the measured Doppler shift of the star 51 Pegasi itself (53 m/sec)?

*15. *The Doppler Formula.* The amount of Doppler shift for light or radio waves can be calculated from this formula:

$$\frac{\text{wavelength shift}}{\text{rest wavelength}} = \frac{v}{c}$$

The rest wavelength is the wavelength of a particular spectral line in an object that is not moving (relative to us), v is the velocity of the star from which we observe a wavelength shift, and c is the speed of light ($c = 300,000,000$ m/sec). Suppose that, in a particular star, a spectral line with a rest wavelength of 600 nanometers is found to be shifted by 0.1 nanometer (toward the blue). How fast is that star moving toward us, in meters per second? (1 nanometer = 10^{-9} meter.)

*16. *Finding a Center of Mass.* In the simple case of a two-body system, for example, a star and a single planet, the position of their center of mass can be determined from:

$$M_* \, r_* = M_p \, r_p$$

In this formula, M_* and M_p are the masses of the star and the planet, respectively, and r_* and r_p are the distances from each body to the center of mass. Consider the Sun and Jupiter, which are separated by 780,000,000 kilometers. The Sun's mass is about 1,000 times that of Jupiter. How far from the center of the Sun is the center of mass for this two-body system? Is this inside or outside the edge of the Sun? (The Sun's radius is 700,000 kilometers.)

Web Projects

1. *New Planets.* Find the latest information about extrasolar planet discoveries. Create a personal "planet journal," complete with illustrations as needed, with a page for each of at least three recent discoveries of new planets. On each journal page, be sure to note the technique that was used to find the planet, give any information we have about the nature of the planet, and discuss how it does or does not fit in with our current understanding of extrasolar planets in general.

2. *The Kepler Mission.* The Kepler mission is the first funded mission designed expressly to look for Earth-size planets around other stars. Go to the Kepler Web site and learn more about the mission. Write a one- to two-page summary of the mission's goals and its current status.

3. *Planet-hunting Interferometers.* Other future missions will use interferometry to learn about extrasolar planets. Go to the Web site for one future interferometry mission under consideration, such as SIM, TPF, or Darwin. For the mission you chose, write a one- to two-page summary of the mission's goals and its current status.

CHAPTER 11

The Search for Extraterrestrial Intelligence

There are approximately 2,500 stars within 50 light-years of Earth. If any of these nearby stellar systems host inhabited planets, they could already know about us through our high-frequency radio, radar, and television transmissions. These broadcasts—unintentional evidence of our technological society—are moving into space at the speed of light and currently are washing over star systems at the rate of almost one new star system a day. By using sufficiently powerful radio telescopes, others could learn that we're here.

We are only beginning to signal our presence, but other civilizations may have been doing something similar for a long time. In the Milky Way Galaxy alone, there could be tens of billions of Earth-like worlds, and some of these worlds could be filling the interstellar voids with their broadcasts. Using both radio and optical telescopes, we are attempting to find such transmissions. These experiments, called the *search for extraterrestrial intelligence* (SETI), are the primary topic of this chapter.

11.1 What Is SETI Searching For?

The search for extraterrestrial intelligence (SETI) differs in a fundamental way from all the other searches for life that we have discussed in this book. Those searches are concerned not just with finding life, but with evidence that *might* point to life elsewhere—such as whether habitable planets are common or rare or whether our understanding of the origin of life would allow for life to arise on Mars. In contrast, SETI seeks clear and conclusive evidence only of technologically advanced life.

Indeed, if SETI is successful in receiving and interpreting a message from a distant civilization, it might give us answers to many or most of the other questions we have discussed. To begin with, the discovery of a distant civilization would immediately prove that life is not unique to Earth. Because it is very likely that any extraterrestrials we might detect with SETI experiments would be more advanced than we are, there's at least a possibility that we might be able to learn a great deal from their transmissions, if they are understandable to us.

The principal goal of this chapter is to explore the methods by which scientists are now searching for evidence of other civilizations. First, however, it's worth asking whether our current understanding of life in the universe gives us any good reason to believe that the search may be successful. To some extent, the answer to this question may not matter, as stated so eloquently in the quotation that opens this chapter. That is, our innate scientific curiosity inspires us to search even if we cannot be certain that the search will ever be successful. But while we may not know the probability of success, recent advances in astrobiology allow us to say a lot about the factors that might influence this probability. As in all of science, knowledge about such factors can provide important guidance to research efforts. In this section, we will identify the major factors that relate to the probability of success of SETI efforts.

The Drake Equation

In 1961, the first scientific conference on the search for extraterrestrial intelligence was held in Green Bank, West Virginia, at the radio observatory where a pioneering search for an alien signal had recently been conducted. (We will discuss this search, called Project Ozma, in Section 11.3.) There were only 10 attendees at the conference—just about the entire world's complement of people with a professional interest in the subject at the time—but their expertise ranged across the disciplines of astronomy, biology, and engineering. While setting the meeting's agenda, astronomer Frank Drake (Figure 11.1) tried to concisely summarize the factors that would determine whether any attempt to detect intelligent extraterrestrials could succeed. He came up with a simple equation that, at least in principle, could be used to calculate the number of civilizations existing elsewhere in our galaxy (or in the universe at large) *from which we could potentially get a signal*. Note that this definition limits what we mean by "civilization" in this context. For example, the ancient Greeks had a remarkable civilization, but their civilization doesn't count under this definition because they never developed radio or other technologies that could be used effectively to communicate across space.

The **Drake equation,** as it is now called, cannot give us a definitive result, that is, a well-defined

FIGURE 11.1 Astronomer Frank Drake, with the equation he first wrote in 1961. He is currently the Chairman of the Board at the SETI Institute in California.

number of civilizations. Rather, it lays out the factors that are important in determining this number. The equation has played an important role in research bearing on life in the universe, since much of it deals with life in general and not just the smaller fraction of life that is intelligent enough to produce signals. Many of the factors in the Drake equation have been discussed earlier in this book as part of our search for biology on other worlds. Consequently, we expect that—with improvements in our knowledge of these factors—our estimate of the number of signaling civilizations will get better. However, it's not essential to await such improvements before we embark on a SETI search.

To keep our discussion of the Drake equation simple, we'll focus only on the number of civilizations in our own galaxy. We can always extend our estimate to the rest of the universe simply by multiplying the result we find for our galaxy by 100 billion, the approximate number of galaxies in our universe [Section 1.2]. We limit our focus for practical reasons, too. Any signals from other galaxies will be severely weakened by distance, making them far harder for us to detect. In addition, to stay focused on key ideas that we have already discussed, we will consider a slight modification of Drake's original equation. (For the original equation, see the box "Frank Drake and His Equation" (on page 276). In modified form, the Drake equation looks like this:

$$\text{Number of civilizations} = N_{\text{HP}} \times f_{\text{life}} \times f_{\text{civ}} \times f_{\text{now}}$$

Let's examine each term to see how this equation tells us the number of civilizations in the Milky Way Galaxy capable of interstellar communication:

- N_{HP} is the number of habitable planets in the galaxy. It is the first term because we assume that a prerequisite to having life or a civilization is having a habitable planet on which that life can evolve.

- f_{life} is the fraction of habitable planets that actually *have* life. For example, if $f_{\text{life}} = 1$, it would mean that all habitable planets have life; if $f_{\text{life}} = 1/1,000,000$, it would mean that only 1 in a million habitable planets has life. Thus, the product $N_{\text{HP}} \times f_{\text{life}}$ tells us the number of life-bearing planets in the galaxy.

- f_{civ} is the fraction of the life-bearing planets upon which a civilization capable of interstellar communication *has at some time* arisen. For example, if $f_{\text{civ}} = 1/1,000$, it would mean that such a civilization has existed on 1 out of 1,000 planets with life, while the other 999 out of 1,000 have not had a species intelligent enough to build radio transmitters, high-powered lasers, or other devices for interstellar conversation. When we mul-

tiply this term by the first two terms to form the product $N_{\text{HP}} \times f_{\text{life}} \times f_{\text{civ}}$, we get the total number of planets upon which intelligent beings have evolved and developed a civilization at some time in the galaxy's history.

- f_{now} is the fraction of the civilization-bearing planets that happen to have a civilization *now*, as opposed to, say, millions or billions of years in the past. This term is important because it tells us how many civilizations we could potentially talk to,[1] since there's no point in listening for civilizations that are long gone. Because the prior three terms told us the total number of civilizations that have *ever* arisen in the galaxy, multiplying by f_{now} tells us how many civilizations we could potentially make contact with today. For example, if the first three terms were to tell us that 10 million planets in the galaxy have at some time had a communicating civilization but f_{now} turns out to be 1 in 5 million, then only two civilizations could be expected to exist today. As we will see shortly, the value of f_{now} must depend largely on how long civilizations survive once they arise.

To summarize, the complete Drake equation gives us a way to calculate the number of civilizations capable of interstellar communication that are currently sharing the Milky Way Galaxy with us. As such, it provides a useful way of organizing our thinking about the problem, because it tells us exactly what numbers we need to know to learn the answer. Indeed, it suffers from only one significant drawback: We don't know the value of any of its terms!

THINK ABOUT IT . . . *To make sure you understand the basic idea behind the Drake equation, try it with some sample numbers. Suppose there are 1,000 habitable planets in the galaxy, that 1 in 10 habitable planets has life, that 1 in 4 planets with life has at some point had an intelligent civilization, and that 1 in 5 civilizations that have ever existed is in existence now. For this case, identify each term in the modified Drake equation above, and find the total number of civilizations. Explain your result. (Note: These numbers are probably unrealistic, but they will still show how the equation works.)*

---

[1] Technically, for purposes of communicating with civilizations, we must factor light-travel times into this term. For example, if a civilization arises around a star 10,000 light-years away, we can talk to it only if the civilization was there 10,000 years ago as opposed to "right now." In the discussion in this book, we will assume that this technicality is properly taken into account, so we need not deal with it explicitly.

If anyone could be called the Father of SETI research, Frank Drake is that person. As a young man, Drake learned electronics in the Navy. He then studied for an advanced degree in astronomy and upon graduation took a job at the new National Radio Astronomy Observatory (NRAO) in Green Bank, West Virginia. In the late 1950s, the observatory was busy with construction of a large radio telescope, a project that would take many years. Consequently, it opted to buy an "off-the-shelf" instrument that could be used right away. This telescope boasted a 26-meter (85-foot) reflector, which at the time was larger than most of the world's operating radio telescopes.

Once this telescope was up and running, the observatory staff was encouraged to come up with interesting experiments for its use. Drake had already been thinking about the possibility of interstellar communication by radio, and he proposed a simple experiment to search for alien signals from two nearby star systems. This became Project Ozma, the first modern SETI experiment. A year later, in 1961, Drake organized a conference at Green Bank to discuss the possibility that such an experiment could actually *find* an extraterrestrial transmission. His "agenda" for that conference became known as the Drake equation.

Drake's original form for his equation is:

$$N = R_* \times f_{\text{planet}} \times n_e \times f_{\text{life}} \times f_{\text{intell}} \times f_{\text{civ}} \times L$$

N is the number of transmitting civilizations in our galaxy. R_* is the galactic birthrate of stars suitable for hosting life, in stars per year. For example, since there are roughly 100 billion stars in the Milky Way and the galaxy is approximately 10 billion years old, R_* is approximately 10 per year (assuming rather crudely that all stars are suitable). The term f_{planet} is the fraction of such stars having planets; n_e is the number of planets per solar system that have an environment favorable for life; f_{life} is the fraction of such planets on which life actually evolved; f_{intell} is the fraction of inhabited worlds that develop intelligent life; f_{civ} is the fraction of planets having intelligent beings that produce a civilization capable of interstellar communication; and L is the lifetime that such civilizations are "on the air," broadcasting signals. You might want to compare these terms with the more compact factors used in our discussion of the equation on p. 275.

Using some admittedly optimistic estimates for the first six terms in the equation, Drake suggested that we could reasonably guess that they would multiply to approximately 1 per year. This is the "birthrate" of civilized societies in the Milky Way. To find out how many are broadcasting now, we need only multiply this rate by L, the number of years during which they broadcast. This is analogous to determining the number of students attending a college: It's the number per year that enter as freshmen (the entrance rate) times the number of years they spend as students (typically four). If the birthrate for civilizations is taken to be 1 per year, the Drake equation becomes simply $N = L$; that is, the number of transmitting civilizations is simply the average lifetime (in years) of a transmitting society.

Frank Drake's personalized license plate.

Unfortunately, L is dependent on sociology rather than on astronomy or biology—making it far more difficult to determine a value for L. Attempts to estimate L usually try to guess what the one technological civilization we know—our own—is likely to do. Only a half-century after inventing radio, we also developed atomic weapons. To some people, this suggests that L might be very short—only a few centuries or less. On the other hand, we can be optimistic and assume that we will survive our own technology and exist as a society for millions of years into the future. Perhaps one of the most important things we could learn from a SETI detection is that not all technologically sophisticated societies are doomed to early self-destruction.

Numbers for the Drake Equation

The only term in the Drake equation for which we can make even a reasonably educated guess is the number of habitable planets, N_{HP}. As we discussed in Chapter 10, the detections of extrasolar planets to date show that planets of some kind are likely to be quite common. While we cannot be sure whether this also means that *habitable* planets are common, so far we have no good reason to think otherwise.

Moreover, the example of our own solar system suggests that it is possible to have more than one habitable planet per system, since we have good reason to believe that Mars was habitable at some point in its past. Given that there are several hundred billion stars in the Milky Way Galaxy, it seems entirely reasonable to suppose that there could be 100 billion or more habitable planets. Nevertheless, the actual number might be far smaller, and we won't know for sure until we begin to get data about habitable planets in the next decade.

The rest of the formula presents more difficulty. For the moment, we have no rational way to estimate the fraction f_{life} of habitable planets upon which life actually arose. The problem is that we cannot generalize when we have only one example to study—our own Earth. Still, we are not completely without guidance. The fact that life apparently arose rapidly on Earth [Section 5.1] suggests that the origin of life was fairly "easy," in which case we might expect that most or all habitable planets would also have life, making the fraction f_{life} close to 1. However, until we have solid evidence that life arose anywhere else, such as on Mars, it is also possible that Earth was somehow very lucky and that f_{life} is so close to 0 that life has never arisen on any other planet in our galaxy.

Similarly, we have little basis on which to guess the fraction f_{civ} of life-bearing planets that eventually develop a civilization. On one hand, the fact that life flourished on Earth for almost 4 billion years before the rise of humans might suggest that it is very difficult to produce a civilization even when there is life. On the other hand, given that roughly half the stars in the Milky Way are older than our Sun, there has been plenty of time for evolution to work on numerous planets. Any evolutionary drive toward intelligence might inevitably lead to huge numbers of civilizations, even if it takes a long time on any given world. This question of whether intelligence is a rare accident or an inevitable result of evolution is so important to the issue of the search for extraterrestrial civilizations that we will devote the next section to investigating it.

The final term in the equation, f_{now}, is particularly interesting because it is related to the survivability of civilizations. Consider our own example. In the roughly 10 billion years during which our galaxy has existed, we have been capable of interstellar communication via radio for only about 50 years, or only about 1 part in 200 million of our galaxy's history. If we were to destroy ourselves tomorrow (saving students the unpleasantness of a final exam), our technological "lifetime"—the length of time we could make ourselves known to other star systems—would be only 50 years. If this is typical of other civilizations, then our chances of finding a signal from any one of them at any random time would be only 1 in 200 million. In that case, even if there have been hundreds of millions of civilizations in the galaxy's history, no more than a few would be detectable now.

Of course, we have not yet destroyed ourselves, and we may be severely underestimating the fraction f_{now}. For example, suppose civilizations routinely survive for a billion years. Then the chance of a signal reaching us from any given civilization at a random time is 1 in 10. If $f_{now} = 1/10$, then there may be numerous communicating civilizations out there now, even if civilizations arise rather infrequently. To take some numbers, suppose that only 1 in 10 million stars ever gets a planet with a civilization. In a galaxy of 100 billion stars, this would mean only about 10,000 civilizations ever arise. But if 1/10 of them are here *now*, then there are some 1,000 civilizations we could potentially find. Four decades ago, considerations like these led Frank Drake to conclude that the typical lifetime of civilizations must be one of the primary factors—perhaps even *the* primary factor—in the potential success of SETI efforts. Moreover, as we'll discuss in Chapter 13, the issue of civilization survivability plays a very important role when we examine the question of why we don't yet know of other intelligences—a mystery sometimes called the *Fermi paradox*.

The Value and Limitations of the Drake Equation

The Drake equation is mathematically simple, but its chain of terms is only as strong as its weakest link. If we know one term poorly, there is no way to improve our estimate of the number of civilizations by knowing other terms well. For example, while we might have better estimates of most of the factors in the equation as our knowledge of astronomy and biology improves, f_{now} depends on sociological factors—that is, the behavior of alien civilizations. Do they quickly self-destruct, or do they survive for long times? The only way we can make realistic estimates of f_{now} is by actually detecting extraterrestrial societies. Thus, as long as we have this uncertainty in f_{now}, we will face a corresponding uncertainty in knowing the total number of signaling worlds even if astrobiology research someday allows us to pin down precise values for terms such as N_{HP} and f_{life}.

As a result of this "weakest link" problem, as well as the fact that all of the terms are still highly uncertain, we cannot draw any definitive conclusions from the Drake equation. Indeed, some of the factors have such a large uncertainty that the numbers we enter into the equation (choosing numbers within the range consistent with our present knowledge) can give us anything from an optimistic view that a large fraction of stars in our galaxy have intelligent, communicating beings to a pessimistic view that we would have to search a very large number of galaxies to find even one other example of intelligence. The main value of the Drake equation, then, is in pointing out what factors are important and underscoring the implications of our lack of knowledge about particular factors. It can be used, therefore, to help us realize what the issues are and where we remain ignorant and thus steer us to areas in which more research is needed.

THINK ABOUT IT . . . *The Drake equation assumes that each transmitting civilization has sprung up independently on its own habitable planet. Can you think of reasons why this assumption might be too limiting?*

11.2 The Question of Intelligence

Our discussion of the Drake equation has pointed out several key factors that we must understand better if we want to know how many other civilizations exist. We have already discussed (in earlier chapters) the uncertainties surrounding the number of habitable planets and the origin of life. However, we have not yet discussed the ideas built into f_{civ}, which describes the probability that life will eventually give rise to intelligence and technologically adept civilization. In this section, we turn our attention to this important topic.

SETI is likely to be successful only if intelligent life is widespread. Is it? Broadly speaking, there are two opposing schools of thought on this question. One school considers intelligence comparable to our own (that is, able to develop both science and technology) to be an unlikely intruder on the stage of life. From this point of view, biology might be widespread but the evolution of technological intelligence extremely rare. Life has existed on our planet for some 3.5 billion years (or more). But only in the last few million years has our genus *Homo* developed the capability to understand the environment, and only within the last half millennium have we come to understand the nature of the Earth and begun our exploration of the cosmos. At the very least, *Homo sapiens'* late appearance on Earth suggests that a long period of evolution must precede the emergence of technologically intelligent creatures.

Moreover, as we discussed in Chapter 5, our existence seems to have resulted from a number of chance events. For example, the Cambrian explosion that gave rise to the "body plans" of all modern animals (including those of our phylum, Chordata) might have been the result of environmental stress introduced by snowball Earth episodes, and the rise of mammals might never have occurred if the K–T impact had not wiped out the dinosaurs (Figure 11.2). The chance nature of these events and the many other forks in the evolutionary road suggest to some people that the appearance of technological intelligence was an enormously improbable event.

The second school of thought holds the opposite view. It proposes that there is evolutionary pressure for intelligence, that is, that various evolutionary

mechanisms consistently encourage an increase in intelligence for a wide range of species. If this is true, then some technologically intelligent species would still have evolved on Earth even if a different sequence of past events had prevented the existence of humans. Because we are the only species on our planet that has ever developed an advanced civilization, no other known species directly supports the view that the evolution of technological intelligence is likely. Instead, those who adopt this viewpoint look more generally at the process of evolution for evidence that some of the workings of natural selection promote greater intellectual capability.

Convergent Evolution

The evolutionary argument in favor of widespread intelligence is based on the phenomenon of **convergent evolution,** the tendency of organisms of different evolutionary backgrounds that occupy similar ecological niches to come to resemble one another. In such cases, natural selection often produces analogous adaptations. One example of convergent evolution is the shape of large marine predators. For example, dolphins and sharks evolved independently from earlier mammals and fish, respectively, but both have the same streamlined body form. The obvious reason for this is that being shaped like a torpedo makes for greater speed underwater—and speed has clear survival value for a predator. We say that the evolution of originally quite different animals has *converged* on this optimized underwater shape.

For another example, consider eyesight. Vision is a useful adaptation, and most animals have some ability to see. However, the eye was not a unique evolutionary invention. Studies of evolutionary relationships show that eyes evolved independently at least eight different times. Indeed, the design of the human eye is by no means the best. Compared to the eyes of some other animals, ours have "flaws," such as the fact that the nerves in our eyes come together in a bundle before exiting through the back of our eyeballs, resulting in a small blind spot. The several independent origins of eyes suggest that evolution tends to *converge* toward developing some kind of eye to provide vision.

THINK ABOUT IT . . . *Recent research has uncovered an example of convergent evolution in crabs. Because all crabs looks pretty much the same, you might expect that they all share a common evolutionary past. But comparison of the DNA of various crab species shows that their backgrounds are quite varied; different species of these crusty critters have separately evolved from, for example, shrimplike or lobsterlike ancestors. Can you think of any reasons why a crablike body, with its unusual*

FIGURE 11.2 The path of evolution on Earth was severely affected by chance events such as the K–T impact 65 million years ago, which wiped out the dinosaurs and most other species. This painting depicts the 10-kilometer wide asteroid as it approached the Earth. If this rock had arrived at Earth's orbit 20 minutes earlier (or 20 minutes later), it would have missed our planet. In that case, would intelligent beings still have eventually arisen?

manner of walking and nearly round shape, might be a natural evolutionary development?

Like speed or eyesight, intelligence—the way an animal processes information—is subject to natural selection. If there are ecological roles for which keener intelligence has survival value, then we would expect a convergent evolution to greater brain power for animals in these niches. Intelligence would be just as likely to emerge as any other generally useful adaptation. In this case, evolution would tend to raise the level of intelligence to at least some degree in a great many species, which in turn would increase the chance that some of these species would evolve even higher intelligence. With more players in the game, the chance of producing human-style intelligence would be greater. Thus, if we could establish the presence of an evolutionary trend favoring intelligence, we would be encouraged to think that the emergence of technological intelligence elsewhere is not enormously improbable.

In principle, we can test whether intelligence is evolutionarily favored by measuring the brain power of a variety of animal species over time. Some researchers are doing exactly that.

Measuring Intelligence

How do we measure the mental ability of animals? Few creatures are eager or able to take IQ tests, but we can resort to a simpler measure of raw brain power based on brain mass.

Figure 11.3 shows the brain weight for a sample of birds and mammals (including primates) plotted against their body weight. There is a clear and expected trend: Heavier animals have heavier brains. By drawing a straight line that fits these data, we define an average brain mass for each body mass. Those sampled animals whose brain mass falls on this line are said to have an **encephalization quotient (EQ)** of 1, which means a typical allotment of mental ability for creatures of their size. They have enough brain mass, we can assume, for a basic set of behaviors. If their brain mass falls above this line, then it seems reasonable to suppose that they are capable of more elaborate behavior. Creatures whose brain mass falls below the line are presumably less mentally agile than average animals in this sample. The EQ, the brain mass relative to the value on the line EQ = 1, serves as an indicator of general intelligence. Although other indicators of intelligence have been suggested (the amount of brain folding, for example, or the number of neural connections), measuring EQ has the advantage of being fast and easy. Moreover, we can compute an EQ for vanished species by measuring the volume of the fossil cranial cavity, from which we can determine the brain mass, and combine this measurement with an estimate of the creature's total body mass.

EQ is an admittedly simple measure, and it might be likened to judging the computational ability of computers by weighing their CPU chips. Nonetheless, EQ seems to correlate well with complex behavior. Carnivorous animals that need to hunt down their meals generally have higher EQs than leaf eaters, and animals that lavish care on their offspring score higher than those that ignore them.

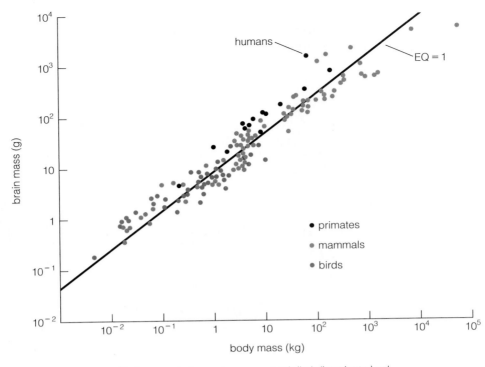

FIGURE 11.3 This graph shows the comparison of brain mass to body mass for some mammals (including primates) and birds. There is a strong and obvious correlation in that the bigger the body, the bigger the brain. The straight line represents an "average" of the ratio of brain mass to body mass, which we define as an encephalization quotient (EQ) of 1. Animals that fall above the line have an EQ greater than 1 and are therefore "smarter" than average. Animals that fall below the line have an EQ less than 1 and are "intellectually challenged." (Adapted from Sagan, 1977.)

What happens when we use EQ to investigate our premise that, on Earth, there has been a trend toward increasing brainpower over time? If we look at contemporary species, we find that while humans don't sport the most massive brains (whales, for instance, have much larger brains than we do), we do have the biggest brains in relation to body mass. Our species' encephalization quotient is 7, meaning that our brains are 7 times more massive than would be expected for an "average" mammal of the same body weight. This EQ not only exceeds that of chimps and dolphins, whose EQs are 2.5 and 4.5, respectively, but it is higher than that of any other known species, alive or extinct. Thus, these numbers confirm what we already know: We're the cleverest critters on the planet. However, a look to the past shows that we were not the only creatures evolving toward greater intelligence. By measuring EQs for a range of mammals, for example, biologists have found that the dolphins of several million years ago had higher EQs than did our primate ancestors of the same period and lower EQs than present-day dolphins. Dolphins, primates, and other animals all appear to have become smarter over time, and we just happen to have come out at the top of the intelligence heap.

Another way to approach the question of the evolution of intelligence is to ask what factors might encourage its appearance and whether those factors are likely to be commonplace. To begin with, a good many "pre-adaptations" seem necessary. High-performance brains like those of birds and mammals need a vigorous metabolism, so intelligence is most likely to arise in warm-blooded animals. In addition, a relatively large body size is necessary to house a large brain (although recent research with insects suggests that, by miniaturizing neurons, nature might in principle be able to pack a lot of intelligence into small packages). Extended parenting is probably also a necessary precondition for intelligence, which wouldn't be very useful without learning. Such preconditions have existed for a wide range of species on Earth during at least the last 50 million years, the same period in which intelligence seems to have risen most dramatically.

Once the necessary preconditions are in place, what evolutionary mechanisms will select for intelligence? One mechanism is the interaction among individuals of social species. Dolphins and primates are highly social animals. Success in a social environment is enhanced by intelligence, because there is survival value in being able to judge the mood and meaning of fellow creatures. Social position in your troop or pod depends on how savvy and canny you are. Also, an elevated social position often allows you to have first choice in mates, so clever, high-ranking individuals will tend to produce clever, high-ranking offspring. We might reasonably expect intelligent aliens, if they exist, also to have evolved in interactive social environments.

Some evolutionary effects cause competing species to ratchet up one another's intellect. For example, large carnivores and their prey encourage improvements in one another's intelligence. When a lioness stalks gazelles for dinner, she's more likely to catch a gazelle that's less aware of its surroundings or less cagey in devising escape maneuvers. The lions get a meal and in the process inadvertently raise the average intelligence of the surviving gazelles. The rise in gazelle intelligence makes getting a meal tougher for less mentally agile lions, who then preferentially drop out of the gene pool. The less alert, less cunning of both species are weeded out, and the smarter members survive.

In summary, several common circumstances in the animal world naturally select for brain power, and there is evidence that more than one group of animals on our planet has been on the track to high general intelligence. This argues against the idea that the appearance of intelligence is some special, extraordinary accident of evolution on Earth and suggests that, given time and a competitive environment, many intelligent species will arise on any inhabited world. The probability of many candidates means at least some possibility that one or more of these species will develop humanlike brainpower. On the other hand, increased intelligence comes at a cost in terms of resources to support it (such as a high metabolism or carrying around a heavy head), and it is unclear that these resources would not be better utilized by, say, evolving the ability to run faster or fight more fiercely. Moreover, many species *didn't* evolve intelligence over the last 50 or 100 million years. In the end, the example of terrestrial life alone does not tell us whether there is an evolutionary imperative toward humanlike intelligence. And, even if there is such an imperative, we still must ask whether high intelligence necessarily leads to technical competence and the ability to communicate across interstellar distances.

The Communication Question

It might seem natural that a species with sufficient brainpower will eventually develop science and technology. However, there are both physiological and sociological counterarguments to this idea.

On the physiological side, suppose dolphins were as intelligent as we are. Their lack of hands and their need to live in water would prevent them from building anything resembling modern technology. They might be very smart and have sophisticated social structures, but without telescopes they wouldn't know much about astronomy, and without radios or

lasers they wouldn't be able to talk to the dolphins of other worlds. Wolves have mouths, elephants have trunks, and ravens sport beaks and feet, but none of these animals have overall body designs that would allow them to manipulate complex tools—no matter how smart they might be.

On the sociological side, remember that *Homo sapiens* emerged in more or less its current form about 100,000 years ago. Certainly, by 20,000 or 30,000 years ago our ancestors had essentially the same level of intelligence as we do. Yet our ability to communicate through space is less than a century old, and it derives from the emergence of science—itself an endeavor that has developed only gradually over the past 2,500 years. Many different cultures developed mathematics and astronomy to varying degrees, but only the cultural line that emerged from ancient Greece ultimately led to modern science. Was this a lucky accident, or was it inevitable that some culture would develop modern technology?

Again, a single example—what happened on our own planet—cannot guarantee that science and technology will be common in the cosmos, even if high intelligence is. The only way we can answer the question of how frequent or rare technological civilizations might be is to find evidence of them on other habitable worlds.

11.3 The SETI Context

Receiving a message from another civilization could be one of the most important events of human history. We would know that we are not alone in the universe. We might learn a great deal about science and sociology. We might even learn how other civilizations have successfully survived periods in their history in which they had the power to destroy themselves. How might we receive such a message? In this section, we'll investigate the context in which we engage in SETI efforts. We'll begin with some historical background, then discuss the types of signals that SETI might be able to detect. We'll also explore a few possible ways of detecting extraterrestrial intelligence that don't involve communication signals.

Early Experiments

Shortly after its invention, radio was recognized as a possible means of extraterrestrial communication. After all, radio travels at the speed of light and can easily bridge the airless voids of space. If extraterrestrials are using radio, then we might detect their presence without anyone having to leave the home planet.

As the twentieth century dawned, two pioneers of wireless technology became convinced (wrongly) that they had heard aliens on the airwaves. One of these pioneers was Guglielmo Marconi (1874–1937), generally celebrated as the man who made radio practical. The other was Nikola Tesla (1856–1943). It was Tesla who first claimed extraterrestrial contact.

Tesla was both eccentric and brilliant. He was a prolific inventor (among other things, he made the first fluorescent lamp), but his most enduring legacy was the use of alternating current (AC) for distributing electrical power. AC—produced by having a voltage that cycles positive and negative 60 times a second—is commonplace today, but it was a radical concept when Tesla first proposed its use for a Niagara Falls generating station. His idea was vigorously opposed by Thomas Edison, who had set up the first commercial power plant in Manhattan. Edison's plant distributed direct current (DC, which has a constant voltage), and Edison tried to frighten consumers by telling them that the rival alternating current would cause electrocution. The advantage of AC is that transformers can easily raise or lower the voltage, increasing it for low-loss transmission over the great distances from the power plant to your neigh-

borhood and then lowering it for final distribution to your home. If you look at the power poles on any residential street that has aboveground electrical lines, you'll see the "step-down" transformers that provide the 115 volts (in the U.S.) feeding the wall sockets of the houses.

Tesla promoted AC because of these benefits, but he tried to go even farther. He thought it might be possible to bring power to the people with "induction," rather than with copper cables, sending the energy through the air as low-frequency radio waves. To demonstrate his idea, he built a 60-meter-tall transmitting tower, wrapped with wire, in Colorado Springs, Colorado. His device, intended to radiate electrical energy, was an outsized example of what is now known as a Tesla coil. One night the inventor noted that his apparatus was picking up electrical disturbances, which he ascribed to interplanetary communication. In 1901, he claimed to be the first to establish communication between worlds. We now believe that he was listening to an atmospheric phenomenon known as "whistlers," electrical noise created by distant lightning discharges.

Movie Madness: Contact

Do the aliens know we're here? They might, if they're not very far away.

Contact, the movie based on Carl Sagan's novel about getting in touch with our celestial pals, starts out with a nifty sequence in which the camera backs away from Earth, slips by the Moon and planets, and eases out into the galaxy and beyond. During this high-speed countermarch, we hear the sounds of radio programs that have reached each of these cosmic outposts. With every step outward, we move back in time. (OK, there's some cinematic license here: as the camera passes Saturn, we've regressed to 1950s rock and roll. Saturn is never more than 88 light-*minutes* from Earth!)

In fact, less than 100 light-years out, the Earth really does go "silent." Easy evidence for the presence of *Homo sapiens* extends only this far— about one-tenth of one percent of the distance to the far edges of the galaxy.

Fortunately for the cash-starved SETI researchers in *Contact,* some friendly aliens have an outpost around the bright star Vega, a mere 25 light-years away. They've tuned in one of our early TV broadcasts and, apparently intrigued, have replied. Their response is picked up by the film's heroine, Ellie Arroway, as she uses a pair of ear-

phones to monitor the cosmic static received by a large radio telescope. (More license here: radio receivers for SETI sport tens or hundreds of millions of channels. Ellie should either have donned a few million headsets or should have left the listening to computers.)

The alien reply signal, which sounds like a pile driver hitting a pod of whales, contains an original 1936 broadcast (that way we know that the extraterrestrials are deliberately beaming to us) interleaved with construction details for . . . well . . . some sort of large device.

Faced with a SETI detection, the government goes nuts, and so do a lot of the citizenry. Some are ecstatic about the possibility of alien company; others see the news as heralding the apocalypse. Meanwhile, the scientists build the device—a multibillion-dollar machine that looks like the ultimate theme park attraction, but is in fact a wormhole transporter (is there any difference?).

It's nice to think that advanced aliens would want to improve our lot by sending us plans for hi-tech hardware. But this seems an unlikely message from space. After all, if we could somehow contact the Neanderthals and give them plans for a personal computer, do you think they could ever, ever build it?

The year 1901 was also when Marconi successfully sent a signal across the Atlantic, proving that radio was useful for communication over large distances. It took little imagination to guess that this new medium might also serve for sending messages into space. In the early 1920s, Marconi stated that he, too, had picked up signals that came from extraterrestrial sources. These experiments culminated in 1924, during one of the periods when Earth and Mars are closest to each other in their orbits. Marconi and others encouraged anyone with a radio set to listen for Martian broadcasts [Section 7.3]. There was considerable optimism that something would be detected, and a cryptographer was standing by in case the signals required decoding. Alas, no emissions of an extraterrestrial nature were heard. In retrospect, the signals that Marconi heard may also have been "whistlers" from distant lightning, or possibly garbled U.S. Navy broadcasts of which he was unaware.

In retrospect, these early experiments were doomed. We now know that there are no Martians sophisticated enough to assemble radio transmitters. In addition, we also realize that these pioneer listening attempts were all made at the wrong spot on the dial. They were conducted at relatively low frequencies that have problems penetrating Earth's **ionosphere**—a layer of the upper atmosphere that consists of particles ionized by sunlight. The ionosphere acts as a radio "mirror," reflecting low-frequency radio emissions. Indeed, it was the ionosphere that bounced Marconi's 1901 transatlantic signal from England to Canada, thus allowing communication despite Earth's curved surface. The apparatus used by both Tesla and Marconi operated at frequencies too low to penetrate the ionosphere and therefore was insensitive to any cosmic broadcasts.

Although the radio pioneers might have mistaken thunderstorms for Martians, their enthusiasm for radio's use as a long-distance communication medium was ahead of its time. After World War II, when high-frequency radio equipment became widely available, it was possible to listen at frequencies that *could* penetrate the ionosphere. In addition, improvements in antennas and receivers made such experiments enormously more sensitive than the attempts made early in the century. The stage was set for a scientific approach to SETI.

SETI Begins

The start of a scientific approach to SETI is generally attributed to two physicists working at Cornell University. In 1959, Giuseppe Cocconi and Philip Morrison wondered how difficult it would be to send radio signals over interstellar distances. They made simple calculations showing that communication between nearby stars was possible using technology no more advanced than our own.

Cocconi and Morrison realized that the galaxy is considerably older than our solar system and consequently that there could be civilizations among the stars that have existed much longer than ours—perhaps surviving for many millions of years. For such long-lived societies, the construction of powerful radio beacons that could be used to send "hailing signals" to other star systems should be a simple matter.

Cocconi and Morrison advocated looking for such beacons using large radio telescopes aimed at nearby stars. The exact arrangement of the planets around the stars under scrutiny was irrelevant, because the area of sensitivity of any reasonable-size radio telescope would encompass all of a star's possible planetary environs. The only other technical question concerned which frequencies to monitor with the radio receivers.

The radio portion of the electromagnetic spectrum (see Figure 4.24) extends over a wide range of frequencies. When we build a radio receiver, it is sensitive only to a particular set of frequencies within this wide range, which we refer to as its **band** of sensitivity. For example, if you look at a typical FM radio, you'll see that it is designed to cover the band of frequencies from about 87 to 108 MHz. (Recall that the basic unit of frequency, hertz (Hz), is equivalent to waves (or cycles) per second. Thus, 1 megahertz (MHz) is a frequency of 1 million cycles per second.) Of course, radio stations do not broadcast their signals over this entire band—if they did, no matter where you tuned your radio dial you'd hear the overlapping sounds of all radio stations that were transmitting. Instead, radio stations transmit their signals only over a very narrow band of frequencies. For example, if your favorite radio station broadcasts at "97.3 on your FM dial," then you must tune your receiver so that it picks up only a very narrow range of radio frequencies centered on 97.3 MHz. That range of transmitted frequencies is called the **bandwidth** of the signal. As a practical matter, the bandwidth is governed by how much information the broadcast contains. For example, a television signal (Figure 11.4) has a bandwidth 500 times wider than that of an AM radio station, because the TV signal contains picture (video) information in addition to sound (audio).

While local broadcasts have very different bandwidths, a hailing signal designed to get the attention of someone light-years away would likely be confined to a very narrow bandwidth, with all the energy of the transmitter concentrated at one spot on the radio dial. This would make the signal easier to pick out against the background noise naturally produced by hot interstellar gas, distant galaxies, and the radio

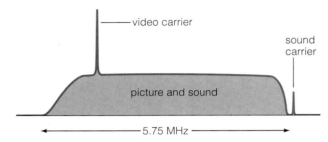

FIGURE 11.4 The spectrum of a television signal, showing how the energy of transmission is spread over a bandwidth of nearly 6 MHz. The large bandwidth is necessary because the signal carries both sound and picture information. However, about one-third of the total power in the TV signal is concentrated in the extremely narrow "spike" known as the *video carrier*. This carrier has a bandwidth of less than 1 Hz, or about one six-millionth of the total bandwidth. Because the video carrier has so much of the total signal power concentrated into such a narrow bandwidth, it would be the easiest part of the signal to detect from a great distance. Hence, it is a good example of the type of narrow-band signal that SETI experiments seek.

receiver itself. But in what part of the radio spectrum would we expect to find such a narrow-band, hailing signal? In truth, we cannot say for certain, since a wide range of radio frequencies is serviceable. Because searching all these frequencies would be a hopelessly difficult task, Cocconi and Morrison proposed that SETI experiments should tune their receivers near a spot on the dial that every scientifically literate society would know: 1,420 MHz. This is the frequency at which neutral hydrogen gas—the major constituent of the thin material that floats between the stars—produces natural radio static. Radio astronomers often use this frequency to study the distribution of interstellar gas in galaxies. Because it is such a useful spot on the dial, astronomers throughout the universe (of whatever species) would have this frequency marked on the receivers of their radio telescopes, and a transmitting beacon tuned near this frequency would be likely to attract the attention of others.

Having set out all the important principles of a radio SETI search, Cocconi and Morrison tried to interest radio astronomers in their idea. They approached astronomers at England's Jodrell Bank Observatory but got little response. However, Frank Drake, then a young astronomer at the Green Bank radio observatory in West Virginia, had independently reached many of the same conclusions as Cocconi and Morrison. In the spring of 1960, using the 26-

meter (85-foot) radio dish at Green Bank, he began a search very much like the one Cocconi and Morrison had in mind (Figure 11.5). He conducted a weeks-long search for 1,420 MHz radio signals from two nearby, Sun-like stars: Epsilon Eridani and Tau Ceti. (These stars are approximately 12 light-years distant.) Drake whimsically named his search Project Ozma, after the fictional princess in the books by Frank Baum. It was the first modern SETI experiment.

Drake's search failed to detect alien signals, but it fired the imagination of many in the science community and ultimately led to a small SETI research program run by the American space agency, NASA. The NASA researchers spent many years studying the feasibility and technology of SETI. Their research led to the construction of specialized receiving systems that could be fitted to large existing radio telescopes. In 1992, the NASA search began. But a year later,

FIGURE 11.5 The 26-meter (85-foot) radio telescope used by Frank Drake in his pioneering 1960 SETI search, Project Ozma. This instrument was the first to be built at the National Radio Astronomy Observatory in Green Bank, West Virginia. Drake scrutinized two nearby star systems for signals near the 1,420 MHz frequency.

with the data collection barely under way, the program was cancelled by the U.S. Congress. Since then, scaled-down SETI programs have continued with private funding in the United States and as small university research efforts elsewhere in the world. (We'll discuss current SETI efforts in Section 11.4.)

Categories of Signals

Project Ozma was searching for a deliberately broadcast signal at a specific frequency that would be known to astronomers anywhere. Drake was hoping to find a deliberate interstellar hailing signal, or beacon. But there are other possible signals we might hope to find. Broadly speaking, alien signals could fall into any one of the following three categories:

1. Signals used for local communication on the world where intelligent beings live. Our own radio and television signals fall into this category, because they are designed for our own use and not for interstellar communication. Another local use for radio signals is radar. Advanced civilizations might use radar to locate comets that pose a potential threat to their planet, for example.

2. Signals used for communication between a civilization's home world and some other site, such as a colony or spacecraft on another world. We have used relatively weak signals of this type to communicate with our interplanetary spacecraft. Such signals would be far stronger if they were being used, say, to communicate between colonies on planets in different star systems light-years apart.

3. Intentional signal beacons (as discussed above), designed to get the attention of other societies.

In principle SETI can search for all three types of signal, but in practice our ability to receive signals depends on the sensitivity of our equipment. To get a rough idea of what we might be able to detect with our current technology, let's consider our own signals as an example. The first commonplace, high-power, high-frequency transmissions from Earth were our early television broadcasts. These transmissions began in earnest during the 1950s, so they are now some 50 light-years away in space—followed by all television broadcasts since. Thus, in principle, any civilization within about 50 light-years could watch our old television shows. However, while television transmitters are fairly powerful, their antenna systems are designed to spread the signal over a wide angle (so that everyone in the area can receive them). This means that the strength of any signal that is by

chance aimed at any given star is quite weak. And, like all light, these signals continue to weaken with distance (following an inverse square law, as shown in Figure 6.1). If another civilization has the same sort of receiving technology that we do, they could detect our television signals only if they were within about 1 light-year of us, nearer than the nearest stars (which are over 4 light-years away). Turning the situation around, we see that we are not yet capable of detecting signals in the first category listed above, unless for some reason they are being broadcast with much more power than we use for our own television signals. (Some of our military radars, which employ large antennas to narrowly focus their transmissions, could be detected as far away as a few tens of light-years.) The situation is no better for signals in the second category, at least if such signals are comparable in strength to those we currently use for communicating with our own interplanetary spacecraft.

Our technology is rapidly improving, and in the future we might be able to detect alien signals in any of the three categories. But for the moment, at least, our best chance of detection is with signals in the third category. Beacon signals should be the easiest to detect because they would deliberately be made strong enough to be heard across interstellar distances. They should also be the least difficult to interpret, because they presumably would be designed for easy decoding.

Of course, signals of the third type will exist only if other societies make a deliberate decision to broadcast them. We ourselves have done very little intentional broadcasting to the stars. The most powerful of these occasional transmissions was made in 1974 and lasted only 3 minutes (Figure 11.6). For this transmission, the powerful planetary radar transmitter on the Arecibo radio telescope was fired up and used to send a simple pictorial message to the object M13, a globular cluster containing a few hundred thousand stars. This target has many possible places where a civilization able to receive the message might be located. However, M13 is about 21,000 light-years from Earth, so it will take our signal some 21,000 years to get there and another 21,000 years for any response to make its way back to Earth.

Should we be broadcasting more beacon signals into space? While this question continues to intrigue both researchers and the public, the consensus has been that we should focus on receiving signals first. After all, as we've seen in our consideration of the Drake equation, the chances that anyone will pick up a signal depend on how long broadcasting civilizations are "on the air." There's little point in transmitting signals for only a few weeks or even a few years. A broadcasting project would require long-term investment and a great deal of patience. Perhaps it is too soon for us to consider such a project; after all,

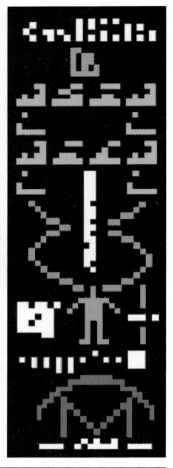

FIGURE 11.6 (**a**) The Arecibo radio telescope in Puerto Rico is the world's largest single radio dish, with a diameter of 305 meters (1,000 feet). (**b**) In 1974, a short message was broadcast to the globular cluster M13 using the Arecibo radio telescope. The message consisted of 1,679 bits, and each bit was represented by one of two radio frequencies. The bits make up a rectangular grid with 73 rows and 23 columns (each of these numbers is a prime number, which hopefully will enable any alien recipients to guess the layout of the grid). The resulting graphic represents the Arecibo radio dish, our solar system, a human stick figure, and a schematic of DNA and the eight simple molecules used in its construction. The colors are shown only to make the components clearer; the actual picture was sent in "black and white."

we developed radio technology only in the past century. There may be galactic civilizations that have had the ability to transmit signals into space for hundreds of millennia or longer. As the new kids on the block, it might make more sense to listen first.

THINK ABOUT IT . . . *Some people consider the 1974 broadcast to have been a dangerous exercise that might attract the unwanted attention of hostile aliens. In general, do you think it is "safe" for us to broadcast messages to the stars? Why or why not?*

Other Ways of Searching

Before we describe current SETI efforts, it's worth noting that there might be other ways of detecting extraterrestrial civilizations besides picking up signals traveling through interstellar space. These possibilities apply if civilizations have advanced far beyond our own capabilities and are able either to travel between the stars or to undertake major "astro-engineering" projects that would be visible at a great distance.

If other civilizations have achieved interstellar travel, then they might have visited our solar system. In that case, they may have left artifacts behind, either accidentally or deliberately. Some people claim that we already have such artifacts from UFOs, but as we discussed in Chapter 2 (and will discuss further in Chapter 12) these claims do not meet the standards of science. There are more plausible ways to imagine aliens leaving artifacts, and many science fiction writers have considered them. For example, in the classic movie *2001: A Space Odyssey,* based on the story by Arthur C. Clarke, highly advanced aliens buried a monolith on the Moon. When humans found the monolith, it notified the extraterrestrials.

It might be a while before we can search the Moon and the planets for artifacts such as buried monoliths, but a few researchers have suggested that aliens wishing to get our attention might leave calling cards in places where we could find them somewhat more easily. In particular, some people believe we should look especially hard at the so-called **Lagrange points** of the Earth–Moon system, named for the eighteenth-century French mathematician Joseph-Louis Lagrange (1736–1813). At these five positions in space, the effects of gravity from the Earth and the Moon "cancel" in such a way that, if you floated weightlessly at one of these five points, you wouldn't be tugged toward either body. Figure 11.7 shows the five Lagrange points for the

In the movie *Contact,* aliens make it easy for us to recognize their message by playing back to us one of our own television transmissions. In reality, our broadcasts have made it only a short distance into space (roughly 50 light-years), and it's unlikely that any alien civilization is near enough to have both found such a signal and returned it to Earth. How would we recognize and decode an alien message?

We've noted that today's radio SETI experiments look for narrow-band signals of the type that only a transmitter could make. The fact that a signal is confined to one spot on the radio dial would strongly suggest that it's artificial, because most natural signals would be expected to have broader bandwidths. If the signal also were flashing on and off or switching between two nearby frequencies, we would suspect that we have a coded message. We would undoubtedly record the pattern and try to analyze what was being "said."

The first thing to do would be to look for repetition in the pattern. Any deliberate message sent our way would be routinely repeated, because the senders couldn't be sure when we would detect it. Knowing the total length of the message might help us greatly in figuring it out. If the total number of flashes or frequency changes was, for example, 1,679 (as it was for the 1974 Arecibo message), then we might note that this is the product of two prime numbers, 23 and 73. (A prime number is one that can be divided only by 1 and itself without any remainder. The prime numbers are 1, 3, 7, 11, . . .) We could then arrange the message in a 23-by-73 grid and look for pictures or other figures.

This simple approach is one we have taken, but alien messages could be far more sophisticated. In the novel and movie *Contact,* the pictures were 3-D, not flat, and in that case we should look for messages whose length is the product of three prime numbers. Alternatively, the message might be encoded in ways similar to the schemes used for sending files on the Internet. Given the enormous variety of possible ways a message could be transmitted, it might be that we would never be able to understand an alien broadcast. However, if advanced societies truly wish to get in touch, then they would undoubtedly go to some trouble to make their messages simple enough to be understandable to any civilization able to build the telescopes necessary to receive them.

Earth–Moon system. Note that three of them are on a line passing through the centers of the Earth and Moon and that the other two are located 60° to either side of the Moon. These latter two positions, known as L4 and L5, are of particular interest. Unlike the other three, they are "stable," which means that if you started at L4 or L5 but then drifted slightly away, the competing effects of gravity from the Earth and the Moon would bring you back to your starting point. This effect is much like keeping a marble in a shallow bowl; if the marble is moved a bit off-center, it still rolls back to the bottom of the bowl. L4 and L5 would be obvious places to leave artifacts in cold storage for possible retrieval after long periods of time. Two decades ago, a limited survey of the Lagrange points was made with small telescopes. The hope was to find bright objects that might indicate the presence of parked artifacts. Although the search didn't turn up anything clearly artificial, these locations might be worth a more thorough reconnaissance in the future.

If interstellar travelers have not left artifacts for us to find, we might still be able to detect their powerful, interstellar spacecraft. As we will discuss in Chapter 12, spacecraft capable of traveling at speeds close to the speed of light will likely have enormous engines powered by energy sources such as nuclear fission, nuclear fusion, or matter-antimatter drives. These engines would leave telltale signs of their operation—signs that might be detectable at distances

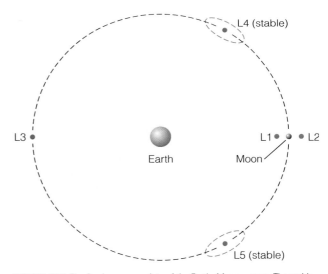

FIGURE 11.7 The five Lagrange points of the Earth–Moon system. The stable points L4 and L5, with the stable regions schematically indicated by the dashed ovals, are most attractive for the long-term parking of artifacts or probes. As the Moon orbits the Earth, the Lagrange points also orbit, so that they remain in the same relative orientation to the Earth and Moon.

of hundreds or even thousands of light-years away. Although no deliberate searches for distant rockets have been made, this type of phenomenon might be inadvertently discovered in the course of more conventional astronomical research.

Sufficiently advanced civilizations might also betray their presence through tremendous feats of "astroengineering." In terms of a civilization's ability to exploit natural energy resources, the twentieth-century Russian physicist Nikolai Kardashev suggested that there might be three distinct categories of civilization:

1. **Planetary (or Type I) civilizations,** which use the resources of their home planet

2. **Stellar (or Type II) civilizations,** which corral the resources of their home star

3. **Galactic (or Type III) civilizations,** which employ the resources of their entire galaxy

We are in the first category, since we exploit (often with abandon) only the meager resources of our home planet. The lights of our cities and the heat from our homes and factories, while considerable, are feeble on a cosmic scale and would be extraordinarily difficult to detect at the distances of the stars. Civilizations in the third category are so advanced that they might be difficult to imagine. As Arthur C. Clarke has stated, "Any sufficiently advanced technology is indistinguishable from magic."

However, we might be able to detect civilizations in the second category. Such a civilization, for example, might decide to fully capitalize on solar energy. A star like the Sun puts out a great deal of power. If we could capture just 1 second's worth of the Sun's total energy output, it would be enough to meet current world demand for energy for approximately the next 1 million years. But nearly all the Sun's energy escapes into space, and the Earth intercepts only the tiny bit that is headed our way. In principle, a technologically adept civilization could capture *all* of its star's energy by fashioning a large, thin-walled sphere (possibly built from a dismantled outer planet) around their solar system and covering the inner surface with solar cells or their equivalent. Such spheres are called **Dyson spheres** after physicist Freeman Dyson, who proposed their possible existence (Figure 11.8).

Because a Dyson sphere would absorb all the light of its interior star, you might think it would be invisible; however, the laws of physics dictate that waste heat must escape from the sphere. This heat would be radiated as infrared light that we could detect with specialized telescopes. Thus, we could discover the presence of a stellar civilization by finding the infrared signature of a Dyson sphere. Limited searches have already been undertaken, so far to no avail.

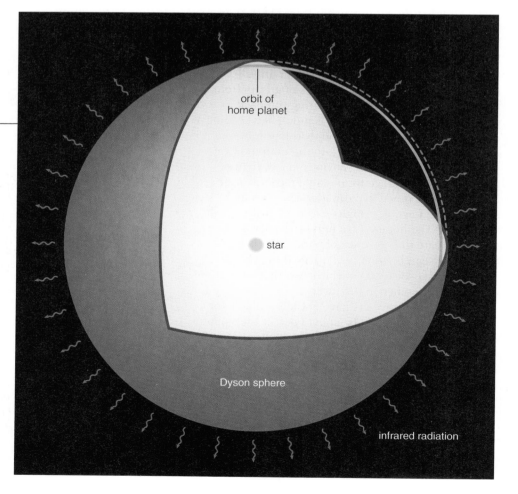

FIGURE 11.8 Schematic representation of a Dyson sphere, constructed exterior to an inhabited planet, with waste heat escaping from its outer surface in the form of infrared light.

11.4 SETI Today

Artifacts and "astroengineering" aside, the best chance we have of detecting another civilization is to eavesdrop on signals broadcast from other worlds. As we've discussed, all of the early SETI searches used radio telescopes. This may seem logical, since we often think of radio as something that we listen to. But remember that radio is a form of light and that we "listen" to it only after the light is converted to sound by a radio receiver. In essence, the radio station uses electronics to convert the information content of the sound (for example, music) into light waves. The light waves travel through the air to your radio receiver, where electronics convert the information back into sound. It just so happens that the wavelengths and frequencies of light we use for this process are in the radio part of the spectrum, which is why we say we are "listening to the radio." We use the same process to encode pictures in radio waves, which is why television also uses radio frequencies.

The use of radio waves for encoding sound or images is a technological choice, not a requirement. In principle, any form of light would do, though radio has some clear advantages. For example, unlike other forms of light, radio waves pass easily through walls, allowing radios and televisions to work both indoors and out. Nevertheless, we frequently use other forms of light for information transmission. For example, if your computer or telephone is hooked up to a fiber optic cable network, information is being transmitted in the form of infrared or visible light waves bouncing through the fiber optics.

In this section, we will describe current efforts to search for other civilizations by looking for messages encoded in light. We'll begin by considering radio searches, which have the longest pedigree in SETI experiments.

Radio SETI

The techniques of radio SETI in use today are elaborations on the earlier schemes proposed by Cocconi, Morrison, and Drake. A large radio dish—almost always a radio telescope constructed for more conventional research purposes—is used to collect the hoped-for cosmic signals. A low-noise amplifier at the antenna's focus boosts the signal levels before they are further processed. The processing usually consists of digitally "slicing" the incoming, wide-band signal (typically tens of megahertz in width) into many narrow-band channels. This processing is based on the assumption that a deliberate extraterrestrial signal will be broadcast only in a narrow band of frequencies.

During the search, the radio telescope may be either pointed in select directions, such as toward individual stars, or swept across the heavens (in either a random or a deliberate pattern) to study a larger section of the sky. The former technique, known as a *targeted search,* proceeds on the assumption that not all locations in space are equally probable sites for intelligent life. For example, Drake put forth the reasonable hypothesis that civilizations were most likely to exist on planets around Sun-like stars, and consequently he targeted his searches to these types of stars. The sweep technique, known as a *sky survey,* makes no assumptions about where intelligent aliens might be located.

Several radio SETI projects are currently under way. Table 11.1 summarizes key features of three major projects. Project Phoenix and SERENDIP both use the Arecibo telescope (see Figure 11.6a),

Table 11.1 Current Radio SETI Projects

Project	Telescope	Search Type	Number of Channels	Individual Channel Widths	Total Band Covered	Detectable Transmitter Power at 100 Light-years (assuming 100 m transmitting antenna)
Project Phoenix (SETI Institute)	Arecibo 305 m	Targeted	56 million	1 Hz	1,200–3,000 MHz	100,000 watts
SERENDIP (University of California, Berkeley)*	Arecibo 305 m	Sky survey	168 million	0.6 Hz	1,370–1,470 MHz	1 million watts
Southern SERENDIP (SETI Australia Centre)	Parkes 64 m	Sky survey	58 million	0.6 Hz	1,418.0–1,420.5 MHz	1 million watts

* About 2.5% of the data collected by the SERENDIP project is being distributed over the Internet for processing on a downloadable screen saver. This project (called SETI@home) has involved more than 3 million home computer users.

FIGURE 11.9 The 64-meter Parkes radio telescope in New South Wales, Australia. A SETI experiment "piggybacks" on this telescope while it is engaged in other astronomical research. Piggyback schemes avoid competition for telescope time and therefore provide SETI experiments with a lot of data. On the other hand, the SETI astronomers —only "along for the ride"—have no say in where the telescope is aimed. Since there are stars in every direction, this type of sky survey could still chance upon a transmitting civilization.

while Southern SERENDIP uses the Parkes radio telescope in Australia (Figure 11.9). For each project, the search type is indicated, along with the number of narrow-band channels being searched and the overall band of frequencies covered in each search. The last column of the table gives an indication of how sensitive each experiment is. It lists the minimum transmitter power required of an alien broadcaster to produce a detectable signal. For the purposes of the table, the aliens are assumed to be 100 light-years away and using a transmitting antenna 100 meters in diameter. Project Phoenix, for example, could detect such an extraterrestrial broadcasting setup if the transmitter had a power of 100,000 watts or more. Many radio stations on Earth are this powerful, although they are not using a 100-meter antenna to narrowly focus their transmissions.

While today's radio SETI experiments are vastly more sensitive and comprehensive than Drake's pioneering Project Ozma, they are limited by two important constraints. First, all current projects use telescopes that are also used for other astronomical research. Thus, they get only limited observing time. In some cases, such as Southern SERENDIP, the SETI studies "piggyback" on other observations,

which means that the choice of where the telescope points is dictated by the needs of other research programs. Second, terrestrial radio interference, especially from radar and orbiting satellites, is becoming an increasing problem for SETI efforts. The first problem has at least a partial solution in sight. In 2005, construction should be complete on the Allen Telescope Array in Northern California (Figure 11.10). This instrument will consist of approximately 350 small (6-meter) radio dishes with a combined collecting area equivalent to that of a single 100-meter telescope. The Allen Telescope Array is a joint venture of the SETI Institute and the University of California, Berkeley, and will be used full-time for SETI observations.

Solving the second problem is more difficult. Terrestrial radio interference greatly hinders SETI searches, because telecommunications satellites and earth-bound radar produce lots of narrow-band signals—exactly the type being sought. These signals are so strong that they are picked up by radio telescopes no matter which direction they are pointed. SETI researchers have devised various tricks to sort out terrestrial signals from possible extraterrestrial ones, but this problem will continue to worsen. Perhaps some time in the future, we will be able to

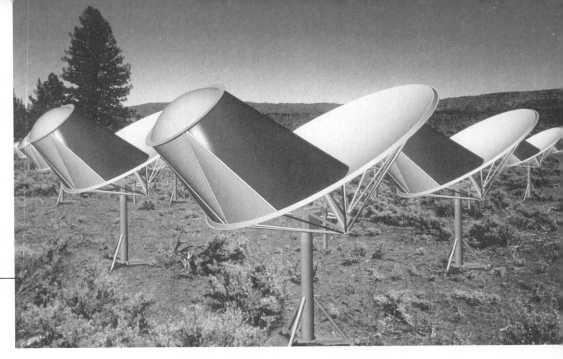

FIGURE 11.10 Artist's impression of the Allen Telescope Array, a telescope that will be used full-time for SETI and simultaneously for conventional radio astronomy studies. The array will comprise 350 radio dishes, each 6 meters in diameter, and will be situated at Hat Creek, California. This instrument is scheduled for completion by 2005. (The "hoods" are structures that minimize the amount of radio noise from the ground below the antennas, noise that would otherwise reduce the telescope's sensitivity.)

place a radio telescope on the back side of the moon, where it would be shielded from Earth's noisy radio presence. Meanwhile, however, SETI can take another approach that is far less sensitive to interference: looking for signals with visible light.

Optical SETI

The idea of communicating between worlds via visual signals has a long and interesting history. In fact, scientists considered using light as a communication medium even before radio's invention. Karl Gauss, the eminent nineteenth-century German mathematician, is said to have suggested planting trees in the form of a right triangle in Siberia. The triangle was to be bounded by clear-cut squares. This greenery graphic presumably would be seen by inhabitants of the Moon, proving that Earthlings are at least smart enough to know Pythagoras' theorem. While this story may be apocryphal, Gauss is known to have proposed using 100 mirrors, each about 1 meter on a side, to shine the light of gas lamps into space. "This would be a discovery even greater than that of America, if we could get in touch with our neighbors on the Moon," he claimed. Another nineteenth-century suggestion was to dig trenches in the Sahara in various geometric shapes, fill them with water and oil, and ignite them at night. Proposals such as these were intended to put us in touch with sophisticated societies that were imagined to live on the Moon or on Mars.

Despite these old ideas, until fairly recently optical SETI did not draw much attention from researchers. Part of the reason had to do with estimates of the energy needed. In general, it requires far less energy to send an interstellar signal via radio than via visible light, so researchers guessed that radio would be the aliens' technology of choice. In addition, while radio waves travel through interstellar space with ease, visible light tends to be absorbed by tiny grains of interstellar dust that float between the stars. This dust creates the dark rift through the center of the Milky Way as we see it in our sky (Figure 11.11). When we look in directions within the galactic plane, the dust completely blocks our view of visible light beyond a few thousand light-years—a relatively short distance compared to the 100,000-light-year diameter of our galaxy. The fact that visible light can be used to send signals only for limited distances through the galaxy seemed to argue against its use as a beacon for interstellar communication.

Researchers no longer believe that either problem is a serious strike against optical signaling. The limitations of interstellar dust are real, but signals that can travel up to a few thousand light-years could still be quite useful. After all, many millions of stars lie within 1,000 light-years of Earth. Moreover, if longer-distance communication is needed, using infrared rather than visible light would largely circumvent the problem, because infrared light penetrates the interstellar dust. The energy-cost issue argues in favor of radio only if we consider transmitters that beam their signals in all directions into space. For focused communication, such as sending a signal to a particular nearby star, visible light can easily be concentrated into a narrow beam (by using a large lens or mirror). Whereas radio is ideal for a radio station that wants its signal to be picked up by

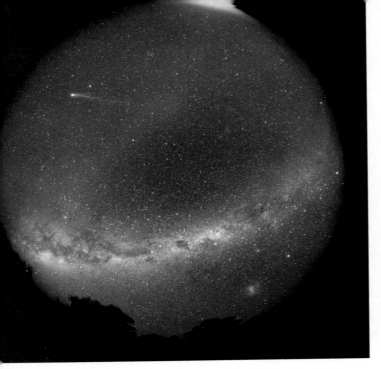

FIGURE 11.11 A "fish-eye" photograph of the Milky Way in the Australian sky. The dark rift running through the center of the Milky Way is dark because starlight is blocked from our view by interstellar dust. (Comet Hyakutake is visible near the upper left in this 1996 photo.)

Optical SETI efforts today are attempting to do just that. Like their radio counterparts, these efforts are entirely passive—that is, they are searching for incoming signals but are not transmitting anything outward—and are systematically checking out the vicinities of local stars. The experiments look for very short pulses of light that are bright enough to outshine the background shower of photons naturally produced by the alien world's host star. Photomultiplier tubes, sensitive light detectors able to see very short flashes, are affixed to conventional mirror telescopes to hunt for these photon bursts. Figure 11.12 shows the setup for an optical SETI experiment at the Lick Observatory, near San Jose, California. This experiment is very sensitive: If someone is sending out laser pulses with a transmitter no more sophisticated than we ourselves could build, the Lick experiment could pick it up from as far away as about 500 light-years.

THINK ABOUT IT . . . *It has been suggested that one benevolent use for our nuclear weapons would be to rocket them into space, line them up in geometric patterns, and then detonate them as a "broadcast signal" to distant alien societies. Can you think of reasons why this would be a largely ineffective SETI signal?*

Other Types of Signals

We have considered both radio and optical SETI, but are these necessarily the only or best ways to communicate? As already noted, we can easily imagine infrared signals being used in place of visible signals in order to penetrate interstellar dust. Similarly, distant civilizations might choose to communicate with ultraviolet light or X rays. However, these other forms of light have no obvious advantages over radio and optical options, and they are more expensive in terms of energy. As a result, SETI researchers feel justified in concentrating at least our current, early searches on radio and optical frequencies.

What about signals using more exotic technologies? For example, some people have suggested that advanced civilizations might send messages via the ghostlike subatomic particles called *neutrinos* or via the so-called *gravity waves* that Einstein predicted but that we have not yet detected directly.[2] Of course, we cannot rule out these possibilities, but we might wonder what the point would be. After all, neither of these schemes would communicate any faster than light, and in fact they would make both

people living in all directions from the station, visible light from a flashlight (or a laser beam) is a better medium if you want to signal someone in a particular place. SETI researchers have concluded that visible light might well be useful for interstellar communication and that optical SETI thus could be a worthwhile enterprise.

How might an alien civilization communicate optically? Simply turning a continuously shining, high-powered laser on someone else's star system isn't particularly effective. A continuous beam could be lost in the glare of starlight from the transmitter's home sun unless it's extremely bright. In addition, a light that is always on doesn't convey much information (think of Morse code). Short bursts of laser light, say a billionth of a second long, would work better, because they can momentarily outshine the starlight even with relatively modest laser power. A series of such bursts—a pulse train—could contain patterns that make up a message.

Such arguments have led us to imagine that advanced alien societies might be using automated, high-powered laser transmitters to send pulsed messages to thousands of nearby stars. The messages, perhaps only a few seconds long, could be repeated every several dozen hours. If a civilization in our galactic neighborhood is sending laser messages, then we might hope to find the laser "pings" by monitoring nearby stars.

---

[2] However, we have strong indirect evidence for the existence of gravity waves, and a new experiment called LIGO may detect them within just a few years.

transmitters and receivers far harder to build. It is also possible that sufficiently advanced civilizations have developed communication technologies that are undetectable by us. These technologies might involve either physics that we have not yet discovered or simply ways of disguising signals so that our detectors are not sensitive enough to pick them out from natural background signals. Thus, if other civilizations *want* to keep their communications secret from us, they probably can. SETI hopes rest with civilizations that either are looking for contact or don't care if their communications are intercepted by others. In either of these cases, sticking with the simplest available technologies would seem to make the most sense.

11.5 Considerations of a SETI Success

As we've discussed, nearly all SETI experiments are looking for either narrow-band radio signals or extremely short bursts of light. A narrow-band radio signal of the right type would immediately betray the presence of intelligence. Similarly, the light pulses sought in optical SETI searches would also be distinctive, since such bursts of light aren't normally produced by stars. Indeed, it is the *artificiality* of signals, rather than any particular message content, that would convince us that we had uncovered an extraterrestrial presence.

So far, no SETI experiment has turned up a confirmed extraterrestrial transmission. Several intriguing signals have been reported, however. Perhaps the best known was found in 1977 by the (now defunct) automated SETI search at the Ohio State Radio Observatory. When an Ohio State astronomer examined the data from the night's observing, in the form of a computer printout, he found a signal so impressively strong that he wrote "Wow" on the printout margin.

FIGURE 11.13 Observing for Project Phoenix, the SETI Institute's radio search program, in the control room at the Arecibo radio telescope in Puerto Rico. The observer, Jill Tarter (the prototype for the fictional Ellie Arroway in the novel and movie *Contact*), monitors the progress of the experiment on computer work stations. It is the computers that do the "listening," since many millions of channels are monitored simultaneously.

The "Wow" signal, made popular by its appealing nomenclature, was never seen again despite repeated attempts at the same frequency and the same sky position. Consequently, SETI researchers do not consider it a detection of extraterrestrial intelligence but rather some sort of terrestrial interference.

Despite their continued failure to find a persistent, verifiable signal, however, those engaged in SETI research remain optimistic (Figure 11.13). The motivation for this optimism is the rapid improvement in the abilities of the telescopes and detectors used in SETI. Because much of the hardware required for SETI is built with digital electronics, SETI research benefits from the rapid growth in computer capability. The density of transistors placed on silicon chips has been doubling approximately every 18 months for decades now (a phenomenon sometimes called Moore's law, after Gordon Moore, a cofounder of Intel Corporation). This exponential improvement in electronics motivates the optimism typical of SETI researchers.

What Happens If a Signal Is Found?

What will happen if we do receive an artificial signal from the stars? A signal detection would need to be thoroughly verified by continued observations both at the discovery telescope and at other observatories. The nature of the signal itself will evidence artificiality, but scientists need to be sure it isn't due to terrestrial interference, equipment failure, or a college prank. It might take many weeks of careful work by excited astronomers before they felt fully confident of a genuine detection.

Once a detection has been verified, the news must be released to the outside world. Given the potential implications of the discovery of an extraterrestrial civilization, this step must be taken with some care. For this reason, SETI research groups have agreed to follow a protocol known as the *Declaration of Principles Concerning Activities Following the Detection of Extraterrestrial Intelligence*. The protocol has no force of law, but it lays out a reasonable course of action. In particular, once a signal had been found and verified as truly extraterrestrial, astronomers around the world would be notified so that they could swing their telescopes in the direction of the signal and learn as much as possible. Governments and the public would also be informed directly. The open nature of research and the necessity to confirm any detection by using other telescopes dictate that the discovery could not be kept secret or covered up. The evidence, after all, is "up in the sky" and accessible to anyone with access to a large telescope. A recent addition to the protocol addresses the matter of sending a reply to any detected signal. The SETI community has suggested that any deliberate response should represent a consensus of the world's population, not just the wishes of whatever group made the detection.

What We Could Learn

Any SETI success in itself would be an astonishing event, because it would prove that we are not alone in the universe. Depending on the nature of the signal, however, we might learn far more.

Suppose we found a message—bits of information—accompanying a radio or an optical signal. Might we be able to decipher it? In fiction, messages from aliens often have a mathematical bent, because we assume that mathematics will be developed by any technological society. The fictional aliens send us the value of pi or numbers from a well-known mathematical series. This is usually done so that we will recognize that the signal is from intelligent beings. But this sort of labeling wouldn't be needed, because we are seeking narrow-band radio signals or pulsed laser light that clearly would be artificial. In addition, using mathematics as a "language" would be a cumbersome way to express things such as political systems, religious beliefs, art, or even physical appearance. Our own messages to space have been pictorial (see Figure 11.6b and Figure 12.2), and pictures might be a better way to communicate to unfamiliar societies.

If a civilization is broadcasting to us, it is likely to be hundreds or even thousands of light-years distant, and two-way communication would be tedious at best. A broadcasting society therefore might not expect a reply and might simply send a one-way message, perhaps some version of their encyclopedia. Such a message would take us a while to figure out but would be worth the trouble. From our discussion of the Drake equation, we learned that the chances of detecting another society increase with the age of that civilization. Consequently, if we do find a signal, the chances are great that it comes from a civilization far older than our own. The knowledge we might gain from such an advanced society would be enormous.

Finally, we should also be aware of the possibility that the signal we find will not be a beacon intended for societies like our own and might contain a message that we can never decode. Even in that case, we would at least have learned that we are neither alone nor particularly special.

THE BIG PICTURE

In this chapter, we have explored the rationale and the methods of the search for extraterrestrial intelligence. As you continue in your study, keep in mind the following "big picture" ideas:

- SETI is both a part of and distinct from other efforts in astrobiology research. Its justifications and methods depend on what we learn more generally about life in the universe. Whereas other astrobiology research makes slow and steady progress, however, SETI offers the potential to give us absolute proof that we are not alone in one fell swoop—but only if we receive a clear signal from another civilization.

- We do not yet know enough about life in the universe to make a reasonable estimate either of the number of civilizations that might exist or of our odds of achieving success in SETI efforts. However, the Drake equation gives us a way to identify areas in which research is needed, and at least some of the evidence from evolution suggests that intelligent beings might be common.

- SETI today is conducted primarily by searching for either radio or optical signals transmitted by distant civilizations. There may be other means of interstellar communication, but it seems reasonable to suppose that radio or optical signals will be used by at least some, if not all, other civilizations.

- The scientific and technological issues of SETI are important but may well pale in comparison to the societal issues if a signal is found. As a result, an important part of SETI work involves thinking about what will happen if the search ultimately proves successful.

Review Questions

1. What is the *Drake equation?* Describe what we know about each of its terms, and briefly discuss both its value and its limitations.

2. What is *convergent evolution?* How does this idea suggest that intelligence would tend to be an evolutionary imperative?

3. Briefly explain the idea of the *encephalization quotient (EQ)*. How does it suggest that humans are indeed intelligent? What does it tell us about intelligence among other animal species?

4. Briefly describe early attempts at interplanetary communication by Marconi and Tesla. Why were these attempts doomed from the start?

5. Briefly discuss early SETI efforts. What do we mean by narrow-band versus wide-band signals?

6. Briefly discuss the possibilities of finding other civilizations via artifacts or "astroengineering."

7. Summarize the current techniques of radio SETI and some of the major current projects.

8. Explain the method behind current optical SETI efforts.

9. Briefly discuss some of the issues that would surround an actual SETI detection.

Discussion Questions

1. *Measuring Intelligence.* In judging the intelligence of animals, we use the *encephalization quotient* (EQ), which depends on the ratio of brain weight to body weight. Can you think of situations for which this might be a poor way to gauge intelligence? For example, could animals have some special processing needs (such as the navigation mechanism of bats) that would make their brains larger without contributing to their intelligence? Alternatively, could animals exist (on Earth or elsewhere) whose brains were relatively lightweight but who were still highly intelligent? Defend your opinion.

2. *Detecting Signals.* SETI scientists are sometimes criticized for using "old technology" in their search for signals. Perhaps extraterrestrials have moved beyond radio and light signaling and are using something much more sophisticated. Discuss (1) the advantages of radio and light for interstellar communication and (2) any reasonable alternatives you can think of. There is always the possibility that "new physics" will provide faster or more efficient methods for signaling. Do you think this is a reason to limit current SETI efforts?

3. *Societal Reaction.* It is frequently said that the detection of a signal by SETI would revolutionize human society. Does this statement seem reasonable? Some researchers have tried to find historical events, such as the Copernican revolution or the publishing of Darwin's theories of evolution, whose impact might compare to that of a SETI detection. Are such examples likely to be accurate in predicting how we would react? How likely do you think it is that a SETI discovery would cause either mass panic or an outbreak of universal brotherhood?

Problems

Evaluate the Opinions. Each of **problems 1–8** makes a clear statement of opinion. Evaluate each statement and write a paragraph or two describing why you agree or disagree with it.

1. Humans are the "crown of creation" and an inevitable result of billions of years of evolution.

2. If, for some reason, we humans were to suddenly wipe out our species, another species—possibly the raccoons—would soon evolve greater intelligence than we possessed.

3. Sea creatures, no matter how clever they are, could never master the technology required to communicate with other worlds.

4. Most of the intelligence in the universe is not biological, but artificial ("machine intelligence").

5. Because SETI researchers are "listening" to star systems that are hundreds of light-years distant, there's a good chance that by the time we hear a signal the civilization that sent it will have disappeared.

6. No advanced society would ever construct a beacon transmitter, because it would inevitably attract attention and might be dangerous. Similarly, we should not make deliberate transmissions to the stars.

7. We should consider including an "artifact hunt" in the space program that would search on the Moon for objects left behind by advanced, extraterrestrial societies.

8. Looking for signals from star systems is a poor approach, because any truly advanced civilization will have moved beyond its home planet and populated interstellar space.

9. *Communication.* Imagine that you have to fashion a short message that would tell extraterrestrials something about human society. What would you "say" using only three simple pictures? What would you "say" if you could use only a half-page of English text?

10. *Talking Back.* Suppose SETI were to find a signal coming from a star system 200 light-years away. Write a one- to two-page essay describing what, if anything, we should do to establish contact. You should think about how quickly we should respond, what the response should be, and what possible dangers might be involved.

11. *Contact.* Watch the movie *Contact,* and pay careful attention to the SETI experiment described in the first third of the film. How accurately does this experiment reflect any of the current SETI search programs? Did you spot any obvious scientific or technical errors? Write a one- to two-page essay comparing *Contact* to the reality of SETI efforts.

*12. *How Many Stars to Search.* The number of star systems that a SETI search would have to investigate before achieving success depends on how common signaling societies are in the galaxy. This is the number estimated by the Drake equation. Suppose this number is 1 million. How many star systems must be checked out by SETI in order to find one signal? What if the number is 10,000? (Assume that there are roughly 400 billion stars in our galaxy.)

*13. *Distance to E.T.* The stars near us are separated by roughly 5 light-years on average and are fairly uniformly distributed in all directions. Suppose the number of signaling societies in our galaxy is 10,000 (assume a total of 400 billion stars in our galaxy). Make an estimate of how far it is to the nearest broadcasting civilization by assuming that the galaxy is a cube filled with stars that are all on a perfect lattice, separated from each other by 5 light-years.

*14. *Power Used by E.T.* Using the Arecibo telescope, which is 305 meters (1,000 feet) in diameter, our SETI receivers could detect a signal from a similar-size transmitter 100 light-years away if its power is 10,000 watts. Assuming that the alien civilization is using this same transmitter setup but is on the other side of the galaxy (roughly 80,000 light-years away), how large an antenna would we need to hear the signal? Keep in mind that with every doubling of distance, signals weaken by a factor of four.

Web Projects

1. *Current SETI Research.* Go to the SETI Institute's Web site and use links listed there to make an inventory of current SETI projects worldwide. Organize these projects according to whether they are radio or optical, and then separate targeted searches from sky surveys. Prepare a one-page summary of this table that discusses how thorough the current searches for extraterrestrial signals are. You should consider how many star systems have been looked at carefully, how wide a band (for radio searches) has been covered, and how sensitive the searches are.

2. *SETI@home.* This is a project organized by researchers at the University of California, Berkeley, to process radio SETI data on home computers. Download the free SETI@home screen saver onto your computer, and use it to analyze data collected by Project SERENDIP. Write a one-page description of the general processing scheme used by SETI@home, as well as the types of signals it is searching for.

3. *What to Do in Case of a Signal Detection?* Download the text of the protocol *Principles Concerning Activities Following the Detection of Extraterrestrial Intelligence* from the SETI Institute's Web site. Write a short discussion of what SETI scientists expect will happen if they detect an extraterrestrial signal and whether this expectation is realistic. In particular, do you think a discovery could be "covered up," or would it leak out to the public before the scientists themselves were sure of the discovery?

CHAPTER 12
Interstellar Travel

Science fiction routinely portrays our descendants hurtling through the galaxy, wending their way from one star system to another as easily as we now travel from one country to the next. We have already used rockets to explore other worlds in our solar system. Could an extension of this technology permit future generations to travel among the stars?

In an age of rapid technological progress, it may seem inevitable that our rockets will soon reach the depths of space. The reality, however, is that interstellar travel is much more challenging than bridging the distances to nearby moons and planets. There are engineering and physical constraints, not least of which is the cosmic speed limit—the speed of light. Nevertheless, it is important for us to consider the possibility of interstellar travel, not only for our own future but for what it means to the search for intelligent life elsewhere. If advanced societies are capable of traveling from star to star, then civilizations might be widespread even if the development of intelligence is itself a rare phenomenon.

In this chapter, we will discuss why rocketing humans to the stars remains well beyond our current capabilities. We will also study both the possibilities and the challenges of interstellar travel; this will allow us to gauge the likelihood that other species might now be traveling among the stars. Finally, we will revisit the question of UFOs as we consider the credibility of claims that extraterrestrials are visiting the Earth.

12.1 The Challenge of Interstellar Travel

The fact that interstellar travel is a daunting enterprise is due to a simple circumstance: the tyranny of distance. The stars are so incredibly remote that only in the nineteenth century did astronomers develop instruments of sufficient precision to measure the distances of even the closest other suns. When it was realized just how far away these pinpoints of light are, the French philosopher Blaise Pascal was moved to write that "the eternal silence of infinite spaces" left him terrified.

We've considered the vast distances to the stars in earlier chapters, using the scale model of the solar system introduced in Chapter 1. Here we consider the distances from the point of view of travel. Four of our past interplanetary probes—*Pioneers 10 and 11, and Voyagers 1 and 2*—are traveling fast enough to escape the solar system. How long will it take them to reach the stars?

The probes are all traveling at similar speeds, so let's take the first, *Pioneer 10,* as an example. This spacecraft, launched in the early 1970s, took 21 months to reach its target, the planet Jupiter, before heading out of the solar system. This might seem speedy enough in view of the fact that the giant planet is never closer than 628 million kilometers to Earth. But our nearest stellar neighbor, the Alpha Centauri star system, is 70,000 times farther away than Jupiter. If *Pioneer 10* were to cover the 4.4 light-years to Alpha Centauri at the same average speed at which it traveled to Jupiter, the journey would take

115,000 years. But *Pioneer 10* was not aimed at Alpha Centauri or at any other deliberate target. Like the other three probes, it is moving into space along a trajectory determined by its final planetary encounter (Figure 12.1). Plotting this trajectory along with the motions of nearby stars, we find that the closest *Pioneer 10* will come to any star in the next million years is 3.3 light-years. In about 2 million years, the probe will reach the general neighborhood of the bright star Aldebaran, in the constellation Taurus. (Anyone who has ever wondered why we did not equip the Voyager or Pioneer spacecraft with instruments for studying other stars now has the answer.)

The Pioneer and Voyager spacecraft might seem unlikely emissaries to the stars, but all four carry simple messages for any extraterrestrial beings that might find them. The Pioneer probes are fitted with a small engraved plaque bearing a drawing of a man and a woman as well as diagrams giving the layout of the solar system and our general location in the galaxy (Figure 12.2). The Voyager craft, launched 4 years after the Pioneer craft, carry a somewhat more sophisticated message consisting of pictures, multilingual greetings, and two dozen musical selections (ranging from Chuck Berry to Bach) on a gold-plated copper record (Figure 12.3). Although you might wonder how intelligible these earthly calling cards might be to any aliens, the chances that they will ever be found are slim. They are like bottled messages thrown into the ocean surf and were intended more as a statement to Earthlings than to extraterrestrials.

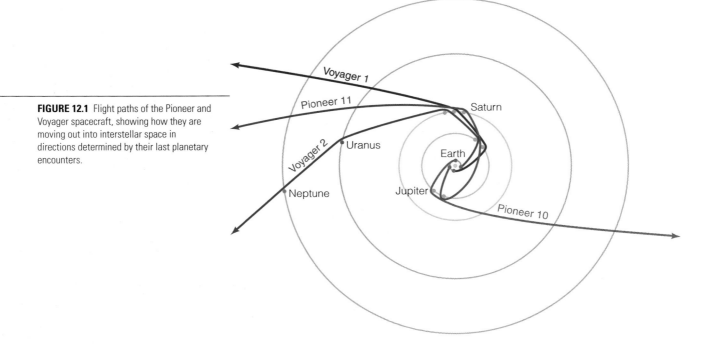

FIGURE 12.1 Flight paths of the Pioneer and Voyager spacecraft, showing how they are moving out into interstellar space in directions determined by their last planetary encounters.

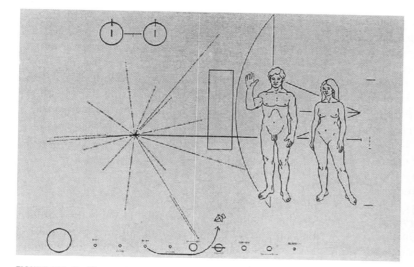

FIGURE 12.2 The Pioneer plaque, designed by Carl Sagan, Linda Sagan, and Frank Drake, shows human figures in front of the spacecraft (which provides them scale) as well as a diagram of our solar system (bottom). The spacecraft's origin, the third planet from the Sun, is schematically indicated, as is the Sun's position relative to nearby pulsars (the "prickly" graphic to the left). The periods of these pulsars, indicated in binary code, will help anyone reading the plaque to determine when the craft was launched, because pulsars slow over time. The radio frequency emitted naturally by neutral hydrogen, also indicated on the plaque, provides the unit of time for the pulsar data. The plaque is about the size of an automobile license plate and was engraved in a California trophy shop.

FIGURE 12.3 *Voyagers 1* and *2* carry a phonograph record—a 12-inch gold-plated copper disk containing music, greetings, and images from Earth.

THINK ABOUT IT . . . *The Pioneer and Voyager "messages" are in the form of sounds, music, and pictures. But this assumes that any aliens finding these craft would have sensory organs similar to ours. Is it possible that our messages are too anthropocentric to even be recognized, or are there good reasons to think that E.T. will have eyes and ears, with characteristics similar to ours? Defend your opinion.*

The Cosmic Speed Limit

The *Pioneer 10* example makes the problem of interstellar travel quite clear. But it also seems to offer an obvious solution: Build spacecraft that can travel a lot faster. If it would take a little more than a hundred thousand years for *Pioneer 10* to reach the nearest stars, then a spacecraft that travels 100,000 times faster should be able to make the trip in only a little over a year.

Unfortunately, this seemingly obvious solution is not allowed by the laws of physics. In particular, we know from Einstein's *special theory of relativity* that it is impossible to travel through space faster than the speed of light. This might not seem too limiting, given that light travels incredibly fast—about 300,000 kilometers per second (186,000 miles per second), fast enough to circle the Earth almost eight times in just 1 second. But even at this remarkable speed, light takes time to travel the vast distances

between the stars, which is why we measure stellar distances in light-years [Section 1.2]. The nearest stars are about 4.4 light-years away, which means it takes their light 4.4 years to reach us. Because that is the fastest possible speed of travel, any spacecraft we build would take longer than 4.4 years for the one-way trip and hence at least 8.8 years for a round-trip. To make a trip across our entire galaxy—a distance of 100,000 light-years—would take any spacecraft a minimum of 100,000 years.

Could it be that Einstein's theory is wrong and that we will someday find a way to break this cosmic speed limit? This is unlikely. Special relativity merits the status of being a scientific theory because it is supported by an enormous body of evidence. Its predictions have been carefully tested and verified in countless experiments, so it cannot simply be "wrong." While it might someday be replaced by a more comprehensive theory, the verified results will not simply disappear; the cosmic speed limit will almost certainly remain in place.

The fact that we cannot exceed the speed of light might at first make distant stars seem forever out of reach. However, this is not entirely the case, thanks to other effects of special relativity. Einstein's theory also tells us that time and space are not as absolute as we think of them in our everyday lives. In particular, time and space differ for objects—or spacecraft or people—moving at different speeds. While it is true that no spacecraft we send to Alpha Centauri could ever return home sooner than 8.8 years after it left, people on the spacecraft would see things very differently if they traveled at a speed sufficiently close to the speed of light. As we'll discuss in Section 12.4, this effect means that, with starships traveling at speeds near the speed of light, we *could* travel to very distant places within our lifetimes—but upon returning home, we would find that much more time had passed for those on Earth than for us.

There might be another way around the cosmic speed limit. Einstein's second famous theory, the *general theory of relativity*, suggests the existence of dimensions outside our normal dimensions of time and space. Thus, while we cannot travel *through* space at a speed faster than the speed of light, it might be possible to find ways either to traverse these other dimensions or to "warp" space within

them, creating "shortcuts" to distant locations. In principle, the shortcuts could make distances to the stars far less for the travelers than they appear to us on Earth, thereby allowing much shorter trips even while still obeying the cosmic speed limit. We'll discuss these possibilities in Section 12.5. For the time being, at least, we have no idea how to escape ordinary time and space—or even if it is possible—and thus are stuck with the travel limitations imposed by the speed of light.

Energy Issues

Another challenge of interstellar travel is the tremendous amount of energy it would require, particularly if we wanted to send people and not just lightweight robotic probes to the stars.

RELATIVITY AND THE COSMIC SPEED LIMIT

Einstein's special theory of relativity is often portrayed as being difficult, but its basic ideas are easy to understand. Let's begin with an example that explains the "relative" part of relativity.

Imagine a supersonic plane trip from Nairobi, Kenya, to Quito, Ecuador. Both cities happen to be located very near the Earth's equator (Figure 12.4). If the plane traveled westward at a speed of 1,650 km/hr, it would precisely match the speed of the Earth's eastward rotation. Thus, if you could watch this trip from, say, the Moon, it would appear that the plane never goes anywhere: It lifts off the ground in Nairobi, stays still while the Earth rotates beneath it, and finally settles back to the ground when Quito arrives at its location. Of course, the travelers on the plane would see things quite differently, believing that they had flown across the face of the Earth at supersonic speed.

We therefore have two different viewpoints for this trip. People on Earth say the plane traveled westward across the Earth's surface at a speed of 1,650 km/hr, while observers on the Moon say the plane stayed still (a speed of zero) while the Earth rotated beneath it. In fact, there are many other equally valid viewpoints. For example, aliens looking from afar at our solar system would see the plane moving at a speed of more than 100,000 km/hr—the Earth's speed in its orbit around the Sun. Observers living in a very distant galaxy would see the plane moving away from them at a significant fraction of the speed of light, since it would be carried along with our galaxy in the expanding universe. The only thing all observers would agree on is that the plane is traveling at a speed of 1,650 km/hr *relative* to the surface of the Earth.

This example shows that questions like "Who is really moving?" and "How fast are you going?" have no absolute answers. Einstein's special theory of relativity gets its name from the fact that it tells us that measurements of time and space, as well as measurements of motion, make sense only when we describe whom or what they are being measured relative to.

The theory of relativity does not, however, say that "everything is relative." In fact, it tell us that two things in the universe are absolute:

1. The laws of nature are the same for everyone.
2. The speed of light is the same for everyone.

It is from these two "absolutes" that all the strange consequences of the theory follow, such as the impossibility of exceeding the speed of light and the fact that travelers moving at high speed relative to Earth will measure a different trip time than people who stay at home. We will not go through all the consequences here, but let's look at the most famous one: the cosmic speed limit.

As we will see, the speed limit follows from the fact that the speed of light is the same for everyone, meaning

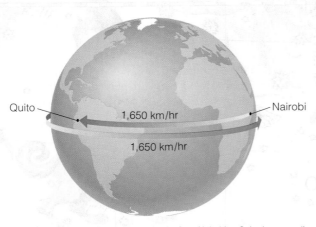

FIGURE 12.4 A plane flying at 1,650 km/hr from Nairobi to Quito (westward) travels precisely opposite the Earth's eastward rotation. Thus, viewed from afar, the plane remains stationary while the Earth rotates underneath it.

Imagine that we wanted to colonize an extrasolar planet. To get a decent-size colony started, we'd need to send a fair number of people with many different sets of skills. For the sake of argument, suppose we wanted to send 5,000 people, meaning we would need a starship of similar capacity to the large starships used in the *Star Trek* television shows and movies. How much energy would such a ship require?

Interestingly, the minimum energy requirement doesn't depend on the fuel source or the ship design at all. Sending a bowling ball flying through the air takes more energy than sending a baseball flying at the same speed, regardless of whether the energy comes from your arm, from a catapult, or from some kind of gas-powered launcher. Similarly, sending either ball flying at a faster speed takes more energy. That is, the energy required to put an object in motion depends on only two things: the object's mass and the speed with which you want it to move.

We can estimate the mass of the starship by comparing it to other ships that transport large numbers of passengers. For example, the *Titanic* weighed about 18,000 kilograms per passenger, but its accommodations for most passengers were hardly roomy and it carried provisions for only a couple of weeks, not many years. If we conservatively adopt the same per-person weight for our starship, we would expect its total mass to be about 100 million kilograms. Let's assume further that our starship travels at a modest 10% of light speed, which means it will take more than 40 years to reach the nearest stars. Now that we

that no matter how we travel we'll always measure light to be traveling at light speed. How do we know this is true? First, it has been verified by many experiments. No matter how objects are moving, the speed of light always turns out to be the same. In addition, our observations of the cosmos prove that this must be true. Some distant galaxies are moving away from us at speeds close to the speed of light, yet their light still arrives here traveling at the speed of light. Perhaps even more convincingly, if the speed of light were not always the same, we could not see distinct stars in binary star systems. Consider the star moving toward us in Figure 12.5. If the speed of light depended on the star's motion, its light would be coming toward us somewhat faster than the "normal" speed of light. Half an orbit later, when it was moving away from us, its light would travel to us at less than the "normal" speed of light. Thus, the light from the "fast" side of the orbit would tend to catch up with the light emitted earlier from the "slow" side, reaching us at the same time and smearing the star's image so that we'd see it in different positions all at once, instead of seeing it as a distinct star. Thus, the simple fact that we see distinct stars in binary star systems proves that the speed of light does not depend on the stars' motions; that is, light always travels at the same speed.

We can now see where the cosmic speed limit comes from. Imagine that you have just built the most incredible rocket possible and are taking it on a test ride. You accelerate away from Earth and just keep going faster and faster and faster. With enough fuel, wouldn't you eventually be moving away from Earth at a speed faster than the speed of light?

No, and here's why. Remember that you will always measure light to be traveling at the speed of light, 300,000 km/s. Thus, *you* will always find that the light from your rocket's headlights moves out ahead of you at the speed of light. This shouldn't be too surprising, because it's just another way of saying that you can't catch up with your

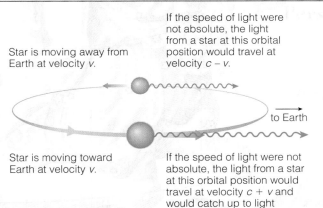

Star is moving away from Earth at velocity *v*.

If the speed of light were not absolute, the light from a star at this orbital position would travel at velocity *c − v*.

to Earth

Star is moving toward Earth at velocity *v*.

If the speed of light were not absolute, the light from a star at this orbital position would travel at velocity *c + v* and would catch up to light emitted from other positions in the orbit.

FIGURE 12.5 If the speed of light were *not* absolute, the speed at which light from a star in a binary system comes toward Earth would depend on its position and velocity toward us in the binary orbit. Thus, we would not see the star as a distinct point of light.

own light. But here's the key: Because everyone always measures the same speed of light, the people back on Earth will *also* say that your headlight beams move through space at 300,000 km/s. And because they will agree with you about your inability to catch up with this light, they will conclude that you must be going *slower* than the light. There is no way around it. In principle, you *could* always keep going faster and faster, but your speed as seen from Earth (or anyplace else) would still never exceed the speed of light because you could never catch up with the light from your own headlights. Instead, you would be seen to be getting ever closer to the speed of light but never quite reaching it. That is why the speed of light is a cosmic speed limit.

have estimated both the mass and the speed of our starship, a simple physics formula allows us to calculate the energy needed. (The required formula is the one used to compute kinetic energy, which is equal to $1/2 \times m \times v^2$, where m and v are the mass and velocity of the moving object.) The energy needed to get this ship to cruising speed is calculated to be 4.5×10^{22} joules, roughly equivalent to 100 times the world's current annual energy use.

In fact, we should double this value, because the amount of energy required to slow down the ship for a soft landing at the colony is the same as that required to accelerate it to cruising speed. Thus, the total energy bill for the trip would be equivalent to at least two centuries' worth of current world energy usage. We can put this in monetary terms. Using a typical price for home electricity (10¢ per kilowatt-hour), the energy cost of sending our craft to another star would be about $2,500,000,000,000,000,000. (To this you can add the cost of food and fresh towels for 40 years.) Clearly, unless we find a way to produce enormously more energy at vastly lower prices, large-scale interstellar travel will be out of reach.

12.2 Leaving Earth

Today, when we think of space travel, we invariably think of rockets. But it wasn't always so. The Greeks spoke of celestial travel in the mythological flights of Phaeton and Icarus, neither of which involved rockets and both of which ended badly. As described by the Roman writer Ovid (43 B.C.–18 A.D.), Phaeton was an impetuous youth who borrowed the Sun god's chariot for a joy ride across the sky. Losing control of the horses on this ill-fated mission, Phaeton inadvertently started to set fire to the Earth. Humanity's home was saved only when Zeus intervened by killing the cheeky charioteer. Icarus, the son of Daedalus, made a similar ill-advised move and became the first fictional flight victim. The feather wings Icarus used to cruise the heavens were held together with wax, and they underwent a fatal meltdown when, despite his father's warnings, he flew too close to the Sun. These stories suggest that, while they were open to the possibility of other worlds, the Greeks regarded thoughts of human flight as hubris.

The Middle Ages saw a continuation of this less-than-enthusiastic view of human flight. Medieval thinkers were not intrigued by practical travel to the heavens, reserving such jaunts for spiritual ventures. Only after the Renaissance, when the philosophy of modern scientific inquiry took hold, did humankind once again consider the possibility of breaking Earth's bonds. By 1687, Issac Newton had produced a treatise on universal mechanics that not only de-scribed the workings of the heavens but explained the physics required to reach them.

Newton's third law of motion states that "for every action there is an opposite and equal reaction." Envision the recoil of a gun when it is fired. The bullet moves in one direction, and the gun moves in the opposite direction. Squids, octopuses, and some other mollusks employ a similar technique. A squid takes in water that it then squirts out at higher speed behind it, thus propelling itself forward. A rocket operates slightly differently, vaporizing on-board fuel that is shot out the back. However, in both cases it is Newton's third law that accounts for the forward motion.

Development of the Rocket

A rocket has been described as the simplest type of engine, and even Newton realized it could work in empty space. Serious thought about travel to other worlds began to surface in the nineteenth century. In the 1860s and 1870s, the French author Jules Verne wrote influential stories describing travel to the moon. His propulsion scheme used an oversize artillery shell specially constructed for the task. While this scheme was hardly practical (the enormous acceleration of the shell when fired would turn the passengers to pancakes), Verne's writings stimulated investigation of space travel by three giants of early rocketry: Konstantin Tsiolkovsky (1857–1935) in Russia, Hermann Oberth (1894–1989) in Germany, and the American Robert Goddard (1882–1945). These fledgling rocket scientists explored many of the theoretical possibilities of this type of propulsion. In particular, they worked out the so-called **rocket equation,** which describes how a vehicle's final speed depends on the propellant velocity. (This equation is given in problem 10 at the end of the chapter.) Both Tsiolkovsky and Goddard realized that it would be difficult for a single rocket to reach **escape velocity,** the speed necessary to overcome gravity and leave Earth behind (about 11 km/s, or 25,000 mi/hr), so they proposed the use of multistage vehicles for space flight. The three pioneers also envisioned space stations, ICBMs, and ion drives.

At first, few people saw much benefit in turning these ideas into working hardware. In a 1920 technical publication, Goddard mused about the possibility that a sufficiently large rocket could reach the Moon. His speculation was immediately ridiculed by the *New York Times,* which claimed that no lunar-bound rocket could ever work since there is no air between Earth and the Moon. A rocket would have nothing to "push against." However, contrary to the *Times'* impression, rockets do not operate by pushing against air—or anything else. They simply employ Newton's

FIGURE 12.6 Early rocketry.

a Robert Goddard stands by his pioneering liquid-fuel rocket in 1926. This craft reached a modest altitude: the height of a four-story building.

b The German V-2 rocket, first launched in 1942, was used against England during World War II. Note that only 16 years separate the V-2 from Goddard's first rocket.

third law, firing hot gas in one direction so that the rocket moves in the opposite direction. In fact, atmospheres hinder the performance of rockets, because they create drag that slows rockets down.

A few years later, in 1926, Goddard launched his first liquid-fueled rocket from a field in Auburn, Massachusetts (Figure 12.6a). It reached a height of 13 meters. This heroic, build-it-in-a-garage phase of rocketry was short-lived. The 1930s brought larger-scale efforts, particularly in Germany, where the military saw value in guided missiles. The German work culminated in the development of the V-2 rocket, used against England during World War II (Figure 12.6b). The rapid development of rocketry that followed the war was driven largely by German scientists who had been recruited by both the Russians and the Americans. The space age truly began in October 1957 with the launch of the Soviet Union's *Sputnik I,* an 84-kilogram beeping metal ball —the world's first artificial satellite.

Although primarily spurred by national rivalries and military considerations, rocket development over the last five decades has allowed humans and their robot proxies to enter those tantalizing realms that had so long been beyond their grasp. What the Greeks could only dream, we can now do. Today's rockets—the direct descendants of those first envisioned nearly a century ago—are fast enough and powerful enough to allow us to explore the nearby worlds of our solar system. But could they ever be improved enough to take us to the stars?

Limitations of Chemical Rockets

Even today, every rocket we use to launch spacecraft works in basically the same way as Goddard's first rocket. The engines ignite and burn a chemical fuel, such as a mixture of oxygen and kerosene. The chemical burning creates very hot gas, which is expelled through a narrow nozzle, propelling the spacecraft into orbit or to other worlds. These chemical rockets serve our current purposes fairly well (though many people dream of new technologies that would allow us to leave Earth at far lower cost). Unfortunately, they are completely inadequate for interstellar travel.

The largest chemical rocket built to date was the Saturn V, the rocket that carried the Apollo astronauts to the Moon (Figure 12.7). This vehicle burned liquid oxygen and kerosene, with water the major combustion product. The hot water vapor was expelled out the back at a speed of about 3 kilometers per second—roughly three times the speed of a rifle bullet. The Saturn V consisted of three separate "stages," that is, three distinct rockets perched atop one another so that each lower stage could drop away after exhausting its fuel. We can see why these multiple stages were useful—and why chemical rockets are limited—by investigating rocket mechanics in a bit more depth.

Let's start by imagining that a rocket like the Saturn V had only a single stage. In order to leave Earth behind, we need to reach the Earth's escape velocity of 11 kilometers per second. Using the

FIGURE 12.7 The Saturn V rocket, which was used to carry the Apollo astronauts to the Moon. The most powerful rocket yet built, this 40-year-old, three-stage design weighed about 30% more than the Space Shuttle at lift-off and was capable of hurling a 45,000-kilogram (50-ton) payload to the moon. With a launchpad mass of 2.8 million kilograms for the Moon trips, its overall mass ratio was 62.

can we ever succeed? As the early rocket pioneers realized, the trick is to use multiple stages. By discarding each stage as its fuel is used up, the upper stages don't need to accelerate the dead weight of those below. The rocket as a whole—with all of its stages—still must weigh 39 times as much as the parts that will actually reach space. But each stage requires a much lower mass ratio, because the weight of the rocket will decrease as stages are discarded. For example, to escape Earth by means of a three-stage rocket (assuming the stages have identical mass ratios and exhaust velocities), each stage need have a mass ratio of only 3.4—well within our capabilities. (Problem 11 at the end of the chapter leads you through this calculation.)

THINK ABOUT IT . . . *The Space Shuttle does not use stacked rockets like the Saturn V, but it still uses staging. Explain how. (Hint: It should be obvious if you look at a picture of the Shuttle just before launch.)*

In principle, adding more stages can propel chemical rockets to higher speeds, but not high enough for convenient interstellar travel. For example, an oxygen-kerosene rocket like the Saturn V consisting of a stack of *100* stages (each with a mass ratio of 3.4) would reach a speed of 370 km/sec—33 times faster than the Saturn V but only a little over 0.1% of the speed of light. A trip to Alpha Centauri would still take some 4,000 years. Using more efficient chemical fuels (such as oxygen and hydrogen, rather than kerosene) can help, but by no more than about a factor of 2. Indeed, no matter what engineering refinements we consider, chemical rockets simply are not powerful enough to deliver large payloads to the stars in a reasonable length of time. Interstellar travel requires a different approach.

12.3 "Conventional" Interstellar Spacecraft

Chemical rockets may be insufficient for travel to the stars, but other technologies have much greater promise. Generally speaking, we can break these technologies into two groups. The first uses "conventional" technology—that is, technology that seems within our grasp (at least if we disregard cost), even if we don't yet have it. The second group involves technologies that are theoretically possible but far beyond our present capabilities. In this section, we'll investigate a few conventional technologies that would allow at least a modest degree of interstellar travel. In the next section, we'll explore the more far-out ideas.

rocket equation developed by the pioneers of rocketry, we can calculate the **mass ratio** required to attain this speed. The mass ratio is defined as the mass of the fully fueled rocket (including any spacecraft it is carrying) divided by the rocket (and spacecraft) mass after all the fuel is burned. We find that reaching escape velocity requires a mass ratio of 39, meaning that the fueled rocket on the launchpad must weigh 39 times more than the empty rocket and spacecraft alone. That is, the fuel weight would be about 38 times the weight of the spacecraft and the engines. This is clearly a discouraging requirement and one that's just about impossible to meet given the weight of tanks, fuel pumps, fins, and astronauts. Indeed, the best single-stage rockets have mass ratios of only 15 or less.

If it takes a mass ratio of 39 to leave the Earth and our best rockets have mass ratios of only 15, how

Nuclear Rockets

Chemical reactions involve shuffling the outer electrons of atoms. While these reactions can seem quite powerful (consider the drama of a Space Shuttle launch), the energy they release is insignificant compared to the amount of energy at least potentially available in the reacting materials. According to Einstein's famous formula, $E = mc^2$, any piece of matter contains an amount of energy equivalent to its mass multiplied by the speed of light squared. (Thus, in the formula, E is the energy content of the mass, m is the amount of mass, and c is the speed of light.) This represents an enormous amount of energy. For example, if you could turn a 1-kilogram (2.2-pound) rock completely into energy, the energy released would be equivalent to that contained in nearly 8 billion liters of gasoline—or as much gasoline as is used by all cars in the United States in a week. However, while this energy is "there" in any piece of matter, it is very difficult to extract. Chemical reactions extract so little of it that we do not notice any change in the mass of the reacting materials.

Nuclear reactions, in contrast, can noticeably affect the mass of reacting materials. They involve changes in the dense atomic nucleus. Two basic types of nuclear reactions can be used to generate power: fission and fusion. Nuclear **fission** involves the splitting of large nuclei such as uranium or plutonium. When a uranium nucleus is split, approximately 0.07% of its mass is turned into energy. Thus, if 1,000 grams of uranium undergo fission, you'd find that the fission products (the material left over after the fission has occurred) would weigh a total of only 999.3 grams, 0.07% less than the starting weight. Although this mass loss may sound fairly small, the energy it releases dwarfs that released by chemical reactions. Nuclear fission bombs are what destroyed the Japanese cities of Hiroshima and Nagasaki at the end of World War II, and all current nuclear power plants get their energy from fission.

Nuclear *fusion*, the power source of the Sun and other stars [Section 1.3], is about ten times more efficient. Fusion of hydrogen into helium converts about 0.7% of the hydrogen fuel mass into energy. The Sun, for example, fuses 600 million tons of hydrogen into helium *each second*. The resulting helium weighs 0.7% less than the original hydrogen, or about 596 million tons. The other 4 million tons of mass simply "disappears" as it becomes the energy that makes our Sun shine. We humans have managed to achieve nuclear fusion here on Earth, but not in a well-controlled, commercially useful way. Although fusion is used as the primary power source in thermonuclear bombs (or "H-bombs"), we have not yet figured out how to harness it for daily use. This is unfortunate; not only is the fuel for fusion (hydrogen) readily available in water, but the efficiency of fusion is so great—at least compared to current energy sources such as oil, coal, and hydroelectric power—that it would seem almost unlimited. For example, if we could somehow hook up a nuclear fusion plant to your kitchen sink, then by continuously fusing the hydrogen in the water flowing from the faucet we could generate more than enough power to meet all the current energy needs of the United States.[1] That is, with your kitchen faucet fusion plant, we could stop the drilling and importing of oil, dismantle all hydroelectric dams, shut down all coal-burning power stations, get rid of all fission power plants, and still have power to spare. And there'd be no more worries about global warming, because fusion does not release any greenhouse gases into the atmosphere.

THINK ABOUT IT . . . *Scientists have been working for decades in hopes of developing the technology for viable nuclear fusion power plants, but so far without success. As a result, governments are reluctant to spend a lot of money on further fusion research, although some is still ongoing. How much effort do you think we should put into fusion power? If we achieved it, how do you think it would change our world?*

The tremendous advantage of nuclear energy over chemical energy was bound to appeal to rocket scientists. In 1955, the U.S. Atomic Energy Commission and the U.S. Air Force (and later NASA) embarked on an experiment called *Project Rover* to develop nuclear fission reactors that could be flown in a rocket (Figure 12.8). The idea was to use the fission reactor to generate enormous heat, which would be used to bring hydrogen gas to a temperature of millions of degrees before expelling it out the engine nozzles. (Note that the hydrogen was being used as a propellant, not for fusion.) At its peak, Project Rover employed 1,800 people and ultimately tested six fission engines. The program made substantial progress and showed that fission-powered rockets could achieve speeds at least two to three times as great as similar-size chemical rockets. By the late 1960s, NASA officials were confident that the Project Rover rockets could be used to send humans to Mars in what they hoped would be an

---

[1] Actual attempts to generate fusion power use deuterium (the isotope of hydrogen with one neutron), which is present naturally in the ratio of about 1 part deuterium to 50,000 parts ordinary hydrogen. Thus, with deuterium, the needed water flow would be about 50,000 times greater than that of your kitchen faucet—but this flow (about 130,000 liters per minute) is still only about that of a small stream.

FIGURE 12.8 President John F. Kennedy departing the Nevada Test Site after viewing a full-scale mock-up of a nuclear powered engine for Project Rover, December 8, 1962.

Alamos Scientific Laboratory realized that one way to get a rocket up to much higher speed would be to toss small nuclear (fusion) bombs out the rear and let the resulting explosions push the craft forward. The bombs, released at a rate of one every few seconds or more, would drop back about 50 meters and then detonate behind a large metal "pusher plate" affixed to the tail of the rocket (Figure 12.9). This would provide an impulse to move the rocket forward. Despite suffering obvious abuse, the pusher plate wouldn't vaporize because it would be exposed to these searing explosions for only a few milliseconds at a time. The Los Alamos scientists calculated that a spaceship 1 mile long accelerated by the rapid-fire detonation of a million H-bombs could reach Alpha Centauri in just over a century. Thus, Project Orion represented the first true "starship" design to be fashioned by humans. No actual construction ever began, though in principle we could build a Project Orion–type starship with existing technology. However, it would be very expensive and would require an exception to the international treaty banning nuclear detonations in space. Project Orion ended in 1965, due both to budget cuts and to the nuclear test ban treaty.

immediate follow-on to the Apollo Moon landings. However, the political climate changed, and the United States abandoned its early plans for a human mission to Mars. Project Rover was officially terminated in 1973.

Another experimental approach, dubbed *Project Orion,* was more radical. Physicists at Nevada's Los

Another nuclear rocket design was developed in the 1970s by the British Interplanetary Society under the name *Project Daedalus* (Figure 12.10). The idea was to shoot frozen fuel pellets of deuterium and helium-3 into a reaction chamber where they would

FIGURE 12.9 Artist's conception of the Project Orion starship, showing one of the small H-bomb detonations that propel it. Debris from the detonation impacts the flat disk, called the pusher plate, at the back of the spaceship. The central sections (enclosed in a lattice) hold the bombs, and the front sections house the crew.

FIGURE 12.10 Artist's conception of a robotic Project Daedalus starship. The front section (right) holds the scientific instruments. The large spheres hold the fuel pellets for the central fusion reactor.

undergo nuclear fusion. The fuel pellets, about the size of gravel, would be shot into the chamber at a rate of 250 pellets per second. There they would be encouraged to fuse by electron beams, producing a rapid-fire series of explosions that would propel the ship. Because we cannot yet build nuclear fusion reactors, this design remains beyond our current technological capabilities. Nevertheless, the proponents of Project Daedalus developed a plan for sending a robotic spacecraft to Barnard's star, a dim, type M star 6 light-years distant and the next closest star to Earth beyond the Alpha Centauri system. After 4 years of firing the engine, the craft would reach about one-tenth the speed of light and then spend the next four decades coasting to its destination. Once there, it would deploy probes and sensors to relay back photos and other data, giving us our first close-up view of another stellar system only about 50 years after its launch.

Nuclear-powered rockets are undoubtedly feasible in some form. Still, at best they would achieve speeds of about one-tenth the speed of light. Interstellar journeys would be possible, but it would take decades for them to reach even the nearest stars.

Ions, Sunlight, and Lasers

The propulsion schemes described thus far involve a relatively quick acceleration of the rocket to high speed, after which the engines shut down and the craft cruises for whatever length of time it takes to reach its target. Another approach is to use a low-powered rocket whose engines keep firing continuously. The **ion engine** is an example of this approach. It works something like a television picture tube in that it accelerates charged particles (ions). In a television tube, electrons are fired from the back of the tube to the phosphor screen that faces the viewer. An ion rocket would do the same, with a nozzle taking the place of the phosphor screen. Although such an electric rocket would need to be started in space (it wouldn't have enough thrust to lift itself off the Earth and needs to work in a vacuum), it could keep firing for long periods of time because the mass expelled per unit time is small. Moreover, the exhaust ions are shot from the craft at tremendous speeds, and such an ion drive could reach speeds approaching a percent or so of the speed of light. NASA has tested a low-power ion drive on a spacecraft called *Deep Space 1*.

Other schemes envision spacecraft that can overcome the limitations of the rocket equation by not

FIGURE 12.11 Artist's conception of a spaceship propelled by a solar sail, shown as it approaches a forming planet in a young solar system. The sail is many kilometers across. The scientific payload is at the central meeting point of the four scaffoldlike structures.

taking along the bulky fuel. An obvious possibility that's been considered for nearly a century is to use sunlight as power. Large, highly reflective, very thin (to minimize mass) **solar sails** could be pushed by the pressure exerted by sunlight. This pressure is so slight that we normally don't notice it, but in the vacuum of space, where friction is absent, the steady pressure of sunlight impinging upon a mirrored surface could build up impressive speed, particularly with sails hundreds of kilometers in size. Solar sailing might well prove to be a fairly inexpensive way of navigating within the solar system. Surprisingly, it might even be useful for interstellar travel (Figure 12.11). Although the push from the Sun would slowly fade once such craft reached the outer solar system (at Saturn, the light intensity is less than 1% its value near Earth), a solar sailing vehicle that was started very near the Sun might achieve speeds of a few percent of light speed. It could then coast to neighboring stars in less than a century.

The fact that sunlight weakens so much with distance limits the ultimate speed of a solar sailing spacecraft. However, we could get around this problem by using a powerful laser, rather than sunlight, as a power source. In principle, the laser could provide a steady and continuous "push" for the solar sail, all the way to its destination if necessary. If building large sails proves too difficult, the laser could be used to vaporize propellant on the rocket that would then move the craft forward in the usual rocketlike manner. As these craft moved light-years away, a large focusing mirror hundreds of kilometers in size would be needed at the laser base in order to concentrate the beam on the pinpoint target that the spacecraft had become. The primary drawback to these schemes is the power requirement. For example, accelerating a ship to half the speed of light within a few years would require a laser that uses 1,000 times more power than all current human power consumption. Nevertheless, this approach would allow us to travel to nearby stars in a decade or two rather than many decades.

A laser-powered rocket would leave the passengers dependent on the efforts of those back on Earth (or wherever the laser installation was located) to keep the laser shining so they could accelerate to their desired final velocity. This might be somewhat risky given the fact that even the fastest of these transports would be en route and accelerating for decades. What if the laser crew went on strike? In addition, with no laser shining in the *opposite* direction, slowing the spacecraft to a halt at its destination (let alone returning home) would be a problem. One possibility is to use on-board propellant heated by the laser. The propellant could then be fired out the front of the craft to slow it down. Alternative braking schemes that use natural magnetic fields in space have also been suggested.

Interstellar Arks

Another, less demanding approach to interstellar travel is often featured in science fiction. Forget the hi-tech rocketry and accept relatively low speeds. Then deal with the resulting long travel times by putting the crew into suspended animation—hibernation, if you will—and letting them doze their way to the stars. A challenging variation on this idea is to somehow reengineer the travelers to live long enough to cruise the galaxy. A third suggestion is to build enormous craft that can accommodate a very large crew: in essence, an "ark." Many generations would live out their lives aboard this slow-moving vehicle before it finally reached its distant destination.

The difficulty with the first suggestion is that no one yet knows how to put humans to sleep for hundreds or thousands of years (and then have them wake up). However, recent work by genetic researchers has identified genes that seem to control hibernation in animals. There is some indication that they may also work in humans. If we were to learn how to activate these genes suitably, future astronauts might be able to sleep away the long travel times.

The second suggestion, to reengineer the crew to live the many thousands of years necessary for interstellar travel at conventional speeds, is an intriguing thought, but it is entirely speculative. Could we somehow stop the aging process, enabling people to live such long lives that centuries or millennia of travel might seem like a walk to the corner store? We simply don't know. A variation on this scheme might involve machine intelligence, if it could be constructed. It is easier to imagine machines that never die than immortal humans.

The idea of building interstellar arks usually gets a skeptical reaction from sociologists. They point out that long voyages on Earth (even those that last only a few months) often end badly. Crews splinter into antagonistic factions and frequently fight for control of the ship. In addition, we might justifiably fear a deterioration in the level of expertise of the crew, with the result that the generation of folk that finally reaches the target star system would neither remember why it journeyed there nor have the technical skills required to land on or colonize a world in the system.

THINK ABOUT IT . . . *One simple way to get around the difficulties of interstellar travel is to radically reduce the size of the payload. If we could cram a robotic probe into a package the size of a tennis ball, existing rockets might be able to send it to the stars. Does such an approach seem feasible or useful? Explain.*

12.4 Relativistic Rocketry

We have seen that the conventional approaches to interstellar travel—using chemical, nuclear, or laser-powered rockets or a solar sail—will not bring us to the stars in anything less than decades. What we really need for interstellar travel are ships that can travel at speeds close to the speed of light. We could then reach the nearest stars in years and explore the space within a few tens of light-years of the Sun in decades. Moreover, such ships would be a boon for any on-board passengers, who, thanks to the effects of special relativity, would experience the march of time more slowly than those left behind.

The Role of Relativity

First, let's look at the benefits of truly rapid travel. Obviously, greater speed reduces trip time, and we would naively expect a rocket moving at 90% of the speed of light ($0.9c$) to require only 1/9 the travel time of a rocket traveling at 10% of the speed of light ($0.1c$). This is true for those who observe the rocket's progress from Earth. But there is an additional factor: the **time dilation** experienced by the crew. According to Einstein's special theory of relativity, when a moving object travels at close to the speed of light, its length becomes shorter in the direction of movement, its mass becomes greater, and time measured aboard proceeds more slowly than time measured by clocks at rest. (Note that time will seem perfectly normal to the passengers on the trip; it runs slow only in comparison to time back on the home planet. For example, if people on the home planet could observe what was going on inside the spaceship, they'd see that spaceship time was running slow— that is, clocks on the ship tick more slowly than clocks on the home planet, hearts of passengers on the ship beat more slowly than hearts on the home planet, and so on. But time would still feel "normal" to the passengers.)

In other words, a year after the launch of such a rocket (as measured on the home planet), the crew will have aged something *less* than a year. Table 12.1 shows the benefits of relativistic travel for a hypothetical trip from Earth to the star Vega, which is 25 light-years away (and which was visited by the character played by Jodie Foster in the movie *Contact*).

Table 12.1 Trip Times to Vega (distance = 25 light-years)

Velocity as Fraction of Light Speed (v/c)	Trip Time as Measured on Earth (years)	Trip Time as Measured on Spacecraft (years)
0.00005 (chemical rocket)	500,000	500,000
0.1 (nuclear, ion, or laser rocket; solar sail)	250	249
0.5	50	43
0.7	36	26
0.9	28	12
0.99	25	3.5
0.999	25	1.1
0.9999	25	0.35

From the crew's standpoint, the trip can be made arbitrarily short simply by increasing the ship's velocity so it is nearer the speed of light. For friends left behind on Earth, the trip is never observed to take less than 25 years.

If you study the table carefully, you might wonder if the crew of a very high velocity rocket would conclude that they're traveling faster than the speed of light—which would violate the cosmic speed limit of relativity. For example, at a speed of 99.99% of the speed of light (0.9999c), their trip takes only 4 months for a distance we said was 25 light-years. This would seem to imply a speed some 75 times the speed of light. However, special relativity also tells us that distances shrink at high speed. Once traveling at high speed, the crew would find that the distance to Vega is not 25 light-years as we measure it on Earth but instead has shrunk to a little under 0.4 light-year. Thus, they can cover this short distance in a short time, and they'll never think they are traveling faster than the speed of light.

Movie Madness: Star Trek

If you want to boldly go where no one has gone before, without spending a few hundred centuries doing it, you need warp drive.

As almost everyone knows, the various incarnations of the *U.S.S. Enterprise, Star Trek*'s famous interstellar transport, high-tail it from one part of the galaxy to another in short order thanks to a futuristic propulsion system. But what *is* warp drive, anyway?

According to the show's technical manuals, the term is merely slang for 'continuum distortion propulsion'—a Latinate mouthful that describes a scheme by which powerful fields are used to distort space and allow speeds faster than light. As we discuss in Section 12.5, rapid travel by warping space is not an entirely nutty idea. It might be possible. And if physics were to allow it in *principle,* couldn't our clever descendants do it in practice?

Maybe yes, maybe no. A major problem is that even if you could warp space, to do so would take enormous amounts of energy. *Star Trek* deals with this small technical detail by fueling the field-generating warp engines with antimatter (antihydrogen, to be precise). Combining antimatter with ordinary matter is the most efficient combustion imaginable, as the entire mass of both is converted to energy.

Of course, there's still the problem of making the antimatter, not to mention shipping it to service stations around the galaxy (being careful to keep it out of the hands of pirates—antimatter is costly). But the truly interesting thing about warp drive is the range of speeds attained. At "Warp 1" you're loping along at the speed of light. By "Warp 9"—near the top of the *Enterprise*'s speedometer—you're streaking through space at 1,000 times lightspeed. This means you can traverse the galaxy in a century, which is short enough to be possible, but long enough to allow you to get effectively stranded and interfere with "Starfleet Command's Prime Directive."

Mind you, you could forget warp drive entirely, and stick with the physics we know by building starships that go at 99+ percent of light speed. Special relativity would guarantee that travel times as perceived by the ship's crew would be short. They could cross the galaxy overnight, according to their own watches. But *Star Trek* has opted out of the relativistic approach for good reason, for otherwise the *Enterprise* crew would return home to find all their family and friends long dead and forgotten. Starfleet headquarters would probably be just an archaeological dig.

Better to call up Scottie in the engine room and tell him to put the pedal to the space-bending metal.

In fact, if we could somehow boost our spacecraft to speeds arbitrarily close to the speed of light, we could go anywhere in the universe within a human lifetime. Astronomer Carl Sagan considered a hypothetical rocket that accelerates at a steady 1g (or "1 gee")—an acceleration that would feel comfortably like gravity on Earth—to the halfway point of its voyage. This constant acceleration would bring the ship closer and closer to the speed of light, though it would never exceed it. (Note that the acceleration of 1g is constant, but the speed is not!) The ship then reverses and decelerates at 1g to its destination. During most of the trip, the ship would be traveling at speeds quite close to the speed of light, so time would pass very slowly on the ship compared to time on Earth. Longer trips would mean longer acceleration, bringing the ship even closer to the speed of light for most of the journey. Calculations show that such a ship could make a trip to a star 500 light-years away in only about 12 years according to those on board the ship. However, 500 years would pass on Earth. If a crew of 20-year-olds left Earth in the year 2100, they would be merely 32 years old when they reached their destination; but it would be the year 2600 on Earth (actually a bit later, since they are traveling at not quite the speed of light). If they sent a radio message back to Earth, the message would take 500 years to arrive here across the 500-light-year distance. More than 1,000 years after the crew left, we'd get a message from people who had aged only 12 years since they'd last been seen on Earth.

Even longer trips would be possible in principle. For example, in about 21 years of ship-board time, a craft with a constant 1g acceleration could bridge the 28,000-light-year distance to the center of the Milky Way Galaxy, where the crew could observe firsthand the mysterious black hole thought to reside there. The 2.5-million-light-year distance to the Andromeda Galaxy could be traveled in only about 29 years of ship's time. Thus, passengers could travel to the Andromeda Galaxy, spend 2 years studying one of its star systems and taking our first pictures of the Milky Way as it appears from afar, and return only 60 years older than when they left. However, they would not exactly be returning "home," since 5 million years would have passed on Earth. In this sense, special relativity offers sufficiently fast travelers a one-way "ticket to the stars." No place is out of reach—but you cannot come home to the same people and places you left behind.

In any event, while such incredible trips are allowed by the laws of physics, the energy costs would be extraordinary. Because special relativity also tells us that an object's mass increases as it approaches the speed of light, the energy cost rises just as much as time slows down. Indeed, that is one explanation for why the speed of light cannot be reached: As the ship gets closer and closer to the speed of light, its mass becomes greater and greater, so the same rocket thrust generates ever-less additional speed. In fact, the mass heads toward infinity as the ship's speed approaches the speed of light—and no force in the universe can give a push to an infinite mass. The ship can never reach the speed of light, no matter what engines it might have, and even getting close to that speed would require amounts of energy far beyond anything we will be able to muster in the near future. Nevertheless, such practical difficulties can't stop us from speculating, and at least two potential ways of approaching the speed of light are known to exist in principle. We discuss these next.

Matter-Antimatter Rocketry

The most efficient energy source we have seen so far is nuclear fusion. But fusion converts only 0.7% of the mass of the fusing hydrogen into energy; 99.3% of the mass still remains, as helium. Is there a way to turn more of the mass, or even all of it, into energy? Yes. It is called **matter-antimatter annihilation.**

Antimatter might sound like the stuff of science fiction, but it really exists. All material things are composed of "ordinary" matter, but physicists have discovered particles that are in some ways the mirror image of normal particles, differing principally in their electrical charge. The first known such particle, christened the *positron,* was discovered in 1932. It is the antimatter twin of the electron; that is, it is identical to an electron except that it has a positive rather than a negative charge. In 1955, the proton's antimatter partner was found: the antiproton. If you were to introduce an antiproton to a positron, they would form an atom of antihydrogen. This antimatter atom would behave chemically just like ordinary hydrogen, except for one thing: You wouldn't want to get near it. When matter and antimatter meet, the result is total annihilation, with 100% of the mass turning into energy. (Note that antimatter still has mass just like ordinary matter; there is no such thing as "antimass.")

The annihilation of matter and antimatter can create energy in a variety of forms. For example, annihilation of positrons with electrons produces energy as a burst of gamma rays. The annihilation of heavier particles, such as antiprotons with protons, produces a gush of particles that soon decay into neutrinos and gamma rays. These might be amenable to powering a rocket, because the flood of reaction particles could be directed out a rearward-facing nozzle. A matter-antimatter rocket could, in principle, achieve speeds of 90% of the speed of light with modest mass ratios.

While such numbers are seductive, the problem lies in rounding up and storing the required antimatter. No practical reservoirs of this material are known. Instead, we would have to manufacture the antimatter, as physicists now do with high-energy particle accelerators (although worldwide production amounts to only a few billionths of a gram per year). Unfortunately, the energy needed to manufacture the antimatter would be enormous. For example, with present technology, manufacturing 1 ton of antimatter—far less than would be needed for an interstellar trip—would take more energy than humankind has used in all of history. Moreover, even if we could make the antimatter, we don't yet know of a good way to store it aboard our rockets, since it would have to be kept in some type of container in which it never touched any walls.

Interstellar Ramjets

Another approach to achieving truly relativistic velocities circumvents the problems involved in carrying highly energetic fuel on board. The idea is that a starship could collect its fuel as it goes, using a giant scoop to sweep up interstellar gas. Because this gas is mostly hydrogen, it could be funneled to a nuclear reactor, fused into helium, and then expelled out the back. Such propulsion systems are known as **interstellar ramjets** (Figure 12.12). In principle, interstellar ramjets can accelerate continuously by collecting and using fuel non-stop, getting arbitrarily close to the speed of light.

Of course, there are practical difficulties. The typical density of the gas between the stars is only a few atoms per cubic centimeter, so the scoop would need to be hundreds of kilometers across to collect adequate supplies of fuel. As Carl Sagan said, we are talking about "spaceships the size of worlds." Another problem facing an interstellar ramjet—or any ship traveling at relativistic speeds—comes from the interstellar gas and dust itself. At 99% of the speed of light ($0.99c$), a particle the size of a sand grain packs energy equivalent to an explosion of about 100 kilograms of TNT. Even individual atoms encountered at this speed are deadly, so the ship would need substantial shielding to protect both its structure and its crew.

Any type of relativistic travel, whether with matter-antimatter engines, interstellar ramjets, or

---

FIGURE 12.12 Artist's conception of a spaceship powered by an interstellar ramjet. The giant scoop in the front (left) collects interstellar hydrogen for use as fusion fuel.

some as-yet-unthought-of method, remains far beyond our current technological capabilities. But at least we have found ways by which a very advanced civilization *might* be able to travel among the stars. Whether anyone has actually achieved this ability remains unknown.

12.5 Wormholes and Hyperspace

If you are a science fiction fan, our discussion of interstellar travel may depress you. Interstellar tourism and commerce seem out of the question, even with ships that travel at speeds close to the speed of light, because of the long times involved (at least as seen from home planets and colonies). If we are ever to travel about the galaxy the way we now travel about the Earth, we will need spacecraft that can somehow get us from here to there much faster than the cosmic speed limit would seem to allow. Could such spacecraft be possible?

No one really knows. However, there just might be a "loophole" in the law limiting cosmic speed. In particular, while Einstein showed that we can't travel *through* space faster than the speed of light, his theory of general relativity suggests that there might be "shortcuts" that, in effect, let us travel *outside* ordinary space in a way that greatly reduces the distances to be traveled. If so, then we might reach far-off places by taking a shortcut that lessens the distance to them. In this section, we investigate these interesting possibilities.

Hyperspace

In 1916, Einstein enlarged his earlier work (special relativity) to produce the theory of general relativity. We live in a four-dimensional universe, Einstein noted, with three dimensions of space and one of time (together called *space-time*). The three spatial dimensions are not rigid and invariant but rather can be warped. Einstein realized that matter can provide the distortion. For example, any mass, such as the Sun, will produce space curvature.

How can we tell that space is "curved"? One simple test is to consider the paths of light beams, which travel through space in what we call straight lines. If space had no curvature, then two parallel light beams—for example, from two laser pointers taped side by side—would never cross. However, general relativity insists that the presence of matter can cause a warping of space that will lead these parallel beams to cross. The matter produces a distortion of the ordinary three dimensions of

space into other, hypothetical, dimensions we can't see or "get into." These additional dimensions are called **hyperspace.**

In order to visualize this idea, physicists often resort to "embedding diagrams," such as that depicted in Figure 12.13. Space is reduced from three dimensions to two, resembling a rubber sheet; if you inhabited the world shown in the figure, you would be flat, infinitely thin, and incapable of appreciating that any dimension exists on either side of the sheet. Embedded in the sheet is a large mass, such as a star, that causes the sheet to distort into hyperspace. You cannot see the hyperspace dimensions, but you can make measurements on the rubber sheet that will tell you whether or not it is curved. In fact, during solar eclipses, we have measured changes in the apparent positions of stars whose light passes near the Sun. As Figure 12.13 shows, these changes are what we would expect if our space is curved through hyperspace.

The warping of space is usually quite small. Even the bending of starlight passing close to the Sun amounts to only a fraction of a thousandth of a degree. But if space could be distorted more dramatically, the distortion might offer us shortcuts to distant destinations.

Black Holes, Wormholes, and Warp Drive

The bizarre objects called black holes are, in fact, holes in space-time—places where space becomes so distorted that it in effect becomes a bottomless pit. As a result, science fiction writers have sometimes imagined using black holes as shortcuts to other places. Unfortunately, this idea suffers from at least two major drawbacks. First, the only known black holes are themselves very far away, so getting to them in the first place would be a problem. Second, we do not know of any way we could survive a close encounter with a black hole.

FIGURE 12.13 This embedding diagram shows how starlight is bent as it passes near the Sun, causing stars to appear slightly offset from their true positions in space. This effect has been measured during solar eclipses, proving that our space really is curved through hyperspace.

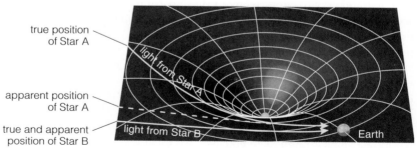

true position of Star A

apparent position of Star A

true and apparent position of Star B

light from Star A

light from Star B

Earth

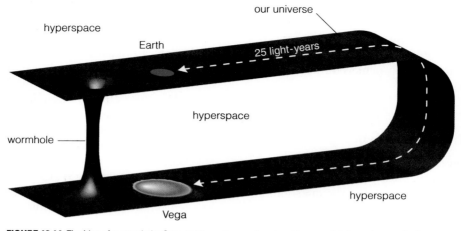

FIGURE 12.14 The idea of a wormhole. Once again, we have reduced our three spatial dimensions to the flat, two-dimensional realm of a rubber sheet. Two distortions of space, one near the Earth and one near Vega, have met up in hyperspace, forming a connecting tunnel. Going from Earth to Vega in ordinary space would be a 25-light-year trip. But the wormhole offers a radically shorter route—one that might be traversed in minutes without ever exceeding the speed of light.

However, a related phenomenon, called a **wormhole,** might be more useful. Just as a worm might shorten its trip from one side of an apple to the other by tunneling through it, so might a wormhole provide a hyperspace shortcut to a distant part of the universe. Imagine a dense mass floating somewhere near Earth, distorting the surrounding space into hyperspace. Now imagine a similar distortion occurring somewhere else in the cosmos, many light-years away. If these two distortions somehow meet up in hyperspace, they could connect two distant places in ordinary space via a short, hyperspace tunnel (Figure 12.14). Traversing this tunnel might take very little time (perhaps minutes) and could short-circuit the necessity of traveling those many light-years.

While this is clearly an appealing idea, could it actually be made to work? In particular, how do we arrange for the wormhole's opportune existence? Quantum physics suggests that on the tiniest scales of the universe, in regions of space far smaller than an atomic particle, space-time is a seething foam, constantly punctured by distortions into hyperspace. In this highly microscopic world, wormholes might be forming (and self-destructing) all the time. It's conceivable that a highly advanced society might have learned how to capture one of these natural wormholes—one that connected two places of interest—and to quickly enlarge it to a size that would permit its use for travel. However, such wormhole construction would seem to carry an impossibly large energy cost—estimated to be a thousand times the energy released by an exploding massive star (a supernova). Moreover, even if the energy could be found, we do not yet know of a way to stabilize a wormhole against immediate collapse. All in all, we do not know enough about physics to say for sure whether wormhole travel is even possible. Given this uncertainty, we can but imagine and hope. Carl Sagan's book and movie *Contact* postulated a network of wormhole tunnels permitting fast travel throughout the universe—but even his fictional characters did not know who had built them.

Another possible way to travel great distances in a short time might be to exploit the warping of space by placing a dense mass in front of a spacecraft. Hanging like bait on a fishing line, this black hole on a stick would allow the shortening of spatial distances in front of the rocket. Much as you might move across a floor by scrunching the carpet in front and straightening it out behind, this highly unusual craft would bend space in front and leave it unaltered behind. You would continually fall into the warped space in front of the craft. This concept comes closest to what we know as the "warp drive" of science fiction. However, it would require either capturing a black hole (difficult, to be charitable) or creating one using enormous amounts of energy. In either case, the black hole would have to be carried along for the ride.

Could there be simpler ways to take advantage of hyperspace? We do not know. This leaves the door wide open for science fiction writers. In the *Star Wars* movies, a simple flip of a lever takes a ship outside our universe and into hyperspace, allowing nearly instantaneous travel to anywhere. In *Star Trek,* a command from the captain sends the ship into warp drive, apparently without the need for a black hole in front of the ship. If such schemes are at all possible, we have no inkling of how they might work. But who knows what an advanced civilization might have discovered? After all, we have been doing physics in earnest for only a few centuries. Others out there may have been doing it for millions or billions of years.

12.6 Are Aliens Already Here?

So far in this chapter, we have discussed interstellar travel as a possibility, not a reality. But a substantial fraction of Americans claim to believe not only that interstellar travel is possible, but that aliens are visiting us on a routine basis. We are referring, of course, to UFOs (unidentified flying objects). In Chapter 2, we discussed why purported UFO sightings have not risen to the level of science. Nevertheless, given the widespread belief in UFOs, it's worth looking a little more deeply into the claims in light of what we have learned about the realities of interstellar travel.

UFOs

Public opinion polls since the 1960s have consistently shown that about half the American public believes that good evidence for intelligent extraterrestrials exists, and many among them suspect the evidence is being covered up by the government. The bulk of the claimed evidence for this remarkable notion consists of UFO sightings—many thousands of UFOs are reported each year.

No one doubts that unidentified objects are being seen. The question is whether they are alien spacecraft. The first modern report of such craft was related by businessman Kenneth Arnold in June 1947. Arnold was flying a private plane near Mount Rainier, in the state of Washington, when he spied nine objects that appeared to be streaking across the sky at nearly 2,000 km/hr. A reporter for the United Press wrote up Arnold's experience as a sighting of "flying saucers," and the story became front-page news throughout America.

Within a decade, "flying saucers" had invaded popular culture, if not our planet. They were often seen in books and movies, and the term is still used today. This is ironic, given that Arnold didn't actually describe the objects he saw as saucer shaped, or even round. Three decades after the sighting, he explained that this impression was the result of a newspaperman's error. When, in 1947, the United Press reporter asked him how the objects moved, Arnold answered that they "flew erratic, like a saucer if you skip it across the water." He was describing their motion, not their shape. In fact, several later investigators have suggested that the objects seen by Arnold were meteors, flashes of light caused by small particles entering the Earth's atmosphere.

Despite the reporter's misunderstanding, the idea of alien disks buzzing the countryside caught the imagination of the public. It also interested the U.S. Air Force, which spent two decades conducting investigations into the nature of UFOs. The military's interest was prompted largely by Cold War concerns that UFOs might represent new types of aircraft being developed by the Soviet Union.

In the 1950s and 1960s, teams of academics met to study the most interesting of the UFO reports. In the overwhelming majority of cases, these experts were able to plausibly identify the UFOs. They included bright stars and planets, aircraft and gliders, rocket launches, balloons, birds, ball lightning, meteors, atmospheric phenomena, and the occasional hoax. For a minority of the sightings, the investigators could not deduce what had been seen, but their overall conclusion was that there was no reason to believe UFOs to be either highly advanced Soviet craft or visitors from other worlds. The Air Force ultimately dropped its investigation of the UFO phenomenon.

However, some believers felt that the inquiries were either incomplete or part of a ruse organized by the government to put people off the scent of alien visitation. The number of UFO sightings increased (of course, the number of human-made objects in the sky was also growing during these years), and countless books, photos, and film clips were offered as evidence (Figure 12.15). Yet little of the photographic material was compelling. Some of it was ambiguous (is that a distant spacecraft or a nearby bird or bug?), and some was clearly faked. While UFO witnesses were frequently credible (they included seasoned pilots), usually what was seen had several possible explanations, only one of which was that the UFO was an alien spacecraft.

The fact that some 90% of the well-documented UFO cases were explainable as earthly phenomena only encouraged some people to point to the 10% that weren't clearly explained. However, this is not a valid argument for the idea that the unexplained cases represent alien spaceships. After all, if a metropolitan police department solves 90% of the murder cases in a large city—all committed by humans against other humans—it doesn't suggest that the other 10% were committed by, say, aliens.

THINK ABOUT IT . . . *If you were given a year off to investigate UFOs, what sorts of experiments would you conduct? What types of evidence might truly suggest that fast-moving lights in the sky are nonterrestrial craft?*

FIGURE 12.15 A UFO. This photo, which shows far more detail than most pictures made of supposed visitors from other worlds, has the familiar saucer shape made popular by a reporter's error. In fact, this photo is faked, and the saucer is only a lamp shade.

Crashed Aliens in Roswell

Although UFOs generally don't seem to leave behind physical evidence of their presence, in at least one celebrated occurrence supposedly alien artifacts were found: the so-called Roswell incident of 1947 [Section 2.3]. It is claimed that the artifacts—supposedly debris from a crashed saucer—were meticulously collected and removed by members of the armed forces. The military is also accused of secreting away the bodies of the saucer's alien occupants. The fact that no evidence from this event is available for scientific scrutiny has been blamed on a nefarious—and surprisingly efficient—cover-up by the U.S. federal government.

The incident began in early July 1947, only a few weeks after the nationwide coverage of Kenneth Arnold's "flying saucers." A rancher found some crash remnants in a pasture close to the city of Roswell, New Mexico. He reported the debris to the local sheriff, who then passed on the information to the Roswell Army Air Field, which was nearby. Several military personnel drove out to the ranch, picked up the debris, and explained to the local papers that they had recovered the remains of a "flying disk."

This dismaying story was quickly quashed. Only a day later, an Air Force officer from Fort Worth, where the debris had been flown, held a press conference in which he dismissed the flying disk notion. He stated that the debris was merely a crashed weather balloon, substantially ending interest in the incident at the time. But in 1978 UFO investigator Stanton Friedman began looking into the events at Roswell. Friedman claimed that the debris was from a spacecraft and that alien occupants had been picked up as well. He believed that the government was covering up both the fact of the mishap and its extraterrestrial victims.

The possibility of alien bodies lent this story an appeal beyond the routine sightings of lights in the sky. However, the credibility of the idea of an alien accident in Roswell hinges on whether you believe the witness testimony. Recent investigators have pointed out that Friedman and others conducted their interviews more than *three decades* after the event and the testimonies were inconsistent. Some supposed witnesses to the crashed "saucer" had originally claimed not to have seen it. Others were caught in flat-out lies. In the end, witness testimony must be judged as "he said, she said" evidence and is not rigorous enough to prove something as profoundly important as alien visitation.

But something *did* crash at Roswell. What was it? Despite the Air Force claim, the debris was not a weather balloon. Instead, it seems to have been the balloon train and some remnants of a device designed to detect Soviet nuclear tests, developed as part of a highly classified operation known as Project Mogul. The scheme was to launch instrumented balloons that would hover at constant altitude near the borders of the Soviet Union and listen for sound waves produced by a distant nuclear blast. In the late 1940s, tests of this device were being made at Roswell. The principal scientist behind Project Mogul is still alive and is dismayed by the insistence that the debris was not from his experiment but rather part of an extraterrestrial craft. Photos taken of the recovered materials by the Air Force (see Figure 2.12) match contemporary photos of Mogul balloon trains.

In short, there is a rational explanation for what happened at Roswell. Claims that the debris was from a downed alien craft are based on witness accounts made long after the fact, and inconsistencies in the accounts make them less than credible. It might be more interesting to believe that small people from another world just happened to make a navigational error a few weeks after flying saucer mania first swept the country, but the evidence suggests only a crashed Air Force project.

Crop Circles, Abductions, and More

UFOs and crashed aliens can nearly always be readily explained in terms that don't involve visitors from other worlds. The same is true for other highly publicized events interpreted as proof of an extraterrestrial presence in the neighborhood, including abductions, crop circles, and miscellaneous phenomena such as mutilated cattle and goat-eating chupacabras.

Crop circles, geometric patterns made in wheat fields, have generated a great deal of interest among the public (Figure 12.16). For several decades, such patterns have appeared each summer in England. According to some, they are the work of visiting extraterrestrials who are using this method to communicate to us. The proof offered for their nonhuman origin is the speed with which they appear (overnight) and the condition of the matted-down wheat, which is claimed to be inconsistent with simple trampling by humans. But there are many indications that these patterns are merely pranks, not alien messages. The continuing increase in complexity of the patterns over the years suggests that their creators are getting better at their craft. The fact that they are made only at night also suggests human subterfuge. Moreover, many of the crop circles have been acknowledged as pranks, and the human pranksters have demonstrated for news reporters how the crop circles can be made quickly and easily. (Competitions in crop circle making have even been held!) In one interesting case reported in a documentary about crop circles, a self-proclaimed "expert" on the alien origins of crop circles acknowledged that some crop circles were pranks but claimed that not all of them could be. He explained to interviewers why a particular set of circles could not possibly have been made by humans—and then watched in dismay as the interviewers showed him a video of experienced pranksters making the very ones he had just labeled "alien."

Cattle mutilations and the reputed killings of various small animals by short, bipedal creatures known as chupacabras have also been attributed to alien activity, despite a centuries-long history of attacks on farm animals by both earthly predators and humans. The only evidence offered to connect these attacks to extraterrestrial beings is uncertainty over who or what is doing the killing.

A more interesting phenomenon involves alien abduction of humans, a widespread curiosity that many people find compelling. Although small in overall percentage of the population, a substantial number of Americans claim to have been alien abductees, often supposedly molested by extraterrestrials while asleep (far fewer are disturbed during waking hours, it seems). The victims are either taken aboard spacecraft for observation and unwholesome

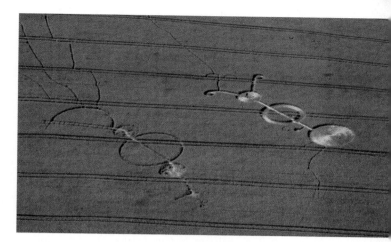

FIGURE 12.16 Crop circles. These relatively simple patterns have been "cropping up" in southern England for several decades and are claimed by some to be manifestations of alien activity. However, the designs are easily produced in a few hours by a small group of motivated students using nothing more than boards and ropes, suggesting a terrestrial origin. In addition, we might wonder why sophisticated extraterrestrials would travel many light-years simply to carve graffiti in our wheat.

experiments or simply watched while they lie immobilized in their beds. However, similar phenomena have been described since ancient times in a multitude of cultures. In the past, the accused culprits have been witches, ghost babies, and goblins. Today, the molester of choice is frequently a visitor from the stars.

According to many psychologists, such abduction experiences can be attributed to **sleep paralysis,** which occurs during REM (rapid eye movement) sleep. During REM sleep, the body is naturally paralyzed so that we don't thrash physically along with the movements we make in our dreams. Sometimes, this paralysis persists for a few minutes after the brain has started waking up—which can give a person the alarming sensation of being awake in a paralyzed body. Visions and other sensations often occur in this state. Sleep paralysis is experienced by approximately half of all people at some point in their lives. It therefore seems quite reasonable to imagine that a small percentage of people might believe they had experienced some type of alien abduction. Proponents of alien abduction dispute this idea, noting that a few victims were fully awake and alert during their experience. However, in some cases daydreams are known to produce sensations similar to those produced by sleep paralysis. Once again, we are left to accept either a very simple and earthly explanation (sleep paralysis) or an extraordinary one (abduction by aliens).

Ancient Visitations

The vast majority of those who believe in extraterrestrial visitation claim that the aliens and their craft are among us now. However, some people have made claims concerning alien visits in the human past or even earlier.

The idea that Earth might have been visited seems plausible, especially if civilizations turn out to be common in our galaxy. Our planet might well be of interest to alien scientists, and we'd have no way of knowing whether alien ships studied the Earth in the billions of years that preceded the evolution of Homo sapiens. Such visits might even have continued after the rise of civilization. Is there any evidence of alien visits to our ancestors? A few people have claimed that there is, but the evidence they offer is far from compelling.

The claimed evidence of past alien visitation consists of things such as ancient drawings (often on rocks or in caves) that supposedly show alien beings, archaeological wonders including the desert markings on the Nazca plains of Peru (Figure 12.17), and even the great pyramids of Egypt. The ancient drawings are ambiguous; what appears to some people to be a space helmet appears to others to be a ceremonial headdress. The Nazca markings—claimed by some to be landing strips for alien craft—could just as well be the work of the Nazca Indian culture that occupied the area about two thousand years ago. No special technology would have been needed to make

FIGURE 12.17 Hundreds of lines and patterns, generally obvious only from the air, are etched in the sand of the Nazca desert in Peru. This aerial photo shows the large figure of a hummingbird. Some people have claimed that these patterns must have been created or inspired by aliens, though it would not have been difficult for the local people to make them on their own.

the markings, and patterns visible only from the air seem more likely to have been intended for gods than for aliens. Claims of alien origin for the Egyptian pyramids are nothing less than an insult to ancient people, because they seem to suggest that the Egyptians were incapable of constructing such impressive structures on their own. In fact, while the pyramids were a remarkable achievement, their construction was well within the technical expertise of the kingdom of the pharaohs. All other known ancient structures—including Stonehenge, the Mayan pyramids of Central America, and the massive stone heads of Easter Island—were also within the technical capability of the cultures that created them.

Is There a Case for Alien Visits?

Despite the lack of evidence of alien visits past or present, many people continue to believe they have occurred. Champions of alien visitation generally explain away the lack of compelling evidence in one of two ways: government cover-ups or a failure of the mainstream scientific community to take the relevant phenomena seriously. Let's consider both of these arguments.

Government conspiracy and cover-up are popular notions, particularly in the United States. Frankly, it's conceivable that a secretive government might try to put a lid on the best spacecraft videos and reconnaissance images from orbiting satellites, though the motivation for doing so is unclear. The usual explanations are that the public couldn't handle the news or that the government is taking secret advantage of the alien materials to design (via "reverse engineering") new military hardware. But both explanations are silly. Half the population already believe in alien visitors and would hardly be shocked if newspapers announced tomorrow that aliens were stacked up in government warehouses. As for the reverse engineering of extraterrestrial spacecraft, we should keep in mind the difficulty of traveling from star to star. Any society that could do so would be technologically far superior to our own. Our reverse-engineering their spaceships is as unlikely as Neanderthals' constructing personal computers just because a laptop somehow landed in their cave. In addition, while a government might successfully hide evidence for a short time, the evidence is unlikely to remain secret for decades (more than five decades, in the case of the Roswell claims). And, unless the aliens landed only in the United States, can we seriously believe that *every* government would cooperate in hiding the evidence?

The alleged disinterest of the scientific community is also an unimpressive claim. Scientists are constantly competing with one another to be first with a great discovery and thus are aggressive in pursuing any important new phenomenon. Clear evidence that aliens exist and are (or have been) on our planet would be hailed as one of the most important discoveries of all time. Countless researchers would work evenings and weekends, without pay, if they thought such a discovery was possible. The fact that few scientists are engaged in such study reflects not a lack of interest but a lack of evidence worthy of study.

Indeed, lack of physical evidence is the foremost reason given by scientists to justify their skepticism regarding UFOs and alien visits. Airport radars and orbiting satellites have never yielded proof of the passage of an alien spacecraft. No one has ever walked into a research lab with an alien artifact, such as a piece of material that we could not manufacture with our current technology—not even a fragment. In Earth orbit, the U.S. Air Force tracks thousands of pieces of "space junk" from our own satellites and shuttles, including discarded rockets, exploding bolts, a Hasselblad camera, paint chips, frozen urine, and the occasional astronaut glove. But no one has ever found a single piece of space junk that can be attributed to aliens. Most scientists are open to the possibility that we might someday find evidence of alien visits, and many would welcome aliens with open arms. But so far the evidence is simply lacking. Extraterrestrial visitors to Earth remain a routine fixture of cinema and television, but in real life they are like ghosts—a pervasive and attractive idea that the vast majority of scientists treat with skepticism.

Of course, it is very important to distinguish between claims that aliens are visiting us now (or visited Earth in the past) and the possibility that alien civilizations might exist. When, after World War II, space travel moved from the theoretical to the practical, it was only natural to assume that what we were trying to do—travel to other worlds—is routinely done by other civilizations. As we've seen in this chapter, there is a great difference between journeys within the solar system and jaunts to the stars. While the former are straightforward, the latter are both enormously difficult and stupefyingly costly in terms of energy. Nonetheless, interstellar travel doesn't violate physics, and advanced societies might have been spurred to voyage from star to star despite the formidable technical obstacles. This possibility leads to an interesting, if hypothetical, problem that we will consider in the next chapter—the Fermi paradox.

THE BIG PICTURE

In this chapter, we have explored the possibilities for and challenges of interstellar travel. As you continue in your study, keep in mind the following "big picture" ideas:

- Interstellar travel may be a staple of science fiction, but it remains well beyond our current capabilities. Nevertheless, it is possible that the challenges can be surmounted and that other civilizations might already have overcome them.

- Although it is not possible to travel faster than the speed of light, the cosmic speed limit might not be as detrimental to interstellar travel as it seems. At speeds close to the speed of light, time dilation would make it possible for the travelers to go great distances in relatively short times—though a long time would pass back home.

- We do not yet know whether it is possible to reach the stars by taking "shortcuts" with science fiction ideas such as wormholes, hyperspace, and warp drive. These shortcuts would effectively lessen the distance to faraway places, allowing us to reach them in short times without ever violating the speed-of-light limit. At present, the laws of physics seem to allow for any or all of these things, at least in principle, but some undiscovered laws may prohibit these forms of travel.

- Despite the many claims of UFOs and other supposed alien visits to Earth, no compelling evidence for such visits has ever presented itself.

Review Questions

1. Briefly describe the voyages of *Pioneers 10* and *11* and *Voyagers 1* and *2*. How do these spacecraft illustrate the challenge of interstellar travel?

2. How does the speed of light affect the possibility of interstellar travel?

3. Explain why the energy needed for interstellar travel also makes it quite challenging.

4. Briefly describe how rockets work and give a short history of rocketry. What is the *rocket equation* used for? What do we mean by a *mass ratio* for a rocket?

5. Briefly explain why chemical rockets are inadequate for sending people or large robotic probes to the stars.

6. Describe the fusion-powered starships of Project Orion and Project Daedalus. How quickly could such ships reach the stars?

7. Discuss a few ways of reaching the stars (other than nuclear rockets) that are, at least in principle, within our current technological reach.

8. How would *time dilation* affect space travel at speeds close to the speed of light? Discuss possible ways of achieving such speeds, including matter-antimatter engines and *interstellar ramjets*.

9. Briefly discuss how Einstein's general theory of relativity might allow "shortcuts" by which we could reach distant stars in shorter times than we would expect from their measured distances.

10. Discuss several types of claims about alien visitation on Earth. Why, so far at least, do they seem unlikely to be true?

Discussion Questions

1. *Distant Dream or Near-Reality.* Considering all the issues surrounding interstellar flight, when, if ever, do you think we are likely to begin traveling among the stars? Why?

2. *Seeding the Galaxy.* If interstellar travel is forever impractical, are there other ways an advanced civilization might spread its culture? Clearly, communication is possible, although the speed of light makes conversations between star systems maddeningly tedious. Could a society send the information required to assemble members of its species (its "DNA," for instance) and therefore spread through the galaxy at the speed of light? Can you imagine other ways of spreading a culture without starships? Explain.

3. *Sociology of Interstellar Travel.* Suppose we somehow built a spaceship capable of relativistic travel and volunteers were being recruited for a journey to a star 15 light-years away. Would you volunteer to go? Do you think others would volunteer? In light of the effects of time dilation, discuss the benefits and drawbacks of such a trip.

4. *Visiting Aliens.* Suppose aliens *are* visiting the Earth (or observing from close range). Based on what you've learned about the challenge of interstellar travel, how would their technology compare to ours? Discuss the implications of your answer to claims of UFO visits and "standard" science fiction stories that imagine wars between visiting aliens and us. Would abductees be prodded and probed with instruments similar to those we might find in our own hospitals?

5. *Dealing with UFO Claims.* Given the large number of people who claim to have seen a UFO, you are likely to know at least one such person, now or in the future. Perhaps *you* have seen a UFO. Suppose someone who has seen a UFO believes deeply that it was an alien spacecraft. What, if anything, would you say to that person? Why?

Problems

Fantasy or Science Fiction? Each of **problems 1–8** describes some futuristic device or discovery. In each case, decide whether it is plausible according to our present scientific understanding or is unlikely to be possible. Explain your reasoning.

1. A brilliant teenager discovers a way to build a rocket that burns coal as its fuel and can travel at half the speed of light.

2. Using beamed energy propulsion from a laser powered by energy produced at a windmill farm in the California desert, NASA engineers are able to send a solar sailing ship on a journey to Alpha Centauri that will take only 50 years.

3. Human colonization of the moons of Saturn occurs using spaceships powered by dropping nuclear bombs out the back of the ships.

4. In the year 2750, we receive a signal from a civilization around a nearby star telling us that the *Voyager 2* spacecraft recently crash-landed on its planet.

5. The General Rocket Corporation (a future incarnation of General Motors) unveils a new personal interstellar spacecraft that works as an interstellar ramjet with a scoop about 10 meters across.

6. Members of the first crew of the matter-antimatter spacecraft *Star Apollo*, which left Earth in the year 2165, return to Earth in the year 2450 looking only a few years older than when they left.

7. In the year 2009, we finally uncover definitive evidence of alien visits to Earth when a flying saucer crashes in the Rocky Mountains and its oxygen-kerosene fuel ignites a forest fire.

8. Traveling through a wormhole apparently constructed by an advanced civilization, explorers journey from our solar system to a star system near the center of the galaxy in just a few hours.

9. *Interstellar Travel in the Movies.* Choose a science fiction movie in which aliens (or future humans) are engaged in some type of interstellar travel. In a one- to two-page essay, briefly describe how they supposedly accomplish the travel and evaluate in depth whether the scheme seems plausible.

*10. *The Rocket Equation.* Because the physics of rockets is quite simple, rocket performance can be described by only a few quantities: M_i, the mass of the rocket (including any spacecraft it is carrying) with all its fuel; M, the mass of the rocket after the fuel has been burned; v_e, the velocity of the exhaust gas expelled out the back; and v, the forward velocity of the rocket. The rocket equation, which tells us how these quantities are related, is:

$$v = v_e \ln\left(\frac{M_i}{M}\right)$$

(Note that ln is the natural logarithm; your calculator should have a key for computing it.) This equation can be solved to find the mass ratio needed for a particular choice of velocities:

$$\frac{M_i}{M} = e^{\left(\frac{v}{v_e}\right)}$$

($e \approx 2.718$; your calculator should also have a key for computing e to any power.)

a. Suppose a rocket with mass ratio $M_i/M = 15$ has engines that produce an exhaust velocity of 3 km/s. What is its final velocity? Is it sufficient to escape Earth?

b. Suppose you want a rocket to achieve escape velocity from Earth (11 km/s) and its engines produce an exhaust velocity of 3 km/s. What mass ratio is required? Briefly explain the meaning of this mass ratio.

*11. *The Multistage Rocket Equation.* The rocket equation takes a slightly different form for a multistage rocket:

$$v = nv_e \ln\left(\frac{M_i}{M}\right)$$

where n is the number of stages.

a. Suppose a rocket has three stages with mass ratio $M_i/M = 3.4$ and engines that produce an exhaust velocity of 3 km/s. What is its final velocity? Is it sufficient to escape Earth?

b. Suppose a rocket has 100 stages with mass ratio $M_i/M = 3.4$ and engines that produce an exhaust velocity of 3 km/s. What is its final velocity? Compare it to the speed of light.

*12. *Relativistic Time Dilation.* The effects of time dilation on a fast rocket can be calculated with a simple formula if we assume that the rocket travels at constant speed (which is somewhat unrealistic, because the rocket would need time to accelerate and decelerate):

$$t_{rocket} = t_{Earth} \times \sqrt{1 - \frac{v^2}{c^2}}$$

where t_{rocket} is the amount of time that passes on the rocket, t_{Earth} is the amount of time that passes on Earth, v is the rocket's velocity (speed), and $c = 3 \times 10^8$ km/s is the speed of light.

a. Suppose a rocket travels at the escape velocity from Earth (11 km/s). Will time on the rocket differ noticeably from time on Earth? Explain.

b. Suppose a rocket travels at a speed of $0.9c$. If the rocket is gone from Earth for 10 years as measured on Earth, how much time passes on the rocket?

c. Suppose a rocket travels at a speed of $0.9999c$. If the rocket is gone from Earth for 10 years as measured on Earth, how much time passes on the rocket?

*13. *Long Trips at Constant Acceleration.* Consider a spaceship on a long trip with a constant acceleration of 1g. Although the derivation is beyond the scope of this book, it is possible to show that, as long as the ship is gone from Earth for many years, the amount of time that passes on the spaceship during the trip is approximately:

$$T_{ship} = \frac{2c}{g} \ln\left(\frac{g \times D}{c^2}\right)$$

where D is the distance to the destination and ln is the natural logarithm. If D is in meters, $g = 9.8$ m/s^2, and $c = 3 \times 10^8$ m/s, the answer will be in units of seconds. (*Hint:* In all cases, be sure you convert the distances from light-years to meters (1 light-year $\approx$ 9.5×10^{15} meters) and convert the final answers from seconds to years (1 year $\approx 3.15 \times 10^7$ seconds).)

a. Suppose the ship travels to a star that is 500 light-years away. How much time will pass on the ship? Approximately how much time will pass on Earth? Explain.

b. Suppose the ship travels to the center of the Milky Way Galaxy, about 28,000 light-years away. How much time will pass on the ship? Compare this to the amount of time that passes on Earth.

c. The Andromeda Galaxy is about 2.2 million light-years away. Suppose you had a spaceship that could constantly accelerate at 1g. Could you go to the Andromeda Galaxy and back within your lifetime? Explain. What would you find when you returned to Earth?

Web Projects

1. *Advanced Spaceship Design.* NASA supports many efforts to incorporate new technologies into spaceships. Although few of these reach the level of being suitable for interstellar colonization, most are innovative and fascinating. Learn about one such NASA project and write a short summary of your findings.

2. *Nuclear Rockets.* Learn more about one of the past government programs to develop a nuclear rocket, such as Project Rover or Project Orion. Write a one- to two-page report about the program.

3. *Alien Visits.* Learn more about a claim of alien visitation to Earth, past or present. Write a one- to two-page report explaining the claim and the evidence that supports it and discussing the plausibility of the claim.

CHAPTER 13
The Fermi Paradox

As difficult as interstellar travel might be, the history of human progress suggests that we might eventually achieve it. If we can rocket to the stars, then other civilizations should also be able to develop that capability. And, if civilizations are common, it's likely that some societies—perhaps many—began colonizing the galaxy long before the earliest humans walked the Earth. Indeed, some civilizations could have begun their spread among the stars even before Earth was born. Yet we have no evidence of any such galactic civilization existing around us.

In this chapter, we'll investigate the astonishing implications of these seemingly simple ideas. We'll discuss why it is reasonable to expect that other civilizations could have had a head start on us by millions or even billions of years. We'll see that, in principle, even a single space-faring civilization that predates us by a few tens of millions of years (or more) could have colonized the entire galaxy. These facts will lead us to consider possible solutions to the so-called *Fermi paradox:* If someone could have colonized the galaxy by now, why don't we see any evidence of colonization? As with many topics in this book, we will not be able to offer a definitive answer. But we will see that the answer, no matter what it turns out to be, may have profound implications for the future of our own civilization.

13.1 Where Is Everybody?

The paradox we are considering in this chapter was first stated in 1950 by the Nobel Prize–winning Italian-American physicist Enrico Fermi (Figure 13.1). During a lunch at the Los Alamos National Laboratory in northern New Mexico, the conversation drifted to the possibility of extraterrestrial intelligence. The physicists present at the lunch were considering the likelihood that sophisticated cosmic societies might exist in great abundance. Fermi replied to these speculations with a disarmingly simple question: "So where is everybody?" Although serious scientific discussion of his query did not get under way for many years, its central idea is now known as the **Fermi paradox.**

The essence of the Fermi paradox is almost as simple as Fermi's original question. It begins with the idea that neither we nor our planet should be in any way special, and thus other Earth-like planets and other advanced civilizations (meaning civilizations capable of space travel) ought to be fairly common in the galaxy. This is more or less what we conclude from the Drake equation [Section 11.1], unless the rare Earth hypothesis turns out to be correct [Section 10.5]. However, as we will discuss in this chapter, a large number of civilizations would necessarily mean many civilizations with the opportunity to develop advanced technology and interstellar travel long before we came on the scene with our rockets and radio telescopes. In that case, it seems that someone else should have colonized the galaxy already. Even using rockets that can travel at only a small fraction the speed of light, reaching every star in the Milky Way could be accomplished in an amount of time that is short compared with the galaxy's age. But we see no evidence of such a galactic colonization effort. Thus, we have two seemingly contradictory ideas:

1. The idea that neither we nor our planet is in any way special suggests that someone should have colonized the galaxy by now.

2. The idea of a galactic civilization implies that we should be surrounded by evidence of this civilization—but aside from unconvincing claims of extraterrestrial UFOs, no such evidence exists.

By definition, the existence of two such seemingly contradictory ideas constitutes a *paradox*. But unlike some logical paradoxes (e.g., a statement such as "This statement is false"), this paradox must have some solution. You can probably already think of ways out of the contradiction, and we will discuss a variety of possible solutions in Section 13.4. The simplest is this: The galaxy is not colonized because we are the first civilization to attain a high level of technological sophistication. If this is true, then all our SETI efforts are a waste of time, because there's no one out there to talk to.

FIGURE 13.1 Enrico Fermi (1901–1954), one of the leading physicists of the twentieth century, received the Nobel Prize in 1938 for work in understanding radioactive decay and predicting the existence of the particles known as neutrinos. By the time of his Nobel Prize, the Fascists had risen in Italy and the Nazis were in power in Germany. Abhorring these ideologies, he chose not to return to Italy after attending the Nobel Prize ceremony in Sweden. Instead, he moved to the United States, where he became a prominent figure in the Manhattan Project, which was working on development of the atomic bomb. Element 100 in the periodic table, fermium, was named in his honor.

To some extent, further consideration of our own future technology makes the paradox seem even more contradictory. We have already built robots to explore the planets, and it seems reasonable to assume that we will continue to do so in the future. Consider the types of robots we might construct with our future technological capabilities. We could program them to go to another world, dig up resources, use the resources to build factories, and use the factories to build spacecraft and more robots. These new robots would then go on to the next world, where they would do the same thing. Thus, these robots would be self-replicating, though in a way quite different from the self-replication of biological beings. The general idea of such self-replicating machines was first proposed by the American mathematician and computer pioneer John Von Neumann (1903–1957); thus, they are often called **Von Neumann machines.**

The use of Von Neumann machines would allow us to explore much farther and wider than we could by going to other worlds ourselves. Moreover, whereas interstellar travel poses huge barriers to us due to our limited life spans, these machines could presumably still function after journeys through space that take centuries or millennia. Once we sent the first wave of these machines to a few nearby star systems, they would gradually spread from star system to star system.

In 1981, physicist Frank Tipler used this idea of "colonization" by self-replicating Von Neumann machines to extend the Fermi paradox. In essence, Tipler argued that civilizations could effectively make their presence felt throughout the galaxy even without achieving the ability to send themselves on interstellar journeys. As soon as a civilization reached a level that allowed them to build Von Neumann machines, these machines would begin to spread through the galaxy. Because such colonization would require technology only slightly beyond our own, Tipler argued that if civilizations are common, then the galaxy would already be overrun by self-replicating machines. Because it isn't, Tipler concluded that we are alone and thus that SETI is a waste of time.

Is there any way around the Fermi paradox—and its extension with self-replicating Von Neumann machines—other than concluding that we are alone? Yes, though we can't be sure that any of these ways are correct. In the rest of this chapter, we'll look more closely at the issues that underlie the paradox and consider possible arguments to resolve it. We'll also discuss the implications of the various solutions.

THINK ABOUT IT . . . *Before reading further, can you think of any possible reasons besides our being alone that would explain why the galaxy is not overrun with either beings or robots? Explain.*

13.2 The Age of Civilizations

The first key premise of the Fermi paradox is that, if civilizations are at all common, many should have arisen long before our own arrival. Although this idea is considered obvious by some people, it bears at least some further investigation. In this section, we'll explore the scale of cosmic time to see why it is unlikely that ours is the first civilization to arrive on the cosmic scene.

The Cosmic Calendar

As we discussed in detail in Chapter 4, the Earth is about 4.6 billion years old. The universe is much older still, with current estimates putting the time since the Big Bang at between about 12 and 15 billion years [Section 1.3]. Our species has been around to study such things for only a tiny fraction of all this time. One of the best ways to get a feel for the scale of cosmic time is through a device, developed by Carl Sagan, called the *cosmic calendar*.

The cosmic calendar represents the entire history of the universe as a single calendar year. Thus, the Big Bang takes place at the first instant of January 1,

and the present day is just at the stroke of midnight on December 31. Although the universe might be anywhere from about 12 to 15 billion years old, we can keep the calendar simple by assuming an age of 12 billion years. Note that with this choice:

- Each month on the cosmic calendar represents 1 billion years in the history of the universe.

- Each day on the cosmic calendar represents about 33 million years.

- Each hour on the cosmic calendar represents about 1.4 million years.

- Each minute on the cosmic calendar represents about 23,000 years.

- Each second on the cosmic calendar represents about 390 years.

Figure 13.2 shows the cosmic calendar, with a few notable events in cosmic history indicated. On this scale, the Milky Way Galaxy—and many or most other galaxies—probably formed sometime in early February. The stellar recycling that created elements heavier than hydrogen and helium was under way. In fact, many generations of stars must have come and gone quite quickly, gradually increasing the proportion of heavy elements to their current value of about 2% (with the other 98% being hydrogen and helium). Remember that massive stars—those that produce most of the heavy elements—burn out in just a few million years or less. On the cosmic calendar, this corresponds to only a few hours. Thus, on the scale of the cosmic calendar, every day since the formation of the Milky Way in February could have seen several generations of massive stars.

Our solar system and the Earth did not form until mid-August on this scale. By early September, life on Earth was flourishing. Note, however, that the great diversification of animal life that we call the Cambrian explosion [Section 5.4] did not occur until mid-December. The dinosaurs appeared on Christmas Day. Then, in a cosmic instant, they disappeared forever with the K–T impact, which occurred around 12:30 A.M. on December 30. The 65 million years that separate us from the dinosaurs is less than 2 full days on the scale of the cosmic calendar.

Perhaps the most astonishing thing about the cosmic calendar is that the entire history of human civilization falls into just the last half-minute. The ancient Egyptians built the pyramids only about 13 seconds ago on this scale. It was only about 1 second ago that Kepler and Galileo first proved that the Earth orbits the Sun rather than vice versa. The average college student was born about 0.05 second ago, around 11:59:59.95 P.M. On the scale of cosmic time, the human species is the youngest of infants, and a human lifetime is a mere blink of an eye.

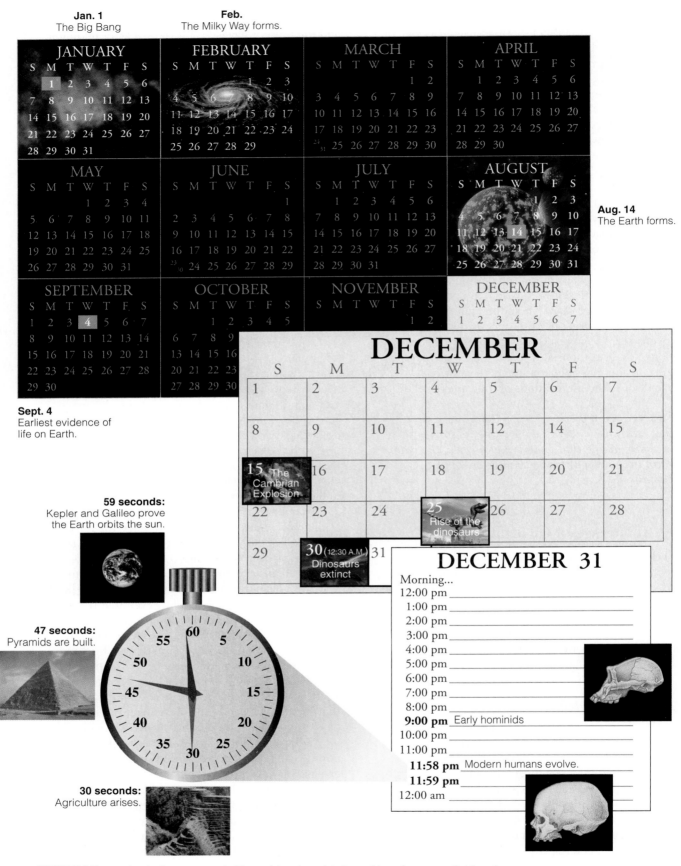

Jan. 1
The Big Bang

Feb.
The Milky Way forms.

Aug. 14
The Earth forms.

Sept. 4
Earliest evidence of life on Earth.

59 seconds:
Kepler and Galileo prove the Earth orbits the sun.

47 seconds:
Pyramids are built.

30 seconds:
Agriculture arises.

DECEMBER

S	M	T	W	T	F	S
1	2	3	4	5	6	7
8	9	10	11	12	14	15
15 The Cambrian Explosion	16	17	18	19	20	21
22	23	24	25 Rise of the dinosaurs	26	27	28
29	30 (12:30 A.M.) Dinosaurs extinct	31				

DECEMBER 31

Morning...
12:00 pm _____
1:00 pm _____
2:00 pm _____
3:00 pm _____
4:00 pm _____
5:00 pm _____
6:00 pm _____
7:00 pm _____
8:00 pm _____
9:00 pm Early hominids _____
10:00 pm _____
11:00 pm _____
11:58 pm Modern humans evolve.
11:59 pm _____
12:00 am _____

FIGURE 13.2 The cosmic calendar compresses the history of the universe into 1 year; this version assumes that the universe is 12 billion years old, so each month represents about 1 billion years. Only within the last few seconds of the last day has human civilization taken shape. (This version of the cosmic calendar is adapted from a version created by Carl Sagan.)

Civilizations on the Cosmic Calendar

We can get a sense of when and how often civilizations might have arisen by considering the age of the universe. We can then put these ideas in perspective with the cosmic calendar. Of course, we do not know how many civilizations have arisen, so we'll have to make some guesses.

Let's start by asking when the first civilizations might have arisen. First, we need to know when Earth-like planets could have formed. This depends on when enough heavy elements were present to make planetesimals that could accrete into terrestrial planets. There is some controversy about this, but given that generations of massive stars were living and dying every few million years (several generations per day on the cosmic calendar), at least some heavy elements must have become available quite soon. Many scientists believe that even stars born quite early in our galaxy's history were able to harbor Earth-like planets. In that case, if we assume that the 4.6 billion years it took from the birth of Earth to modern humans is typical, the first civilizations could have arisen about $4\frac{1}{2}$ billion years after our galaxy formed—around mid- to late-June on the cosmic calendar. In real time, this means civilizations could have begun to arise in the Milky Way Galaxy as early as 6 billion years ago. (Similar considerations apply to other galaxies.)

Let's further suppose that about one in a million stars has a habitable planet that eventually gets a civilization. A conservative estimate of 100 billion stars in our galaxy (the actual number is probably several hundred billion) then gives 100,000 civilizations. If we assume that the first of these 100,000 civilizations arose 6 billion years ago, then civilizations could have been arising ever since at an average rate of one civilization about every 60,000 years. On the cosmic calendar, this means a new civilization approximately every 3 minutes.

We are led to similar conclusions with almost any other assumptions we make, unless we are willing to believe that the Earth is so unique and so rare that no other planet in the galaxy is suitable for civilization. For example, suppose we take the more pessimistic assumption that only *one in a billion* stars gives rise to a civilization. That would still mean at least 100 civilizations over the past 6 billion years, or about 1 civilization every 60 million years on average. This corresponds to the rise of a new civilization approximately every 2 days on the cosmic calendar.

What if Earth-like planets could not have formed as early as we have assumed? Let's suppose instead that planets like ours were not possible until the proportion of heavy elements attained the level it had when our solar system was born. Even with this constraint, we know that this proportion of heavy

elements was not reached in an instant. If we conservatively assume that Earth-like planets could have begun to form only about 5 billion years ago—the beginning of August on the cosmic calendar—then other planets could still have had a half-billion-year head start on evolution compared to the Earth. Maintaining our assumption that it takes $4\frac{1}{2}$ billion years for an intelligent species to evolve, then possible homes for civilizations would include only planets around stars that formed during the half-billion years before our Sun was born. But this is not really much of a constraint. The rate of star formation in our galaxy is more or less constant, and a half-billion years amounts to roughly 1/20 of the galaxy's history. Thus, in a half-billion-year period, about 1/20 of the galaxy's stars were born, which means at least 5 billion stars. If one in a million of these stars gives rise to a civilization, we would still find some 5,000 civilizations that had arisen over the past half-billion years—an average of one civilization every 100,000 years. On the cosmic calendar, this would mean that the first civilization arose around mid-December and civilizations have arisen about every 5 minutes since then.

No matter how we look at it, we are drawn to a clear conclusion: If civilizations are at all common, then plenty of them should predate our own civilization by thousands, millions, or even billions of years.

The Youngest Civilization

Another way to look at the issue of civilizations is to consider the likelihood that other civilizations are at a technological level close to ours. To do this, we'll use the more conservative assumption about when Earth-like planets could have formed; that is, we'll suppose that civilizations have been possible for only about the past half-billion years. As we found above, spreading 5,000 civilizations over this period—or one for every 1 million stars of the right age—would mean that civilizations arise an average of 100,000 years apart. In other words, we would expect the *most recent* civilization besides ours to have come into being some 100,000 years ago. If members of this civilization did not destroy themselves but managed to continue to progress, then their level of science and technology should be some 100,000 years ahead of ours. Even if they arose on the far side of the galaxy from us, some 80,000 light-years away, radio signals now arriving from their home planet would be coming from a civilization 20,000 years ahead of ours (since the signals would take 80,000 years to get here). And every other civilization would be much older.

There's a galactic club out there, and Earth has been asked to join. Actually, membership is not optional. You *will* join, and you *will* behave.

That's the stern message delivered in *The Day the Earth Stood Still,* a 1951 film that suggests that the specter of cold-war nuclear holocaust might best be eliminated with the help of some extraterrestrial meddling. The movie begins as well-intentioned aliens decide to visit the nation's capital and save us from ourselves. Citizens watch in dismay as a glowing saucer cruises by some famous landmarks and eventually sets down on the White House lawn. Help is at hand.

The landed craft has few occupants: Klaatu, a well-spoken alien who resembles a New York lounge lizard, and his robot pal, Gort, who looks like the Tin Man on steroids. These two characters step out of their disk-shaped craft to give us "the word." Klaatu declares, with some pride, that they've traveled 250 million miles to do this. That sounds like a long haul, but of course it's not: within that paltry distance all you'll find are the Sun, Moon, Mercury, Venus, and Mars. These aliens supposedly represent some sort of cosmic organization put together to keep the peace, but the low mileage on their odometer suggests that they're just members of the inner-solar-system neighborhood association.

Klaatu wants to speak to the United Nations, but the politicians won't agree to listen, so the debonair alien opts for second best: He'll pitch his spiel to a conclave of the world's top scientists. After a brief demo of extraterrestrial capabilities to get everyone's attention (Klaatu and Gort shut down all earthly machinery for a half-hour—an amazing feat routinely, and annoyingly, managed by your local power company), they get to the point: Stop threatening everyone in the universe with atomic weaponry, or Gort and his robotic kin will wipe out our planet. Apparently Gort is a "death star" on legs.

So it boils down to this: Mom has told you to behave, and if you don't, Dad is going to really punish you. Needless to say, this isn't quite the agenda of any cosmic empire imagined by scientists, nor even that of the galactic federation in *Star Trek.* If there are real aliens out there who've mastered both interstellar travel and the colonization of other worlds, they undoubtedly regard A-bombs as mere firecrackers.

And besides, would anyone really want to trust a bunch of nickel-plated robots to keep the peace? After all, what if they get wet and seize up?

THINK ABOUT IT . . . *Take a moment to contemplate what the technology might look like for a civilization 100,000 years ahead of ours. Discuss with friends or classmates some of the possibilities. What about a civilization that arose earlier, say, 1 million or 100 million years ago?*

13.3 Galactic Colonization

We have found that if civilizations are at all common, then not only have many preceded us in the Milky Way Galaxy, but they have preceded us by at least thousands or millions of years. Some civilizations may predate us by hundreds of millions or even billions of years. As stated in the second premise of the Fermi paradox, this conclusion implies that our galaxy should already have been colonized. In this section, we'll see why.

Colonization Models

Suppose a civilization arose a long time before us and started sending out spacecraft to colonize other hab-

itable planets. How long would it take them to colonize the entire galaxy?

The answer clearly depends on the civilization's technological capabilities. For example, if it has the technology to build spacecraft that can travel at speeds close to the speed of light, then it could add colonies throughout the galaxy fairly quickly. It would take just a few years to send settlers to planets around stars only a few light-years away. These colonies could then grow and eventually send emigrants to more distant stars. Because no place in the galaxy is more than 100,000 light-years (the diameter of the galaxy) away from any other place, this civilization could reach and settle every star system within just a few hundred thousand years. If many civilizations have preceded us by millions of years, then there has been plenty of time for the galaxy to have been colonized in this way.

Interestingly, the conclusion is not that much different if we assume much slower speeds. For example, suppose a civilization has nuclear rockets of the type discussed in Chapter 12 (such as the Project Orion or Project Daedalus rockets). Although we do not yet have such rockets, we are already capa-

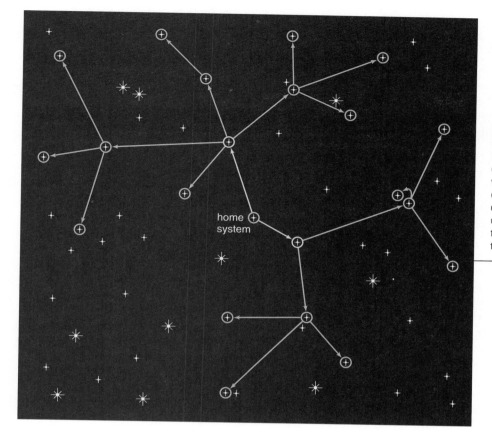

FIGURE 13.3 The coral model of galactic colonization. Colonization begins when the inhabitants of one star system send a few craft to nearby stars. After a time during which the new colonies grow and mature, each new colony sends a few ships with colonists to yet more distant stars, and so on. The colonization "frontier" expands at the edges, much like the growth of coral in the sea.

ble of building some of them in principle, and others do not seem far beyond our technological grasp. Such rockets might attain speeds of about 10% of the speed of light (0.1*c*).

Imagine a colonization effort begun by a civilization that has emerged around a star somewhere in the Milky Way Galaxy. It initiates its project by firing off interstellar spacecraft to nearby stars, with a suitable number of colonists aboard each craft. Let's assume that the average distance between star systems is about 5 light-years, as it is in our part of the galaxy. With a nuclear spacecraft traveling at 10% of the speed of light, the 5-light-year journey from one star system to the next would take about 50 years. This trip would be possible within a human lifetime and might even be easy if the colonizers have found ways to hibernate during the voyage or if they have somewhat longer life spans than we do (either naturally or through medical intervention).

After arriving at a new star system, the colonists establish themselves, doing whatever is required to make their adopted star system viable and increasing their numbers. Let's assume that 150 years after reaching the star system the population has grown sufficiently to allow the colonists to send their own pilgrims into space, adding yet more star systems to the growing civilization.

Figure 13.3 shows how such colonization would gradually spread through the galaxy. The process starts at the home star system. The first few colonies are located within just a few light-years. These colonies then lead to other colonies at greater distances, as well as at unexplored locations in between. The growth tends to expand the empire around the edges of the existing empire, much like the growth of coral in the sea. For this reason, this type of colonization model is often called a **coral model.**

The overall result is a gradually expanding region in which all habitable planets are colonized. For the case we have described here, the colonized region moves outward from the home star system at about 1% of the speed of light. (Problem 11 at the end of the chapter leads you through this calculation.) Thus, if the home star is near one edge of the galactic disk so that colonizing the entire galaxy means inhabiting star systems 100,000 light-years away, the civilization could expand through the entire galaxy in about 10 million years. The required time would be a few million years less if the home star is in a more central part of the galaxy.

For an even more conservative estimate, suppose the colonists have rockets that travel at only 1% of the speed of light and it takes each new colony 5,000 years until it is ready to send out additional colonists. Even in this case, the region occupied by this civilization would grow at a rate of roughly 1/1000 (0.1%) the speed of light, and the entire galaxy would be colonized in 100 million years. This is still a very short time compared to the time that has been available for civilizations to arise. Again, it seems that someone should have colonized the galaxy by now.

Motives for Colonization

In developing our colonization model, we assumed that other civilizations would *want* to send out colonists and colonize the galaxy. Is this a reasonable assumption?

We can address this question by considering ourselves as an example, since one of the premises of the Fermi paradox is that we are not special in any way. That is, if we would be likely to colonize the galaxy, then we should assume that others would probably act in the same way. Because we have not yet reached the technological level needed to start interstellar colonization, it's impossible to know with certainty whether we would try it if and when we achieved the capability. However, the history of the human species strongly suggests a predisposition to colonize any new territory available to us.

In many ways our entire history has been one of colonization. Modern humans arose in Africa about 100,000 years ago and almost immediately began expanding around the world. Indeed, the expansion of the human species on Earth probably looked much like the coral model we have described for galactic colonization. Early humans moved outward, gradually encompassing a larger and larger region of our planet. By about 10,000 years ago, our ancestors already lived in almost every habitable place on Earth. Even after humans had effectively colonized the entire planet, attempts at colonization did not stop. For example, Europeans colonized the Americas, with devastating consequences for the people already living there.

Might our inclination to colonize subside in the future? Possibly, but recent history suggests otherwise. Already there are organizations dedicated to colonizing Mars. It doesn't matter if most people would have no interest in going, because a tiny fraction of the human population would be more than sufficient to start a new colony on another planet. In summary, it seems that if other civilizations are at all like us, they would take advantage of technological opportunities to colonize the galaxy.

Moreover, even if other civilizations don't have an inherited predisposition toward colonization, many other motives might serve to encourage it. For example, some members of an alien civilization might choose to leave their home to escape war or persecution (as did many Europeans and others coming to America, for example). A society might deliberately send out colonists in an attempt to make its civilization "extinction proof." While our civilization, for example, could easily be wiped out in a variety of ways—from nuclear warfare to environmental catastrophe—a civilization spread among star systems would have a much more difficult time self-destructing. A particular environmental problem affects only one planet, and the long travel times between stars would make it almost impossible to wage war on multiple planets at once. Finally, if a civilization survived long enough for its star to reach an age at which life would be extinguished [Section 9.5], its members might have no choice but to move on in search of a new home.

THINK ABOUT IT . . . *Consider these and other possible reasons why a civilization might choose to start colonizing other star systems. In general, do you think it is reasonable to assume that other civilizations would attempt galactic colonization? Defend your opinion.*

Nonmotives for Colonization

Before we leave this topic, it's worth looking at a few ideas that are sometimes suggested as motives for colonization but that break down on closer examination. The most notable example of such ideas concerns the alleviation of population pressure.

After growing quite slowly for thousands of years, human population began a dramatic upward swing a few hundred years ago. Figure 13.4 shows human population over the past 12,000 years. If the current trend were to continue, today's human population of about 6 billion (a threshold reached in 1999) would double to 12 billion by about 2050, double again to 24 billion by 2100, and double again to 48 billion by 2150. Indeed, within just a few more centuries of such growth, we would not fit on the Earth even if we all stood elbow-to-elbow. Clearly, our rapid population growth on Earth must stop soon, or we will face an unparalleled catastrophe. Could colonization be the answer to our population problem?

Not a chance. Let's suppose we wanted to stabilize the population at its current size, but through colonization rather than changes in the population growth rate. Currently, we add about 100 million people to the Earth each year. Thus, to keep our population stable, we'd need to move 100 million people per year off the planet. A typical Space Shuttle launch costs about $100 million and takes only six or seven people into space (and the Shuttle can't reach other planets, let alone other stars). Even if we somehow found the resources to build enough Space

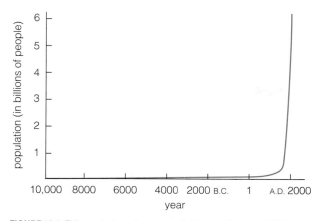

FIGURE 13.4 This graph shows human population over the past 12,000 years. Note the tremendous population growth that has occurred in just the past few centuries.

Shuttles—and the fuel to launch them—the cost of sending 100 million people into space each year would be more than a quadrillion dollars, some 1,000 times the gross national product of the United States. There are not too many things that we can say are outright impossible, but solving population problems through colonization is one of them.

THINK ABOUT IT . . . *Note that, at our current growth rate, the world adds about 300 million people—equivalent to the entire population of the United States—every 3 years. Discuss the consequences of this growth. How serious a problem is population growth for our species? Discuss possible solutions to this problem.*

Another less-than-viable colonization motive is conquest. As we've seen, even if we do find other inhabited planets with intelligent beings, we are likely to be at least either many thousands of years ahead of them or many thousands of years behind them. If we are thousands of years ahead, conquering them would be like the United States conquering cavemen—there hardly seems much to gain. If we are behind them, we would not be likely to prevail.

Some people suggest that we would colonize not so much through a motivation for direct conquest but rather as the consequence of our species' general tendency toward aggression. This is harder to rule out as a viable motive, but many science fiction writers (including Gene Roddenberry, creator of the *Star Trek* series) have pointed out a potential flaw in this idea: If we continue to be as aggressive and warlike as we have been in the past, we are unlikely to survive long into the future because our capacity for destruction has risen along with our level of technology. Thus, these writers argue that our surviving long enough to achieve the technology for interstellar

travel will necessarily mean that we have found ways to overcome our aggressive and warlike tendencies. In that case, colonization will occur because of curiosity and a desire to explore, not because of any desire for empire building.

13.4 Possible Solutions to the Paradox

We have now seen why the Fermi paradox is real; that is, it really does seem that if civilizations are at all common, then the galaxy should have been colonized long ago. So why don't we see any evidence of a galactic civilization? There are many possible explanations, but broadly speaking we can group them into three categories:

1. We are alone. There is no galactic civilization because civilizations are extremely rare—so rare that we are the first to have arisen on the galactic scene.

2. Civilizations are common, but no one has colonized the galaxy. There are at least three possible reasons why this might be the case:
   (i) Technological difficulties. Interstellar travel is much harder or vastly more expensive than we have guessed, so civilizations are unable to venture far from their home worlds.
   (ii) Sociological considerations. Our desire to explore is unusual, and other societies *choose* not to leave their home star systems. Also, for one reason or another, colonizers might run out of steam before they've conquered large tracts of galactic real estate.
   (iii) Self-destruction. Many civilizations have arisen, but they have all destroyed themselves before achieving the ability to colonize the stars.

3. There *is* a galactic civilization, but it has deliberately avoided revealing its existence to us.

Let's examine each of these categories in more depth.

We Are Alone

The idea that we are alone is certainly the simplest solution to the Fermi paradox. Many people object to this solution on philosophical grounds, because it would suggest that our circumstances are very special compared to those that have arisen around any other of the more than 100 billion star systems in the galaxy. If this is true, it would go against almost everything else we have learned since the time of Copernicus. That is, while our ancestors might have imagined our planet to be the center of the universe, more recent astronomical discoveries all seem to suggest that we are not particularly special. The Earth is merely a planet orbiting the Sun, our Sun is a rather ordinary star in the Milky Way Galaxy, and our galaxy is much like many other galaxies in the universe.

However, while the idea that we are alone might be philosophically unappealing, we cannot rule it out on scientific grounds. Indeed, proponents of the rare Earth hypothesis [Section 10.5] would not be surprised to learn that we are alone. Recall that, according to this hypothesis, the combination of circumstances that allowed for intelligent life on Earth is so rare that we are likely to be the only civilization in the galaxy. The ideas that underlie the rare Earth hypothesis are quite controversial, but even if they prove incorrect there may be other reasons why we are alone. For example, perhaps some undiscovered law of nature has rendered civilizations impossible until quite recently. In that case, it would not be so strange to imagine that we are the first civilization, even if many others may follow.

THINK ABOUT IT . . . *The rare Earth hypothesis and other similar ideas are generally used to suggest that we might be alone in the Milky Way Galaxy. Recall from Chapter 1 that the total number of stars in the universe is larger than the number of grains of dry sand on all the beaches on Earth. With that in mind, do you think it is also possible that we are alone in the universe? Why or why not? (Hint: consider the factors in the Drake equation.)*

Civilizations but No Colonization

The second category of explanations offers the possibility that civilizations are common but colonization is not. Three possible reasons why this might be the case were listed: technological difficulties, sociological considerations, and self-destruction. Let's consider each of these.

Technological Difficulties In Chapter 12, we found that while interstellar travel is far beyond our current capabilities, there is no reason why it could not be achieved by a more advanced civilization. But perhaps it is more difficult than we have imagined—so difficult as to be essentially impossible.

The energy cost of interstellar travel is often suggested as an impasse. Recall that a large interstellar starship traveling at only about 10% of the speed of light would require energy comparable to what the world currently uses in a hundred years [Section 12.1]. Clearly, this requirement is prohibitive for us today, and it might be so high compared to the costs of building habitats in our own solar system that migra-

tion to other stars might always seem untenable. However, it's certainly possible that an advanced civilization could overcome this problem. The ability to produce power through nuclear fusion, for example, might allow us to generate the needed energy with relative ease. In an extreme case, a civilization capable of building a Dyson sphere [Section 11.3] would have access to all the energy produced by its star. It seems unlikely that the energy requirement alone could preclude all interstellar colonization.

A related possibility is that some other unknown biological or physical barrier to interstellar travel exists. For example, we have assumed that in the future we'll be able to find a way to keep crews alive for the decades required to go from one star system to the next, but perhaps this is actually much more difficult than we have imagined. Possibly, some type of unknown danger lurking in space prevents intelligent beings from traveling among the stars. Science fiction writers have certainly considered such possibilities—for example, a mysterious effect that causes interstellar travelers to go insane—but they seem far fetched in light of what we presently know about interstellar space.

It's worth noting that neither energy considerations nor lurking dangers would be enough to stop a civilization from sending out self-replicating Von Neumann machines. But there might be other reasons why no such machines are out there. For example, such machines would tend to grow in number at a rapid (exponential) rate and could in principle use up all the resources in the galaxy in just a few million years. Carl Sagan (in a paper cowritten with William Newman) addressed this idea by suggesting that any civilization smart enough to build such machines would also be smart enough to recognize their dangers and therefore would not construct them in the first place.

Sociological Considerations As we've discussed, it seems quite likely that, given the technological and economic opportunity, we will choose to engage in interstellar travel and galactic colonization. But could it be that we are somehow exceptional in having this desire and that other civilizations are perfectly content to stay at home? Like the "we are alone" idea, this idea is philosophically unappealing. It too suggests that we are somehow special rather than typical of intelligent beings. After all, we are products of the competitive forces that drive evolution by natural selection, and these forces ought to be similar on any world with life. Moreover, our colonization models show that it would take only *one* other civilization to colonize the entire galaxy in a few million years, so the lack of interest in space travel would have to apply to *every other* civilization that has ever arisen. If civilizations are common—say, if there are the

100,000 civilizations expected if one in a million stars has one—it's very difficult to believe that not one other civilization has had any interest in interstellar travel.

On the other hand, even with fast rockets, colonization of the entire galaxy would still take millions of years. Perhaps no civilization can maintain enthusiasm for an effort that lasts this long. The individual colonies, separated by many light-years, might evolve along different lines (either biological or cultural), shattering the unity of the empire and bringing further expansion to a halt. While these possibilities are not unreasonable, it still remains true that only one civilization needs to persevere with its colonization efforts in order to bring the entire galaxy under its wing.

One other sociological consideration suggests that advanced societies might start out like us but then "engineer" themselves in such a way as to shut down their drive to colonize. On our planet, and presumably on others, the development of rocketry occurred at roughly the same time as the invention of nuclear weapons, chemical weapons, and other methods of mass destruction. Societies that remain aggressive are in constant danger of self-destruction. Therefore, civilizations might be motivated to find ways to reduce or channel their aggressive tendencies, perhaps through some type of genetic engineering. The oldest alien cultures, according to this line of reasoning, have managed to rid themselves of dangerous aggression. It's conceivable that in the process they have also chosen to focus on improving life on their home planet rather than on moving out into space.

Self-destruction A much more sobering possibility for why the galaxy might remain uncolonized even if many civilizations have arisen—one that assumes we are completely typical of intelligent beings—is that societies inevitably self-destruct before attaining the capability for interstellar travel. While this idea is horribly tragic, it is not farfetched. Nuclear weapons provide clear proof that the very technology needed for interstellar travel can also be used to destructive ends. Similarly, any society that learns to tap energy resources would almost certainly use the most accessible energy first, which on any Earth-like planet is likely to be fossil fuels. Thus, like us, other civilizations must face the dangers posed by global warming and other environmental problems. Population growth probably also poses similar problems for all civilizations. Rapid population growth is a natural consequence of biological reproduction for a species that is no longer subject to the whims of predators or childhood diseases, which tend to hold population growth in check. From this perspective, a society can survive long enough to achieve interstellar travel only if it successfully navigates what amounts to a very difficult obstacle course—one in which each obstacle would mean the end of its civilization. Could it be that it simply can't be done?

THINK ABOUT IT . . . *What odds would you give for humanity's surviving long enough to achieve interstellar travel? Why?*

There Is *a Galactic Civilization*

The third category of explanations for the Fermi paradox in essence suggests that there is no paradox at all: The galactic civilization is out there, but we do not yet recognize it. Indeed, UFO buffs might claim that scientists are blind to the obvious proof of this suggestion. While we can't rule out the possibility, no evidence of alien visitation has yet withstood scientific scrutiny. Nevertheless, there are many ways by which a galactic civilization could avoid our detection.

One idea simply assumes that a galactic civilization would have no particular interest in us. After all, to a society that is millions or billions of years ahead of us, we might seem no more interesting than the bacteria living between your toes seem to you. Of course, if civilizations are communicating or traveling among the stars, we might be able to discover them. Signaling is precisely what SETI experiments look for. The fact that we have not yet received a clearly extraterrestrial broadcast may simply be a consequence of not yet having looked at a sufficient number of stars.

Another possibility is that civilizations are aware of our presence but have deliberately chosen to keep us in the dark. This idea is sometimes called the **zoo hypothesis,** although it might better be called the "wildlife refuge" hypothesis. Just as we set aside nature reserves that are supposed to be left alone as places where wildlife can thrive without our intervention, a galactic civilization might declare planets like ours off limits to exploration. (*Star Trek* fans might think of this as the "Prime Directive" solution to the Fermi paradox.) One objection to the zoo hypothesis is that even if civilizations wanted to hide from us, we would still be able to intercept their communications among themselves. On the other hand, as we've already noted, SETI searches for signals have so far carefully investigated only a small amount of cosmic real estate and could easily have missed such communications. Alternatively, the communication might involve a technology that we have not yet developed and that therefore is undetectable to us at present.

A closely related idea suggests that a sophisticated galactic civilization might reveal itself to new societies only after they reach a certain level of technology. Perhaps the extraterrestrials place monitoring devices near star systems that show promise of emerging intelligence and patiently wait until these devices record the presence of civilization. This idea is sometimes called the **sentinel hypothesis,** after a science fiction story by Arthur C. Clarke titled "The Sentinel." The story became the basis of the book and movie *2001: A Space Odyssey,* in which a monolith buried on the Moon signals our presence when we finally dig it up. Carl Sagan used a similar idea for the book and movie *Contact,* in which a signaling station around the star Vega amplifies and beams back our own television broadcasts to us, leading us to our first glimpse of a galactic civilization.

Summary: Answering the Fermi Paradox

We have discussed a variety of possible solutions to the Fermi paradox, but we currently have no way of knowing which, if any, of them is correct. Some of the possible solutions would indeed make SETI efforts a waste of time, but others offer the hope of a SETI success at any time. Thus, we cannot use the Fermi paradox to draw any conclusions about the viability of SETI. Given the uncertainties, it seems worthwhile to continue the search. The reason for doing so was eloquently stated by one of SETI's pioneers, Philip Morrison[1]:

> *It is fine to argue about [the number of civilizations]. After the argument, there is no easy substitute for a real search out there. . . We owe the issue more than mere theorizing.*

13.5 The Astonishing Implications of the Fermi Paradox

The Fermi paradox may have its origins in a simple question—Where is everybody?—but we have seen that finding an answer is much more complex than asking the question. In fact, if we consider our possible answers in more depth, we find that each leads to astonishing implications for our own species.

Consider the first solution—that we are alone. If this is true, then our civilization is a remarkable achievement. It implies that through all of cosmic evolution, among countless star systems, we are the first piece of the universe ever to know that the rest of the universe exists. Through us, the universe has attained self-awareness. Some philosophers and many religions argue that the ultimate purpose of life is to become truly self-aware. If so, and if we are alone, then the destruction of our civilization and the loss of our scientific knowledge would represent an inglorious end to something that took the universe 12 billion years or longer to achieve. From this point of view, humanity becomes all the more precious, and the collapse of our civilization would be all the more tragic. Knowing this to be the case might help us learn to put petty bickering and wars behind us so that we might preserve all that is great about our species.

The second category of solutions has much more terrifying implications. If thousands of civilizations before us have all failed to achieve interstellar travel on a large scale, what hope do we have? Unless we somehow think differently than all previous civilizations, we will never go far in space. Given that we have always explored when the opportunity arose, this solution almost inevitably leads to the conclusion that failure will come about because we destroy ourselves. Let's hope this answer is wrong.

The third solution is perhaps the most intriguing. It says that we are newcomers on the scene of a galactic civilization that has existed for millions or billions of years before us. Perhaps this civilization is deliberately leaving us alone for the time being and will invite us to join it when we prove ourselves worthy. If so, our entire species might be on the verge of beginning a journey every bit as incredible as that of a baby emerging from the womb and coming into the world.

You can probably see why the Fermi paradox involves far more than a simple question. No matter what the answer turns out to be, learning it is sure to mark a turning point in the brief history of our species. Moreover, this turning point is likely to be reached within the next few decades or centuries. We already have the ability to destroy our own civilization. If we do so, then our fate is sealed. But if we survive long enough to develop technology that can take us to the stars, the possibilities seem almost limitless.

Imagine for a moment the grand view, a gaze across the centuries and millennia from this moment forward. Picture our descendants living among the stars, having created or joined a great galactic civilization. They will have the privilege of experiencing ideas, worlds, and discoveries far beyond our wildest imagination. Perhaps, in their history lessons, they will learn of our generation—the generation that history placed at the turning point and that managed to steer its way past the dangers of self-destruction and onto the path to the stars.

---

1 As quoted in S. J. Dick, *The Biological Universe,* Cambridge, 1996, p. 453.

THE BIG PICTURE

We began this chapter with a simple question, explored it in some depth, and followed it through to its profound implications for our future. As you continue in your study, keep in mind the following "big picture" ideas:

■ The Fermi paradox is not easily dismissed. Based on an examination of the cosmic time scale and of the seeming ease with which an advanced civilization ought to be able to colonize the galaxy, it is quite surprising that we have found no evidence of a galactic civilization.

■ People have thought of numerous possible solutions to the Fermi paradox, but we do not yet know whether any one of them is correct. The only way we can ever hope to reconcile the paradox is to continue our study of life in the universe and our search for extraterrestrial intelligence.

■ We live at a unique moment in the history of the human species. We have the ability to destroy our civilization and perhaps even to drive our species to extinction. But if we survive, our descendants might have a boundless future. From this perspective, no generation has ever borne such great responsibility.

Review Questions

1. Briefly describe the two seemingly contradictory ideas that underlie the *Fermi paradox*.

2. What are *Von Neumann machines?* How do they affect the Fermi paradox?

3. Briefly describe the cosmic calendar and how it puts time into perspective.

4. Describe several scenarios that give varying answers to the question of the average time between civilizations in the galaxy. Why do all of these scenarios seem to imply that we must be the youngest civilization around?

5. Describe the *coral model* of galactic colonization. Why do we conclude that civilizations could have colonized the galaxy by now even with technology not much more advanced than ours?

6. Briefly discuss possible motives for galactic colonization, as well as several nonmotives.

7. Summarize the three general categories of possible solutions to the Fermi paradox, and discuss each category in some detail.

8. Briefly discuss the profound implications of the Fermi paradox.

Discussion Questions

1. *Beyond the Milky Way.* We have focused on the question of other civilizations in the Milky Way Galaxy. Do any of the key points or implications change if we consider civilizations in other galaxies? Why or why not?

2. *Human Significance.* Consider what you have learned about the size of the universe and the extent of cosmic time. Some people say that these render us utterly insignificant in the universe. Others say that the mere fact that we know them gives us great cosmic significance. What is your opinion? Defend it.

3. *The Turning Point.* Discuss the idea that our generation has acquired a greater responsibility to future humans than any previous generation. Do you agree with this assessment? If so, how should we deal with this responsibility? Defend your opinions.

Problems

Science Fiction Scenarios. Each of **problems 1–8** describes a science fiction scenario that, while perhaps common and entertaining, may or may not be plausible. For each statement, write a paragraph explaining why it seems either plausible or unlikely. Explain your reasoning clearly.

1. Aliens from a distant star system invade the Earth in order to steal our technology.

2. Aliens from a distant star system invade the Earth with the intent of destroying us and occupying our planet, but we successfully fight them off with a great effort by our best scientists and engineers.

3. Aliens arrive on Earth but virtually ignore our presence, finding the diversity of earthly bacteria to be much more scientifically interesting.

4. The galaxy is divided into a series of empires, each having arisen from a different civilization, that hold each other at bay through the threat of military action.

5. All members of the galactic empire, being much more advanced than our civilization, have become peaceful and sympathetic. They have not made contact with us because they are protecting us from a rush of advanced knowledge that would make our own strivings to learn about the universe seem puny and uninteresting.

6. The species that created a great galactic civilization is inherently warlike and never misses an opportunity for a fight, either with others or among themselves.

7. There is a single, great galactic civilization that originated on a single planet long ago, but it is now made up of beings from many different planets that were each assimilated into the galactic culture in turn.

8. Aliens are already living among us, cleverly disguised as humans so that they can learn more about our cultural and social institutions.

9. *Solution to the Fermi Paradox.* Among the various possible solutions we have discussed for the Fermi paradox, which do think is most likely? (Or, if you have no opinion as to their likelihood, which do you like best?) Write a one- to two-page essay in which you explain why you favor this solution.

10. *Aliens in the Movies.* Choose a science fiction movie (or television show) that involves encounters between different alien species. In light of what we have discussed about cosmic time, are the encounters portrayed realistically? Write a one- to two-page critical review of the movie in which you focus primarily on the question of whether the encounters are realistic.

*11. *The Coral Model of Colonization.* It's relatively easy to create a formula for estimating the time it would take for a civilization to colonize the galaxy. Imagine that a civilization sends colonists to stars that are an average distance D away and sends them in spacecraft that travel at speed v. Then the time required for travel, which we'll call t_{travel}, is $t_{travel} = D/v$. Now suppose that the colonists build up their colony for a time t_{col}, at which point they send out their own set of colonists to other star systems (with the same average distance and same spacecraft speed). Then the speed at which the civilization expands outward from the home star, which we'll call v_{col} (for the speed of colonization), is $v_{col} = D/(t_{travel} + t_{col})$. However, this is true only if the colonization is always directed straight outward from the home star. In reality, the colonists will sometimes go to uncolonized star systems in other directions, so we will introduce a constant k that accounts for this zigzag motion. Now our equation for the speed at which the civilization expands outward from the home star is:

$$v_{col} = k \frac{D}{\left(t_{travel} + t_{col}\right)}$$

$$= k \frac{D}{\left(\dfrac{D}{v} + t_{col}\right)}$$

For the purposes of this problem, assume that $k = 1/2$ and that the average distance between star systems is $D = 5$ light-years.

a. How fast (as a fraction of the speed of light) does the civilization expand if its spacecraft travel at $0.1c$ and each colony builds itself up for 150 years before sending out the next wave of colonists? How long would it take the colonists to expand a distance of 100,000 light-years from their home star at this rate?

b. Repeat part (a), but assume that the spacecraft travel at $0.01c$ and that each colony builds itself up for 1,000 years before sending out more colonists.

c. Repeat part (a), but assume that the spacecraft travel at $0.25c$ and that each colony builds itself up for 50 years before sending out more colonists.

Web Projects

1. *Solutions to the Fermi Paradox.* Learn more about someone's pet solution to the Fermi paradox. Write a short summary of that solution, and discuss how it integrates with the ideas we have discussed in this chapter.

2. *SETI and the Fermi Paradox.* Learn more about how the debate about the Fermi paradox has affected support for and funding of SETI efforts. Write a short summary of what you learn.

3. *Von Neumann Machines.* Learn more about Von Neumann machines, including how they work, what they might be used for, and what dangers they might pose. Based on what you have learned, do you believe we will ever build such machines? Defend your opinion in a one- to two-page essay.

The known is finite, the unknown is infinite; intellectually we stand on an islet in the midst of an illimitable ocean of inexplicability. Our business in every generation is to reclaim a little more land.
—Thomas H. Huxley

CHAPTER 14
Contact—Implications of the Search and Discovery

We've covered a lot of ground up to this point. We've studied life on Earth and learned about the prospects for life elsewhere in our solar system and on planets that reside among the stars. Our scientific discussion mirrors a broader cultural phenomenon—an intense interest in ideas relating to extraterrestrial life. This interest shows up in many aspects of modern life, including TV shows and the movies, the supermarket tabloids (which regularly contain articles discussing UFOs and alien abductions), and a notable public interest in the exploration of space and the search for life beyond Earth.

In this closing chapter, we discuss some possible reasons why humans are so deeply interested in the question of whether life is present elsewhere. We'll begin with a brief review of why the idea of life beyond Earth seems scientifically reasonable and then explore how the search for it helps us revisit age-old questions about the nature of humanity. We'll also discuss the philosophical and cultural consequences of finding life elsewhere—life of any kind, whether microbial or intelligent or somewhere in between.

14.1 Is There Life Elsewhere?

We know of only one example of life existing in the universe—life here on Earth. In this book we have talked about whether life might exist elsewhere—either on planets in our own solar system or on planets orbiting other stars—but no matter how reasonable and plausible the idea that life exists elsewhere might seem, we still don't know for sure that it does. Because of this uncertainty, all we can do today is discuss the issues that could determine whether extraterrestrial life is likely or unlikely to exist and how we might search for evidence of it.

Despite these limitations, we are at a unique point in the long history of the debate over the possibility of extraterrestrial life. We have the technological capability to explore Mars and much of the rest of our solar system, and we are rapidly developing technology that might allow us to find evidence about whether life exists on planets around other stars. After millennia of speculation about life beyond Earth, we have the potential to discover life on another planet within perhaps the next one or two decades. This remarkable prospect calls us to discuss the philosophical and cultural consequences of finding life elsewhere. But first let's summarize the key issues in our discussion about the prospects of finding life and the search for life in the universe.

THINK ABOUT IT . . . *To some degree, scientists have been saying for at least the last three centuries that we may be on the verge of discovering extraterrestrial life. For example, Kepler speculated about life on the Moon, Percival Lowell was convinced he saw evidence of a civilization on Mars, and the Viking scientists designed experiments to look for microbial life on Mars. Are we being overly optimistic when we claim now to be on the verge of possible discoveries? Why or why not?*

Why Life Seems Likely

Why do we think life might exist elsewhere? Although we don't know for sure whether it does, current science offers reasons for optimism. We can examine the nature of life on Earth—its building blocks and how it originated—and understand the environmental conditions in which life can exist. We can look at the other planets and satellites in our solar system and determine whether the conditions conducive to life exist there. And we can look for planets around other stars, learn how abundant they are and how they form, and determine whether some of them might be Earth-like planets (that is, rocky planets in the inner regions of their solar systems

that might have liquid water at the surface) that could be capable of supporting life.

When we do these things, we find three key pieces of evidence that point to the idea that life should be common in the universe. We'll list them and then discuss each briefly in turn:

1. The chemical elements that comprise life are common throughout the universe, and complex, carbon-bearing molecules important to life on Earth appear to form easily and naturally under conditions that should be common on many planets.

2. Life on Earth thrives under a wide range of environmental conditions that we once considered too extreme to be capable of supporting life, and many of these types of environments are likely to be found on other planets in our own solar system and beyond.

3. It seems that life appeared on Earth quite quickly once the conditions became conducive to supporting life (that is, after the end of the heavy bombardment), implying that the origin of life took place rapidly and making it seem plausible that life would arise quickly elsewhere when the right conditions exist.

Let's consider the first item in our list. The elements from which life is constructed (carbon, hydrogen, oxygen, nitrogen, and almost two dozen others) are found nearly everywhere in the universe. Hydrogen is the most abundant element by far, but the other elements used for life exist in at least modest quantities, because they have been created by nuclear fusion in earlier generations of stars or in supernova explosions of these stars. They were ejected by the supernova explosions into interstellar space where they were incorporated into the gas and dust clouds out of which later stars and planets formed. Moreover, experiments in laboratories on Earth and spectroscopic observations of distant objects show that the elements utilized by life combine readily into the molecular building blocks of life. For example, we have found molecules as complex as amino acids in meteorites, and we have detected numerous organic molecules even in interstellar space. We therefore expect that organic molecules should form abundantly and naturally under a range of conditions that includes those that were present on the early Earth and are likely to be present on other geologically active planets. A wide availability of organic molecules would make it seem likely that the starting points for an origin of life exist on many worlds.

If the starting points for life are widely available, the next question concerns whether available environments can allow life to arise and thrive. This is the issue addressed by the second point on our list. We have learned that life on Earth is incredibly robust and can thrive in a wide variety of environments that seem "extreme" to humans. For example, we have found life on Earth in the hot water near deep-sea vents, in the dry and frigid deserts of Antarctica, and inside rocks deep underground. The diversity of environments in which we find life on Earth suggests that life could survive on any planet that meets relatively simple environmental requirements—the presence of liquid water (or possibly another liquid), access to the requisite elements and molecules, and an energy source to drive metabolism. These conditions are likely to be met on any geologically active, rocky planet. Such planets would have the necessary elements and the potential for energy to be available via water-rock chemical reactions.

The first two points on our list tell us that the starting points for life are widely available and that life, once started, can survive in a wide range of environments. The third point tells us that getting life to start may not be difficult. The early history of the Earth, as recorded in the oldest rocks, indicates that life appeared very soon after the end of the heavy bombardment, which is when it first became possible for life to exist continuously. Even with the present uncertainties in the interpretation of the rock record, life must have originated on a time scale that is very short compared to the age of the Earth and to the expected ages of other habitable planets. (Note that even if life originated on Venus or Mars and was transferred to Earth, the origin and transfer must have happened very quickly.) Thus, unless the Earth was somehow atypical, it seems that life might be the natural, straightforward consequence of the types of chemical reactions that can occur in planetary environments. When we put all three key pieces of evidence together, it seems reasonable to imagine life existing in at least a few other places in our own solar system and on many similar worlds throughout the universe.

Prospects for Finding Life in Our Solar System

If life is indeed as common and wide-ranging as the evidence suggests it could be, then the first place to look for life is within our own solar system. Here in the Sun's neighborhood, Mars and Europa seem the most likely places besides Earth to harbor life.

Mars shows evidence for liquid water having been present at its surface early in its history and within the crust throughout its history. The martian atmosphere contains several of the key elements of life—notably carbon (in the form of gaseous carbon dioxide), hydrogen and oxygen (in the form of water), and nitrogen. The other necessary elements are found in surface and near-surface rocks. Energy to drive metabolism could come from chemical reactions between the water and the rocks, for example, allowing the possibility of organisms much like those that live within rocks on the Earth.

Jupiter's moon Europa probably also has large amounts of liquid water. Although we are not yet absolutely certain, it seems likely that Europa is even more of a "water world" than Earth, with a global, 100-kilometer-thick ocean lying beneath an ice surface. We also expect that Europa has heavier elements in its rocky interior, so all the elements needed for life should be abundant. Energy for life again could be supplied by chemical reactions between the water and the underlying rock, or by chemical compounds created by the impact of high-energy particles onto surface ice (driven by Jupiter's rotating magnetic field). It seems plausible that at least microbial life could exist on Europa if its ocean proves to be real.

Other places in our solar system could potentially support life. Ganymede and Callisto, for example, show evidence of subsurface liquid water, similar to Europa's ocean though at greater depth beneath the surface. Saturn's moon Titan, although too cold to have liquid water at the surface today, may have pools or lakes of other liquids. It might also have deep-underground seas of liquid water and ammonia, and pockets of liquid water might persist for thousands of years following large impacts on the surface. At a minimum, Titan has abundant organic molecules in its atmosphere and on its surface.

In summary, at least two worlds in our solar system—Mars and Europa—seem to be good candidates for having life today or for having had it at some time in the past, and several other worlds seem to be possible candidates for life. As a result, NASA and other space agencies have embarked on a program to try to determine whether life might actually exist elsewhere in our solar system. Spacecraft are being developed for missions to Mars, with launches planned approximately every 2 years (at the times when Earth and Mars line up to make the trip relatively easy). Numerous other spacecraft are exploring or are being developed to explore other worlds, including the Cassini spacecraft currently en route to Saturn and Titan. Such missions will help us learn not only whether life can exist on various worlds, but also how interior, surface, and atmospheric processes play out on different planets and moons. This knowledge will help us understand why some worlds end up habitable and others don't. It will also provide guidance to us as we seek to answer the question of what planets in other solar systems might be like.

Prospects for Finding Life Among the Stars

Efforts that will allow us to search for life beyond our own solar system have begun. We have already discovered many dozens of Jupiter- and Saturn-size planets orbiting other stars, and in a few cases we have found stars orbited by more than one planet. However, we don't expect these planets to harbor life, and our technology is not yet up to the task of detecting Earth-size planets. Nevertheless, the formation of planets seems to be a natural consequence of the processes that lead to star formation. This makes it seem likely that a substantial fraction of stars would have Earth-size or Earth-like planets.

However, very few of the newly discovered solar systems look like ours. Most known extrasolar planets have characteristics different from those of the gas giants in our own solar system. We refer to them as "hot Jupiters" because they are Jupiter-like in size (and presumably in composition) but orbit very close-in to their central star. These planets probably formed in the outer parts of their solar systems and reached their current orbits by migrating inward. Such migration would be devastating for any Earth-like planets that already existed in these systems. Thus, if this type of planetary migration is common, then Earth-like planets might be rare. On the other hand, we have not yet studied enough other solar systems to be able to determine whether or not Earth-like planets are common. Scientists are rapidly discovering more extrasolar planets and are developing technology that will soon allow us to look directly for Earth-size planets. Within one to two decades we not only might know how common such planets are, but also might be able to characterize them and determine from telescopic observations (of their atmospheric composition, for example) whether they could or do hold life.

If rocky, Earth-like planets are indeed common, life could be widespread throughout the galaxy. The possible widespread existence of at least microbial life begs the question of the potential for more complex or intelligent life. Our one example of intelligence here on Earth does not allow us to extrapolate or even hazard a guess as to whether intelligence should be widespread or rare. Searches for radio or light signals from possible extraterrestrial civilizations have been going on for about four decades. Although these SETI efforts have not yet met with success, only a small sample of the possible homes for civilization has been searched so far. As with the question of extraterrestrial life in general, the question of whether intelligent extraterrestrial life exists remains open.

14.2 Extraterrestrial Life and the Human Condition

We have reviewed the key scientific issues pertaining to life elsewhere. Now we turn our attention to issues that we have for the most part neglected up to this point in the book, including philosophical and societal issues that touch on why the search for life beyond Earth is of such broad intellectual interest to scientists and the public alike. We've saved these topics for this final chapter not because they are less interesting or less important, but rather because they are the capstone of all the discussion up to this point.

One exciting aspect about the potential for life elsewhere is that everybody seems interested in it. Some people are interested in understanding the scientific debate about possible microbes on Mars, while others may be interested in the discoveries of planets orbiting other stars and the potential for life on those planets, the possibility of intelligent life elsewhere in our galaxy, or UFOs and alien abductions. Regardless of what drives the interest, a large fraction of the population seems to be engaged by such topics.

Why are people interested in whether life exists elsewhere in our own solar system or beyond? What would it mean to us to find convincing evidence for such life? What would it mean to search and not find evidence? We'll discuss a few possible answers to these questions, but you should recognize that the ideas we discuss may not resonate the same way with everybody. Think about what issues might be driving your own interest and especially about how and why these personal issues might differ from the ideas we discuss here.

Our Changing Perspective on the World

One key reason why many people are caught up in the search for life in the universe is the way it could change our perspective on the world. Much of what we have learned about the world around us over the past several thousand years has changed the way we interact with it. For example, a mere 10,000 years ago—a blink of an eye in the history of our planet—humans were primarily a hunter-gatherer society, living out their lives with little knowledge about what was beyond the next mountain or valley. Today, people in nearly every corner of the world recognize themselves as part of a tremendous cosmos. We have learned about the Earth in an integrated way, in that what we do in one place affects the environment, people, and societies in all other places around the globe. In addition, we see the Earth as one planet of many in our solar system, our solar system and the

Sun as one of more than 100 billion star systems in our galaxy, and our galaxy as just one of billions of similar galaxies in the visible universe. This expansion of our world (or contraction, depending on how you perceive it!) has brought with it a need to rethink our views both of ourselves and of our relationship to the rest of the world.

A few key events in the last millennium have fundamentally changed how we perceive our relationship to the world around us. We have already discussed most of these events in the context of the nature of science and the history of scientific thought. Here we discuss them in the context of how they affect our understanding of humanity.

Perhaps the first major transition toward a modern scientific view of ourselves (and maybe the most significant one) began nearly 500 years ago with the Copernican revolution. The work of Copernicus, Galileo, Kepler, and Newton allowed us to recognize that the Earth is not at the center of the universe but instead orbits the Sun. This displacement of the Earth from the center of the universe had profound philosophical and psychological effects. The Earth, and by extension humanity, could no longer be viewed as the center of everything. And the world could no longer be viewed as obeying only the laws of providence rather than the laws of physics. This shift in thinking had very little impact on people's day-to-day lives—we would be hard-pressed even today to think of a way in which whether the Earth goes around the Sun or vice versa matters to our daily activities. However, it made a fundamental difference to our world view. As a result, the idea initially met widespread resistance in Western society and was not fully accepted for at least a couple of hundred years following Copernicus.

A second major shift occurred in the mid-1800s with the recognition by Charles Darwin and Alfred Russel Wallace of the processes that drive the evolution of species. The basic idea of evolution was not new. By the time of Darwin, scientists already recognized that the Earth was old, that fossils represented organisms that had lived in prior times, that older fossils were different from younger ones, and that both were different from living organisms. These discoveries had already convinced many scientists that the nature of living organisms had changed over time. What Darwin and Wallace discovered was a way to explain why and how species change: through descent with modification and competition for survival, a process we refer to today as natural selection. We now recognize that all of the species present on Earth represent the end product of some 4 billion years of evolution, traceable back to the earliest history of life on Earth.

Like the Copernican revolution, the recognition of the nature of evolution sparked a change in our view of ourselves and of our world. Humans could no longer be considered as the center of the biological universe. Rather, we represent just one more species on Earth, one more end product of the same 4 billion years of evolution. This shift in perspective is so profound that many people still do not accept the idea of evolution by natural selection. Indeed, while essentially all biologists consider the theory of evolution central to an understanding of the history and nature of life on Earth, much of the public, at least in the United States, does not accept it. Opinion polls indicate that about half of the American public hold various creationist views of the origin and history of life. About half of the rest, while accepting the evidence of evolution, see it as guided by the power of a deity rather than as the seemingly random, unguided process envisioned by biologists. The intensity of the public controversy only underscores the tremendous impact that the theory of evolution has had on Western society and on our world view.

THINK ABOUT IT . . . *Which perspective shift do you think has had a bigger impact on humanity: the shift brought on by the Copernican revolution or the shift brought on by the theory of evolution? Defend your opinion.*

A third shift in perspective is taking place today with the discovery of other solar systems and the modern understanding of the potential for life to be widespread in the universe. Although many astronomers had long suspected that planets ought to be common around other stars, the first definitive evidence of extrasolar planets is only about a decade old. The discoveries to date already show us that Jupiter-size planets must be fairly common. If we ultimately confirm our guess that worlds like ours exist or even are common, we will no longer be able to regard the Earth as special in any essential way. Throughout the history of scientific thought, new discoveries have increasingly displaced us from the center of the universe, both physically and metaphorically. Finding that life is common throughout the universe would lead to yet another major shift in perspective. In the rest of this section, we'll briefly explore why.

The Impact of Finding Life on Human Perspective

People have long speculated about life beyond Earth, and many people—including many scientists—have at times been convinced that life exists on the Moon, on Mars, or on other worlds. Why, then, would the discovery of extraterrestrial life have a major impact

When we explore the societal implications of discovering life beyond Earth, whether microbial or intelligent, one of the most common questions that arises is whether this discovery might have any impact on religion. The answer depends largely on which religion we consider. Historically, some religions have had a great deal of conflict with science, while others have had none. Here we will focus in a fairly general way on monotheistic religions such as Judaism, Christianity, and Islam.

One view is that such a discovery, with its implications for how life originated on Earth, would be inconsistent with the views of these modern religions. According to this view, if life is common and there is nothing special about life on Earth or the human species, there is no longer any reason to think there is a God.

However, questions concerning the implications of the existence of extraterrestrial life have been debated by theologians for centuries. Most have concluded that religion should not be unduly affected by a discovery of life beyond Earth. Religions have adapted to many scientific discoveries in the past and likely would be able to adapt to a discovery of life or intelligence elsewhere. We need only look back to the issues raised by the Copernican revolution to see how religious thinking can evolve and adapt. Galileo's belief in a Sun-centered solar system, though initially branded as heresy, eventually held sway in the same religious hierarchy that tried and convicted him. And for all the controversy that still surrounds evolution in some circles, most religions have found ways to adapt to

this profound scientific discovery. For example, Pope John Paul II argued recently that the evidence in favor of Darwinian evolution is overwhelming and there is no longer any reason to doubt its occurrence. He asserts, however, that at some point God infused humanity with a soul that separates us from the rest of life on Earth.

Since the time when the Copernican revolution raised issues about the nature of the physical world, people have been trying to understand the relationship between science and religion. Some have argued that the two are fundamentally intertwined, given that there is only a single existence that must encompass both the physical and the spiritual world. Others have argued that the two are separated relatively easily, with the physical world being governed by the laws of nature, as inferred from the application of logic to our observations of the world around us, and the spiritual world existing independent of the physical world, and beyond our ability to understand from our physical-world perspective. According to this view, religion is dominated by a faith that is independent of how we interpret our own physical existence.

In summary, it might not be difficult to reconcile science and religion. Discoveries in astrobiology certainly will affect our views of religion, but they are unlikely to affect the degree of interaction dramatically. Thus, we expect that both science and religion will continue to play important roles in our society if and when we discover extraterrestrial life.

on human perspective? The answer lies in the difference between guessing and knowing. As long as there is uncertainty about the existence of life on other worlds, people are free to hold a wide range of opinions. People living before the time of Copernicus could continue to believe in an Earth-centered world, but it was very hard to continue to do so after Galileo, Kepler, and Newton offered convincing proof to the contrary. An actual discovery of life beyond Earth would force us, both as individuals and as a society, to reconsider the place of our planet and our species in the cosmos.

In contemplating the significance of finding life elsewhere, let's begin by considering what would happen if we found microbial life on Mars. The first question we would probably ask is whether the life was genetically related to terrestrial life (suggesting that it had migrated between planets on meteorites) or instead represented an independent origin of life on Mars. We could answer this question by determining the structure of the molecules that make up

the martian life. Does it use DNA and RNA molecules similar to those used by terrestrial life? Does it use the same amino acids or the same proteins to carry out enzymatic reactions? Do the molecules that participate in life have the same "handedness" to them? It seems unlikely that there would be only one solution to the problems of containing and passing on the genetic information required for life to reproduce, of catalyzing the chemical reactions that comprise life, and of storing and using energy in metabolism. We would expect life that had an origin independent from life on Earth to have a different chemical structure.

While a discovery of life on Mars that was genetically related to terrestrial life would be exciting, it would not have as great an impact as a discovery of life that showed evidence of an independent origin. If and when such a discovery occurs, it likely will seem to be the final step in recognizing that we on Earth are not so special. Life would be seen as just another example of the types of chemistry that can occur in a planetary environment, albeit an especially interesting one. A discovery of nonterrestrial-based life on Mars would be consistent with the views that have been put forward about the ease of formation of life and would suggest that at least microbial life was common throughout the galaxy.

Would a discovery of alien microbial life have the same philosophical impact as a discovery of extraterrestrial intelligence? In informal polls, many people indicate that only the discovery of alien intelligence would have truly profound significance for them. Many scientists working in the nascent field of astrobiology disagree, however, feeling that the discovery of even the simplest single-celled organism on another planet would have profound significance for us. If this life had an origin independent from terrestrial life, it would tell us that the origin of life was not a unique event. With proof that life has originated twice, we would have every reason to think it has originated many times and thus that life is widespread in the universe. A discovery of alien microbial life would also help us better understand life in general. We would learn more about the conditions under which life can arise and persist, as well as the conditions under which it can evolve into more complex forms. That knowledge, in turn, would have implications for whether intelligent life might be common.

While many people who have studied the issue of life in the universe expect that we will find life to be widespread, we cannot truly envision the consequences of an actual discovery unless and until it occurs. Moreover, the history of science tells us to be prepared for surprises. For example, the real discovery of extrasolar planets proved surprising, despite their predicted existence: We learned that other solar systems can be very different from our own. If and when we do find life elsewhere, we should expect to be equally surprised about its nature.

Extraterrestrial Intelligence and the Nature of Humanity

What would finding intelligent life elsewhere mean to us? We discussed some of the philosophical implications in Chapter 13; here we focus on how it might affect our lives more directly. Of course, it is difficult to predict how we would respond to such a discovery, and it is likely that people would exhibit a wide range of responses. We can, however, look to science fiction as a guide to understanding how people might react, because the theme of such discoveries and contact is a staple of science fiction.

At one extreme, some people might look to extraterrestrial intelligence as a source of help for solving

FIGURE 14.1 A 1906 drawing used to illustrate H. G. Wells's book *The War of the Worlds,* in which Earth is invaded by hostile martians. This illustration appeared in newspapers in 1938, on the day following the radio broadcast based on the book that caused a panic among people who did not realize that it was fictional. *War of the Worlds* was also a 1953 movie.

our problems. This view has been portrayed in many books and movies, including *Contact* and *2001—A Space Odyssey.* Interestingly, many of these stories suggest, first, that we are not able to solve our own problems without outside help and, second, that extraterrestrials will want to help us!

At the other extreme, some people imagine that extraterrestrials would come to Earth and destroy our civilization, either deliberately or by accident. The intentional destruction of our civilization is the scenario in the movies *Independence Day* and *War of the Worlds,* in which other-world civilizations wreak havoc on Earth, destroying cities and killing people with abandon (Figure 14.1). The accidental destruction of the Earth's culture was the theme of the early years of the TV show *Earth: Final Conflict.* Alien

When cinema aliens come to Earth, it's usually wise policy to head for the storm cellar. But when wrinkly little ET is accidentally left behind by his planetary pals, it turns out to be good news—at least for a suburban kid or two.

ET is the quintessential alien film—a movie that long held the record for being the most successful picture of all time. There's good reason. Appealing little ET, who has the stature and gait of a penguin, is every kid's dream. After all, he's a friend (and really useful because with his super powers he can help you outsmart adults), and he's exclusively *your* friend. Other kids will stop kicking verbal sand in your face when you show up at the playground with a guy from another galaxy, even if he has a face like a polished redwood burl.

Aside from this childlike wish fulfillment, *ET* encapsulates everything we hope or think is true about intelligent extraterrestrials. To begin with, ET is benign and nonthreatening. Unlike evil aliens, who look like reptiles or insects, ET resembles a baby, with his short nose, big eyes, and wrinkled skin. He's only two feet high, weighs 35 pounds, and has a 25 watt fingertip. He clearly comes from a kinder, gentler planet, since his only interest in Earth's biota is its plants. (No insects had to die for this film.) Scientists who have given any thought to the true nature of advanced aliens (and those that can come to Earth are clearly advanced) have often equated technological prowess with peaceful behavior. The aliens will be friendly. This doesn't square very well with our experience on Earth, but one can hope.

In addition, ET not only looks like us; he acts like us, functions like us (he can get blotto on beer), and is interested in our personal lives. None of this is likely to be true, of course. He's also well adapted to terrestrial conditions, waddling around without a space suit and, indeed, without any clothes at all. In addition, nefarious federal agents wearing jackets, ties, and drab personalities are busily trying to keep the cuddly creature's visit under wraps, something that many people believe is happening in real life.

All of this may be in keeping with the public's perception of what aliens would be like. But if SETI succeeds and we eventually learn the true nature of extraterrestrials, it is far more likely that they will be of a construction and temperament that are far beyond our most fevered imaginings.

---

beings in that show integrated themselves into Earth's culture, which then was in danger of being subsumed by the aliens and lost entirely. We see analogies to this type of inadvertent destruction in our own history. In many cases, when modern technological societies have come into contact with nontechnological societies, the latter have been subsumed into the former, with their culture essentially being lost.

Neither possibility has any empirical or observational data to back it up, and holding one or the other view may mean little more than that the person holding that view is an optimist or a pessimist, respectively. If and when intelligent life is discovered, the reality may lie somewhere in between—or it might be completely different.

Significance of the Search Itself

A discovery of extraterrestrial life would undoubtedly bring important practical benefits. For example, studying it will help us understand what characteristics of terrestrial life are unique to Earth and what characteristics apply generally to life anywhere. If we find intelligent life elsewhere, we would learn much more about the nature of intelligence and might be exposed to cultures and societies very different from those of humans. If we could communicate with more advanced beings, we could possibly learn the secrets of the nature of the universe, the nature of consciousness and the mind, and technological marvels that could dramatically change our life here on Earth.

However, for many people excited by the scientific search for life in the universe, the search itself is much more than a means to an end. For them the search is just one more critical component in our exploration of the world around us. Other components in astronomy and space science include exploring the planets and moons in our solar system as a way to understand how planets work and exploring stars and galaxies as a way to determine the nature of our universe. Other components in biology include exploring the origin and evolution of life on Earth so that we can understand how we ourselves came to exist.

FIGURE 14.2 The Earth as viewed in space from the *Apollo 17* spacecraft. Seen from this perspective, we recognize the strong connections between the terrestrial ecosystem and the planet itself, and we recognize that life, indeed, is a planetary phenomenon. As we explore the universe, we will learn whether the formations of planets around stars and the occurrence of life on planets are rare or commonplace. We may then finally answer the question of whether we are alone.

In all these cases, our exploration does not seem to be driven solely by the desire to find specific answers to the scientific questions we are asking, because in each case we end up asking more questions. Instead, we seem to be driven by our inherent curiosity, our desire to understand the world around us (Figure 14.2). Sometimes our curiosity leads to discoveries with practical applications, while at other times it simply helps us understand how or why the world is as it is. Understanding the world around us means learning about the broader-scale environment in which humans exist. Understanding the occurrence of planets orbiting other stars helps us understand the significance of the occurrence of planets orbiting the Sun, including Earth. Understanding the occurrence of life elsewhere allows us to understand the significance of the occurrence of life on Earth. And understanding the potential for intelligent life beyond Earth brings with it an understanding of the meaning of the occurrence of intelligent life here on Earth. In essence, by learning about the world around us we are learning about ourselves and about what it means to be human.

Exploring the universe is but one activity among many that help us learn about ourselves, including searching for new forms of life in extreme environments on the Earth, exploring the Earth itself, exploring the nature of human consciousness and the human mind, and exploring the human condition through literature and the arts. Each of these activities helps us learn about what it means to be alive and to be human. Thus, the exploration of the universe and the search for extraterrestrial life can be seen as nothing less than an attempt to determine the nature of humanity and our place in the universe. While it is possible that this view is simply an attempt by scientists to feel that their life's work has value and importance on a grand scale, the ideas seem to resonate with many other people as well. Again, it is important for each individual who addresses the issue of life in the universe to think personally about the significance of the search and whether it is worth the costs involved.

THE BIG PICTURE

Throughout this book, we have integrated concepts from a wide variety of distinct disciplines. A study of the potential for life in the universe includes observations and theories from geology and geophysics, planetary science and astronomy, atmospheric physics, chemistry and biochemistry, and molecular and evolutionary biology. These disciplines together allow us to construct a framework for understanding a topic of broad intellectual interest.

The issues related to life in the universe go beyond the purely scientific, however, crossing the disciplinary boundaries to include the humanities. The issue of life elsewhere has substantial connections to human existence, and we can ponder the societal and philosophical implications. Only when we take a perspective that spans disciplinary boundaries can we really appreciate the broader significance of the potential for life beyond Earth. However, we don't yet know how widespread life might be, and we must remember that we presently know of only one planet that supports life—the Earth. Continuing to explore the Earth, the solar system, the galaxy, and the universe may eventually enable us to answer fundamental questions about whether we are unique.

NASA administrator Sean O'Keefe summarized NASA's mission by emphasizing our need "to understand and protect our home planet, to explore the universe and search for life, and to inspire the next generation of explorers." We hope that you might be inspired to continue the exploration you have begun by reading this book. Most important, we hope that you will continue to have an interest in understanding life on Earth, in our solar system, and beyond and that you will recognize and value the connections among exploration, science, and society.

Essay or Discussion Questions

1. *Is There Life Elsewhere?* After considering all the evidence to date about the likelihood of extraterrestrial life, do you believe it is likely that we'll find microbial life elsewhere? Do you believe it is likely that we'll find intelligent life elsewhere? Defend your opinions, using arguments based on the full range of scientific issues discussed in this book.

2. *Microbial or Intelligent?* Do you think the implications of discovering microbial life elsewhere would be any more or less profound than those of discovering extraterrestrial intelligence? Explain your reasoning.

3. *Extraterrestrial Life and Your Religion.* Would the discovery of extraterrestrial life have any important implications for your own personal religious beliefs? Would it affect the current "official" beliefs (if any) of your religion? Explain.

4. *Extraterrestrial Life and Religion in the United States.* Do you think the discovery of extraterrestrial life would affect the current status of the debate over science and religion in the United States? For example, would it alter the controversy surrounding the teaching of evolution in the schools? Why or why not?

5. *Aliens and Everyday Life.* While the discovery of extraterrestrial life would surely be profound, do you think it would alter any aspect of our everyday lives? If so, how? If not, why not?

6. *The Search Itself.* Suppose we spend a fair amount of money and effort searching for life over the next few decades and ultimately find no evidence for life beyond Earth. Will the search have been a waste, a success, or something in between? Defend your opinion.

APPENDIX A

USEFUL NUMBERS

Astronomical Distances

1 AU $\approx 1.496 \times 10^8$ km

1 light-year $\approx 9.46 \times 10^{12}$ km

1 parsec (pc) $\approx 3.09 \times 10^{13}$ km ≈ 3.26 light-years

Universal Constants

Speed of light: $\qquad c = 3 \times 10^5$ km/s $= 3 \times 10^8$ m/s

Gravitational constant: $\qquad G = 6.67 \times 10^{-11} \dfrac{m^3}{kg \times s^2}$

Mass of a proton: $\qquad m_p = 1.67 \times 10^{-27}$ kg

Mass of an electron: $\qquad m_e = 9.1 \times 10^{-31}$ kg

Useful Reference Values

Luminosity of the Sun: $\qquad 1 L_{Sun} = 3.8 \times 10^{26}$ watts

Mass of the Earth: $\qquad 1 M_{Earth} = 5.97 \times 10^{24}$ kg

Radius (equatorial) of the Earth: $\qquad 1 R_{Earth} = 6{,}378$ km

Acceleration of gravity on Earth: $\qquad g = 9.8$ m/s^2

Escape velocity from surface of Earth: $\qquad v_{escape} = 11.2$ km/s $= 11{,}200$ m/s

A FEW MATHEMATICAL SKILLS

This appendix reviews the following mathematical skills: powers of 10, scientific notation, working with units, the metric system, and finding a ratio. You should refer to this appendix as needed while studying the textbook.

B.1 Powers of 10

Powers of 10 simply indicate how many times to multiply 10 by itself. For example:

$$10^2 = 10 \times 10 = 100$$

$$10^6 = 10 \times 10 \times 10 \times 10 \times 10 \times 10 = 1,000,000$$

Negative powers are the reciprocals of the corresponding positive powers. For example:

$$10^{-2} = \frac{1}{10^2} = \frac{1}{100} = 0.01$$

$$10^{-6} = \frac{1}{10^6} = \frac{1}{1,000,000} = 0.000001$$

Table B.1 lists powers of 10 from 10^{-12} to 10^{12}. Note that powers of 10 follow two basic rules:

1. A positive exponent tells how many zeros follow the 1. For example, 10^0 is a 1 followed by no zeros, and 10^8 is a 1 followed by eight zeros.

2. A negative exponent tells how many places are to the right of the decimal point, including the 1. For example, $10^{-1} = 0.1$ has one place to the right of the decimal point; $10^{-6} = 0.000001$ has six places to the right of the decimal point.

Table B.1 Powers of 10

Zero and Positive Powers			Negative Powers		
Power	Value	Name	Power	Value	Name
10^0	1	One			
10^1	10	Ten	10^{-1}	0.1	Tenth
10^2	100	Hundred	10^{-2}	0.01	Hundredth
10^3	1,000	Thousand	10^{-3}	0.001	Thousandth
10^4	10,000	Ten thousand	10^{-4}	0.0001	Ten thousandth
10^5	100,000	Hundred thousand	10^{-5}	0.00001	Hundred thousandth
10^6	1,000,000	Million	10^{-6}	0.000001	Millionth
10^7	10,000,000	Ten million	10^{-7}	0.0000001	Ten millionth
10^8	100,000,000	Hundred million	10^{-8}	0.00000001	Hundred millionth
10^9	1,000,000,000	Billion	10^{-9}	0.000000001	Billionth
10^{10}	10,000,000,000	Ten billion	10^{-10}	0.0000000001	Ten billionth
10^{11}	100,000,000,000	Hundred billion	10^{-11}	0.00000000001	Hundred billionth
10^{12}	1,000,000,000,000	Trillion	10^{-12}	0.000000000001	Trillionth

Multiplying and Dividing Powers of 10

Multiplying powers of 10 simply requires adding exponents, as the following examples show:

$$10^4 \times 10^7 = \underbrace{10,000}_{10^4} \times \underbrace{10,000,000}_{10^7} = \underbrace{100,000,000,000}_{10^{4+7} = 10^{11}} = 10^{11}$$

$$10^5 \times 10^{-3} = \underbrace{100,000}_{10^5} \times \underbrace{0.001}_{10^{-3}} = \underbrace{100}_{10^{5+(-3)} = 10^2} = 10^2$$

$$10^{-8} \times 10^{-5} = \underbrace{0.00000001}_{10^{-8}} \times \underbrace{0.00001}_{10^{-5}} = \underbrace{0.0000000000001}_{10^{-8+(-5)} = 10^{-13}} = 10^{-13}$$

Dividing powers of 10 requires subtracting exponents, as in the following examples:

$$\frac{10^5}{10^3} = \underbrace{100,000}_{10^5} \div \underbrace{1,000}_{10^3} = \underbrace{100}_{10^{5-3} = 10^2} = 10^2$$

$$\frac{10^3}{10^7} = \underbrace{1,000}_{10^3} \div \underbrace{10,000,000}_{10^7} = \underbrace{0.0001}_{10^{3-7} = 10^{-4}} = 10^{-4}$$

$$\frac{10^{-4}}{10^{-6}} = \underbrace{0.0001}_{10^{-4}} \div \underbrace{0.000001}_{10^{-6}} = \underbrace{100}_{10^{-4-(-6)} = 10^2} = 10^2$$

Powers of Powers of 10

We can use the multiplication and division rules to raise powers of 10 to other powers or to take roots. For example:

$$(10^4)^3 = 10^4 \times 10^4 \times 10^4 = 10^{4+4+4} = 10^{12}$$

Note that we can get the same end result by simply multiplying the two powers:

$$(10^4)^3 = 10^{4 \times 3} = 10^{12}$$

Because taking a root is the same as raising to a fractional power (e.g., the square root is the same as the 1/2 power, the cube root is the same as the 1/3 power, etc.), we can use the same procedure for roots, as in the following example:

$$\sqrt{10^4} = (10^4)^{1/2} = 10^{4 \times (1/2)} = 10^2$$

Adding and Subtracting Powers of 10

Unlike with multiplication and division, there is no shortcut for adding or subtracting powers of 10. The values must be written in longhand notation. For example:

$$10^6 + 10^2 = 1{,}000{,}000 + 100 = 1{,}000{,}100$$

$$10^8 + 10^{-3} = 100{,}000{,}000 + 0.001 = 100{,}000{,}000.001$$

$$10^7 - 10^3 = 10{,}000{,}000 - 1{,}000 = 9{,}999{,}000$$

Summary

We can summarize our findings using n and m to represent any numbers:

- To *multiply* powers of 10, *add* exponents: $10^n \times 10^m = 10^{n+m}$

- To *divide* powers of 10, *subtract* exponents: $\dfrac{10^n}{10^m} = 10^{n-m}$

- To *raise* powers of 10 to other powers, multiply exponents: $(10^n)^m = 10^{n \times m}$

B.2 Scientific Notation

When we are dealing with large or small numbers, it's generally easier to write them with powers of 10. For example, it's much easier to write the number 6,000,000,000,000 as 6×10^{12}. This format, in which a number *between* 1 and 10 is multiplied by a power of 10, is called *scientific notation*.

Converting a Number to Scientific Notation

We can convert numbers written in ordinary notation to scientific notation with a simple two-step process:

1. Move the decimal point to come after the *first* nonzero digit.

2. The number of places the decimal point moves tells you the power of 10; the power is *positive* if the decimal point moves to the left and *negative* if it moves to the right.

   Examples:

$$3{,}042 \xrightarrow[\text{3 places to left}]{\text{decimal needs to move}} 3.042 \times 10^3$$

$$0.00012 \xrightarrow[\text{4 places to right}]{\text{decimal needs to move}} 1.2 \times 10^{-4}$$

$$226 \times 10^2 \xrightarrow[\text{2 places to left}]{\text{decimal needs to move}} (2.26 \times 10^2) \times 10^2 = 2.26 \times 10^4$$

Converting a Number from Scientific Notation

We can convert numbers written in scientific notation to ordinary notation by the reverse process:

1. The power of 10 indicates how many places to move the decimal point; move it to the *right* if the power of 10 is positive and to the *left* if it is negative.

2. If moving the decimal point creates any open places, fill them with zeros.

   Examples:

$$4.01 \times 10^2 \xrightarrow[\text{2 places to right}]{\text{move decimal}} 401$$

$$3.6 \times 10^6 \xrightarrow[\text{6 places to right}]{\text{move decimal}} 3{,}600{,}000$$

$$5.7 \times 10^{-3} \xrightarrow[\text{3 places to left}]{\text{move decimal}} 0.0057$$

Multiplying or Dividing Numbers in Scientific Notation

Multiplying or dividing numbers in scientific notation simply requires operating on the powers of 10 and the other parts of the number separately.

   Examples:

$$(6 \times 10^2) \times (4 \times 10^5) = (6 \times 4) \times (10^2 \times 10^5) = 24 \times 10^7 = (2.4 \times 10^1) \times 10^7 = 2.4 \times 10^8$$

$$\frac{4.2 \times 10^{-2}}{8.4 \times 10^{-5}} = \frac{4.2}{8.4} \times \frac{10^{-2}}{10^{-5}} = 0.5 \times 10^{-2-(-5)} = 0.5 \times 10^3 = (5 \times 10^{-1}) \times 10^3 = 5 \times 10^2$$

Note that, in both these examples, we first found an answer in which the number multiplied by a power of 10 was *not* between 1 and 10. We therefore followed the procedure for converting the final answer to scientific notation.

Addition and Subtraction with Scientific Notation

In general, we must write numbers in ordinary notation before adding or subtracting.

   Examples:

$$(3 \times 10^6) + (5 \times 10^2) = 3{,}000{,}000 + 500 = 3{,}000{,}500 = 3.0005 \times 10^6$$

$$(4.6 \times 10^9) - (5 \times 10^8) = 4{,}600{,}000{,}000 - 500{,}000{,}000 = 4{,}100{,}000{,}000 = 4.1 \times 10^9$$

When both numbers have the *same* power of 10, we can factor out the power of 10 first.

   Examples:

$$(7 \times 10^{10}) + (4 \times 10^{10}) = (7 + 4) \times 10^{10} = 11 \times 10^{10} = 1.1 \times 10^{11}$$

$$(2.3 \times 10^{-22}) - (1.6 \times 10^{-22}) = (2.3 - 1.6) \times 10^{-22} = 0.7 \times 10^{-22} = 7.0 \times 10^{-23}$$

B.3 Working with Units

Showing the units of a problem as you solve it usually makes the work much easier and also provides a useful way of checking your work. If an answer does not come out with the units you expect, you probably did something wrong. In general, working with units is very similar to working with numbers, as the following guidelines and examples show.

Five Guidelines for Working with Units

Before you begin any problem, think ahead and identify the units you expect for the final answer. Then operate on the units along with the numbers as you solve the problem. The following five guidelines may be helpful when you are working with units:

1. Mathematically, it doesn't matter whether a unit is singular (e.g., meter) or plural (e.g., meters); you can use the same abbreviation (e.g., m) for both.

2. You cannot add or subtract numbers unless they have the *same* units. For example, 5 apples + 3 apples = 8 apples, but the expression 5 apples + 3 oranges cannot be simplified further.

3. You *can* multiply units, divide units, or raise units to powers. Look for key words that tell you what to do.

   ■ *Per* suggests division. For example, we write a speed of 100 kilometers per hour as:

   $$100\frac{\text{km}}{\text{hr}} \quad \text{or} \quad \frac{100 \text{ km}}{1 \text{ hr}}$$

   ■ *Of* suggests multiplication. For example, if you launch a 50-kg space probe at a launch cost *of* $10,000 per kilogram, the total cost is:

   $$50 \text{ k\!\!/g} \times \frac{\$10,000}{\text{k\!\!/g}} = \$500,000$$

   ■ *Square* suggests raising to the second power. For example, we write an area of 75 square meters as 75 m^2.

   ■ *Cube* suggests raising to the third power. For example, we write a volume of 12 cubic centimeters as 12 cm^3.

4. Often the number you are given is not in the units you wish to work with. For example, you may be given that the speed of light is 300,000 km/s but need it in units of m/s for a particular problem. To convert the units, simply multiply the given number by a *conversion factor*: a fraction in which the numerator (top of the fraction) and denominator (bottom of the fraction) are equal, so that the value of the fraction is 1; the number in the denominator must have the units that you wish to change. In the case of changing the speed of light from units of km/s to m/s, you need a conversion factor for kilometers to meters. Thus, the conversion factor is:

   $$\frac{1,000 \text{ m}}{1 \text{ km}}$$

   Note that this conversion factor is equal to 1, since 1,000 meters and 1 kilometer are equal, and that the units to be changed (km) appear in the denominator. We can now convert the speed of light from units of km/s to m/s simply by multiplying by this conversion factor:

   $$\underbrace{300,000 \frac{\text{k\!\!/m}}{\text{s}}}_{\substack{\text{speed of light} \\ \text{in km/s}}} \times \underbrace{\frac{1,000 \text{ m}}{1 \text{ k\!\!/m}}}_{\substack{\text{conversion from} \\ \text{km to m}}} = \underbrace{3 \times 10^8 \frac{\text{m}}{\text{s}}}_{\substack{\text{speed of light} \\ \text{in m/s}}}$$

   Note that the units of km cancel, leaving the answer in units of m/s.

5. It's easier to work with units if you replace division with multiplication by the reciprocal. For example, suppose you want to know how many minutes are represented by 300 seconds. We can find the answer by dividing 300 seconds by 60 seconds per minute:

   $$300 \text{ s} \div 60 \frac{\text{s}}{\text{min}}$$

   However, it is easier to see the unit cancellations if we rewrite this expression by replacing the division with multiplication by the reciprocal (this process is easy to remember as "invert and multiply"):

   $$300 \text{ s} \div 60 \frac{\text{s}}{\text{min}} = 300 \text{ s\!\!/} \times \underbrace{\frac{1 \text{ min}}{60 \text{ s\!\!/}}}_{\substack{\text{invert} \\ \text{and multiply}}} = 5 \text{ min}$$

We now see that the units of seconds (s) cancel in the numerator of the first term and the denominator of the second term, leaving the answer in units of minutes.

More Examples of Working with Units

Example 1. How many seconds are there in 1 day?

Solution: We can answer the question by setting up a *chain* of unit conversions in which we start with 1 *day* and end up with *seconds*. We use the facts that there are 24 hours per day (24 hr/day), 60 minutes per hour (60 min/hr), and 60 seconds per minute (60 s/min):

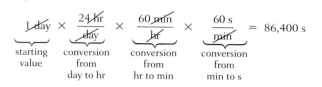

Note that all the units cancel except *seconds,* which is what we want for the answer. There are 86,400 seconds in 1 day.

Example 2. Convert a distance of 10^8 cm to km.

Solution: The easiest way to make this conversion is in two steps, since we know that there are 100 centimeters per meter (100 cm/m) and 1,000 meters per kilometer (1,000 m/km):

$$10^8 \text{ cm} \times \underbrace{\frac{1 \text{ m}}{100 \text{ cm}}}_{\substack{\text{conversion} \\ \text{from} \\ \text{cm to m}}} \times \underbrace{\frac{1 \text{ km}}{1,000 \text{ m}}}_{\substack{\text{conversion} \\ \text{from} \\ \text{m to km}}} = 10^8 \text{ cm} \times \frac{1 \text{ m}}{10^2 \text{ cm}} \times \frac{1 \text{ km}}{10^3 \text{ m}} = 10^3 \text{ km}$$

Alternatively, if we recognize that the number of kilometers should be smaller than the number of centimeters (because kilometers are larger), we might decide to do this conversion by dividing as follows:

$$10^8 \text{ cm} \div \frac{100 \text{ cm}}{\text{m}} \div \frac{1,000 \text{ m}}{\text{km}}$$

In this case, before carrying out the calculation, we replace each division with multiplication by the reciprocal:

$$10^8 \text{ cm} \div \frac{100 \text{ cm}}{\text{m}} \div \frac{1,000 \text{ m}}{\text{km}} = 10^8 \text{cm} \times \frac{1 \text{ m}}{100 \text{ cm}} \times \frac{1 \text{ km}}{1,000 \text{ m}}$$

$$= 10^8 \text{ cm} \times \frac{1 \text{ m}}{10^2 \text{ cm}} \times \frac{1 \text{ km}}{10^3 \text{ m}}$$

$$= 10^3 \text{ km}$$

Note that we again get the answer that 10^8 cm is the same as 10^3 km, or 1,000 km.

Example 3. Suppose you accelerate at 9.8 m/s^2 for 4 seconds, starting from rest. How fast will you be going?

Solution: The question asked "how fast?" so we expect to end up with a speed. Therefore, we multiply the acceleration by the amount of time you accelerated:

$$9.8 \, \frac{\text{m}}{\text{s}^2} \times 4 \text{ s} = (9.8 \times 4) \, \frac{\text{m} \times \text{s}}{\text{s}^2} = 39.2 \, \frac{\text{m}}{\text{s}}$$

Note that the units end up as a speed, showing that you will be traveling 39.2 m/s after 4 seconds of acceleration at 9.8 m/s^2.

Example 4. A reservoir is 2 km long and 3 km wide. Calculate its area, in both square kilometers and square meters.

Solution: We find its area by multiplying its length and width:

$$2 \text{ km} \times 3 \text{ km} = 6 \text{ km}^2$$

Next we need to convert this area of 6 km² to square meters, using the fact that there are 1,000 meters per kilometer (1,000 m/km). Note that we must square the term 1,000 m/km when converting from km² to m²:

$$6 \text{ km}^2 \times \left(1{,}000 \frac{\text{m}}{\text{km}}\right)^2 = 6 \text{ km}^2 \times 1{,}000^2 \frac{\text{m}^2}{\text{km}^2} = 6 \text{ km}^2 \times 1{,}000{,}000 \frac{\text{m}^2}{\text{km}^2}$$

$$= 6{,}000{,}000 \text{ m}^2$$

The reservoir area is 6 km², which is the same as 6 million (or 6×10^6) m².

B.4 The Metric System (SI)

The modern version of the metric system, known as *Système Internationale d'Unites* (French for "International System of Units") or *SI*, was formally established in 1960. Today, it is the primary measurement system in nearly every country in the world with the exception of the United States. Even in the United States, it is the system of choice for science and international commerce.

The basic units of length, mass, and time in the SI are:

■ The *meter* for length, abbreviated m

■ The *kilogram* for mass, abbreviated kg

■ The *second* for time, abbreviated s

Multiples of metric units are formed by powers of 10, using a prefix to indicate the power. For example, *kilo* means 10^3 (1,000), so a kilometer is 1,000 meters; a microgram is 0.000001 gram, because *micro* means 10^{-6}, or one millionth. Some of the more common prefixes are listed in Table B.2.

Metric Conversions

Table B.3 lists conversions between metric units and units used commonly in the United States. Note that the conversions between kilograms and pounds are valid only on Earth, because they depend on the strength of gravity.

Example 1. International athletic competitions generally use metric distances. Compare the length of a 100-meter race to that of a 100-yard race.

Table B.2 SI (Metric) Prefixes

| Small Values | | | Large Values | | |
Prefix	Abbreviation	Value	Prefix	Abbreviation	Value
Deci	d	10^{-1}	Deca	da	10^1
Centi	c	10^{-2}	Hecto	h	10^2
Milli	m	10^{-3}	Kilo	k	10^3
Micro	μ	10^{-6}	Mega	M	10^6
Nano	n	10^{-9}	Giga	G	10^9
Pico	p	10^{-12}	Tera	T	10^{12}

Table B.3 Metric Conversions

To Metric	From Metric
1 inch = 2.540 cm	1 cm = 0.3937 inch
1 foot = 0.3048 m	1 m = 3.28 feet
1 yard = 0.9144 m	1 m = 1.094 yards
1 mile = 1.6093 km	1 km = 0.6214 mile
1 pound = 0.4536 kg	1 kg = 2.205 pounds

Solution: Table B.3 shows that 1 m = 1.094 yd, so 100 m is 109.4 yd. Note that 100 meters is almost 110 yards; a good "rule of thumb" to remember is that distances in meters are about 10% longer than the corresponding number of yards.

Example 2. How many square kilometers are in 1 square mile?

Solution: We use the square of the miles-to-kilometers conversion factor:

$$(1 \text{ mi}^2) \times \left(\frac{1.6093 \text{ km}}{1 \text{ mi}} \right)^2 = (1 \text{ mi}^2) \times \left(1.6093^2 \frac{\text{km}^2}{\text{mi}^2} \right) = 2.5898 \text{ km}^2$$

Therefore, 1 square mile is 2.5898 square kilometers.

B.5 Finding a Ratio

Suppose you want to compare two quantities, such as the average density of the Earth and the average density of Jupiter. The way we do such a comparison is by dividing, which tells us the *ratio* of the two quantities. In this case, the Earth's average density is 5.52 g/cm^3 and Jupiter's average density is 1.33 g/cm^3, so the ratio is:

$$\frac{\text{average density of Earth}}{\text{average density of Jupiter}} = \frac{5.52 \text{ g/cm}^3}{1.33 \text{ g/cm}^3} = 4.15$$

Notice how the units cancel on both the top and bottom of the fraction. We can state our result in two equivalent ways:

- The ratio of the Earth's average density to Jupiter's average density is 4.15.

- The Earth's average density is 4.15 times Jupiter's average density.

Sometimes, the quantities that you want to compare may each involve an equation. In such cases, you could, of course, find the ratio by first calculating each of the two quantities individually and then dividing. However, it is much easier if you first express the ratio as a fraction, putting the equation for one quantity on top and the equation for the other on the bottom. Some of the terms in the equation may then cancel out, making any calculations much easier.

Example 1. Compare the kinetic energy of a car traveling at 100 km/hr to that of a car traveling at 50 km/hr.

Solution: We do the comparison by finding the ratio of the two kinetic energies, recalling that the formula for kinetic energy is $\frac{1}{2} mv^2$. Because we are not told the mass of the car, you might at first think that we don't have enough information to find the ratio. However, notice what happens when we put the equations for each kinetic energy into the ratio, calling the two speeds v_1 and v_2:

$$\frac{\text{K.E. car at } v_1}{\text{K.E. car at } v_2} = \frac{\frac{1}{2} m_{car} v_1^2}{\frac{1}{2} m_{car} v_2^2} = \frac{v_1^2}{v_2^2} = \left(\frac{v_1}{v_2} \right)^2$$

All the terms cancel except those with the two speeds, leaving us with a very simple formula for the ratio. Now we put in 100 km/hr for v_1 and 50 km/hr for v_2:

$$\frac{\text{K.E. car at 100 km/hr}}{\text{K.E. car at 50 km/hr}} = \left(\frac{100 \text{ km/hr}}{50 \text{ km/hr}} \right)^2 = 2^2 = 4$$

The ratio of the car's kinetic energies at 100 km/hr and 50 km/hr is 4. That is, the car has four times as much kinetic energy at 100 km/hr as it has at 50 km/hr.

Example 2. Compare the strength of gravity between the Earth and the Sun to the strength of gravity between the Earth and the Moon.

Solution: We do the comparison by taking the ratio of the Earth–Sun gravity to the Earth–Moon gravity. In this case, each quantity is found from the equation of Newton's law of gravity. The ratio is:

$$\frac{\text{Earth–Sun gravity}}{\text{Earth–Moon gravity}} = \frac{G \dfrac{M_{\text{Earth}} M_{\text{Sun}}}{(d_{\text{Earth–Sun}})^2}}{G \dfrac{M_{\text{Earth}} M_{\text{Moon}}}{(d_{\text{Earth–Moon}})^2}} = \frac{M_{\text{Sun}}}{(d_{\text{Earth–Sun}})^2} \times \frac{(d_{\text{Earth–Moon}})^2}{M_{\text{Moon}}}$$

Note how all but four of the terms cancel; the last step comes from replacing the division with multiplication by the reciprocal (the "invert and multiply" rule for division). We can simplify the work further by rearranging the terms so that we have the masses and distances together:

$$\frac{\text{Earth–Sun gravity}}{\text{Earth–Moon gravity}} = \frac{M_{\text{Sun}}}{M_{\text{Moon}}} \times \frac{(d_{\text{Earth–Moon}})^2}{(d_{\text{Earth–Sun}})^2}$$

Now it is just a matter of looking up the numbers (see Appendix E) and calculating:

$$\frac{\text{Earth–Sun gravity}}{\text{Earth–Moon gravity}} = \frac{1.99 \times 10^{30}\ \text{kg}}{7.35 \times 10^{22}\ \text{kg}} \times \frac{(384.4 \times 10^3\ \text{km})^2}{(149.6 \times 10^6\ \text{km})^2} = 179$$

In other words, the Earth–Sun gravity is 179 times stronger than the Earth–Moon gravity.

THE PERIODIC TABLE OF THE ELEMENTS

Key

12	← Atomic number
Mg	← Element's symbol
Magnesium	← Element's name
24.305	← Atomic mass*

*Atomic masses are fractions because they represent a weighted average of atomic mass numbers of different isotopes—in proportion to the abundance of each isotope on Earth.

1																	2
H Hydrogen 1.00794																	**He** Helium 4.003
3 **Li** Lithium 6.941	4 **Be** Beryllium 9.01218											5 **B** Boron 10.81	6 **C** Carbon 12.011	7 **N** Nitrogen 14.007	8 **O** Oxygen 15.999	9 **F** Fluorine 18.988	10 **Ne** Neon 20.179
11 **Na** Sodium 22.990	12 **Mg** Magnesium 24.305											13 **Al** Aluminum 26.98	14 **Si** Silicon 28.086	15 **P** Phosphorus 30.974	16 **S** Sulfur 32.06	17 **Cl** Chlorine 35.453	18 **Ar** Argon 39.948
19 **K** Potassium 39.098	20 **Ca** Calcium 40.08	21 **Sc** Scandium 44.956	22 **Ti** Titanium 47.88	23 **V** Vanadium 50.94	24 **Cr** Chromium 51.996	25 **Mn** Manganese 54.938	26 **Fe** Iron 55.847	27 **Co** Cobalt 58.9332	28 **Ni** Nickel 58.69	29 **Cu** Copper 63.546	30 **Zn** Zinc 65.39	31 **Ga** Gallium 69.72	32 **Ge** Germanium 72.59	33 **As** Arsenic 74.922	34 **Se** Selenium 78.96	35 **Br** Bromine 79.904	36 **Fr** Krypton 83.80
37 **Rb** Rubidium 85.468	38 **Sr** Strontium 87.62	39 **Y** Yttrium 88.9059	40 **Zr** Zirconium 91.224	41 **Nb** Niobium 92.91	42 **Mo** Molybdenum 95.94	43 **Tc** Technetium (98)	44 **Ru** Ruthenium 101.07	45 **Rh** Rhodium 102.906	46 **Pd** Palladium 106.42	47 **Ag** Silver 107.868	48 **Cd** Cadmium 112.41	49 **In** Indium 114.82	50 **Sn** Tin 118.71	51 **Sb** Antimony 121.75	52 **Te** Tellurium 127.60	53 **I** Iodine 126.905	54 **Xe** Xenon 131.29
55 **Cs** Cesium 132.91	56 **Ba** Barium 137.34		72 **Hf** Hafnium 178.49	73 **Ta** Tantalum 180.95	74 **W** Tungsten 183.85	75 **Re** Rhenium 186.207	76 **Os** Osmium 190.2	77 **Ir** Iridium 192.22	78 **Pt** Platinum 195.08	79 **Au** Gold 196.967	80 **Hg** Mercury 200.59	81 **Ti** Thallium 204.383	82 **Pb** Lead 207.2	83 **Bi** Bismuth 208.98	84 **Po** Polonium (209)	85 **At** Astatine (210)	86 **Rn** Radon (222)
87 **Fr** Francium (223)	88 **Ra** Radium 226.0254		104 **Rf** Rutherfordium (261)	105 **Db** Dubnium (262)	106 **Sg** Seaborgium (263)	107 **Bh** Bohrium (262)	108 **Hs** Hassium (265)	109 **Mt** Meitnerium (266)	110 **Uun** Ununnilium (269)	111 **Uuu** Unununium (272)	112 **Uub** Ununbium (277)						

Lanthanide Series

57	58	59	60	61	62	63	64	65	66	67	68	69	70	71
La Lanthanum 138.906	**Ce** Cerium 140.12	**Pr** Praseodymium 140.908	**Nd** Neodymium 144.24	**Pm** Promethium (145)	**Sm** Samarium 150.36	**Eu** Europium 151.96	**Gd** Gadolinium 157.25	**Tb** Terbium 158.925	**Dy** Dysprosium 162.50	**Ho** Holmium 164.93	**Er** Erbium 167.26	**Tm** Thulium 168.934	**Yb** Ytterbium 173.04	**Lu** Lutetium 174.967

Actinide Series

89	90	91	92	93	94	95	96	97	98	99	100	101	102	103
Ac Actinium 227.028	**Th** Thorium 232.038	**Pa** Protactinium 231.036	**U** Uranium 238.029	**Np** Neptunium 237.048	**Pu** Plutonium (244)	**Am** Americium (243)	**Cm** Curium (247)	**Bk** Berkelium (247)	**Cf** Californium (251)	**Es** Einsteinium (252)	**Fm** Fermium (257)	**Md** Mendelevium (258)	**No** Nobelium (259)	**Lr** Lawrencium (260)

APPENDIX D

CHEMICAL ENERGY FOR LIFE

In the text, we discuss many cases where life obtains energy from chemical reactions, such as reactions that occur at the interface between rock and water or reactions involving minerals in the hot water near deep-sea vents. In this appendix, we discuss the nature of these reactions in a bit more depth so that interested readers can understand how chemistry can provide energy for metabolism.

D.1 Chemical Energy from Disequilibrium

Any mixture of atoms and molecules can naturally undergo chemical reactions that may rearrange the atoms in such a way as to form or break chemical bonds. Left to themselves, chemical reactions will ultimately come to an **equilibrium** that represents a balance between the reacting atoms and molecules and the product atoms and molecules. For example, molecular hydrogen and oxygen can react together to make water. We can write this chemical reaction as:

$$H_2 + 1/2\ O_2 \leftrightarrow H_2O$$

The double arrow indicates that the reaction can proceed in both directions, and the 1/2 in front of the O_2 indicates that the reaction requires only half as many oxygen molecules as hydrogen molecules.

If we begin by mixing hydrogen and oxygen, at first the reaction will proceed only toward making water molecules, because there is no water present initially. But eventually the reaction will proceed at equal rates in both directions, at which time we will have chemical equilibrium. In some cases nearly all the hydrogen and oxygen will be converted to water, while in other cases very little of it may be converted to water. The relative amounts of hydrogen, oxygen, and water at equilibrium depend on the external circumstances (such as pressure, temperature, and the presence of other chemicals). For example, the reaction between H_2 and O_2 needs a "push" to get it started. In a room filled with these two gases, this push might come from lighting a match. The energy from the match gets the reaction started; as H_2 and O_2 combine to form H_2O, additional energy is released that can then trigger more molecules to combine. This sequence can occur extremely rapidly, and the amounts of energy released can cause an explosion.

Imagine that our reaction between hydrogen and oxygen is occurring in a small flask that is inside a large room and that the reaction has come to equilibrium at room temperature. Now, suppose we suddenly do something that disturbs the equilibrium; for example, imagine that we add excess hydrogen and oxygen molecules. The rate of water formation will temporarily speed up until the equilibrium is restored. In the meantime, however, the excess amounts of hydrogen and oxygen mean that the reaction is not in equilibrium; we say that it is in **disequilibrium.** The reaction rate will change automatically in such a way as to bring the relative amounts of hydrogen, oxygen, and water back toward their equilibrium values. Under the right circumstances, these reactions back toward equilibrium can release chemical energy that might be used by life to fuel metabolism.

The key to making such chemical energy available for life lies in having some natural set of circumstances that maintains a state of disequilibrium. In that case, the ongoing disequilibrium means that the reactions are always trying to move back toward equilibrium, thereby offering a continuous source of chemical energy. Of course, this state of affairs is possible only when an external energy source somehow maintains the disequilibrium, but such circumstances may be common on geologically active worlds. For example, chemical disequilibrium occurs near deep-sea volcanic vents because of mixing between high-temperature vent water and the surrounding low-temperature ocean water. Because the vents are continually releasing hot water, the disequilibrium can be maintained over long periods of time. Another place where we commonly find ongoing disequilibrium is at interfaces between rock and water. The rock and water will naturally undergo chemical reactions, but as long as new water (that is, water that has not already undergone reactions with the rock) continuously circulates and comes in contact with the rock, the reactants and products will remain out of equilibrium. Thus, both deep-sea vents and rock-water interfaces offer places where chemical disequilibrium can provide ongoing energy that could be utilized to create complex molecules, to string complex molecules together into complicated structures, or to support metabolism of living organisms.

D.2 Redox Reactions

Chemical disequilibrium has the potential to provide chemical energy for life, but realizing this potential requires chemical reactions that life can actually make use of. A particular class of chemical reactions turns out to be especially important for life—reactions that go by the rather odd-sounding name **redox reactions.** Redox reactions involve an exchange or reshuffling of electrical charge (which occurs through movement of electrons) between the reacting atoms or molecules.

To understand both the idea of a redox reaction and the origin of its bizarre name, consider again what happens when hydrogen and oxygen combine to make water. Viewed on a molecular level, the reaction involves two steps. First, a hydrogen molecule decomposes into two protons (hydrogen nuclei) and two electrons:

$$H_2 \rightarrow 2H^+ + 2e^-$$

(H^+ represents a single proton, which is positively charged, and e^- represents a single negatively charged electron.) Next, the two protons and two electrons combine with an oxygen atom (half of an oxygen molecule) to make a water molecule:

$$1/2\ O_2 + 2H^+ + 2e^- \rightarrow H_2O$$

Viewed in this way, the production of water is a redox reaction, because electrons are effectively transferred from hydrogen to oxygen. Because hydrogen gives up the electrons, we say that it is the **electron donor** for the overall reaction. Because the oxygen takes on the electrons, we say that it acts as the **electron acceptor.**

Now, with just a little more terminology, we can see why this type of reaction is called a redox reaction. In accepting electrons, the electrical charge of the oxygen is *reduced* (because electrons are negatively charged); hence, the first three letters in *redox* refer to this process of **reduction** of electrical charge. The last two letters are a bit more subtle. The charge of the hydrogen is increased, but because this increase occurs as the result of action by oxygen, we say that the hydrogen has become *oxidized*. In fact, oxygen is so efficient at grabbing electrons from other chemicals (atoms or molecules) that chemists have come to use the term **oxidation** to describe the process of losing electrons in general, even when the electrons are lost to something besides oxygen. The overall process of making water from hydrogen and oxygen, $H_2 + 1/2 \, O_2 \rightarrow H_2O$, is called a *redox reaction* because the oxygen gets reduced while the hydrogen gets oxidized.

More generally, a redox reaction always involves the transfer of one or more electrons from an electron donor (which becomes oxidized) to an electron acceptor (which becomes reduced). The transfer of electrons gives off energy that can then drive other chemical reactions, including the biochemical reactions of life.

Redox Reactions Used by Life on Earth

Most of the key energy-generating chemical reactions used by life on Earth are redox reactions. For example, the basic process of aerobic respiration in animals involves combining a sugar acquired by eating, such as glucose ($C_6H_{12}O_6$) with oxygen acquired by breathing to make carbon dioxide and water, releasing energy in the process:

$$C_6H_{12}O_6 + 6O_2 \rightarrow 6CO_2 + 6H_2O$$

This is a redox reaction, because the glucose donates electrons (it is oxidized) while the oxygen accepts electrons (it is reduced). (We can often recognize what is being oxidized or reduced in redox reactions even without knowing precisely how electrons are rearranged. In the above reaction, for example, the C in glucose is being oxidized, because it ends up being combined with a greater number of O atoms. Glucose has equal numbers of C and O atoms, but CO_2 has twice as many O as C atoms and so is more oxidized. The oxygen is being reduced, because it ends up being combined with more hydrogen; it has no H atoms on the left side of the reaction but has two H atoms per O atom on the right side. Thus, for example, the C in CH_4 is more reduced than the C in CO_2, which is more oxidized.)

Many cellular energy-generating processes proceed through chains of redox reactions, sometimes called *electron transport chains* because a series of redox reactions means a series of electron transfers. In photosynthesis, for example, the chain begins when chlorophyll absorbs sunlight. The energy from the sunlight creates disequilibrium in the cell, and this disequilibrium then offers energy that the cell utilizes through a chain of redox reactions. Redox reactions are especially important when we consider the prospects for life in extreme environments, either on Earth or on other worlds. Many Earth organisms use fairly simple redox reactions as their primary source of energy. For example, bacteria known as *Thiobacillus ferrooxidans*, which can thrive in highly acidic conditions such as mine tailings, obtain energy by oxidizing iron:

$$2Fe^{+2} + 1/2 \, O_2 + 2H^+ \rightarrow 2Fe^{+3} + H_2O$$

(Fe^{+2} and Fe^{+3} represent iron atoms missing two and three electrons, respectively.) In this case, the iron is the electron donor (becoming oxidized), and the oxygen is the electron acceptor (becoming reduced). Note that neither preexisting organic molecules nor sunlight is needed for this reaction, which means reactions like this could have been used by early life-forms.

Many other redox reactions produce energy for various microbes on Earth, including reactions involving molecular hydrogen and sulfur. These are especially important when we consider possible energy sources for an origin of life. For example, both iron and sulfur are common in the disequilibrium environments of hot springs and deep-sea vents and thus could be involved in redox reactions that might ultimately lead to life. At rock-water interfaces, chemical reactions between water and iron in rock can produce molecular hydrogen, which can then be used in redox reactions for biochemistry.

In summary, any geologically active world with liquid water may have places where chemical disequilibrium persists for long periods of time, such as near underwater volcanic vents or anywhere where rock and water come into contact. At these places, redox reactions can provide energy that could power biochemical reactions that might ultimately lead to life and that could support life once it arises.

APPENDIX E

PLANETARY DATA

Table E.1 Physical Properties of the Sun and Planets

Name	Radius (Eq[a]) (km)	Radius (Eq) (Earth units)	Mass (kg)	Mass (Earth units)	Average Density (g/cm^3)	Surface Gravity (Earth = 1)
Sun	695,000	109	1.99×10^{30}	333,000	1.41	27.5
Mercury	2,440	0.382	3.30×10^{23}	0.055	5.43	0.38
Venus	6,051	0.949	4.87×10^{24}	0.815	5.25	0.91
Earth	6,378	1.00	5.97×10^{24}	1.00	5.52	1.00
Mars	3,397	0.533	6.42×10^{23}	0.107	3.93	0.38
Jupiter	71,492	11.19	1.90×10^{27}	317.9	1.33	2.53
Saturn	60,268	9.46	5.69×10^{26}	95.18	0.71	1.07
Uranus	25,559	3.98	8.66×10^{25}	14.54	1.24	0.91
Neptune	24,764	3.81	1.03×10^{26}	17.13	1.67	1.14
Pluto	1,160	0.181	1.31×10^{22}	0.0022	2.05	0.07

[a]Eq = equatorial.

Table E.2 Orbital Properties of the Sun and Planets

Name	Distance from Sun[a] (AU)	(10^6 km)	Orbital Period (years)	Orbital Inclination[b] (degrees)	Orbital Eccentricity	Sidereal Rotation Period (Earth days)[c]	Axis Tilt (degrees)
Sun	—	—	—	—	—	25.4	7.25
Mercury	0.387	57.9	0.2409	7.00	0.206	58.6	0.0
Venus	0.723	108.2	0.6152	3.39	0.007	−243.0	177.4
Earth	1.00	149.6	1.0	0.00	0.017	0.9973	23.45
Mars	1.524	227.9	1.881	1.85	0.093	1.026	23.98
Jupiter	5.203	778.3	11.86	1.31	0.048	0.41	3.08
Saturn	9.539	1,427	29.46	2.49	0.056	0.44	26.73
Uranus	19.19	2,870	84.01	0.77	0.046	−0.72	97.92
Neptune	30.06	4,497	164.8	1.77	0.010	0.67	29.6
Pluto	39.54	5,916	248.0	17.15	0.248	−6.39	118

[a]Semimajor axis of the orbit.

[b]With respect to the ecliptic.

[c]A negative sign indicates rotation relative to other planets.

Table E.3 Satellites of the Solar System[a]

Planet Satellite	Radius or Dimensions[b] (km)	Distance from Planet (10³ km)	Orbital Period[c] (Earth days)	Mass[d] (kg)	Density[d] (g/cm³)	Notes About the Satellites
Earth						**Earth**
Moon	1,738	384.4	27.322	7.349×10^{22}	3.34	*Moon:* Probably formed in giant impact.
Mars						**Mars**
Phobos	13×11×9	9.38	0.319	1.3×10^{16}	2.2	*Phobos, Deimos:* Probable captured asteroids.
Deimos	8×6×5	23.5	1.263	1.8×10^{15}	1.7	
Jupiter						**Jupiter**
Small inner moons (5 moons)	10 to 135×82×75	128–222	0.295–0.6745	—	—	*Metis, Adrastea, Amalthea, Thebe, 1999 J1:* Small moonlets within and near Jupiter's ring system.
Io	1,821	421.6	1.769	8.933×10^{22}	3.57	*Io:* Most volcanically active object in the solar system.
Europa	1,565	670.9	3.551	4.797×10^{22}	2.97	*Europa:* Possible oceans under icy crust.
Ganymede	2,634	1,070.0	7.155	1.482×10^{23}	1.94	*Ganymede:* Largest satellite in solar system; unusual ice geology.
Callisto	2,403	1,883.0	16.689	1.076×10^{23}	1.86	*Callisto:* Cratered iceball.
2000J1	4	7,411	129.8	—	—	Probable captured moon with highly inclined orbit.
Irregular group 1 (5 moons)	2–80	11,100–12,700	240–289	—	—	*Leda, Himalia, Elara, Lysithea, 2000 J11:* Probable captured moons with inclined orbits.
Irregular group 2 (25 moons)	1–29	18,900–24,200	−534 to −767	—	—	*Ananke, Carme, Pasiphae, Sinope, 1999J1, 2000 J2–J10, 2001 J1–J11:* Probable captured moons in inclined backward orbits.
Saturn						**Saturn**
Small inner moons (6)	10 to 97×95×77	134–151	0.574–0.695	—	—	*Pan, Atlas, Prometheus, Pandora, Epimetheus, Janus:* Small moonlets within and near Saturn's ring system.
Mimas	199	185.52	0.942	3.70×10^{19}	1.17	*Mimas, Enceladus, Tethys:* Small and medium-size iceballs, many with interesting geology.
Enceladus	249	238.02	1.370	1.2×10^{20}	1.24	
Tethys	530	294.66	1.888	6.17×10^{20}	1.26	
Calypso	15×8×8	294.66	1.888	4×10^{15}	—	*Calypso, Telesto:* Small moonlets sharing Tethys's orbit.
Telesto	15×13×8	294.67	1.888	6×10^{15}	—	
Dione	559	377.4	2.737	1.08×10^{21}	1.44	*Dione:* Medium-size iceball, with interesting geology.
Helene	18×?×15	377.4	2.737	1.6×10^{16}	—	*Helene:* Small moonlet sharing Dione's orbit.
Rhea	764	527.04	4.518	2.31×10^{21}	1.33	*Rhea:* Medium-size iceball, with interesting geology.
Titan	2,575	1,221.85	15.945	1.3455×10^{23}	1.88	*Titan:* Dense atmosphere shrouds surface; ongoing geological activity possible.

Satellite	Dimensions or radius (km)[b]	Orbital radius (10³ km)	Period (days)[c]	Mass (kg)[d]	Density[d]	Notes
Hyperion	180×140×112	1,481.1	21.277	2.8 × 10¹⁹	—	Hyperion: Only satellite known not to rotate synchronously.
Iapetus	718	3,561.3	79.331	1.59 × 10²¹	1.21	Iapetus: Bright and dark hemispheres show greatest contrast in the solar system.
Phoebe	110	12,952	−550.4	1 × 10¹⁹	—	Phoebe: Very dark; material ejected from Phoebe may coat one side of Iapetus.
Irregular group 1 (4 moons)	7–22	11,400–17,100	453–829	—	—	2000 S2, S3, S5, S6: Probable captured moons with highly inclined orbits.
Irregular group 2 (3 moons)	5–15	17,400–18,000	854–901	—	—	2000 S4, S10, S11: Probable captured moons in inclined orbits.
Irregular group 3 (5 moons)	4–10	15,600–23,400	−723 to −1,325	—	—	2000 S1, S7, S8, S9, S12: Probable captured moons in inclined backward orbits.
Uranus						**Uranus**
Small inner moons (11 moons)	10 to 97×95×77	134–151	0.574–0.695	—	—	Cordelia, Ophelia, Bianca, Cressida, Desdemona, Juliet, Portia, Rosalind, Belinda, Puck, 1986 U10: Small moonlets within and near Uranus's ring system.
Miranda	236	129.8	1.413	6.6 × 10¹⁹	1.26	Miranda, Ariel, Umbriel, Titania, Oberon: Small and medium-size iceballs, with some interesting geology.
Ariel	579	191.2	2.520	1.35 × 10²¹	1.65	
Umbriel	584.7	266.0	4.144	1.17 × 10²¹	1.44	
Titania	788.9	435.8	8.706	3.52 × 10²¹	1.59	
Oberon	761.4	582.6	13.463	3.01 × 10²¹	1.50	
Irregular group (5 moons)	???–60	7,170–25,000	580–2,280	—	—	Caliban, Sycorax, Stephano, Prospero, Setebos: Too recently discovered for accurate determination of their properties; several in backward orbits.
Neptune						**Neptune**
Small inner moons (5 moons)	29 to 104×?×89	48–74	0.296–0.554	—	—	Naiad, Thalassa, Despina, Galatea, Larissa: Small moonlets within and near Neptune's ring system.
Proteus	218×208×201	117.6	1.121	6 × 10¹⁹	—	
Triton	1,352.6	354.59	−5.875	2.14 × 10²²	2.0	Triton: Probable captured Kuiper belt object—largest captured object in solar system.
Nereid	170	5,588.6	360.125	3.1 × 10¹⁹	—	Nereid: Small, icy moon; very little known.
Pluto						**Pluto**
Charon	635	19.6	6.38718	1.56 × 10²¹	1.6	Charon: Unusually large compared to its planet; may have formed in giant impact.

[a]*Note:* Authorities differ substantially on many of the values in this table.

[b]$a \times b \times c$ values for the Dimensions are the approximate lengths of the axes for irregular moons.

[c]Negative sign indicates backward orbit.

[d]Masses and densities are most accurate for those satellites visited by a spacecraft on a flyby. Masses for the smallest moons have not been measured but can be estimated from the radius and an assumed density.

SELECTED ASTROBIOLOGY WEB SITES

The Web contains a vast amount of astronomical information. For all your astrobiology Web surfing, the best starting point is the Web site for this textbook:

The Astronomy Place
www.astronomyplace.com

The book Web site contains both general links and links organized by chapter. Tables F.1–F.6 list some other sites that may be of particular use; if any of the links change, in most cases you can find live links at The Astronomy Place.

Primary Astrobiology Sites

General astrobiology sites that serve as great starting points for all kinds of astrobiology information, including the latest news and summaries of key research topics.

Site	Description	Web Address
NASA Astrobiology Institute	NASA's main site for astrobiology	http://nai.arc.nasa.gov
NASA Science News	Great site for all science news related to NASA research, including astrobiology; offers sign-up for e-mail news alerts	http://science.nasa.gov
The SETI Institute	Devoted to the search for life off the Earth	http://www.seti.org
Astrobiology Web	Discovery Channel site dedicated to astrobiology	http://www.astrobiology.com/

More Astrobiology News and Resources

Additional sites offering the latest astrobiology news and other astrobiology resources.

Site	Description	Web Address
Astrobiology: The Living Universe	Thinkquest student-created site for astrobiology	http://library.thinkquest.org/C003763/index.php?page=about00
Astrobiology Central	Astrobiology news and resources maintained by Dr. David Darling	http://www.angelfire.com/on2/daviddarling/
Astronomy Now	General astronomy news, including astrobiology news	http://www.astronomynow.com
EurekAlert	The American Association for the Advancement of Science (AAAS) general news site organized by discipline	http://www.eurekalert.org
JPL News	News releases from NASA's Jet Propulsion Laboratory	http://www.jpl.nasa.gov/news
Marsbugs Newsletter	An astrobiology newsletter	http://www.lyon.edu/webdata/users/dthomas/marbugs/marsbugs.html
Science Online	*Science Magazine*'s online site	http://www.scienceonline.org/
Space.com	Lots of space-related news, including areas for SETI and astronomy	http://www.space.com
SpaceScience.com	General space science news	http://spacescience.com
Universe Today	General space news	http://www.universetoday.com

Key Mission Sites

The Web pages for major astrobiology-related space missions.

Site	Description	Web Address
NASA's Office of Space Science Missions Page	Direct links to all past, present, and planned NASA space science missions	http://spacescience.nasa.gov/missions
Cassini/Huygens	Mission scheduled to arrive at Saturn in 2004	http://www.jpl.nasa.gov/cassini
Hubble Space Telescope	Latest discoveries, educational activities, and other information from the Hubble Space Telescope	http://hubble.stsci.edu
Mars Exploration Program	Information on current and planned Mars missions	http://mars.jpl.nasa.gov
Kepler	Mission to search for extrasolar planets	http://www.kepler.arc.nasa.gov/
Space Interferometry Mission (SIM)	Astronomical observatory that will search for extrasolar planets	http://sim.jpl.nasa.gov/
Terrestrial Planet Finder (TPF)	Proposed mission to search for extrasolar terrestrial planets	http://planetquest.jpl.nasa.gov/TPF/tpf_index.html
Darwin Mission	Proposed European mission to search for extrasolar terrestrial planets	http://www.estec.esa.nl/spdwww/future/darwin/index.html

Sites Concerning the Search for Extrasolar Planets

Site	Description	Web Address
The Extrasolar Planets Encyclopedia	Information about the search for and discoveries of extrasolar planets	http://cfa-www.harvard.edu/planets
PlanetQuest	NASA site devoted to the search for Earth-like planets	http://planetquest.jpl.nasa.gov/index.html
California and Carnegie Planet Search	Site developed by leading extrasolar planet hunters	http://exoplanets.org/

More Useful Web Sites

Some of the authors' favorites among many other noncommercial resources for astronomy and astrobiology.

Site	Description	Web Address
Astronomical Society of the Pacific	An organization for both professional astronomers and the general public, devoted largely to astronomy education	http://www.aspsky.org
Astronomy Picture of the Day	An archive of beautiful pictures, updated daily	http://antwrp.gsfc.nasa.gov/apod
AstroWeb	Listing of major resources for astronomy on the Web	http://www.cv.nrao.edu/fits/www/astronomy.html
Canadian Space Agency	Home page for Canada's space program	http://www.space.gc.ca
European Space Agency (ESA)	Home page for this international agency	http://www.esa.int
NASA Home Page	Almost anything you want to learn about NASA	http://www.nasa.gov
The Nine Planets (University of Arizona)	A multimedia tour of the solar system	http://seds.lpl.arizona.edu/nineplanets/nineplanets/nineplanets.html
The Planetary Society	An organization with more than 100,000 members who are interested in planetary exploration and the search for life in the universe	http://planetary.org
Voyage Scale Model Solar System	The National Mall exhibit discussed in Chapter 1	http://www.voyageonline.org

Astrobiology Society Web Sites

The Web sites of large groups or institutions that currently carry out astrobiology-related research, education and outreach.

Group	Web Address
Arrhenius Laboratory at Scripps Institution of Oceanography	http://www.mrd.ucsd.edu/ga/
Astrobiology Australasia (Australia and New Zealand)	http://www.aao.gov.au/local/www/jab/astrobiology/
Astrobiology: The Living Universe	http://library.thinkquest.org/C003763/index.php?page=about00
Biosphere 2 Astrobiology Launch Pad	http://www.astro.bio2.edu/astrobiology/
Caltech: Geobiology and Astrobiology	http://www.gps.caltech.edu/options/geobiology/geobio.html
Cardiff Center for Astrobiology	http://www.astrobiology.cf.ac.uk/
Center for Astrobiology, Spain (in Spanish)	http://www.cab.inta.es/pagina/home_cab.htm
Arizona State University Astrobiology Institute	http://astrobiology.asu.edu/
UCLA Center for Astrobiology	http://astrobiology.ucla.edu/
UCLA Astrobiology Society	http://www.studentgroups.ucla.edu/abs/
Johnson Space Center Astrobiology Institute for the Study of Biomarkers	http://www-sn.jsc.nasa.gov/astrobiology/biomarkers/
Penn State Astrobiology Research Center	http://psarc.geosc.psu.edu/
University of Colorado Center for Astrobiology	http://argyre.colorado.edu/life/CAB.html
University of Washington Astrobiology Program	http://depts.washington.edu/astrobio/
Woods Hole Oceanographic Institute Astrobiology Program	http://www.mbl.edu/Astrobiology/
Night Eyes: The Auckland Astronomical Society Junior Section	http://www.astronomy.org.nz/events/juniors/more/nighteye.htm
SETI and Astrobiology in Australia	http://www1.tpgi.com.au/users/tps-seti/seti.html
Swedish Astrobiology Network	http://www.oru.se/nat/astrobio/
Southwest Research Institute	http://www.boulder.swri.edu/scinews/astrobio.html
University College London, Astrobiology Institute	http://www.astrobiology.ucl.ac.uk/
UK Astrobiology Forum and Network	http://astrobiology.rl.ac.uk

Glossary

absolute zero The coldest possible temperature, which is 0 Kelvin or about −273.15°C.

absorption (of light) The process by which matter absorbs radiative energy.

absorption-line spectrum A spectrum that contains absorption lines.

acceleration The rate at which an object's velocity changes, given by velocity divided by time; standard units are m/s².

acceleration of gravity The acceleration of a falling object. On Earth, the acceleration of gravity, designated by g, is 9.8 m/s².

accretion (of planets) The process of building larger objects from collisions and impacts of smaller ones. As accretion proceeded in the early solar system, it first built up *planetesimals,* some of which later grew into larger *protoplanets.*

adaptive optics A technique in which telescope mirrors flex rapidly to compensate for the bending of starlight caused by atmospheric turbulence.

aerobic organisms Organisms that cannot survive without molecular oxygen from the atmosphere.

albedo Describes the fraction of sunlight reflected by a surface; albedo = 0 means no reflection at all (a perfectly black surface), while albedo = 1 means all light is reflected (a perfectly white surface).

Amazonian era The present era on Mars, which began about 1.0 billion years ago.

amino acids The molecules that form the building blocks of proteins. Organisms construct proteins from a particular set of 20 amino acids, although several dozen other amino acids can be found in nature. (More technically, an amino acid is a molecule containing both an *amino group,* NH or NH_2, and a *carboxyl group,* COOH.)

anaerobic organisms Organisms that do not require (and may even be poisoned by) molecular oxygen from the atmosphere.

Andromeda Galaxy (M 31; the Great Galaxy in Andromeda) The nearest large spiral galaxy to the Milky Way.

angular resolution (of a telescope) The smallest angular separation that two point-like objects can have and still be seen as distinct points of light (rather than as a single point of light).

angular size (or **angular distance**) A measure of the angle formed by extending imaginary lines outward from our eyes to span an object (or between two objects).

Antarctic Circle The circle on the Earth with latitude 66.5°S.

antimatter Refers to any particle with the same mass as a particle of ordinary matter but whose other basic properties, such as electrical charge, are precisely opposite those of the ordinary matter.

apparent retrograde motion The apparent motion of a planet, as viewed from Earth, during the period of a few weeks or months when it moves westward relative to the stars in our sky.

Archaea One of the three domains of life; the others are Eukarya and Bacteria.

arcminutes (or **minutes of arc**) One arcminute is 1/60 of 1°.

arcseconds (or **seconds of arc**) One arcsecond is 1/60 of an arcminute, or 1/3,600 of 1°.

Aristotelians Ancient Greek followers of Aristotle, who held that there could be only one Earth and the heavens were a realm distinct from Earth.

asteroid A relatively small and rocky object that orbits a star; asteroids are sometimes called *minor planets* because they are similar to planets but smaller.

asteroid belt The region of our solar system between the orbits of Mars and Jupiter in which asteroids are heavily concentrated.

astrobiology The study of life on Earth and beyond; emphasizes research into questions of the origin of life, the conditions under which life can survive, and the search for life beyond Earth.

astrometric technique A technique used to look for extrasolar planets that involves making very precise measurements of stellar positions in the sky (*astrometric* means "measurement of the stars").

astronomical unit (AU) The average (semimajor-axis) distance of the Earth from the Sun, about 150 million km.

atmospheric pressure The surface pressure resulting from the overlying weight of an atmosphere.

atom Consists of a nucleus made from protons and neutrons (except for ordinary hydrogen, which has only a proton) encircled by a cloud of electrons.

atomic mass number The combined number of protons and neutrons in an atom.

atomic number The number of protons in an atom.

atomists Ancient Greek scholars who held that the universe is made from an infinite number of indivisible atoms.

ATP (adenosine triphosphate) The molecule that stores and releases energy for nearly all cellular processes among life on Earth.

AU See *astronomical unit.*

aurora Dancing lights in the sky caused by charged particles entering our atmosphere; called the *aurora borealis* in the Northern Hemisphere and the *aurora australis* in the Southern Hemisphere.

autotroph An organism that gets its carbon directly from the atmosphere in the form of carbon dioxide.

Bacteria One of the three domains of life; the others are Eukarya and Archaea.

band (of sensitivity) The set of frequencies that a particular radio receiver can pick up.

bandwidth (of a transmitted signal) The range of frequencies over which a communication signal is transmitted.

bar The standard unit of pressure, equal to the Earth's atmospheric pressure at sea level.

basalt A dark, dense igneous rock that is commonly produced by undersea volcanoes.

Big Bang The event that gave birth to the universe.

binary star system A star system that contains two stars.

biochemistry The chemistry of life.

biosphere Refers to the "layer" of life on Earth.

black smokers Structures around seafloor volcanic vents that support a wide variety of life.

blueshift A Doppler shift in which spectral features are shifted to shorter wavelengths, caused when an object is coming toward the observer.

brown dwarf An object that forms much like a star from a spinning cloud of gas but that is not large enough to sustain nuclear fusion like a star.

Cambrian explosion The dramatic diversification of animal life on Earth that occurred between about 545 and 505 million years ago.

carbohydrates Molecules, such as sugars and starches, that provide energy to cells and make important cellular structures.

carbon dioxide (CO$_2$) cycle The process that cycles carbon dioxide between the Earth's atmosphere and surface rocks.

carbonate rock A carbon-rich rock, such as limestone, that forms underwater from chemical reactions between sediments and carbon dioxide. On Earth, most of the outgassed carbon dioxide currently resides in carbonate rocks.

carbon-based life Life that uses molecules containing carbon for its most critical functions. All life on Earth is carbon-based.

catalysis The process of causing or accelerating a chemical reaction by involving a substance or molecule that is not permanently changed by the reaction.

catalyst The unchanged substance or molecule involved in catalysis.

cell The basic structure of all life on Earth, in which the living matter inside is separated from the outside world.

Celsius (temperature scale) The temperature scale commonly used in daily activity internationally. Defined so that, on Earth's surface at sea level, water freezes at 0°C and boils at 100°C.

center of mass The point at which a system of objects would balance; the point around which gravitational forces are centered.

chemical bond The linkage between atoms in a molecule.

chemical element A substance made from individual atoms of a particular atomic number.

chemoautotroph An organism that gets its carbon directly from the atmosphere and its energy from chemical reactions involving inorganic molecules.

chemoheterotroph An organism that gets both its energy and its carbon by consuming preexisting organic molecules; all animals are chemoheterotrophs.

chloroplasts Structures in plant cells that produce energy by photosynthesis.

civilization types A way of categorizing civilizations by whether they use resources of their planet, their star, or their galaxy. See also *galactic civilization, planetary civilization, stellar civilization.*

climate Describes the long-term average of weather.

comet A relatively small and icy object that orbits a star.

compound Any substance (crystal or molecule) made from more than one chemical element.

continental crust The thicker, lower-density crust that makes up the Earth's continents. It is made when remelting of

seafloor crust allows lower-density rock to separate. Continental crust ranges in age from very young to as old as about 4.0 billion years.

continental drift The idea that the continents slowly move around on Earth, now known to be a result of plate tectonics.

continuously habitable zone The region around a star in which conditions could allow for surface habitability throughout the history of the star system.

convection The energy transport process in which warm material expands and rises, while cooler material contracts and falls.

convection cells Individual "loops" (rising and falling) of convecting material.

convergent evolution The tendency of organisms of different evolutionary backgrounds to come to resemble one another because they occupy similar ecological niches.

Copernican revolution The dramatic change in human perspective that occurred when the ancient geocentric belief was replaced by the idea that the Earth is a planet orbiting the Sun. It began with the work of Copernicus.

coral model (of colonization) A model of how a civilization might colonize the galaxy, based on growth much like that of coral in the sea.

core (of a planet) The dense central region of a planet that has undergone differentiation.

core (of a star) The central region of a star, in which nuclear fusion can occur.

cosmic rays Particles, including electrons, protons, and atomic nuclei, that zip through interstellar space at close to the speed of light.

cosmos See *universe.*

crust (of a planet) The low-density surface layer of a planet that has undergone differentiation.

crystal A substance made from atoms arranged in precise geometrical patterns, such as in a mineral.

cultural evolution Changes that arise from the transmission of knowledge accumulated over generations.

cyanobacteria Photosynthetic bacteria thought to have been responsible for making most of the oxygen that gradually built up in the Earth's atmosphere.

cycles per second Units of frequency for a wave; describes the number of peaks (or troughs) of a wave that pass by a given point each second. Equivalent to *hertz.*

deuterium A form of hydrogen in which the nucleus contains a proton and a neutron, rather than only a proton (as is the case for most hydrogen nuclei).

differentiation The process in which gravity separates materials according to density, with high-density materials sinking and low-density materials rising.

discovery science The process of science that involves going out and looking at nature in a general way in hopes of learning something new and unexpected.

disequilibrium (chemical) A state in which a mixture undergoing chemical reactions is not in equilibrium.

DNA (deoxyribonucleic acid) The basic hereditary molecule of life on Earth. A DNA molecule consists of two strands, twisted in the shape of a double helix, along each of which lies a long sequence of *DNA bases.*

DNA bases The four DNA bases are adenine (A), cytosine (C), guanine (G), and thymine (T), and they can be paired across the two DNA strands only so that A goes with T and C goes with G.

DNA replication The process of copying DNA molecules.

domain (of life) The highest level by which we currently classify life; the three domains are Eukarya, Bacteria, and Archaea.

Doppler effect The effect that shifts the wavelengths of spectral features in objects that are moving toward or away from the observer.

Doppler technique A technique for looking for extrasolar planets by studying a star's spectrum to look for Doppler shifts indicating that the star is moving around a center of mass.

Drake equation An equation that lays out the factors that play a role in determining the number of communicating civilizations in our galaxy.

Dyson sphere A hypothesized type of large, thin-walled sphere built to surround a star so that an advanced civilization could capture all the energy flowing out from the star. (Named after physicist Freeman Dyson, who proposed its possible existence.)

Earth-orbiters (spacecraft) Spacecraft designed to study the Earth or the universe from Earth orbit.

eccentricity A measure of how much an ellipse deviates from a perfect circle, which has zero eccentricity. Greater eccentricity means the ellipse is more stretched out (elongated).

ecliptic plane The plane of the Earth's orbit around the Sun.

electrical charge A fundamental property of matter that is described by its amount and as either positive or negative. Protons have positive electrical charge, and electrons have negative electrical charge; neutrons have no electrical charge.

electromagnetic spectrum The complete spectrum of light, including radio waves, infrared, visible light, ultraviolet light, X rays, and gamma rays.

electromagnetic wave A synonym for *light,* which consists of waves of electric and magnetic fields.

electron acceptor (in a redox reaction) The chemical (atom or molecule) that gains electrons in an overall chemical reaction.

electron donor (in a redox reaction) The chemical (atom or molecule) that gives up electrons in an overall chemical reaction.

electrons Fundamental particles with negative electrical charge; the distribution of electrons in an atom gives the atom its size.

element See *chemical element*.

ellipse A type of oval that happens to be the shape of planetary orbits. An ellipse can be drawn by moving a pencil along a string whose ends are tied to two tacks; the locations of the tacks are the foci (singular, *focus*) of the ellipse.

encephalization quotient (EQ) A rough measure of animal intelligence based on the ratio of an animal's brain size to its body mass.

endospore A special "resting" cell that allows some organisms to remain dormant for long periods of time.

energy Broadly speaking, energy is what can make matter move. The three basic types of energy are kinetic, potential, and radiative.

enzyme A protein that serves as a catalyst.

eons (geological) The largest divisions of time in Earth's geological history. The four eons are the Hadean, Archaean, Proterozoic, and Phanerozoic.

EQ See *encephalization quotient*.

equilibrium (chemical) A state of balance between the reacting atoms and molecules and the product atoms and molecules in a mixture undergoing chemical reactions.

eras (geological) The second largest divisions of time in Earth's geological history, after eons. The Phanerozoic eon is subdivided into three eras: the Paleozoic, Mesozoic, and Cenozoic.

erosion The wearing down or building up of geological features by wind, water, ice, and other phenomena of planetary weather.

eruption The process of transferring a planet's heat outward by releasing hot lava on the planet's surface.

escape velocity The speed necessary for an object to completely escape the gravity of a large body such as a moon, planet, or star.

Eukarya One of the three domains of life and the one in which all plants and animals are found; the other domains are Bacteria and Archaea.

eukaryote A living organism that is a member of the domain Eukarya and therefore made from one or more eukaryotic cells.

eukaryotic cell A cell that contains a distinct nucleus that is separated from the rest of the cell by its own membrane.

evaporation The phase change from liquid to gas.

evolution (biological) The gradual change in populations of living organisms responsible for transforming life on Earth from its primitive origins to the great diversity of life today.

evolution (technological) Change driven by the rapid development of technology.

evolutionary adaptation An inherited trait that enhances an organism's ability to survive and reproduce in a particular environment.

expansion (of universe) We say that the universe is expanding because the average distance between galaxies appears to be increasing with time.

experimental science Science that uses experiments to seek out general principles of physics, chemistry, or other sciences.

extrasolar planet A planet orbiting a star other than our Sun.

extremophile An organism that thrives under conditions that are extreme by human standards.

Fahrenheit (temperature scale) The temperature scale commonly used in daily activity in the United States. Defined so that, on Earth's surface at sea level, water freezes at 32°F and boils at 212°F.

fault (geological) A place where plates slip sideways relative to each other.

feedback relationships Processes in which one property amplifies (positive feedback) or counteracts (negative feedback) the behavior of other properties.

Fermi paradox The question posed by Enrico Fermi about extraterrestrial intelligence—"So where is everybody?"—which asks why we have not observed other civilizations even though it seems some ought to have spread throughout the galaxy.

field An abstract concept used to describe how a particle would interact with a force. For example, the idea of a *gravitational field* describes how a particle would react to the local strength of gravity, and the idea of an *electromagnetic field* describes how a charged particle would respond to forces from other charged particles.

flybys (spacecraft) Spacecraft that fly past a target object (such as a planet), usually just once, as opposed to entering a bound orbit of the object.

fossil A relic of an organism that lived and died long ago.

fossil record (and **rock record**) The information about Earth's past that is recorded in fossils (fossil record) and rocks (rock record). Note that the terms are often used synonymously.

frequency Describes the rate at which peaks of a wave pass by a point; measured in units of 1/s, often called *cycles per second* or *hertz*.

galactic (Type III) civilization A civilization that employs the resources of its entire galaxy.

galaxy A great island of stars in space, containing from a few hundred million to a trillion or more stars, all orbiting a common center.

Galilean moons The four moons of Jupiter discovered by Galileo: Io, Europa, Ganymede, and Callisto.

gamma rays Light with very short wavelengths (and hence high frequencies), shorter than those of X rays.

gas phase The phase of matter in which atoms or molecules can move essentially independently of one another.

gas pressure Describes the force (per unit area) pushing on any object due to surrounding gas. See also *pressure*.

gene The basic functional unit of an organism's heredity; a single gene consists of a sequence of DNA bases (or RNA bases, in some viruses) that provide the instructions for a single cell function (such as for building a protein).

genetic analysis The analysis of an organism's genes or genome.

genetic code The specific set of rules by which the sequence of bases in DNA is "read" to provide the instructions that make up genes.

genetic engineering Making deliberate changes to an organism's genome.

genome The complete sequence of DNA bases in an organism, encompassing all of the organism's genes.

geocentric universe (ancient belief in) The idea that the Earth is the center of the universe.

geological processes The four basic geological processes are impact cratering, volcanism, tectonics, and erosion.

geological time scale The time scale used to measure the history of the Earth. It is divided into four *eons* (the Hadean, Archaean, Proterozoic, and Phanerozoic). The last (Phanerozoic) eon is subdivided into three *eras* (the Paleozoic, Mesozoic, and Cenozoic), which in turn are subdivided into several *periods*. (The periods are further subdivided into *epochs* and *ages*.)

geology The study of surface features (on a moon, planet, or asteroid) and the processes that create them.

global average temperature The average surface temperature of a planet.

global warming Usually refers to the current warming of the Earth being caused by human input of greenhouse gases into the atmosphere.

gravitational encounters Occur when two (or more) objects pass near enough so that each can feel the effects of the other's gravity and therefore can exchange energy.

Great Red Spot A large, high-pressure storm on Jupiter.

greenhouse effect The effect that makes the surface and lower atmosphere warmer than they would be in the absence of an atmosphere.

greenhouse gases Gases, such as carbon dioxide (CO_2), water vapor (H_2O), and methane (CH_4), that can absorb infrared light emitted by the planetary surface (after the surface is heated by sunlight).

habitable planet (or habitable world) A planet (or world) with environmental conditions under which life could *potentially* arise or survive.

habitable zone The region around a star in which a planet of suitable size could *potentially* have surface temperatures at which liquid water could exist.

Hadean eon The earliest eon in Earth's history, corresponding to times before about 3.8 billion years ago.

half-life The time it takes for half of the atoms to decay in a sample of a radioactive substance.

handedness The property of some molecules, such as amino acids, that allows them to come in two distinct forms that are mirror images of each other.

heavy bombardment The period of time during which the planets in our solar system were heavily bombarded by leftover planetesimals, starting from the time the planets first formed and ending some 3.8–4.0 billion years ago.

heavy elements In astronomy, generally refers to all chemical elements *except* hydrogen and helium.

heredity The characteristics of an organism passed on to it by its parent(s), which it can pass on to its offspring. The term can also apply to the transmission of these characteristics from one generation to the next.

Hesperian era The middle history of Mars, dating from about 3.7 to 1.0 billion years ago.

heterotroph An organism that gets its carbon by consuming preexisting organic molecules.

historical science Science that involves looking at present-day evidence to try to discover something about past events.

hot spot (geological) See *mantle plume.*

hyperspace Hypothetical dimensions beyond our ordinary dimensions of space and time that we can't see or "get into."

hyperthermophile An organism that thrives under conditions of very high temperature compared to what most organisms tolerate.

hypothesis An "educated guess," or proposed explanation (for some set of observed facts), that has not yet been rigorously tested and confirmed.

hypothesis-driven science The process of science that involves proposing an idea and performing experiments or making observations that put it to the test.

ice ages Periods of time during which the Earth becomes unusually cold, so that water from the oceans freezes out as ice and covers a substantial portion of the continents.

ices (in solar system theory) Materials that are solid only at low temperatures, such as the hydrogen compounds water, ammonia, and methane.

igneous rock Rock made when molten rock cools and solidifies.

image A picture of an object made by focusing light.

impact The collision of a small body (such as an asteroid or a comet) with a larger object (such as a planet or a moon).

impact crater A bowl-shaped depression left by the impact of an object that strikes a planetary surface (rather than burning up in the atmosphere).

impact sterilization The process of sterilizing a planet as a result of a large impact.

impactor The object responsible for an impact.

infrared (light) Light with wavelengths that fall in the portion of the electromagnetic spectrum between radio and visible light.

inner solar system Generally considered to encompass the region of our solar system out to about the orbit of Mars.

inorganic Not pertaining to life or the chemistry of carbon molecules.

interferometry A telescopic technique in which two or more telescopes are used in tandem to produce much better angular resolution than the telescopes could achieve individually.

interstellar cloud A cloud of gas and dust between the stars.

interstellar ramjet A hypothesized type of spaceship that uses a giant scoop to sweep up interstellar gas for use in a nuclear fusion engine.

inverse square law Followed by any quantity that decreases with the square of the distance between two objects.

ion engine (rocket) A rocket engine that works by accelerating charged particles and expelling them out its back.

ionosphere A layer of the Earth's upper atmosphere that consists of particles ionized by sunlight. The ionosphere acts as a radio "mirror," reflecting low-frequency radio emissions.

ions Atoms with a positive or negative electrical charge.

isotopes Each different isotope of an element has the *same* number of protons but a *different* number of neutrons.

joule The international unit of energy, equivalent to about 1/4,000 of a calorie.

jovian moons The moons of jovian planets.

jovian nebulae The clouds of gas that swirled around the jovian planets, from which the jovian moons formed.

jovian planets Gaseous planets similar in overall composition to Jupiter.

Kepler's first law States that the orbit of each planet about the Sun is an ellipse with the Sun at one focus.

Kepler's laws of planetary motion Three laws discovered by Kepler that describe the motion of the planets around the Sun.

Kepler's second law States that as a planet moves around its orbit it sweeps out equal areas in equal times. This tells us that a planet moves faster when it is closer to the Sun (near perihelion) than when it is farther from the Sun (near aphelion) in its orbit.

Kepler's third law States that the square of a planet's orbital period is proportional to the cube of its average distance from the Sun (semimajor axis), which tells us that more distant planets move more slowly in their orbits. In its original form, written $p^2 = a^3$.

kingdoms (biological) Except for the three domains, kingdoms (such as the plant kingdom and the animal kingdom) are the highest classification grouping of living organisms.

K–T boundary The thin layer of dark sediments that marks the division between the Cretaceous and Tertiary periods in the fossil record (the "K" comes from the German word for Cretaceous, *Kreide*).

K–T event The mass extinction at the end of the Cretaceous period, thought to have been caused by an asteroid or comet impact.

Kuiper belt The comet-rich region of our solar system that spans distances of about 30–100 AU from the Sun; Kuiper belt comets have orbits that lie fairly close to the plane of planetary orbits and travel around the Sun in the same direction as the planets.

Kuiper belt objects The cometlike objects located in the Kuiper belt.

Lagrange points (of the Earth–Moon system) The five positions in space where the effects of gravity from the Earth and Moon "cancel" in such a way that, if you floated weightlessly at one of these five points, you wouldn't be tugged toward either body.

lander (space mission) A spacecraft that lands on the surface of another world.

latitude The angular north-south distance between a planet's equator and a location on its surface.

light-year The distance that light can travel in 1 year, or 9.46 trillion kilometers.

lipids Complex molecules in cells, also known as fats, that have a variety of functions, including being key components of membranes.

liquid phase The phase of matter in which atoms or molecules are held together but move relatively freely.

lithophile An organism that thrives inside of rock.

lithosphere The relatively rigid outer layer of a planet; generally encompasses the crust and the uppermost portion of the mantle.

Local Group The group of about 30 galaxies to which the Milky Way Galaxy belongs.

Local Supercluster The supercluster of galaxies to which the Local Group belongs.

longitude The angular east-west distance between the prime meridian (which passes through Greenwich, England) and a location on the Earth's surface.

luminosity The total power output of an object, usually measured in watts or in units of solar luminosities ($L_{Sun} = 3.8 \times 10^{26}$ watts).

lunar maria The regions of the Moon that look smooth from Earth and actually are impact basins.

magma Underground molten rock.

magnetic field Describes the region surrounding a magnet in which it can affect other charged magnets or charged particles in its vicinity.

mantle (of a planet) The rocky layer that lies between a planet's core and crust.

mantle convection The flow pattern in which hot mantle material expands and rises while cooler material contracts and falls.

mantle plume (or **hot spot**) A place where a plume of hot material rises up from deep within the mantle, not at the boundary between plates.

martian meteorites Meteorites found on the Earth's surface that apparently were chipped off the surface of Mars.

mass extinction An event in which a large fraction of the species living on Earth go extinct, such as the K–T event in which the dinosaurs died out about 65 million years ago.

mass increase (in relativity) Refers to the effect in which an object moving past you seems to have a mass greater than its rest mass.

mass ratio (of a rocket) The ratio of the initial (launch) mass of the rocket M_i, including its fuel and any spacecraft it is carrying, to its mass M after all the fuel is burned.

matter-antimatter annihilation The result when particles of matter contact their corresponding particles of antimatter, with all their mass turned into energy according to $E = mc^2$.

membrane (cell) A barrier that separates the inside of a cell (or cell nucleus) from the outside.

metabolism The many chemical reactions that occur in living organisms.

metamorphic rock Rock made from igneous or sedimentary rock that gets transformed (but not melted) by high heat or pressure.

meteor A flash of light caused when a particle from space burns up in our atmosphere.

meteorite A rock from space that lands on Earth.

Milky Way Galaxy The galaxy of which our solar system is a member. Our solar system is located about midway out from the center and is one of several hundred billion star systems in the Milky Way Galaxy.

Miller–Urey experiment An experiment first performed in the 1950s designed to learn how organic molecules might have formed naturally on the early Earth.

mineral A rocky substance with a particular chemical composition and crystal structure.

mitochondria The cellular organs in eukaryotic cells in which oxygen helps produce energy (by making molecules of ATP).

model (scientific) A representation of some aspect of nature that can be used to explain and predict real phenomena without invoking myth, magic, or the supernatural.

moist greenhouse effect A process by which a planet could lose water when the atmospheric circulation allows water vapor to rise high enough to be broken apart by ultraviolet light from the Sun.

molecule Consists of two or more atoms chemically bound together in a way that gives the molecule properties distinct from those of its individual atoms.

moon (or **satellite**) An object that orbits a planet. The term *satellite* is also used more generally to refer to any object orbiting another object.

multiple star system A star system that contains two or more stars.

mutation Any change to the base sequence of an organism's DNA.

natural selection The primary mechanism by which evolution proceeds. More specifically, natural selection refers to the process in which, over time, advantageous genetic traits naturally win out (are "selected") over less advantageous traits because they are more likely to be passed down through succeeding generations.

nebula A cloud of gas in space, usually one that is glowing.

neutrons Particles found in atomic nuclei that have no electrical charge.

Noachian era The era on Mars before 3.7 billion years ago.

nonscience Any way of searching for knowledge that does not exhibit the hallmarks of science.

nuclear fission The process in which a larger nucleus splits into two (or more) smaller nuclei.

nuclear fusion The process in which two (or more) smaller nuclei slam together and make one larger nucleus; it is the primary process by which stars generate their energy.

nucleic acids DNA and RNA.

nucleus (of an atom) The compact center of an atom, made from protons and neutrons.

nucleus (of a cell) The membrane-enclosed region of a eukaryotic cell that contains the cell's DNA.

nucleus (of a comet) The solid portion of a comet and the only portion that exists when the comet is far from the Sun.

observable universe The portion of the entire universe that, at least in principle, can be seen from Earth.

Occam's razor The idea that when two (or more) models are capable of explaining the same set of observations, we prefer the model that is simpler and based on fewer assumptions; after William of Occam (1285–1349).

orbital resonance Describes any situation in which one object's orbital period is a simple ratio of another object's period, such as 1/2, 1/4, or 5/3. In such cases the two objects periodically line up with one another, and the extra gravitational attraction at these times can affect the objects' orbits.

orbiters (of other worlds) Spacecraft that go into orbit of another world for long-term study.

organic chemistry The chemistry of organic molecules (whether or not the molecules are involved in life).

organic molecule Generally refers to any molecule containing carbon, but with some exceptions among molecules that are most commonly found independent of life. For our purposes in this book, the important exceptions are carbon dioxide (CO_2) and carbonate minerals, which we do not consider to be organic.

outer solar system Generally considered to encompass the region of our solar system beginning at about the orbit of Jupiter.

outgassing The process of releasing gases from a planetary interior.

oxidation reactions Chemical reactions that remove oxygen from the atmosphere or, more generally, the process of losing electrons in a chemical reaction.

ozone The molecule O_3, which is a particularly good absorber of ultraviolet light.

ozone depletion Refers to the declining levels of atmospheric ozone found worldwide on Earth in recent years, especially in Antarctica.

ozone hole A place where the concentration of ozone in the stratosphere is dramatically lower than is the norm, as has been the case each spring over Antarctica for the past two decades.

paradigm (in science) Refers to general patterns of thought that tend to shape scientific beliefs during a particular time period.

paradox A situation that, at least at first, seems to violate common sense or contradict itself. Resolving paradoxes often leads to deeper understanding.

parallax The apparent shifting of an object against the background due to viewing it from different positions. See also *stellar parallax*.

periodic table of the elements A table that lists properties of all the known elements in an organized way.

periods (geological) The third largest divisions of time in Earth's geological history, after eons and eras.

phase (of matter) Describes the way in which atoms or molecules are held together; the common phases are solid, liquid, and gas.

photoautotroph An organism that gets its carbon directly from the atmosphere and gets its energy from sunlight through photosynthesis; plants are photoautotrophs.

photoheterotroph An organism that gets its carbon by consuming preexisting organic molecules and gets its energy from sunlight through photosynthesis.

photon An individual particle of light, characterized by a wavelength or frequency.

phyla (singular, *phylum*) The next level of biological classification below kingdoms.

pixel An individual "picture element" in a digital picture.

planet A moderately large object that orbits a star. There are no official minimum or maximum sizes for planets, which can lead to disagreement about what counts as a planet. On the minimum side, for example, some astronomers argue that Pluto is too small to count as a planet. On the maximum side, astronomers disagree on the dividing line between large planets (up to several times the size of Jupiter) and "failed stars" (called *brown dwarfs*) that are starlike but too small to sustain significant nuclear fusion in their cores.

planetary (Type I) civilization A civilization that uses the resources of its home planet; we are a planetary civilization by this definition.

planetary nebula The glowing cloud of gas ejected from a low-mass star at the end of its life.

planetesimals The objects formed by accretion in the solar nebula.

plasma A gas consisting of ions and electrons.

plate tectonics The geological process in which lithospheric plates move around the surface of the Earth. It acts like a conveyor belt, with new seafloor crust erupting and spreading outward from mid-ocean ridges and then being recycled back into the mantle by subduction at ocean trenches. It also explains *continental drift,* because plates carry the continents with them as they move.

plates (on a planet) Pieces of a lithosphere that apparently float upon the denser mantle below.

pressure Describes the force (per unit area) pushing on an object; in astronomy, we are generally interested in pressure applied by surrounding gas.

probe (space) A spacecraft designed to explore (probe) some specific object in space.

prokaryote A living organism made from cells in which DNA is *not* confined to a distinct, membrane-enclosed nucleus. Most prokaryotes are single-celled. Prokaryotes include all the organisms in two of the three domains of life: Bacteria and Archaea.

prokaryotic cell A cell that lacks a distinct nucleus.

protein A large molecule assembled from amino acids according to instructions encoded in DNA. Proteins play many roles in cells; a special category of proteins, called *enzymes,* catalyzes nearly all of the important biochemical reactions that occur within cells.

protons Particles found in atomic nuclei that have positive electrical charge.

protoplanetary disk A spinning disk of gas in which planets may eventually form, as they did in our solar system.

protoplanets A term used to describe planetesimals that have grown quite large, to planet size.

pseudoscience Something that purports to be science or may appear to be scientific but that does not adhere to the hallmarks of science; literally, "false science."

Ptolemaic model The Earth-centered model of the universe created by Ptolemy in about A.D. 150 and used until the Copernican revolution.

radio (waves or light) Light with the longest wavelengths in the electromagnetic spectrum; that is, with wavelengths longer than those of infrared light.

radioactive decay The change that occurs in a radioactive nucleus.

radioactive element (or **radioactive isotope**) A substance whose nuclei tend to fall apart spontaneously.

radioactive isotope See *radioactive element.*

radiometric dating The process of determining the age of a rock (i.e., the time since it solidified) by comparing the present amount of a radioactive substance to the amount of its decay product.

rare Earth hypothesis A hypothesis holding that the specific circumstances that have made it possible for complex creatures (such as birds or humans) to evolve on Earth might be so rare that ours may be the only inhabited planet in the galaxy that has anything but the simplest life.

redox reactions Chemical reactions that involve an exchange or reshuffling of elec-

tric charge between the reacting atoms or molecules. A redox reaction always involves the transfer of one or more electrons from an electron donor (which becomes oxidized) to an electron acceptor (which becomes reduced).

redshift A Doppler shift in which spectral features are shifted to longer wavelengths, caused when an object is moving away from the observer.

reduction (chemical) The process of gaining electrons—which reduces the electrical change (because electrons carry negative charge)—in a chemical reaction.

RNA (ribonucleic acid) A molecule closely related to DNA, but with only a single strand and a slightly different backbone and set of bases, that plays critical roles in carrying out the instructions encoded in DNA.

rock record See *fossil record.*

rocket equation An equation that describes how a rocket's final speed depends on its propellant velocity and *mass ratio.*

runaway greenhouse effect A positive feedback cycle in which heating caused by the greenhouse effect causes more greenhouse gases to enter the atmosphere, which further enhance the greenhouse effect until all the greenhouse gases are present in the atmosphere.

rybozymes RNA molecules that function as catalysts.

sample return mission A space mission designed to return to Earth a sample of another world.

satellite See *moon.*

science The search for knowledge that can be used to explain or predict natural phenomena in a way that can be confirmed by rigorous observations or experiments. See also *discovery science, experimental science, historical science,* and *hypothesis-driven science.*

scientific method An organized approach to explaining observed facts through science. It is an idealization of how science is actually performed and corresponds most closely to hypothesis-driven science.

seafloor crust The relatively dense, thin, and young crust found on Earth's seafloors, composed largely of the igneous rock called *basalt.*

seafloor spreading The separation of plates that occurs along mid-ocean ridges, where new crust rises up and pushes existing seafloor crust apart.

search for extraterrestrial intelligence (SETI) The name given to observing projects designed to search for signs of intelligent life beyond Earth.

sedimentary rock Rock that builds up through time as sediments accumulate, often on the seafloor, and become compressed into solid rock. The sediments tend to build up in distinct layers, or *strata.*

seismic waves Earthquake-induced vibrations that propagate through a planet.

semimajor axis Half the distance across the long axis of an ellipse; in this text it is usually referred to as the *average* distance of an orbiting object.

sentinel hypothesis A possible explanation for the Fermi paradox that suggests that extraterrestrials place monitoring devices near star systems that show promise of emerging intelligence and patiently wait until these devices record the presence of civilization.

SETI See *search for extraterrestrial intelligence.*

silicate rock A silicon-rich rock.

sleep paralysis The natural paralysis of the body that occurs during REM sleep; it may occasionally persist for a few minutes after the brain has started waking up, giving a person the alarming sensation of being awake in a paralyzed body. Visions and other sensations often occur in this state.

snowball Earth Refers to recently identified periods of extreme ice ages that may have occurred several times between 750 and 580 million years ago.

solar luminosity The luminosity of the Sun, approximately 4×10^{26} watts.

solar nebula The piece of interstellar cloud from which our own solar system formed.

solar sail A large, highly reflective (and very thin, to minimize mass) piece of material that "sails" using pressure exerted by sunlight.

solar system The Sun and all the material that orbits it, including the planets. Note that the term *solar system* technically refers to our own star system (because solar means "of the Sun") but is sometimes used to describe other star systems. See also *star system.*

solar wind A stream of charged particles ejected from the Sun.

solid phase The phase of matter in which atoms or molecules are held rigidly in place.

spectral lines Bright or dark lines that appear in an object's spectrum, which we can see when we pass the object's light through a prismlike device that spreads the light out like a rainbow.

spectral type A way of classifying a star by the lines that appear in its spectrum; it is related to surface temperature. The basic spectral types are designated by a letter (one of OBAFGKM, with O for the hottest stars and M for the coolest) and are subdivided with numbers from 0 through 9.

spectroscopy The process of obtaining spectra from astronomical objects.

spectrum (of light) See *electromagnetic spectrum.*

speed The rate at which an object moves, given by distance divided by time. Common units are m/s or km/hr.

speed of light The speed at which light travels, about 300,000 km/s.

star Our Sun and other ordinary stars are large, glowing balls of gas that generate heat and light through nuclear fusion—the smashing together of light nuclei to make heavier nuclei—in their cores. (The term *star* is also applied to objects that are in the process of becoming true stars, such as protostars, and to the remains of stars that have died, such as a neutron star.)

star system One or more stars and any planets and other material that orbit them. See also *solar system.*

stellar (Type II) civilization A civilization that employs the resources of its home star (that is, not only the resources available on its home planet).

stellar parallax The apparent shift in the position of a nearby star against distant objects that occurs as we view the star from different positions in the Earth's orbit of the Sun each year.

strata (rock) Layers in sedimentary rock.

stratosphere An intermediate-altitude layer of the atmosphere in which ultraviolet light from the Sun is absorbed.

stromatolites Large bacterial "colonies."

subduction (of tectonic plates) The process in which one plate slides under another.

subduction zones Places where one plate slides under another.

sublimation The phase change going directly from solid to gas.

supernova The explosion of a massive star.

symbiotic relationship A relationship in which both an invading organism and a host organism benefit from living together.

synchronous rotation Describes the rotation of an object that always shows the same face to an object that it is orbiting because its rotation period and orbital period are the same.

technological evolution Change driven by the rapid development of technology.

tectonics The disruption of a planet's surface by internal stresses.

temperature A measure of the average kinetic energy of particles in a substance.

terraforming Changing a planet in such a way as to make it more Earth-like.

terrestrial planets Rocky planets similar in overall composition to Earth.

theory (in science) A comprehensive explanation for some aspect of nature that is supported by abundant evidence.

theory of evolution The theory, first advanced by Charles Darwin, that explains *how* evolution occurs through the process of natural selection. (Also referred to as "Darwinian evolution.")

thermophile An organism that thrives under conditions of high temperature compared to what most organisms tolerate.

tidal force A force that is caused when the gravity pulling on one side of an object is larger than that pulling on the other side, causing the object to stretch.

tidal friction Friction within an object caused by a tidal force.

tidal heating A source of internal heating created by tidal friction.

time dilation (in relativity) The effect in which you observe time running slower in reference frames moving relative to you.

transit When a planet appears to move across the face of the star as viewed from Earth.

troposphere The lowest atmospheric layer, in which weather occurs.

ultraviolet light Light with wavelengths that fall in the portion of the electromagnetic spectrum between visible light and X rays.

uniformitarianism The idea in geology that past events can be understood in terms of the same geological processes that we observe operating today.

universe (or **cosmos**) The sum total of all matter and energy; that is, everything within and between all galaxies.

viscosity Describes the "thickness" of a liquid in terms of how rapidly it flows; low-viscosity liquids flow quickly (e.g., water), while high-viscosity liquids flow slowly (e.g., molasses).

visible light The light our eyes can see, ranging in wavelength from about 400 to 700 nm.

volatile Refers to a substance that evaporates at relatively low temperatures.

volcanism The eruption of molten rock, or *lava,* from a planet's interior onto its surface.

Von Neumann machines Self-replicating machines first proposed by the American mathematician and computer pioneer John Von Neumann (1903–1957).

wavelength The distance between adjacent peaks (or troughs) of a wave.

weather Describes the ever-varying combination of winds, clouds, temperature, and pressure in a planet's troposphere.

wormhole A hypothetical tunnel through hyperspace that might connect two distant places in the universe.

X rays Light with wavelengths that fall in the portion of the electromagnetic spectrum between ultraviolet light and gamma rays.

zoo hypothesis A possible explanation for the Fermi paradox holding that alien civilizations are aware of our presence but have chosen to deliberately avoid contact with us.

REFERENCES AND BIBLIOGRAPHY

Indicates articles or books that are particularly accessible to students not majoring in science.

General (books covering a wide range of topics relevant to many chapters in this text)

* Bennett, J., M. Donahue, N. Schneider, and M. Voit. *The Cosmic Perspective.* Addison Wesley, 2002.

* Darling, D. *The Extraterrestrial Encyclopedia.* Three Rivers Press, 2000.

Delsemme, A. *Our Cosmic Origins.* Cambridge University Press, 1998.

* Dick, S. J. *Life on Other Worlds.* Cambridge University Press, 1998.

* Jakosky, B. *The Search for Life on Other Planets.* Cambridge University Press, 1998.

Lunine, J. I. *Earth: Evolution of a Habitable World.* Cambridge University Press, 1999.

* Sagan, C. *Cosmos.* Random House, 1980.

* Shostak, S. *Sharing the Universe.* Berkeley Hills Books, 1998.

Chapter 1

The Big Bang, Origin of the Elements, Formation of the Solar System

Beckwith, S. V. W. and A. I. Sargent. Circumstellar disks and the search for neighboring planetary systems. *Nature* **383,** 139–144, 1996.

* Kirshner, R. P. The Earth's elements. *Scientific American* **271,** 58–65, 1994.

Lissauer, J. J. Planet formation. *Annu. Rev. Astron. Astrophys.* **31,** 129–174, 1993.

Rees, M. J. Piecing together the biggest puzzle of all. *Science* **290,** 1919–1925, 2000.

Approaches to Astrobiology

Lederberg, J. Exobiology: Approaches to life beyond the Earth. *Science* **132,** 393–400, 1960.

* Morrison, D., and G. K. Schmidt, eds. *Astrobiology Roadmap.* NASA, 1999.

Chapter 2

History of Science and Astrobiology

* Asimov, I. *Asimov's Biographical Encyclopedia of Science and Technology.* Doubleday, 1982.

Crowe, M. J. *The Extraterrestrial Life Debate, 1750–1900.* Dover, 1999.

Crowe, M. J. *Theories of the World: From Antiquity to the Copernican Revolution.* Dover, 1990.

* Dick, S. J. *The Biological Universe.* Cambridge University Press, 1996.

The Nature of Science

* Boorstin, D. J. *The Discoverers.* Vintage, 1985.

Cleland, C. E. Historical science, experimental science, and the scientific method. *Geology* **29,** 987–990, 2001.

Kuhn, T. S. *The Nature of Scientific Revolutions.* University of Chicago Press, 1996.

Popper, K., and D. Miller, (eds.). *Popper Selections.* Princeton University Press, 1985.

* Sagan, C. *The Demon-Haunted World: Science as a Candle in the Dark.* Ballantine, 1996.

Chapter 3

Basic Biology and Evolution

* Campell, N., and J. Reece. *Biology,* 6th ed. Addison Wesley, 2003.

Gould, S. J. Showdown on the Burgess Shale: The reply. *Natural History,* Dec. 1998–Jan. 1999.

Koshland, D. E., Jr. The seven pillars of life. *Science,* **295,** 2215–2216, 2002.

Morris, S. C. Showdown on the Burgess Shale: The challenge. *Natural History,* Dec. 1998–Jan. 1999.

Sagan, C., Life. *Encyclopaedia Britannica,* 1970.

* Zimmer, C. *Evolution: The Triumph of an Idea.* Harper Collins, 2001.

The Three Domains and the Tree of Life (also relevant to Ch. 5)

Brown, J. F., and W. F. Doolittle. Root of the universal tree of life based on ancient aminoacyl-tRNA synthetase gene duplications. *Proc. Nat'l Acad. Sci.* **92,** 2441–2445, 1995.

* Doolittle, W. F. Uprooting the tree of life. *Scientific American* **282,** 90–95, 2000.

Pace, N. R. The early branches in the tree of life. Preprint for *Proceedings of a Symposium on Assembling the Tree of Life.* American Museum of Natural History, 2002.

Pace, N. R. A molecular view of microbial diversity and the biosphere. *Science* **276,** 734–740, 1997.

Woese, C. Interpreting the universal phylogenetic tree. *Proc. Nat'l Acad. Sci.* **97,** 8392–8396, 2000.

Woese, C. The universal ancestor, *Proc. Nat'l Acad. Sci.* **95,** 6854–6859, 1998.

Life in Extreme Environments

* Madigan, M. T., and B. L. Marrs. Extremophiles. *Scientific American* **276,** 82–87, 1997.

Nealson, K. H. The limits of life on Earth and searching for life on Mars. *J. Geophys. Res.* **102,** 23675–23686, 1997.

Sharma, A., et al. Microbial activity at gigapascal pressures. *Science* **295,** 1514–1516, 2002.

Stevens, T. O., and J. P. McKinley. Lithoautotrophic microbial ecosystems in deep basalt aquifers. *Science* **270,** 450–454, 1995.

Van Dover, C. L., et al. Biogeography and ecological setting of Indian Ocean hydrothermal vents. *Science* **294,** 818–823, 2001.

Chapter 4

Radiometric Dating and the Age of the Earth

Dalrymple, G. B. *The Age of the Earth.* Stanford, 1991.

Impact History of the Earth, the Heavy Bombardment, Impact Sterilization

Chyba, C. F., T. C. Owen, and W. H. Ip. Impact delivery of volatiles and organic molecules to the Earth. In *Hazards Due to Comets and Asteroids*, T. Gehrels, ed. U. Arizona Press, 9–58, 1994.

Hartmann, W. K., G. Ryder, L. Dones, and D. Grinspoon. The time-dependent intense bombardment of the primordial Earth/Moon system. In *Origin of the Earth and Moon*, R. M. Canup and K. Righter, eds. U. Arizona Press, 493–512, 2000.

Sleep, N. H., K. J. Zahnle, J. F. Kasting, and H. J. Morowitz. Annihilation of ecosystems by large asteroid impacts on the early Earth. *Nature* **342**, 139–142, 1989.

Geology and Plate Tectonics

Anderson, D. L. Planet Earth. In *The New Solar System*, J. K. Beatty, C. C. Peterson, and A. Chaikin, eds. Cambridge University Press: Sky Publ. Corp., 111–124, 1999.

Long-term Climate Change

* Hoffman, P. F., and D. P. Schrag, Snowball Earth. *Scientific American* **282**, 68–75, 2000.

Kasting, J. F. and D. H. Grinspoon. The faint young sun problem. In *The Sun in Time*, C. P. Sonett, M. S. Giampapa, and M. S. Matthews, eds. U. Arizona Press, 447–462, 1991.

Chapter 5

Early Evidence for Life

Brasier, M., et al. Questioning the evidence for Earth's oldest fossils. *Nature* **416**, 76–81, 2002.

Mojzsis, S. J., G. Arrhenius, K. D. McKeegan, T. M. Harrison, A. P. Nutman, and C. R. L. Friend. Evidence for life on Earth before 3,800 million years ago. *Nature* **384**, 55–59, 1996.

Mojzsis, S. J., T. M. Harrison, and R. T. Pidgeon. Oxygen-isotope evidence from ancient zircons for liquid water at the Earth's surface 4,300 Myr ago. *Nature* **409**, 178–181, 2001.

Schopf, J. W. The oldest fossils and what they mean. In *Major events in the history of life*, J. W. Schopf, ed. Boston: Jones and Bartlett Publishers, 29–63, 1992.

Schopf, J. W., et al. Laser-Raman imagery of Earth's earliest fossils. *Nature* **416**, 73–76, 2002.

Origin of Life on Earth

Cody, G. D., et al. Primordial carbonylated iron-sulfur compounds and the synthesis of pyruvate. *Science* **289**, 1337–1340, 2000.

Huber, C., and G. Wachtershauser. Activated acetic acid by carbon fixation on (Fe,Ni)S under primordial conditions. *Science* **276**, 245–247, 1997.

Pace, N. R. Origin of life—facing up to the physical setting. *Cell* **65**, 531–533, 1991.

Shock, E. L., J. P. Amend, and M. Y. Zolotov. The early Earth vs. the origin of life. In *Origin of the Earth and Moon*, R. M. Canup and K. Righter, eds. U. Arizona Press, 527–543, 2000.

Miller–Urey Experiment

Miller, S. L. The prebiotic synthesis of organic compounds as a step toward the origin of life. In *Major events in the history of life*, J. W. Schopf, ed. Boston: Jones and Bartlett Publishers, 1–28, 1992.

Urey, H. C. On the early chemical history of the Earth and the origin of life. *Proc. Nat'l Acad. Sci.* **38**, 351–363, 1952.

Where Did Life Originate?

Baross, J. A., and S. E. Hoffman. Submarine hydrothermal vents and associated gradient environments as sites for the origin and evolution of life. *Origins of Life* **15**, 327–345, 1985.

D'Hondt, S., S. Rutherford, and A. J. Spivak. Metabolic activity of subsurface life in deep-sea sediments. *Science* **295**, 2067–2070, 2002.

Ferris, J. P., C. H. Huang, and W. J. Hagan, Jr. Montmorrilonite: A multifunctional mineral catalyst for the prebiological formation of phosphate esters. *Orig. Life Evol. Biospheres* **18**, 121–133, 1988.

RNA World and Early Life

Cech, T. R. A model for the RNA-catalyzed replication of RNA. *Proc. Nat'l Acad. Sci.* **83**, 4360–4363, 1986.

Doudna, J. A., and J. W. Szostak. RNA-catalyzed synthesis of complementary-strand RNA. *Nature* **339**, 519–522, 1989.

Joyce, G. F. The rise and fall of the RNA world. *New Biologist* **3**, 399–407, 1991.

Lazcano, A., and S. L. Miller. On the origin of metabolic pathways. *J. Mol. Evol.* **49**, 424–431, 1999.

Morris, S. C. The early evolution of life. In *Understanding the Earth*, G. Brown, C. Hawkesworth, and C. Wilson, eds. Cambridge U. Press, 436–457, 1992.

Impacts and Extinction Events

Alvarez, L. W., W. Alvarez, F. Asaro, and H. V. Michel. Extraterrestrial cause for the Cretaceous-Tertiary extinction. *Science* **208**, 1095–1108, 1980.

Rampino, M. R., and B. M. Haggerty. Extraterrestrial impacts and mass extinctions of life. In *Hazards Due to Comets and Asteroids*, T. Gehrels, ed. U. Arizona Press, 827–857, 1994.

Chapter 6

Environmental Requirements for Life

Chyba, C. F., and K. P. Hand. Life without photosynthesis. *Science* **292**, 2026–2027, 2001.

Nealson, K. H. Post-Viking microbiology: New approaches, new data, new insights. *Origins Life Evol. Biosphere* **29**, 73–93, 1999.

Pace, N. R. The universal nature of biochemistry. *Proc. Nat'l Acad. Sci.* **98**, 805–808, 2001.

Wald, G. The origins of life. *Proc. Nat'l Acad. Sci.* **52**, 595–611, 1964.

Life in the Solar System

Chang, S. Planetary environments and the conditions of life. *Phil. Trans. R. Soc. Lond. A* **325**, 601–610, 1988.

Soffen, G. A. Life in the solar system. In *The New Solar System*, J. K. Beatty, C. C. Peterson, and A. Chaikin, eds. Cambridge: Sky Publ. Corp., 365–376, 1999.

Chapter 7

Introduction to Mars

Carr, M. H. Mars. In *The New Solar System*, J. K. Beatty, C. C. Peterson, and A. Chaikin, eds. Cambridge: Sky Publ. Corp., 141–156, 1999.

Zahne, K., Decline and fall of the martian empire. *Nature* **412**, 209–213, 2001.

Geological and Geochemical Potential for Life on Mars

Baker, V. R. Water and the martian landscape. *Nature* **412**, 228–236, 2001.

Gladman, B. J., et al. The exchange of impact ejecta between terrestrial planets. *Science* **271**, 1387–1392, 1996.

Jakosky, B. M., and R. J. Phillips. Mars' volatile and climate history. *Nature* **412**, 237–244, 2001.

Jakosky, B. M., and E. L. Shock. The biological potential of Mars, the early Earth, and Europa. *J. Geophys. Res.* **103**, 19359–19364, 1998.

Weiss, B. P., Y. L. Yung, and K. H. Nealson. Atmospheric energy for subsurface life on Mars? *Proc. Nat'l Acad. Sci.* **97**, 1395–1399, 2000.

Viking Search for Life on Mars

Klein, H. P. Did Viking discover life on Mars? *Origins Life Evol. Biosphere* **29**, 625–631, 1999.

Klein, H. P., N. H. Horowitz, and K. Biemann. The search for extant life on Mars. In *Mars*, H. H. Kieffer, B. M. Jakosky, C. W. Snyder, and M. S. Matthews, eds. U. Arizona Press, 1221–1233, 1992.

ALH84001 and Implications for Life on Mars

Barber, D. J., and E. R. D. Scott. Origin of supposedly biogenic magnetite in the martian meteorite Allan Hills 84001. *Proc. Nat'l Acad. Sci.* **99**, 6556–6561, 2002.

Friedmann, E. I., J. Wierzchos, C. Ascaso, and M. Winklhofer. Chains of magnetite crystals in the meteorite ALH84001: Evidence of biological origin. *Proc. Nat'l Acad. Sci.* **98**, 2176–2181, 2001.

Golden, D. C., et al. A simple inorganic process for formation of carbonates, magnetite, and sulfides in martian meteorite ALH84001. *Amer. Mineralogist* **86**, 370–375, 2001.

McKay, D. S., E. K. Gibson, Jr., K. L. Thomas-Keprta, H. Vali, C. S. Romanek, S. J. Clemett, X. D. F. Chillier, C. R. Maechling, and R. N. Zare. Search for past life on Mars: possible relic biogenic activity in martian meteorite ALH84001. *Science* **273**, 924–930, 1996.

Treiman, A. H. A short, critical evaluation of proposed signs of ancient martian life in Antarctic meteorite ALH84001. In *Bioastronoomy '99: A New Era in Bioastronomy*, G. A. Lemarchand and K. J. Meech, eds. San Francisco: Astron. Soc. Pac., 303–314, 2000.

Treiman, A. H., J. D. Gleason, and D. D. Bogard. The SNC meteorites are from Mars. *Planet. Space Sci.* **48**, 1213–1230, 2000.

Searching for Life on Mars

Carr, M. H., and J. Garvin. Mars exploration. *Nature* **412**, 250–253, 2001.

DeVincenzi, D. L., M. S. Race, and H. P. Klein. Planetary protection, sample return missions, and Mars exploration: History, status, and future needs. *J. Geophys. Res.* **103**, 28577–28585, 1998.

Mojzsis, S. J., and G. Arrhenius. Phosphates and carbon on Mars: Exobiological implications and sample return considerations. *J. Geophys. Res.* **103**, 28495–28511, 1998.

Westall, F., et al. An ESA study for the search for life on Mars. *Planet. Space Sci.* **48**, 181–202, 2000.

* Zorpette, G. Why go to Mars? *Scientific American*, March 2000.

Chapter 8

Life on Moons

Irwin, L. N., and D. Schulze-Makuch. Assessing the plausibility of life on other worlds. *Astrobiology* **1**, 143, 2001.

* LePage, A. Habitable moons. *Sky and Telescope*, p. 51, December 1998.

* Rothery, D. A. *Satellites of the Outer Planets*. Clarendon Press, 1992.

Europa, Ganymede, and Callisto

Chyba, C. F., and C. B. Phillips. Possible ecosystems and the search for life on Europa. *Proc. Nat'l Acad. Sci.*, **98**, 801, 2001.

Kivelson, M. G., et al. Galileo magnetometer measurements: A stronger case for a subsurface ocean at Europa. *Science* **289**, 1340–1343, 2000.

Pappalardo, R. T., J. W. Head, and R. Greeley. The hidden ocean of Europa. *Scientific American* **281**, 54–63, 1999.

Stevenson, D. J. Jupiter and its moons. *Science* **294**, 71–72, 2001.

Titan

Lunine, J. I. Does Titan have an ocean? A review of current understanding of Titan's surface. *Rev. Geophys.* **31**, 133–149, 1993.

Sagan, C., W. R. Thompson, and B. N. Khare. Titan: a laboratory for prebiological organic chemistry. *Accounts Chem. Res.* **25**, 286–292, 1992.

Chapter 9

Venus

* Grinspoon, D. *Venus Revealed*. Addison Wesley, 1997.

Habitable Zones and Beyond

Doyle, L., ed. *Circumstellar Habitable Zones*. Travis House, 1996.

Kasting, J. F., D. P. Whitmire, and R. T. Reynolds. Habitable zones around main sequence stars. *Icarus*, **101**, 108–128, 1993.

Stevenson, D. J. Life-sustaining planets in interstellar space? *Nature* **400**, 32, 1999.

Wetherill, G. W. The formation and habitability of extra-solar planets. *Icarus* **119**, 219–238, 1996.

Williams, D. M., J. F. Kasting, and R. A. Wade. Habitable moons around extrasolar giant planets, *Nature* **385**, 234–236, 1997.

Chapter 10

Discovery of Extrasolar Planets, Implications for Solar System Evolution

* Croswell, K. *Planet Quest*. The Free Press, 1997.

* Doyle, L. R., H.-J. Deeg, and T. M. Brown. Searching for Shadows of Other Earths. *Scientific American* **283**, 38, 2000.

* Goldsmith, D. *Worlds Unnumbered*. University Science Books, 1997.

* Malhotra, R., Migrating planets. *Scientific American* **281**, 56–63, 1999.

Marcy, G. W. and R. P. Butler. A planetary companion to 70 Virginis. *Astrophy. J.* **464**, L147–L151, 1996.

Marcy, G. W. and R. P. Butler. Planets orbiting other suns. *PASP* (Publications of the Astronomical Society of the Pacific) **112**, 137, 2000.

Marcy, G. W., W. D. Cochran, and M. Mayor. Extrasolar planets around main-sequence stars. In *Protostars and Protoplanets IV*, V. Mannings, A. P. Boss, and S. S. Russell, eds. U. Arizona Press, 1285–1311, 2000.

How to Search for Evidence of Life on Extrasolar Planets

Angel, J. R. P., and N. J. Woolf. Searching for life on other planets. *Scientific American* **274**, 60–66, 1966.

Ford, E. B., S. Seager, and E. I. Turner. Characterization of extrasolar terrestrial planets from diurnal photometric variability. *Nature* **412,** 885, 2001.

Sagan, C., et al. A search for life on Earth from the Galileo spacecraft. *Nature* **365,** 715–721, 1993.

Schindler, T. L., and J. F. Kasting. Synthetic spectra of simulated terrestrial atmospheres containing possible biomarker gases. *Icarus* **145,** 262–271, 2000.

The Rare Earth Hypothesis

Darling, D. Rare Earths and hidden agendas. In *Life Everywhere,* Basic Books, 2001.

Gonzalez, G., D. Brownlee, and P. D. Ward. Refuges for life in a hostile universe. *Scientific American* **285,** 60–67, 2001.

* Ward, P. D., and Brownlee, D. *Rare Earth: Why Complex Life Is Uncommon in the Universe.* Copernicus, 2000.

Chapter 11

Nature of Intelligence and Potential for Its Occurrence

* Calvin, W. H. The emergence of intelligence. *Scientific American* **271,** 100–107, 1994.

Jerison, H. J. *Evolution of the Brain and Intelligence.* Academic Press, 1973.

Marino, L. Brain-behavior relations in primates and cetaceans: Implications for the ubiquity of factors leading to the evolution of complex intelligence. In *Astronomical and Biochemical Origins and the Search for Life in the Universe,* C. B. Cosmovici, S. Bowyer, and D. Werthimer, eds. Editrice Compositori, 553, 1997.

Mayr, E. The probability of extraterrestrial intelligent life. In *Extraterrestrials: Science and Alien Intelligence,* E. Regis, Jr., ed. Cambridge University Press, 23, 1985.

Artifacts

Dyson, F. J. Search for artificial stellar sources of infrared radiation. *Science* **131,** 1667, 1960.

Freitas, R. A., Jr., and F. Valdes. The search for extraterrestrial artifacts (SETA). *Acta Astronautica* **12,** 1027–1034, 1985.

Zubrin, R. Detection of extraterrestrial civilizations via the spectral signature of advanced interstellar spacecraft. In *Progress in the Search for Extraterrestrial Life,* ASP Conference Series, vol. 74, G. S. Shostak, ed. Astronomical Society of the Pacific, 487, 1995.

SETI

Cocconi, G., and P. Morrison. Searching for interstellar communications. *Nature* **184,** 844–846, 1959.

* Davies, P. *Are We Alone?* Basic Books, 1996.

Drake, F. Project Ozma. *Physics Today* **14,** 40, 1961.

* Drake, F. D., and D. Sobel. *Is Anyone Out There?* Delacorte Press, 1992.

* Harrison, A. *After Contact.* Plenum Press, 1997.

Mayr, E. The search for extraterrestrial intelligence: Scientific quest or hopeful folly? The improbability of success. *The Planetary Report* **16,** 4–7, 1996.

* McConnell, B. *Beyond Contact.* O'Reilly and Associates, 2001.

Sagan, C. The search for extraterrestrial intelligence: Scientific quest or hopeful folly? The abundance of life-bearing planets. *The Planetary Report* **16,** 8–13, 1996.

* Shostak, S. The future of SETI. *Sky and Telescope,* 42, April 2001.

Chapter 12

The Voyager Record

* Sagan, C., F. D. Drake, A. Dryan, T. Ferris, J. Lomberg, and L. S. Sagan. *Murmurs of Earth: The Voyager Interstellar Record.* Random House, 1978.

Interstellar Travel

Mallove, E. F., and G. L. Matloff. *The Starflight Handbook: A Pioneer's Guide to Interstellar Travel.* Wiley, 1989.

Mauldin, J. *Prospects for Interstellar Travel.* Science and Technology Series: A Supplement to Advances in the Astronautical Sciences, vol. 80. Univelt, 1992.

Relativity, Black Holes and Worm Holes

Thorne, K. S. *Black Holes and Time Warps: Einstein's Outrageous Legacy.* Norton, 1994.

Scientific Study of UFOs and Alien Visitation Claims

* Alschuler, W. R. *The Science of UFOs.* St. Martin's Press, 2001.

* Korff, K. *The Roswell UFO Crash: What They Don't Want You to Know.* Prometheus Books, 1997.

* Kristoff, N. D. Alien abduction? Science calls it sleep paralysis. *New York Times,* 6 July 1999.

* Sturrock, P. A. *The UFO Enigma.* Warner Books, 1999.

Chapter 13

Will We Try to Colonize Space?

* Bracewell, R. *The Galactic Club.* W. H. Freeman, 1975.

* Dyson, F. *Disturbing the Universe.* Basic Books, 1979.

* Dyson, F. *Infinite in All Directions.* Harper and Row, 1988.

* Heppenheimer, T. A. *Toward Distant Suns.* Stackpole Books, 1979.

* O'Neill, G. K. *The High Frontier: Human Colonies in Space.* Wm. Morrow, 1977.

* Sagan, C. *Pale Blue Dot: A Vision of the Human Future in Space.* Random House, 1994.

Fermi Paradox

Ball, J. A. The zoo hypothesis. *Icarus* **19,** 347, 1973.

* Finney, B. and E. M. Jones eds. *Interstellar Migration and the Human Experience.* University of California Press, 1985.

Hart, M. An explanation for the absence of extraterrestrials on Earth. *Quarterly Journal of the Royal Astronomical Society* **16,** 128, 1975.

Sagan, C., and W. I. Newman. The solipsist approach to extraterrestrial intelligence. *Quarterly Journal of the Royal Astronomical Society* **24,** 113, 1983.

Tipler, F. J. Extraterrestrial intelligent beings do not exist. *Quarterly Journal of the Royal Astronomical Society* **21,** 267, 1980.

Chapter 14

Philosophical and Societal Issues
Related to the Search for Life Elsewhere

Jakosky, B. M. Philosophical aspects of astrobiology. In *Bioastronomy '99: A New Era in Bioastronomy,* G. A. Lemarchand and K. J. Meech, eds. San Francisco: Astron. Soc. Pac., 661–666, 2000.

Jakosky, B. M. and M. P. Golombek. Planetary science, astrobiology, and the role of science and exploration in society. *EOS, Trans. Amer. Geophys. Union* **81,** 58, 2000.

CREDITS

CHAPTER 1 **1.1** © Corbis **1.2** NASA/ Voyager **1.3** © NASA/Voyager 2/NSSDC **1.4** © Seth Shostak **1.5** Illustration by Joe Bergeron **1.6** © Jeff Bennett **1.7** © Jeff Bennett **1.9** (left) Andrea Dupree (Harvard-Smithsonian CFA), Ronald Gilliland (STScI), ESA, and NASA (right) © Anglo-Australian Observatory. Photo by David Malin **1.10, 1.11, 1.14** Illustrations by Joe Bergeron **1.15** NASA/STScI, courtesy of Alfred Schultz and Helen Hart **1.18** (photo) © Steven Frisch/Benjamin Cummings (artwork) Courtesy of Neil A. Campbell, from *Biology*, 5ed, by Neil A. Campbell, Jane B. Reece, and Lawrence G. Mitchell © 1999 by Benjamin Cummings **1.20** © ESO (European Southern Observatory)

CHAPTER 2 **2.1** © David Nunuk/Science Scource/Photo Researchers **2.2, 2.7** © Bettman/Corbis **2.11** Courtesy of Neil A. Campbell, from *Essential Biology*, by Neil A. Campbell and Jane B. Reece, © 2001 by Benjamin Cummings **2.12** © Bettmann/Corbis **2.13** © Richard J. Wainscoat **2.14** NASA

CHAPTER 3 **3.1a** © David M. Phillips/Visuals Unlimited **3.1b** © Biophoto Associates/Photo Researchers, Inc. **3.1c** © Michael Fogden/Bruce Coleman **3.1d** © Woods Hole Oceanographic Institution **3.1e** © Tom & Pat Leeson/Photo Researchers, Inc. **3.1f** © Tom & Pat Leeson/Photo Researchers, Inc. **p. 56** © English Heritage Photo Library **3.2, 3.3** Courtesy of Neil A. Campbell, from *Essential Biology*, by Neil A. Campbell and Jane B. Reece, © 2001 by Benjamin Cummings **3.4** © Jack Wilburn/Earth Sciences/Animals/ Animals **3.5a** © Lester V. Bergman/Corbis **3.5b** © Lester V. Bergman/Corbis **3.5c** © Science Pictures Limited/Corbis **3.5d** © Frank Lane Picture Agency/Corbis **3.6, 3.8, 3.9, 3.12, 3.14, 3.15,** Courtesy of Neil A. Campbell, from *Essential Biology*, by Neil A. Campbell and Jane B. Reece, © 2001 by Benjamin Cummings **3.11** CNRI Science Photo Library **3.16a** © Lester V. Bergman/ Corbis **3.16b** © Lester V. Bergman/Corbis **3.17** © Woods Hole Oceanographic Institution **3.18** © George Steinmetz **3.19** Photo by Arnie Bomblies **3.20** © E. Imre Friedmann **3.21** © H. S. Pankratz, T. C. Beaman/ Biological Photo Service

CHAPTER 4 **4.1a** © Tim Thompson/Corbis **4.1b** © David Muench/Corbis **4.1c** © Layne Kennedy/Corbis **4.2a** © George Gerster/PhotoResearchers **4.2b** © Tom Till **4.2c** © Walter H. Hodge/Peter Arnold, Inc. **4.2d** © Manfred Kage/Peter Arnold, Inc. **4.2e** Courtesy, Dr. David A. Grimald. Photo by

Jacklyn Eckett/The American Museum of Natural History, NY **4.2f** © F. Latreille/Cerpolex **4.2g** © Dr. Martin Lockley, University of Colorado **4.3** Courtesy of Neil A. Campbell, from *Essential Biology*, by Neil A. Campbell and Jane B. Reece, © 2001 by Benjamin Cummings **4.4** © Jeff Greenberg/Visuals Unlimited **4.9** Photo by John W. Valley **4.10** Tom Van Sant/The Geosphere Project, Corbis Stock Market **4.11** © William K. Hartmann **4.12a** © Roger Ressmeyer/Corbis **4.12b** NASA/USGS **4.13** USGS Photo Library, Denver, Colorado **4.15** Tom Van Sant/The Geosphere Project, Corbis Stock Market **4.18** © Earth Satellite Corp./Science Photo Library/Photo Researchers, Inc. **4.20** Digital Image by Dr. Peter W. Sloss (NOAA/NESDIS/ NGDCO)

CHAPTER 5 **5.1a** Biological Photo Services **5.1b** © S. M. Awramik, University of California/Biological Photo Service **5.1c** © S. M. Awramik, University of California/Biological Photo Service **5.2** Biological Photo Services **5.3** S. J. Mojzsis **5.5** © Roger Ressmeyer/Corbis **5.6 5.7** Courtesy of Neil A. Campbell, from *Essential Biology*, by Neil A. Campbell and Jane B. Reece, © 2001 by Benjamin Cummings **5.8b** Courtesy of F. M. Menger and Kurt Gabrielson, Emory University **5.7a** © Sidney Foz, University of Miami/BPS **5.10a** Courtesy of Frederick Atwood, Flint Hill School **5.10b** © T. E. Adams/Visuals Unlimited **5.11** Courtesy of Neil A. Campbell, from *Biology*, 5ed, by Neil A. Campbell, Jane B. Reece, and Lawrence G. Mitchell © 1999 by Benjamin Cummings **5.12** © The Field Museum **5.13** Kirk Johnson/Denver Museum of Natural History **5.14** Brad Snowder **5.15** Image courtesy of Dr. Virgil Sharpton, University of Alaska-Fairbanks **5.16** Quade Paul/fiVth.com **5.18b** © Hinrich Basemann (*www.polarfoto.com*) **5.19a** Geological Survey of Canada **5.19b** Courtesy of Calvin Hamilton **5.20** TASS/Sovfoto **5.22, 5.24** Courtesy of Neil A. Campbell, from *Essential Biology*, by Neil A. Campbell and Jane B. Reece, © 2001 by Benjamin Cummings **5.23** © David Gifford/SPL/Photo Researchers

CHAPTER 6 **6.3** NASA **6.4** (top) © Richard Wainscoat **6.4** (bottom) Russ Underwood (W. M. Keck Observatory) **6.5** NASA **6.6** © Joel Gordon Photography **6.7** Jodi Schoemer **6.8a** Galileo Spacecraft, JPL, NASA, Calvin J. Hamilton/NSSDC **6.8b** NASA/NSSDC **6.9** © Runk/Schoenberger/ Grant Heilman Photography **6.10** STScI and R. Beebe and A. Simon (NMSU) **6.12** © Michael Carroll **6.14** (Moon) Akira Fujii,

(others) NASA **6.15** Los Alamos National Laboratory, NASA, NSSDC **6.16** NASA/ NSSDC **6.17** NASA/JPL/NSSDC

CHAPTER 7 **7.1** Lowell Observatory Photographs **7.2** Copyright © 1938 The New York Times Co. Reprinted by Permission. **7.3** The Viking Project, NASA, NSSDC **7.4a** The Viking Project, NASA, NSSDC **7.4b** Mars Global Surveyor, MLS, NASA **7.5** Viking Orbiter, MLS, NASA **7.6a** Viking Lander, NASA **7.6b** Viking 2 Lander, NASA, NSSDC **7.7** NASA **7.10** NASA, P. James (University Toledo), T. Clancy (Space Science Inst.), S. Lee (University Colorado), NSSDC **7.11** NASA/JPL/Malin Space Science Systems **7.12** J. Bell (Cornell), M.Wolff (Space Science Inst.), Hubble Heritage Team (STScI/ AURA), NASA **7.13** NASA/JPL **7.14** NASA/Mars Global Surveyor/NSSDC **7.16** NASA/JPL **7.17** U.S. Geological Survey **7.18** NASA/JPL/Malin Space Science Systems **7.19** NASA/NSSDC **7.20** NASA and Michael Mellon **7.21** Dr. David E. Smith/ NASA/MOLA Science Team **7.22** NASA/JPL/Malin Space Science Systems **7.25** NASA/JPL **7.26** Radio Age, October 1924 **7.27, 7.28** Viking Lander, NASA/ NSSDC **7.29** © Space Adventures **7.30a,b** NASA, Photo courtesy of Johnson Space Center/NSSDC **7.31** NASA/JPL **7.32a** NASA **7.32b** Photo by R. L. Folk and F. L. Lynch **7.33** (inset) NASA/JPL **7.34a** NASA/James Garvin

CHAPTER 8 **8.2** NASA/JPL/Galileo **8.3** Courtesy Tim Parker/JPL **8.5** NASA/JPL **8.11** PIRL/Lunar & Planetary Laboratory (University of Arizona) and NASA **8.12** DLR/NASA/JPL **8.1** NASA/JPL/Galileo/ Artist's conception by Pam Engebretson. Drawings rendered by Pam Engebretson, from a design by Eric M. DeJong, and Zareh Gorjian of JPL **8.14** University of Arizona and NASA **8.15, 8.16, 8.17b, 8.20a, 8.20b, 8.23** NASA/JPL/Galileo **8.17a** NASA/JPL/ Arizona State University **8.18** Joe Bergeron **8.19** NASA/Goddard Space Flight Center Scientific Visualization Studio **8.21, 8.22** NASA/JPL/Galileo/DLR (German Aerospace Center)/NSSDC **8.24, 8.25** NASA/JPL/ Galileo/NSSDC **8.26a** Seran Gibbard, Bruce Macintosh, Don Gavel, Claire Max (Lawrence Livermore National Laboratory) **8.26b** © Don Davis **8.27** NASA/USGS

CHAPTER 9 **9.2** NASA/JPL **9.6** Copyright © Subaru Telescope, NAOJ. All rights reserved **9.7** J. P. Harington, K. J. Borkowski (University of Maryland)/NASA

INDEX